DICTIONARY OF Advanced Manufacturing Technology

DICTIONARY OF Advanced Manufacturing Technology

V. Daniel Hunt

President, Technology Research Corporation

Elsevier

New York • Amsterdam • London

This book was prepared as an account of work sponsored by Elsevier Science Publishing Co., Inc. Neither Elsevier Science Publishing Co., Inc. nor Technology Research Corporation, nor any of its employees, nor any of their contractors, subcontractors, consultants, or their employees, makes any warranty, expressed or implied, or assumes any legal liability or responsibility for the accuracy, completeness, or usefulness of any information, apparatus, product or process disclosed, or represents that its use would not infringe on privately owned rights.

The views, opinions, and conclusions in this book are those of the author and do not necessarily represent those of the United States Government.

Public domain information and those documents abstracted or used in full, edited, or otherwise used are noted in the acknowledgments or on specific pages or illustrations.

Cover photograph courtesy of Allen-Bradley, a Rockwell International company.

Elsevier Science Publishing Co., Inc.
52 Vanderbilt Avenue, New York, New York 10017

Distributors outside the United States and Canada:
Elsevier Science Publishers B.V.
P.O. Box 211, 1000 AE Amsterdam, the Netherlands

Library of Congress Cataloging in Publication Data

Hunt, V. Daniel.
Dictionary of advanced manufacturing technology.

Bibliography: p.
1. Computer integrated manufacturing systems – Dictionaries. 2. CAD/CAM systems – Dictionaries. I. Title.
TS155.6.H85 1987 670'.28'5 87-20123
ISBN 0-444-01208-7

Current printing (last digit)
10 9 8 7 6 5 4 3 2 1

Manufactured in the United States of America

DEDICATION

To the staff and consultants who make Technology Research Corporation and its products a reality.

Table of Contents

Preface

Advanced Manufacturing Technology development and demonstration have expanded rapidly over the past several years; as a result, new terminology is appearing at a phenomenal rate.

This book provides an overview of advanced manufacturing technology along with brief definitions for CIM, CAD/CAM, robotics, machine vision, and software interface terminology applicable to advanced manufacturing technology. The extensive support material includes points of contact for additional information, acronyms, a detailed bibliography, and other reference data.

The definitions chapter includes advanced manufacturing technology terminology for:

- Computer-Aided Design
- Computer-Aided Manufacturing
- Computer Integrated Manufacturing
- Computer-Aided Process Planning
- Industrial Robots
- Machine Vision
- Artificial Intelligence
- Communication Interfaces
- Standardization Software and
- Production Processes

The Dictionary of Advanced Manufacturing Technology is compiled from information acquired from numerous books, journals, and authorities in the field of advanced manufacturing technology.

I hope this compilation of information will help clarify the terminology for your advanced manufacturing technology activities. Your comments, revisions, or questions are welcome.

V. Daniel Hunt
Springfield, Virginia

Acknowledgments

The information in *The Dictionary of Advanced Manufacturing Technology* has been compiled from a wide variety of authorities who are specialists in their respective fields.

The following publications were used as the basic technical resources for this book. Portions of these publications may have been used in the book. Those definitions or artwork used have been reproduced with the permission to reprint of the respective publisher.

Computerized Manufacturing Automation: Employment, Education, and the Workplace (Washington, D.C.: U.S. Congress, Office of Technology Assessment, OTA-CIT-235, April 1984)

A Competitive Assessment of the U.S. Flexible Manufacturing Systems Industry, U.S. Department of Commerce, International Trade Administration, July 1985.

Industrial Robots—A Summary and Forecast, Copyright 1986, Tech Tran Corporation, Naperville, Illinois.

Machine Vision Systems—A Summary and Forecast, Copyright 1986, Tech Tran Corporation, Naperville, Illinois.

AIM Glossary of Terms, Automatic Identification Manufacturers, Pittsburgh, Pennsylvania, 1981.

APICS Dictionary, Copyright by American Production and Inventory Control Society, Inc., Falls Church, Virginia, 1980.

Glossary of CAM Terms, Computer-Aided Manufacturing-International, Inc., Arlington, Texas.

MAP Reference Specification, Based on GM MAP 2.1 Specification, Developed by MAP/TOP Users Group of SME, Published by The Computer and Automated Systems Association of SME, May 1986, Dearborn, Michigan.

Technical and Office Protocols, Specification Version 1.0 November 1985, Published by The Boeing Company, Seattle, Washington.

The CAD/CAM Handbook, Carl Machover and Robert E. Blauth, Computervision Corporation, 1980, Bedford, Massachusetts.

CAD/CAM Handbook, Eric Teicholz, McGraw-Hill Book Company, 1985.

Introduction to CAD, Donald D. Voisinet, McGraw-Hill Book Company, 1983, New York, New York.

Robotics and Automated Manufacturing, Richard C. Dorf, Reston Publishing Company, 1983, Reston, Virginia.

The preparation of a book of this type is dependent upon an excellent staff, and I have been fortunate in this regard. Special thanks are extended to Janet C. Hunt, Audrey and Vernon Hunt, Donald W. Keehan, Kellie-Jo Anne Ackerman, James Wise, Jr., and Carla Wise for research assistance. Special thanks to Margaret W. Alexander and Janelle Orrico for the word processing of the manuscript.

DICTIONARY OF
Advanced Manufacturing Technology

Introduction

This section provides a brief introduction to advanced manufacturing technology. It gives an overview of the current technology, describes how the technology basically works, and summarizes current developments and growth opportunities.

DEFINITION

There is a great deal of confusion in industry regarding the maze of terms currently in use to describe advanced manufacturing technology. Terms such as mechatronics, factory of the future, CAD/CAM, flexible manufacturing systems, factory automation, and programmable automation all seem to have similar, identical, or overlapping meanings—and these can vary depending on the audience. In order to clarify the emphasis of this book, advanced manufacturing technology (AMT) is defined as systems providing flexibility for, and data-driven computer integration of, the total manufacturing operation. Advanced manufacturing technology includes computer-aided design (CAD) and computer-aided manufacturing (CAM)—for example, robots, numerically controlled (NC) machine tools, flexible manufacturing systems (FMSs), and automated materials-handling (AMH) systems. In addition to machine tools for design and production, advanced manufacturing technology encompasses computer-aided techniques for plant and production management, including management information systems (MISs), computer-aided process planning (CAPP), material requirements planning (MRP), and artificial intelligence (AI). These flexible integrated systems can

Figure 1
Elements of an advanced manufacturing technology system

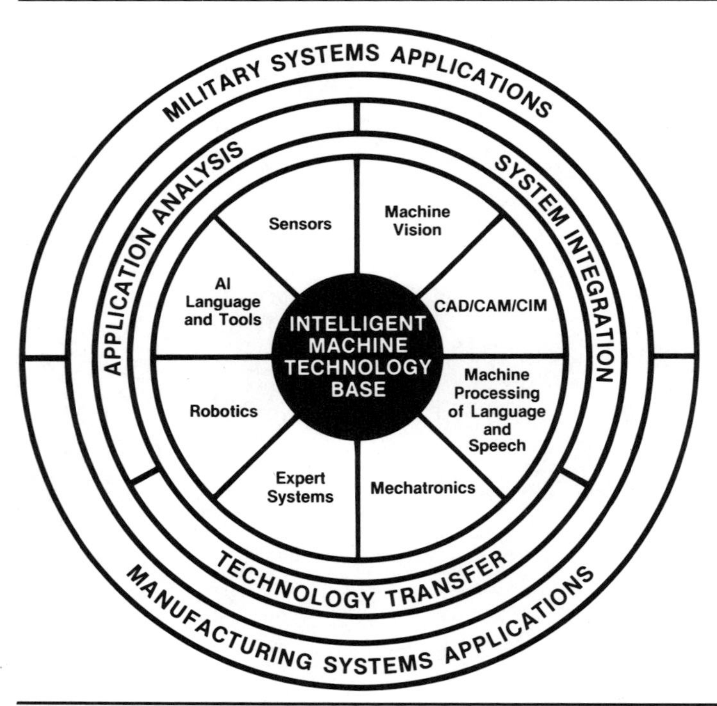

include all aspects of design, manufacturing, and management in a coordinated data-driven computer-assisted system. Figure 1 shows the elements of an advanced manufacturing technology system.

OVERVIEW OF ADVANCED MANUFACTURING TECHNOLOGY

Many manufacturers are now beginning to implement portions of advanced manufacturing technology. Usually proceeding by making evolutionary improvements to their facility, manufacturers move from standalone machines to flexible manufacturing cells (FMCs) to flexible manufacturing systems (FMSs). Today in the United States, an increasing number of companies are investing in computer-integrated manufacturing (CIM) to learn about and demonstrate the productivity and flexibility possible with advanced manufacturing technology. Although there are dozens of FMS installations in the United States, we can only identify a few companies that have fully implemented CIM. Companies such as Allen-Bradley Inc., Frost Inc., and LTV Aerospace & Defense have made major commitments to CIM and have taken the first step toward total data-driven computer-integrated manufacturing.

One of the essential differences between conventional factory machines and computer-integrated manufacturing is the latter's extensive use of data-driven information technology to provide design, manufacturing, and management systems control and communications. The use of data-driven computer-aided technology allows these machines to perform a greater variety of tasks than can fixed automation (as shown in Fig. 2) and to automate some tasks that previously necessitated direct human control.

Advanced manufacturing technology can respond to some of the needs currently facing American manufacturing firms. These

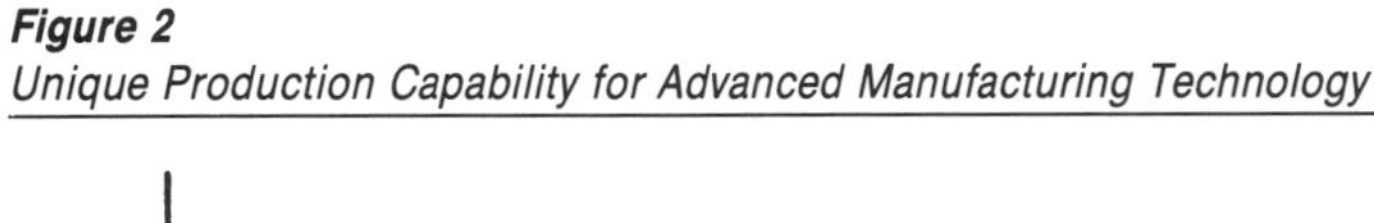

Figure 2
Unique Production Capability for Advanced Manufacturing Technology

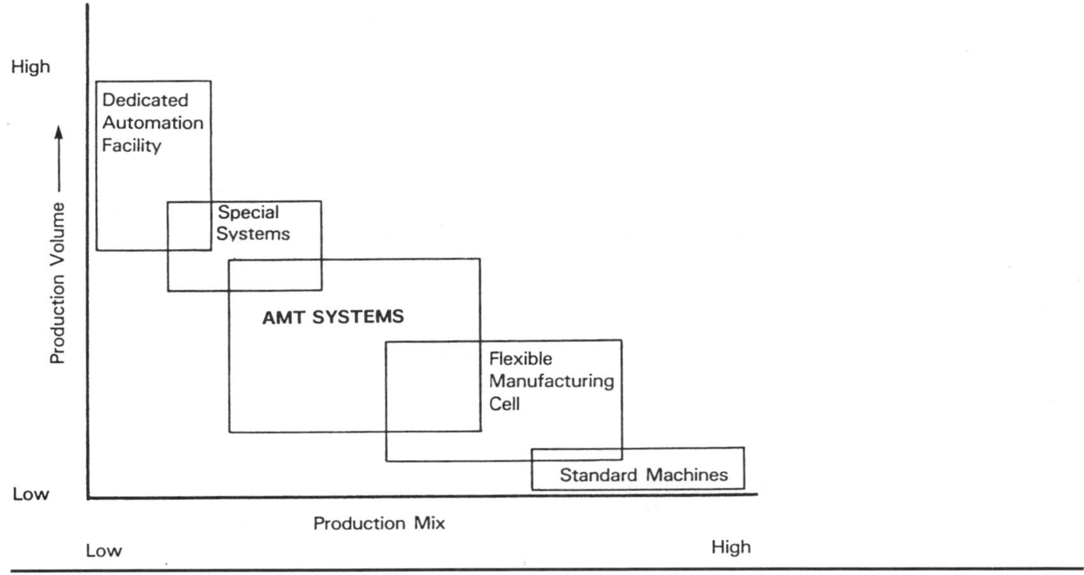

include: (1) improved data and information flow, (2) better engineering and manufacturing coordination, (3) increased efficiency and production flexibility, (4) improved competitiveness, (5) decreased labor cost, (6) greater capital utilization, and (7) improved quality. Implementation of advanced manufacturing technology not only magnifies the productivity and efficiency of a production facility, but it also creates changes in all parts of the factory. Changes in the way parts are designed, procured, and produced can improve productivity significantly and help manufacturers make better use of capital-intensive production facilities.

ELEMENTS OF ADVANCED MANUFACTURING TECHNOLOGY SYSTEMS

This section describes the technologies that together comprise advanced manufacturing technology and evaluates their usefulness as part of data-driven advanced manufacturing technology. In addition, this section examines how the technologies are evolving and what can be expected regarding their capabilities and applications.

This overview presentation is based on the Office of Technology Assessment (OTA) report, *Computerized Manufacturing Automation.** We appreciate the information provided by OTA and Ms Marjory S. Blumenthal, who is currently with General Electric Company. Some charts and data have been revised to reflect the latest information available. The following material is provided to enlighten those end-users who are not familiar with the current development of advanced manufacturing technology.

As mentioned initially, advanced manufacturing technology refers to a family of technologies that lie at the intersection of computer science and manufacturing engineering. Advanced manufacturing technology systems can be switched from one production task to another with relative ease by changing the (usually) computerized instructions. The common element in these tools that makes them different from traditional manufacturing tools is their use of the computer to manipulate and store data and the use of related microelectronics technology to allow communication of data to other machines in the factory.

Advanced manufacturing tools perform four general categories of functions. They are used to (1) design products, (2) help manufacture (both fabricate and assemble) products on the factory floor, (3) assist in management of many factory operations, and (4) provide artificial intelligence/expert system advisory capabilities. Table 1 outlines the principal technologies included in these categories, each of which will be described in the following material.

The first three categories of advanced

Table 1.
Principal Advanced Manufacturing Technologies

I. Computer-aided design (CAD)
  A. Computer-aided drafting and design (CADD)
  B. Computer-aided engineering (CAE)
II. Computer-aided manufacturing (CAM)
  A. Robots
  B. Numerically controlled (NC) machine tools
  C. Flexible manufacturing cells (FMCs)
  D. Flexible manufacturing systems (FMSs)
  E. Automated materials-handling (AMH) systems
  F. Automated storage and retrieval systems (AS/RSs)
III. Tools and strategies for manufacturing management
  A. Management information systems (MISs)
  B. Computer-aided planning (CAP)
  C. Computer-aided process planning (CAPP)
  D. Material requirements planning (MRP)
IV. Artificial intelligence (AI)
  A. Expert systems (ESs)
  B. Natural language processing (NLP)
  C. Smart robots (SRs)
  D. Machine vision (MV)

* Computerized Manufacturing Automation: Employment, Education, and the Workplace (Washington, D.C.: U.S. Congress, Office of Technology Assessment, OTA-CIT-235, April 1984).

manufacturing technology—tools for design, manufacturing, and management—are not mutually exclusive. In fact, the goal of much current research (e.g., mechatronics) is to break down the barriers between these systems so that design and manufacturing are inextricably linked. However, the three categories are useful as a framework for the discussion, particularly since they correspond to the organization of a typical manufacturing firm.

Figure 3
Computer-Aided Design Workstation. Source: Calcomp.

ADVANCED MANUFACTURING TECHNOLOGIES

This section briefly describes the operation of each advanced manufacturing technology area and its applications in manufacturing.

Computer-Aided Design (CAD)

In its simpler forms, CAD is an electronic drawing board for design engineers and draftspersons. Instead of drawing a detailed design with pencil and paper, designers and drafters work at a computer terminal, as shown in Figure 3, instructing the computer to combine various lines and curves to produce a drawing of a part and its specifications. In its more complex forms, CAD can be used to communicate to manufacturing equipment the specifications and process for making a product. Finally, CAD is also the core of computer-aided engineering, in which engineers analyze a design and maximize a product's performance using the computerized representation of the product.

The roots of computer-aided design technology are primarily in computer science. CAD evolved from research carried out in the late 1950s and early 1960s on interactive computer graphics—simply, the use of computer screens to display and manipulate lines and shapes instead of numbers and text. SKETCHPAD, funded by the Department of Defense (DOD) and demonstrated at the Massachusetts Institute of Technology in 1963, was a milestone in CAD development. Users could draw pictures on a screen and manipulate them with a *light pen*, a pen-shaped object wired to the computer that locates points on the screen (see Fig. 4). Such early systems were expensive prototypes and required most of the computing power of the then-largest computers. As a consequence, most of the early users of CAD were aerospace, automobile, and electronics manufacturers.

Several key developments in the 1960s and 1970s facilitated the development of CAD technology. These developments included the continuing decrease in the cost of computing power. A major factor influencing the cost decrease was the development of powerful minicomputers and microcomputers that primarily resulted from electronics manufacturers learning to squeeze more and more circuitry into an integrated circuit chip. Another important technological advance was the development of cheaper, more efficient display screens. In addition, computer scientists began to develop very powerful programming tech-

Figure 4
Light Pen Used to Input Design Information into CAD System. Source: Calcomp.

niques for manipulating computerized images.

How CAD Works

There are various schemes for input of a design to the computer system, and each scheme has its advantages and disadvantages. Every CAD system is equipped with a keyboard, though other devices are often more useful for entering and manipulating shapes. The operator can point to areas of the screen with a light pen or use a graphics tablet, which is an electronically touch-sensitive drawing board; a device called a *mouse* can be traced on an adjacent surface to move a pointer around on the screen. If there is already a rough design or model for the product, the operator can use a *digitizer* to read the contours of the model into computer memory and then manipulate a drawing of the model on the screen. Finally, if the part is similar to one that has already been designed using the CAD system, the operator can recall the old design from computer memory and edit the drawing on the screen.

CAD systems typically have a library of stored shapes and commands to facilitate the input of designs. CAD systems perform four basic functions that can enhance the productivity of a designer or draftsperson. First, CAD allows *replication*, the ability to take part of the image and use it in several other areas of the design when a product has repetitive features. Second, the system can "translate" parts of the image from one location on the screen to another. Third is *scaling*, in which CAD can "zoom in" on a small part or change the size or proportions of one part of the image in relation to the others. Finally, *rotation* allows the operator to see the design from different angles or perspectives. Using such commands, operators can perform sophisticated manipulations of the drawing, some of which are difficult or impossible to achieve with pencil and paper.

Repetitive designs or designs in which one part of the image is a small modification of a previous drawing can be done much more quickly through CAD. On the other hand, CAD can be cumbersome, especially for inexperienced users. Drawing an unusual shape may be fairly straightforward with a pencil but quite complex to accomplish using the basic lines and curves in the system's library.

The simplest CAD systems, as shown in Figure 5, are *two-dimensional* (*2-D*). Like pencil-and-paper drawings, they can be used to model *three-dimensional* (*3-D*) objects if several 2-D drawings from various perspectives are combined. For some applications, such as electronic circuit design, 2-D drawings are sufficient. In the past few years, more sophisticated CAD systems have been developed that allow the operator to construct a 3-D image on the screen—a capability that is particularly useful for complex mechanical products.

Figure 5
2-D Computer-Aided Design System. Source: Calma.

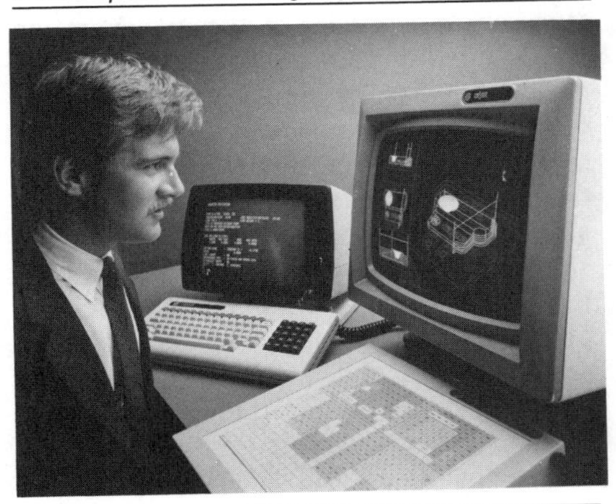

Most CAD systems include several CAD terminals connected to a central mainframe or minicomputer, although some recently developed systems use stand-alone microcomputers. As the operator produces a drawing, it is stored in computer memory, typically on a magnetic disk. The collection of digitized drawings in computer storage becomes a design data base, and this data base is then readily accessible by other designers, managers, or manufacturing staff.

CAD operators have several options for output of their design. All systems have a plotter capable of producing precise and often multicolor paper copies of the drawing. Some systems can generate copies of the design on microfilm or microfiche for compact storage. Others are capable of generating photographic output. In most cases, however, the paper output from CAD is much less important than the pen-and-paper drawings in the manual design process. More important is the fact that the design is stored on a computer disk; it is this version that is most up-to-date and accessible and that will be modified as design changes occur.

The CAD systems described above are essentially drafters' versions of word processors: they allow operators to create and easily modify an electronic version of a drawing. However, more sophisticated CAD systems can go beyond computer-aided drafting in two important ways.

First, such systems increasingly allow the physical dimensions of the product and the steps necessary to produce it to be developed via the computer and communicated electronically to computer-aided manufacturing equipment. Some of these systems present a graphic simulation of the machining process on the screen and guide the operator step by step in planning the machining process. The CAD system can then produce a tape that is fed into the machine tool controller and used to guide the machine tool path. Such connections from computer-aided design equipment to computer-aided manufacturing equipment shortcut several steps in the conventional manufacturing process. They cut down the time required for a manufacturing engineer to interpret design drawings and establish machining plans; they facilitate process planning by providing a visualization of the machining process; and they reduce the time necessary for machinists to interpret process plans and guide the machine tool through the process.

Second, these more sophisticated CAD systems serve as the core technology for many forms of computer-aided engineering (CAE). Beyond using computer graphics merely to facilitate drafting and design changes, CAE tools permit interactive design and analysis. Engineers can, for example, use computer graphic techniques for simulation and animation of products, to visualize the operation of a product, or to obtain an estimate of its performance, as shown in Figure 6. Other CAE programs can help engineers perform finite element analysis, which, essentially, is the breaking down of complex mechanical objects into a network of hundreds of simpler elements to determine stresses and deformations. Computerization in general made finite element

Figure 6
Computer-Aided Engineering System Provides Simulation and Animation Function. Source: Calma.

analysis feasible for the engineer's use, while CAD systems now make it significantly less cumbersome by assisting the engineer in breaking down the object into "elements."

Many of these analytical functions are dependent on 3-D CAD systems that can not only draw the design but also perform *solid modeling*—that is, the machine can calculate and display such solid characteristics as the volume and density of the object. Solid-modeling capabilities are among the most complex features of CAD technology.

Applications

At the end of 1986, there were an estimated 47,000 CAD workstations in the United States. Aerospace and electronic uses of CAD have always led the state of the art. For example, the Boeing Commercial Airplane Company, which began using CAD in the late 1950s, employed the technology extensively in the design of its new-generation 757 and 767 aircraft. Boeing uses CAD to design families of similar parts, such as wing ribs and floor beams. CAD allows designers to make full use of similarities between parts so that redesign and redrafting are minimized. Moreover, CAD has greatly simplified the task of designing airplane interiors and cargo compartments, which are often different for each plane. Moving seats, galleys, and lavatories is relatively simple with CAD, and the system is then used to generate instructions for the machines that later drill and assemble floor panels according to the layout. Finally, Boeing uses CAD and related interactive computer graphics systems as the basis for computer-aided engineering applications, such as checking mechanism clearances and simulating flight performance of various parts and systems.

Computer-aided engineering has also become important in the automobile and aerospace industries, where weight can be a critical factor in product design. These industries have developed CAE programs that can optimize a design for minimum material used while maintaining strength.

Applications for the design of integrated circuits are similarly advanced. Very large-scale integrated (VLSI) circuits, for example, have become so complicated that it is virtually impossible for a person to keep track of the circuit paths manually and to make sure the patterns are correct. There is less need here for geometrically sophisticated CAD systems—integrated circuit designs are essentially a few layers of two-dimensional lines—and more need for computer-aided engineering systems to help the designer cope with the intricate arrangement of the circuit pattern. Such CAE programs are used to simulate the performance of a circuit and check it for "faults," as well as to optimize the use of space on the chip.

CAD is also beginning to be used for nonaerospace mechanical design and in smaller firms; these developments are being spurred on by the marketing of relatively low-priced *turnkey* systems—complete packages of software and hardware that, theoretically, are ready to use as soon as they are delivered and installed. While a standard and reasonably powerful system based on a minicomputer is typically in the $500,000 range, many smaller microcomputer-based systems have been introduced in the past year for under $40,000—in some cases, for as low as $15,000. Very low-cost systems that run on common microcomputers (see Fig. 7) have been introduced, and these have potential uses in a wide variety of firms that otherwise might not consider CAD. The cost of custom-developed, specialized systems, such as those described above for aerospace and electronics applications, is harder to gauge but runs well into the millions of dollars.

The potential advantage of CAD for both large and small mechanical manufacturing firms is that it addresses several of the needs of manufacturing referred to at the beginning of this section. It facilitates use of previous designs and allows design changes to be processed more quickly. Because CAD reduces the time necessary for many design tasks, it also improves design by allowing designers to "try out" a dozen or a hundred different variations, where previously they might have been limited to

Figure 7
ANVIL-1000MD CADD Software Operates on IBM PC AT System. Source: Manufacturing and Consulting Services, Inc.

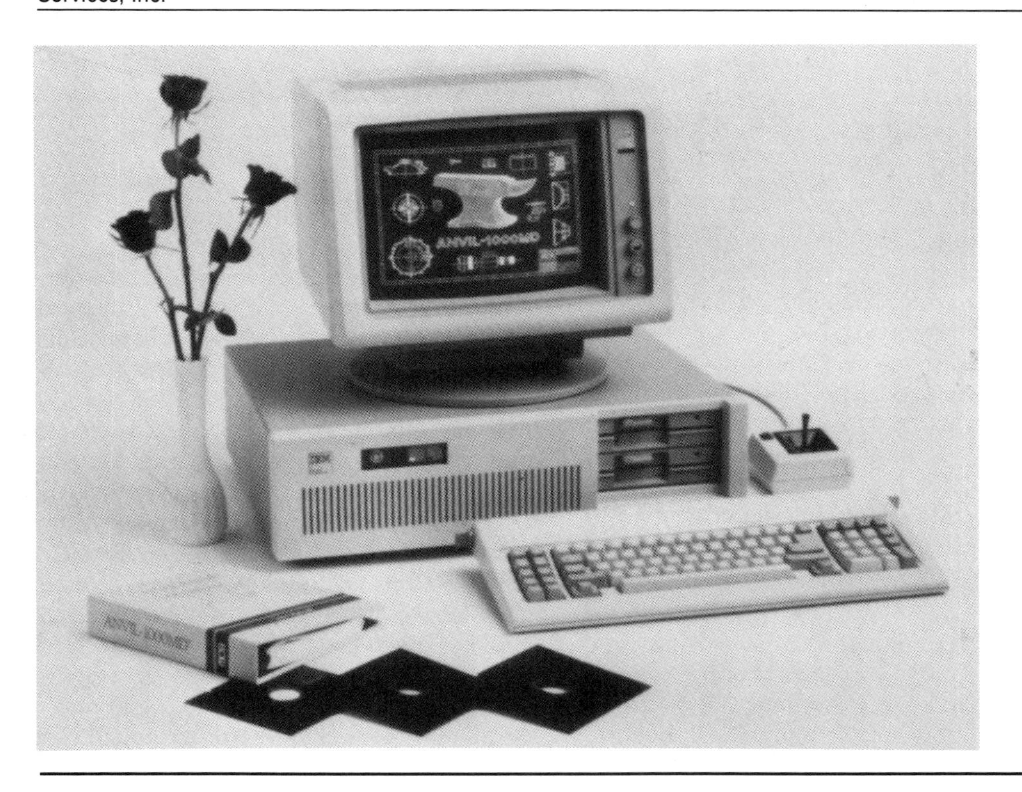

building perhaps three or four prototype models. It also allows many drawings to be constructed more quickly, especially by an experienced CAD operator. Comparisons of CAD design time with manual systems show CAD ranging from 0.5 to 100 times as fast, with 2 to 6 times as fast being typical. For instance, Prototype and Plastic Mold Corporation in Middletown, Connecticut, is a small firm that uses CAD to design short-lived metal molds for plastic parts. The firm's president reported that designs could be produced with CAD roughly twice as fast as previously. For example, they received specifications for a plastic part mold by air express one Saturday morning and planned to return the design drawings by air express that evening—a feat that, they reported, would have been impossible without CAD.

Other applications of CAD, though not directly connected to manufacturing, include architectural drawing and design, graphics for technical publishing, and animation in cinematography.

Figure 8
Cincinnati-Milicron Robot. Source: Cincinnati Milacron.

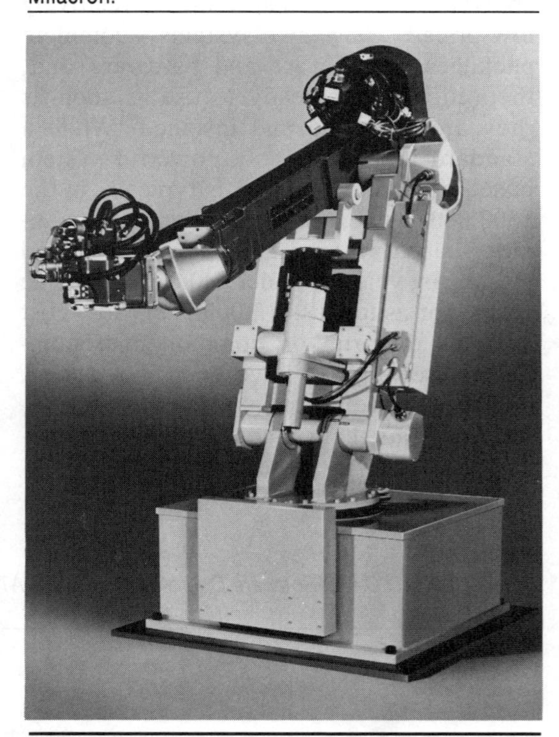

Computer-Aided Manufacturing (CAM) Technologies

Computer-aided manufacturing (CAM) is a widely used term in industrial literature, and it has various meanings. Here it is defined simply as those types of CAM systems used primarily on the factory floor to help produce products. The following sections provide functional descriptions of four CAM tools: 1) robots, 2) numerically controlled machine tools, 3) flexible manufacturing systems, and 4) automated materials handling systems.

Robots

Robots are manipulators that can be programmed to move workpieces or tools along various paths. Most dictionary definitions describe robots as "humanlike," but industrial robots bear little resemblance to a human, as shown in Figure 8.

There is some controversy over the definition of a robot. The Japan Industrial Robot Association, for example, construes almost any machine that manipulates objects to be a robot (this includes most of their very simple robot systems), while the often quoted Robotics Industries Association (RIA) definition emphasizes that the robot must be flexible or relatively easily changed from one task to another. The RIA definition thus excludes preset part-transfer machines—whose path can be changed only by mechanically reworking or rearranging the device—that have been used for decades as a part of large-batch and mass-production systems. Also excluded are *manual manipulators* or *teleoperators*—devices like those for remote handling of radioactive material—that are directly controlled by a human.

As OTA observed in an earlier report on this subject, industrial robots have a dual technological ancestry, emerging from 1) industrial engineering automation technology, a discipline that stretches historically over a century and 2) computer science and artificial intelligence technology, which is only a few decades old. Indeed, there is still disagreement among experts regarding the applications and research directions for robotics. Some emphasize the need for anthropomorphic capabilities in robots such as "intelligence," vision, and mobility, while others view robots simply as more versatile extensions of other manufacturing tools.

Although it is uncertain to what extent artificial intelligence researchers will succeed in developing intelligent machines in the next few decades, it is certain that robots currently available, and those likely to be available in the next decade, neither look like humans nor have more than a fraction of the dexterity, flexibility, or intelligence of humans. Some believe a more accurate term for these machines might be *programmable manipulators*. Nevertheless, it is clear that much of the great popular interest in robotics is rooted in the prevailing vision (or nightmare) of intelligent robots with humanlike characteristics.

How Robots Work

There are three main parts of a typical industrial robot: 1) the computer controller, 2) the power drive unit, and 3) the manipulator/end-effector (see Fig. 9). The con-

Figure 9
The Computer Controller, Power Drive Unit, and Manipulator/End-Effector are the Basic Elements of a Robot System. Source: Cincinnati Milacron.

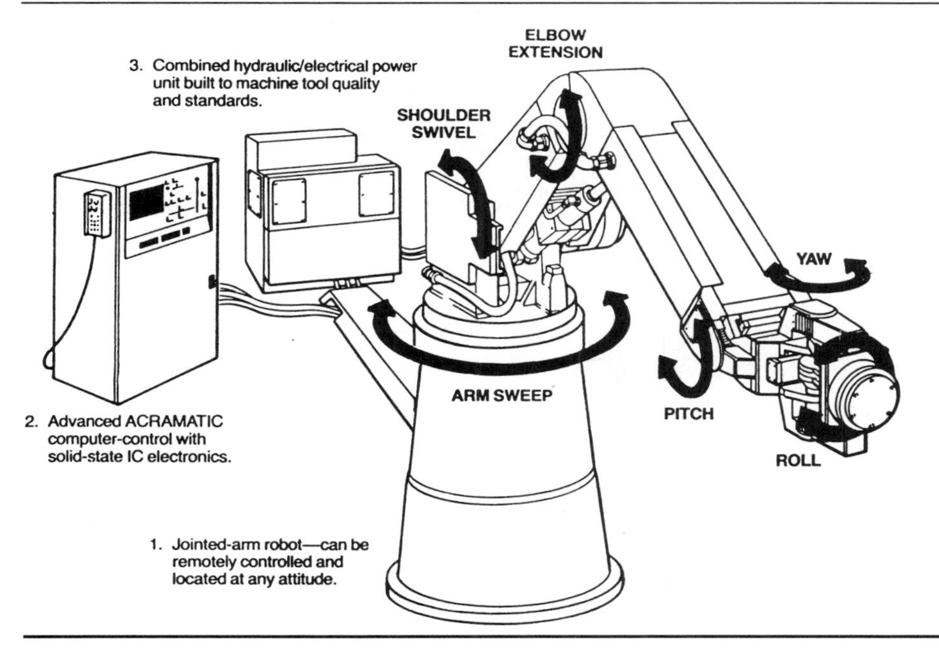

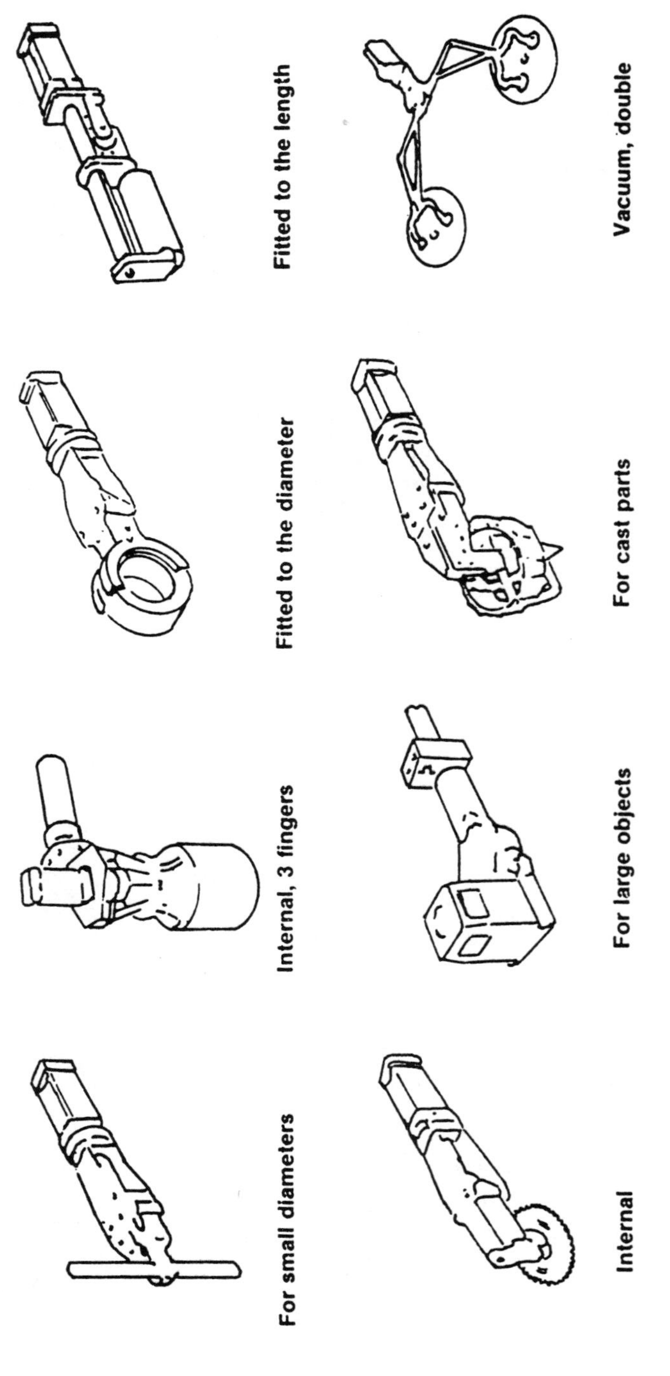

Figure 10
Sample Robot Grippers. Source: Tech Tran Corp., *Industrial robots: a summary and forecast.*

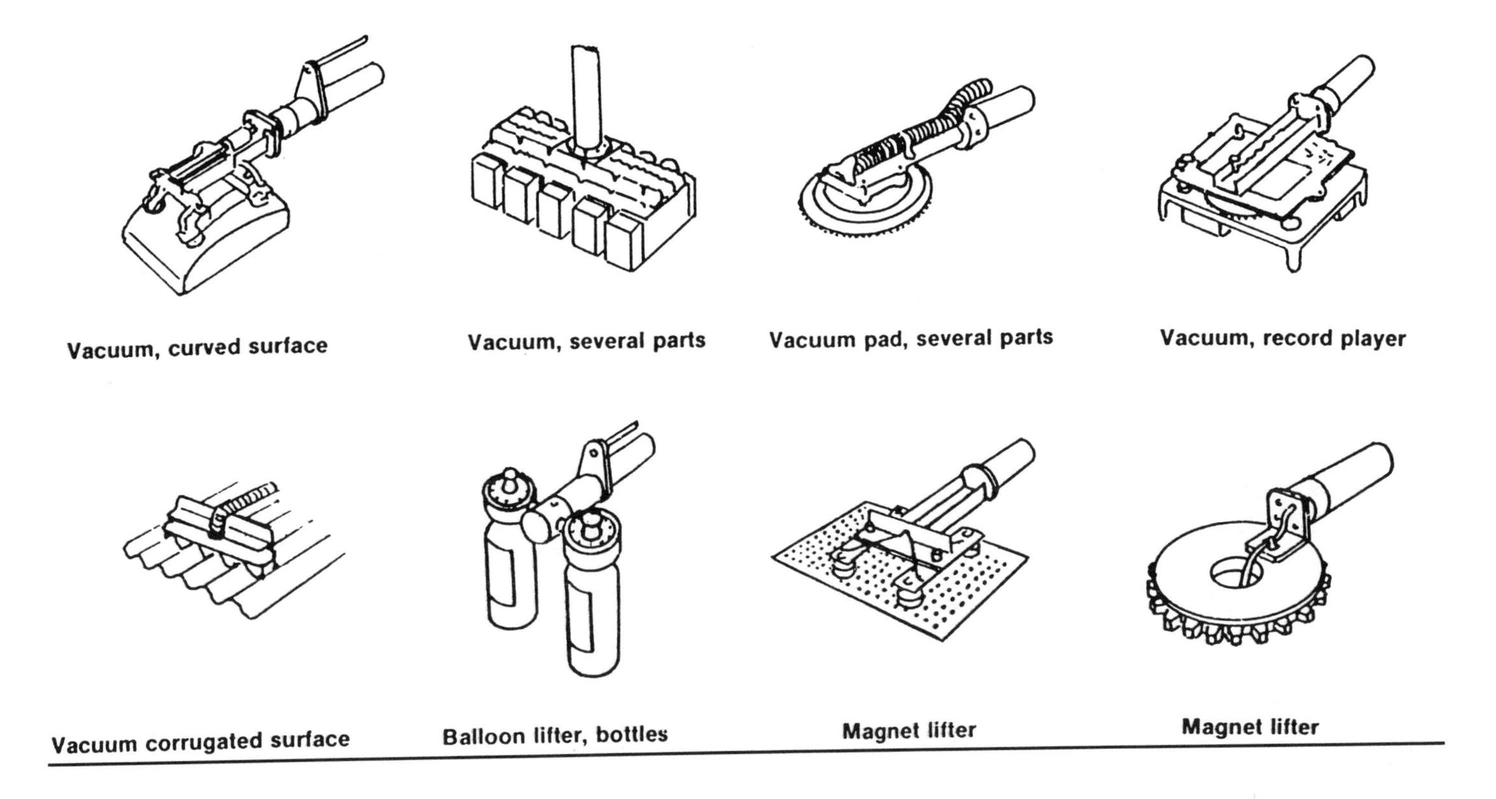
Vacuum, curved surface
Vacuum, several parts
Vacuum pad, several parts
Vacuum, record player
Vacuum corrugated surface
Balloon lifter, bottles
Magnet lifter
Magnet lifter

troller consists of the hardware and software—usually involving a microcomputer or microelectronic components—that guide the motions of the robot and through which the operator programs the machine. The manipulator consists of a base, usually bolted to the floor, an actuation mechanism—the electric, hydraulic, or pneumatic apparatus that moves the arm—and the arm itself, which can be configured in various ways to move through particular patterns. In the arm, *degrees of freedom*—basically, the number of different joints—determine the robot's dexterity, as well as its complexity and cost. Finally, the end-effector, usually not sold as part of the robot, is the gripper, weld gun, spray painting nozzle, or other tool used by the robot to perform its task.

The structure, size, and complexity of the robot varies, depending on the application and the industrial environment. Robots designed to carry lighter loads tend to be smaller and are operated electrically; many heavier units move their manipulator hydraulically. Some of the simpler units are pneumatic. Some of the heaviest material-handling robots and the newer light-assembly robots are arranged gantry-style—that is, with the manipulator hanging from an overhead support. A few robots are mobile to a limited degree, for example, they can roll along fixed tracks in the floor or in their gantry supports.

Similarly, there is a great variety of end-effectors, particularly grippers, most of which are customized for particular applications. Grippers are available to lift several objects at once or to grasp a fragile object without damaging it (Fig. 10, pp. 12–13).

Programming

There are essentially two methods of programming a robot. The most commonly used method is *teaching by guiding*. The worker either physically guides the robot through its path or uses switches on a control panel to move the arm. The controller records that path as it is "taught." This process is rather slow and ties up valuable production equipment for programming. Now beginning to emerge is *offline programming*, where an operator writes a program in computer language at a computer terminal and directs the robot to follow the program.

Each method of programming has advantages that depend on the application. Teaching by guiding is the simplest and is actually superior for certain operations; in spray painting, for example, it is useful to have the operator guide the robot arm through its path because of the continuous curved motions usually necessary for even paint coverage. However, teaching by guiding offers minimal ability to "edit" a path, that is, to modify a portion of the path without rerecording the entire path. Offline programming is useful for several reasons: (1) production need not be stopped while the robot is being programmed; (2) the factory floor may be an inhospitable environment for programming, and offline programming can be done at a computer terminal in an office; (3) as computer-aided manufacturing technologies become more advanced and integrated, they will increasingly be able to generate robot programs automatically from design and manufacturing data bases; and (4) an offline written program can better accommodate more complex tasks, especially those in which *branching* is involved (e.g., "if the part is not present, then wait for the next cycle"). These branching decisions require some kind of mechanism by which the robot can sense its external environment. However, the vast majority of robotic devices are unable to sense their environment, although they may have internal sensors to provide feedback to their controller on the position of the arm joints.

Sensors

Devices for sensing the external environment, while often used in conjunction with

Table 2.
International Robot Installations

Country	Number
Japan	74,000
United States	22,000
West Germany	6,600
France	3,380
Italy	2,700
United Kingdom	2,623
Sweden	2,400
Belgium	860
Poland	285
Canada	273
Czechoslovakia	154
Finland	98
Switzerland	73
Netherlands	71
Denmark	63
Austria	50
Singapore	25
Korea	10
Total	97,265

robots, are a growing technology in themselves. The simplest sensors answer the question, Is something there or not? For example, a light detector mounted beside a conveyor belt can signal when a part has arrived because the part breaks a light beam. Somewhat more complex are proximity sensors that, by bouncing sound off objects, can estimate how far away they are. The technology for these devices is fairly well-established. The most powerful sensors, however, are those that can interpret visual or tactile information; these have just begun to be practical.

Ideally, vision sensors could allow a robot system to respond to changes in its environment and to inspect products as well as or better than a human could. However, using computers to process images from a video camera has proven to be a difficult programming task. Routine variations in lighting the complexity of the everyday environment, common variations in shape or texture, and the difference between a 2-D camera image and a 3-D world all complicate the task of computer processing of a video image.

Other kinds of sensing devices, from proximity sensors to touch and force sensors, have received much less attention than machine vision, but they also play an important role in the factory environment, and are especially significant in assembly applications.

Applications

Table 2 displays some of the most recent estimates of the number of international robot installations by country. Figure 11 estimates the sale and total use of robots in the United States from 1980 through 1986. Such statistics should be interpreted with caution, however. In particular, the number of robots in use is a highly imperfect measure of the level of automation and modernization in an industry or country. Process changes in manufacturing that increase productivity may or may not include robots.

It is also important that robots be viewed as part of the *overall* changes taking place

Figure 11
Actual and Projected U.S. Annual Robot Sales

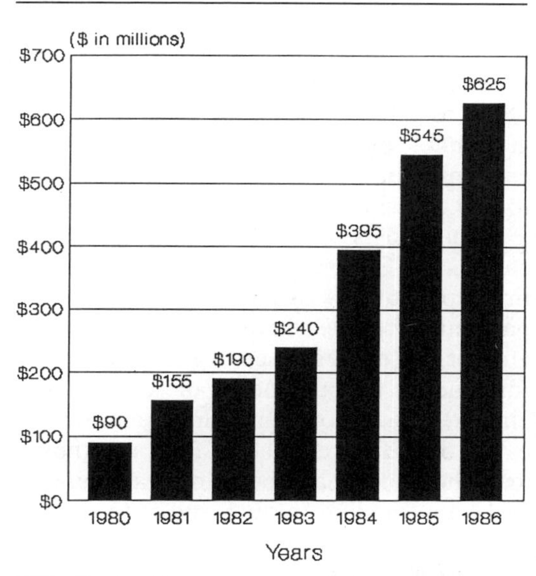

in manufacturing concepts, as automated systems for design and production are installed in greater numbers. The impact of new production concepts, equipment, and systems on production control and machine utilization, inventory control, and management efficiency will—all taken together—have a much greater productivity impact than will the industrial robot alone.

As noted earlier in this section, international comparisons of robot "populations" are also plagued by inconsistencies in the definition of a robot, particularly between the United States and Japan. Regardless of the definition of robot used, Japan leads the world in number of robots in use. The reasons for Japan's emphasis on robot technology include a historical shortage of labor and a tendency to devote more skilled engineering expertise to manufacturing processes than does the United States. In addition, the United States had labor surpluses throughout the 1970s, a situation that tended to induce manufacturers to use labor instead of equipment in production.

Sophistication in reprogrammability, size, and degree of freedom, are some of the key cost factors for an industrial robot. A simple *pick-and-place* machine with two or three degrees of freedom costs roughly $5000 to $30,000, while more complex programmable models, often equipped with microcomputers, cost approximately $25,000 to $90,000 and up.

Table 3 lists some of the potential applications for industrial robots. Many of the first applications of robots have been for particularly unpleasant or dangerous tasks. One of the earliest uses, for example, was for loading and unloading die-casting machines, a hazardous and unpleasant job because of the extreme heat. The best-known uses, however, have been in spray painting and spot welding in auto and auto-related industries, where robots have proved useful for performing particularly hazardous and monotonous jobs, while offering enough

Table 3.
Examples of Current Robot Applications

Material Handling
Depalletizing wheel spindles onto conveyors
Transporting explosive devices
Packaging toaster ovens
Stacking engine parts
Transfer of auto parts from machine to overhead conveyor
Transfer of turbine parts from one conveyor to another
Loading transmission cases from roller conveyor to monorail
Transfer of finished auto engines from assembly to hot test
Processing of thermometers
Bottle loading
Transfer of glass from rack to cutting line

Machine Loading/Unloading
Loading auto parts for grinding
Loading auto components into test machines
Loading gears onto CNC lathes
Orienting/loading transmission parts onto transfer machines
Loading hot-form presses
Loading transmission ring gears into vertical lathes
Loading of electron beam welder
Loading cylinder heads into transfer machines
Loading a punch press
Loading die-casting machines

Spray Painting
Painting of aircraft parts on automated line
Painting of truck bed
Painting of underside of agricultural equipment
Application of prime coat to truck cabs
Application of thermal material to rockets
Painting of appliance components

Welding
Spot welding of auto bodies
Welding front-end loader buckets
Arc welding of hinge assemblies on agricultural equipment
Braze alloying of aircraft seams
Arc welding of tractor front weight supports
Arc welding of auto axles

Machining
Drilling aluminium panels on aircraft
Metal flash removal from castings
Sanding missile wings

Assembly
Assembly of aircraft parts (used with auto-rivet equipment)
Riveting small assemblies
Drilling and fastening metal panels
Assembling appliance switches
Inserting and fastening screws

Other
Application of two-part urethane gasket to auto part
Application of adhesive
Induction hardening
Inspecting dimensions on parts
Inspection of hole diameter and wall thickness

Source: Tech Tran Corp., *Industrial robots: A summary and forecast,* Naperville, Ill., 1986.

flexibility to be easily adapted to changes in car models or body styles.

There are a number of motivations for using robots for such unpleasant jobs. Improvement of job conditions (and, consequently, worker morale) is one of them, although it may not be the primary one. The unpleasant conditions often create high worker turnover and inconsistent product quality. Furthermore, compliance with occupational safety and health regulations that protect people performing these tasks adds to production costs. In addition, tasks like spray painting and spot welding are often relatively easy to automate because the paths the robot is to follow are predictable, and the tasks are repetitive and require little sensing capability.

While spot welding, spray painting, and loading/unloading applications have been the primary uses for robots to date, increasing sophistication in programmability and in sensing is now providing the basis for applications such as arc welding and assembly.

As an example of such an application, a welder at Emhart Corporation's United Shoe Manufacturing plant in Beverly, Massachusetts, uses a robot to arc-weld frames for shoemaking machinery. Several dozen identical frame units are welded at a time, and each frame unit requires perhaps a dozen 2-inch welds to attach reinforcing bars to a steel sheet. The welder clamps the first sheet and reinforcing bars onto a table. Using directional buttons on a *teach pendant*—a portable panel attached to the robot's controller—the welder directs the robot to the spot where it is to begin the first weld. Then, pushing a button to record that location, and still using the teach pendant, the welder moves the robot to the end of the weld and records that location. At that point, the welder presses a button that instructs the machine to "weld a straight line from the first point to the second." Once the welder repeats that entire process for all twelve welds, the welder gives the command for the robot to begin welding, and the robot follows the path it has been led through—this time with its welding gun on. For each subsequent identical frame unit, the only requirements are to clamp down the parts in the same location as the original set on which the machine was "taught," signal the machine to begin, and then inspect the welds after the machine completes its program. The robot controller can store several programs, so that the operator can use the robot to weld different types of frames in any order, as long as the steel plates and reinforcing bars are set up in the appropriate positions.

Note that this application of a robot for arc welding does not use sensors, even though extensive work has been done on developing vision sensors that allow the robot to "see" the seam formed by the two pieces of metal and to follow it automatically. For the fairly simple straightline applications at Emhart, sensors are not necessary. However, if the frame units were out of position by a half inch, the welding robot would put a useless blob of metal where it expected the joint to be. If a clamp were in the way of a programmed weld, the robot would attempt to weld through the clamp, damaging the clamp and itself in the process.

As shown in Figure 12, robots offer different advantages, depending on whether they are compared to hard automation devices or to human workers. Clearly, the flexibility and programmability of robots is prominent in the first case; in comparison with humans, the advantages are likely to be the robot's greater consistency in producing quality work, its endurance, and its ability to tolerate hostile environments.

The disadvantages of robots also depend on what they are compared to: other automation or humans. In the former case, a robotic device is sometimes more expensive than a hard automation device that is not programmable; furthermore, a robot is not as fast—a typical robot moves about as

Figure 12
Optimal Employment of Robots Is in Programmable and Flexible Performance in Mid-Range Applications

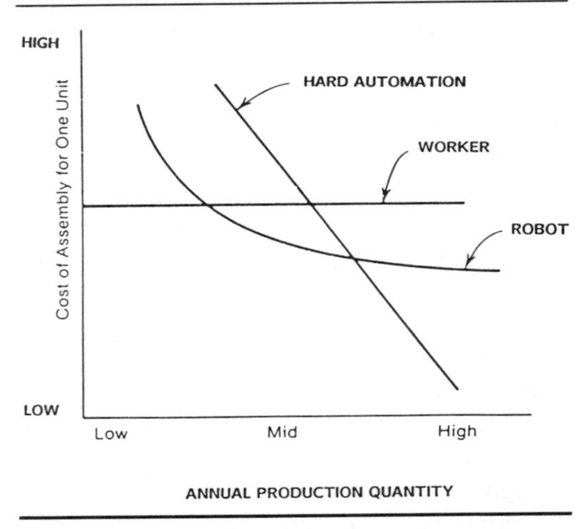

fast as a human, while dedicated automatic part-transfer devices can operate at considerably greater speed. The clear advantage of human workers over robots, on the other hand, comes in situations where extensive sensing, judgment, or intelligence is required and/or where circumstances change so frequently that the expense of programming a robot is uneconomical. For these reasons, it is often suggested that humans, robots, and hard automation devices are best suited for low, medium, and high production volumes, respectively, although there are many exceptions to this, for example, automotive spot welding. Each situation must be evaluated individually.

The design of automated production processes involves determining which tasks are most suitable for a machine and which are most suitable for a human. Several technology experts have argued that some manufacturers' visions of robots as replacements for human workers will prevent the best allocation of tasks between human and machine.

General-purpose robots are already evolving toward special-purpose programmable devices for a particular task (assembly machines and painting machines, for example); if this evolution continues, few robots in the future will look like the general-purpose "arm" of today.

Numerically Controlled Machine Tools

Numerically controlled (NC) machine tools are devices that cut a piece of metal according to programmed instructions concerning the desired dimensions of a part and the steps for the machining process. These devices consist of a machine tool, specially equipped with motors to guide the cutting process, and a controller that receives numerical control commands, as shown in Figure 13.

The U.S. Air Force developed NC technology in the 1940s and 1950s, in large part to help produce complex parts for aircraft that were difficult to make reliably and economically with a manually guided machine tool.

How They Work. Machine tools for cutting and forming metal are the heart of the metalworking industry. Using a conventional, manual machine tool, a machinist guides the shaping of a metal part by hand, moving either the workpiece or the head of the cutting tool to produce the desired shape of the part. The speed of the cut, the flow of coolant, and all other relevant aspects of the machining process are controlled by the machinist.

In ordinary NC machines, programs are written at a terminal that, in turn, punches holes in a paper or Mylar plastic tape. The tape is then fed into the NC controller. Each set of holes represents a command that is transmitted to the motors, guiding the machine tool by relays and other electromechanical switches. Although these machines are not computerized, they are programmable in the sense that the machine can easily be set to making a different part by feeding it a different punched tape; they are automated in that the machine moves its cutting head, adjusts its coolant, and so forth, without direct human intervention. However, most of these machines still require a human operator, though in some cases one person operates two or more NC machine tools. The operator supervises several critical aspects of the machine's operation, as follows:

1. He or she has override control to modify the programmed speeds (rate of motion of the cutting tool) and feeds (rate of cut). These rates will vary depending on the type of metal used and the condition of the cutting tool.
2. The operator watches the quality and dimensions of the cut and listens to the tool, ideally replacing worn tools before they fail.
3. He or she monitors the process to avoid accidents or damage, such as a tool cutting into a misplaced clamp or a blocked coolant line.

Typically, NC programs are written in a language called APT (Automatically Programmed Tools), which was developed during the initial Air Force research on NC. A number of modified versions of APT have been released in the last decade, and some of these are easier to use than the original. The essential concept and structure of the numerical codes, however, has remained the same. APT has become a de facto standard for NC machine tools, in large part because of the momentum it gained from its initial DOD support.

Since 1975, machine tool manufacturers have begun to use microprocessors in the controller, and some NC machines—called computerized numerically controlled (CNC) machines—come equipped with a dedicated minicomputer. CNC machines are equipped with a screen and keyboard for writing or editing programs at the machine; closely related to CNC is direct numerical control (DNC), in which a larger

Figure 13
Block Diagram of Numerically Controlled Machine Tool. Reprinted with Permission of Reston Publishing Company, Prentice-Hall, from Robotics and Automated Manufacturing *by Richard Dorf.*

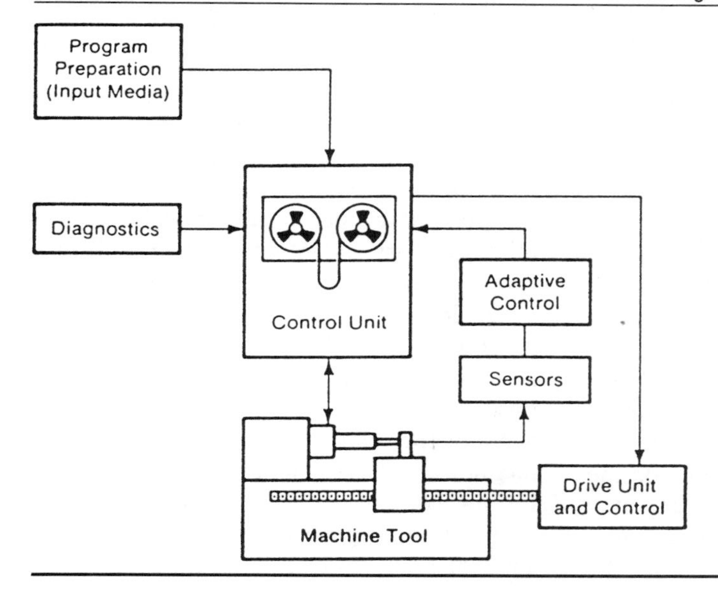

minicomputer or mainframe computer is used to program and run more than one NC tool simultaneously. As the price of small computers has declined over the past decade, DNC has evolved both in meaning and concept into *distributed numerical control*, in which each machine tool has a microcomputer of its own, and where the systems are linked to a central controlling computer. One of the advantages of such distributed control is that the machines can often continue working for some time even if the central computer "goes down."

In all types of NC machine tools, the machining processes are essentially the same: the difference is in the sophistication and location of the controller. CNC controllers allow the operator to edit the program at the machine, rather than sending a tape back to a programmer in a computer room for changes. In addition, because they do not require the use of paper or Mylar tape, CNC and DNC machines are substantially more reliable than ordinary NC machines. The tape punchers and readers and the tape itself have been notable trouble spots; in addition, CNC and DNC machines, through their computer screens, offer the operator more complete information about the status of the machining process. Apart from those features associated with CNC and DNC, some NC tools are equipped with a feature called *adaptive control* that tries to automatically optimize the rates of cut to produce the part as fast as possible, while avoiding tool failure.

Figure 14
Total Number of Numerically Controlled Machine Tools in U.S. Metalworking. Source: 13th American Machinist inventory, *American Machinist.*

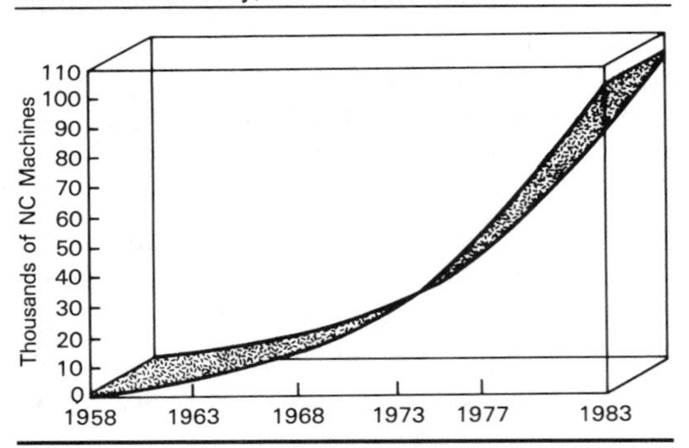

Applications. The diffusion of NC technology into the metalworking industry proceeded very slowly in the 1950s and 1960s, though it has accelerated somewhat over the past 10 years. Figure 14 and Table 4 detail the U.S. population of machine tools. In 1983, numerically controlled machine tools represented only 4.7 percent of the total population, although this figure may be somewhat misleading; the newer NC machine tools tend to be used more than the older equipment, and firms often keep old equipment even when they buy new machines. Some industry experts have estimated that as many as half of the parts made in machine shops are made using NC equipment. Nevertheless, the applications still tend to be concentrated in large firms and in smaller subcontractors in the aerospace and defense industries.

The U.S. machine tool population is significantly older than that of most other countries (see Table 5), and this situation, suggesting relatively low levels of capital investment, has been a source of concern for many in industry and government. In 1983, for the first time in several decades, the percentage of metalcutting tools less than 10 years old increased by 3 percent, although the percentage of metalforming tools less than 10 years old remained at an all-time low of 27 percent.

DOD has encouraged diffusion of NC technology, which has moved beyond the aerospace industry—but not as fast as most observers expected. There are several reasons for the relatively slow diffusion of NC technology. These include high capital cost for an NC machine (perhaps $80,000 to $150,000 and up, as opposed to $10,000 to $30,000 for a conventional machine tool).

Table 4.
Estimated Total Machine Tools in the United States

	Total Units	Metalcutting	Metalforming
Metalworking	2,192,754	1,702,833	489,921
Other industries	380,000	275,000	105,000
Training	74,000	70,000	4,000
In storage and surplus	250,000	200,000	50,000
Total	2,896,754	2,247,833	648,921

Source: 13th American Machinist inventory, *American Machinist.*

In addition, the successful application of NC machine tools requires technical expertise that is in short supply in many machine shops. Training is also a problem, as some users report requiring as much as 2 years to get an NC programmer "up to speed"; small machine shops typically do not have the resources or expertise to train staff to use or maintain computerized equipment. Finally, according to one source, "APT proved to be too complicated for most users outside the aerospace industry. . . . Most machine jobs could be specified in a considerably less complex world."

In spite of the roadblocks to implementing NC, there are some clear advantages. Intricate shapes, such as those now found in the aerospace industry, are nearly impossible for even the most experienced machinist using conventional machine tools. With NC, the parts can be more consistent, because the same NC program is used to make the part each time it is produced. A manually guided machine tool is more likely to produce parts with slight variations, because the machinist is likely to use a slightly different procedure each time he or she makes a part. This may not be a concern in one-of-a-kind or custom production, but it can create problems in batch production. The advantages in consistency brought by NC are seen by many manufacturers as an increase in their control over the machining process.

NC machines tend to have a higher *throughput* than conventional machine tools and hence are more productive. They operate (i.e., cut metal) more of the time than do conventional machine tools, because all the steps are established before the machining begins and are followed methodically by the machine's controller. Further,

Table 5.
Age of Machine Tools in Seven Industrial Nations

		Metalcutting Machines			**Metalforming Machines**		
	Year	Units	≥15 years	≥20 years	Units	≥15 years	≥20 years
United States	83	1,703,000	—	32%	490,000	—	37%
Canada	78	149,400	—	37	61,400	—	26
Federal Republic of Germany	80	985,000	48%	—	265,000	48%	—
France	80	584,000	—	32	177,000	—	32
Italy	75	408,300	—	29	133,000	—	25
Japan	81	707,000	37	—	211,000	31	—
United Kingdom	82	627,900	—	27	146,800	—	28

Source: 13th American Machinist inventory, *American Machinist.* (Note: *American Machinist* used a variety of foreign sources for this table.)

on a complex part that takes more than one shift of machining on a conventional machine tool, it is very difficult for a new machinist to take over where the first left off. The part may remain clamped to the machine and the part and machine tool may lie idle until the original machinist returns. On NC machines, operators can substitute for each other relatively easily, allowing the machining to continue uninterrupted.

As discussed previously, the capability of guiding machine tools with numeric codes opens up possibilities for streamlining the steps between design and production. The geometric data developed in drawing the product on a CAD system can be used to generate the NC program for manufacturing the product.

Flexible Manufacturing Systems

A flexible manufacturing system (FMS), shown in Figure 15, is a production unit capable of producing a range of discrete products with a minimum of manual intervention. It consists of production equipment cells or workstations (machine tools or other equipment for fabrication, assembly, or treatment) linked by a materials-handling system to move parts from one workstation to another, and it operates as an integrated system under full programmable control.

An FMS is often designed to produce a family of related parts, usually in relatively small batches—in many cases less than 100 and even as low as one. Most systems appropriately considered to be FMS include at least four workstations, while some have up to 32. Smaller systems of two or three machine tools served by a robot, which are also called flexible manufacturing systems in some circumstances, are more appropriately termed flexible *machining cells*.

How an FMS Works.

Using NC programs and, often, computer-aided process planning, workers develop the process plan (i.e., the sequence of production steps) for each part that the FMS produces. Then, based on inventory, orders, and computer simulations of how the FMS could run most effectively, the FMS managers establish a schedule for the parts that the FMS will produce on a given day. Next, operators feed the material for each part into the system, typically by clamping a block of metal into a special carrier that serves both as a fixture to hold the part in place while it is being machined and as a pallet for transporting the workpiece. Once loaded, the system itself essentially takes over. Robots, conveyors, or other automated materials-handling devices transport the workpiece from workstation to workstation, according to the process plan. If a tool is not working, many FMSs can reroute the part to other tools that can be substituted for the defective unit.

Maching tools are not the only workstations in an FMS; other possible stations include washing or heat-treating machines and automatic inspection devices. While most current FMSs consist of groups of machine tools, other systems anticipated or in operation involve machines for grinding, sheet metal working, plastics handling, and assembly.

The amount of flexibility necessary to deserve the label "flexible" is arguable. Some FMSs can produce only three or four parts of very similar size and shape, for example, three or four engine blocks for different configurations of engines. One FMS expert argues, however, that in the current state of the technology, a system that cannot produce at least 20 to 25 different parts is not flexible. Indeed, some are being designed to manufacture up to 500 parts.

The essential features that constitute a workable "part family" for an FMS are:

A common shape. In particular, prismatic (primarily flat surfaces) and rotational parts cannot be produced by the same set of machines.

Figure 15
Block Diagram of a Flexible Manufacturing System. Reprinted with Permission of Reston Publishing Company, Prentice-Hall, from Robotics and Automated Manufacturing *by Richard Dorf.*

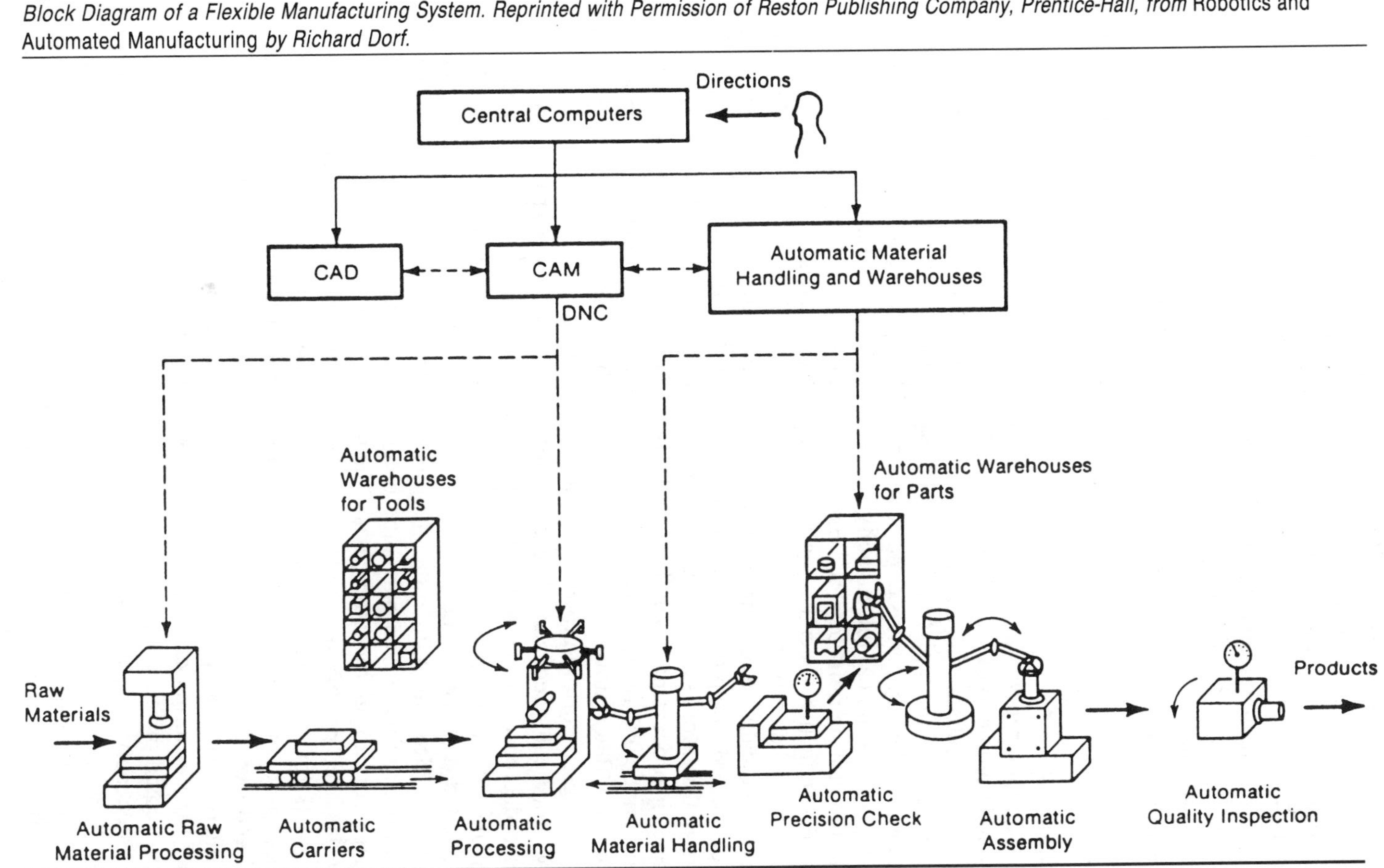

Size. An FMS will be designed to produce parts of a certain maximum size, for instance, a 36-inch cube. Parts that are larger or very much smaller may not be handled.

Material. Titanium and common steel parts cannot be effectively mixed, nor can metal and plastic.

Tolerance. The level of precision necessary for the set of parts must be in a common range.

Applications.

For a manufacturer with an appropriate part family and volume to use an FMS, the technology offers substantial advantages over stand-alone machine tools. In an ideal FMS arrangement, the company's expensive machine tools work at near full capacity. Turnaround time for manufacture of a part is reduced dramatically, because parts move from one workstation to another quickly and systematically, and computer simulations of the FMS help to determine optimal routing paths. Most systems have some redundancy in processing capabilities, and this can automatically reroute parts around a machine tool that is down. Because of these time savings, work-in-process inventory can be drastically reduced. The company can also decrease its inventory of finished parts, since it can rely on the FMS to produce needed parts on demand.

Finally, FMSs can reduce the *economic order quantity* for a given part—the batch size necessary to justify setup costs. When a part has been produced once on an FMS, setup costs for later batches are minimal, because process plans are already established and stored in memory, and materials handling is automatic. In the ultimate vision of an FMS, in terms of cost per unit, the machine could produce a one-part batch almost as cheaply as it could produce 1000. While there are, in practice, unavoidable setup costs for a part, the FMS's capability to lower the economic order quantity is particularly useful in an economy where manufacturers perceive an increased demand for product customization and smaller batch sizes.

A midwestern agricultural equipment manufacturer, for example, uses an FMS to machine transmission case and clutch housings for a family of tractors. The company had considered *hard automation*—a transfer line—to manufacture the parts but expected a new generation of transmissions within 5 years that would render the transfer line obsolete. They chose an FMS instead, because it could be more easily adapted to other products. In the system, a supervisory computer controls 12 computerized machining centers and a system of chain-driven carts that shuttle the fixtured parts to the appropriate machines. The supervisory computer automatically routes parts to those machines with the shortest queue of workpieces waiting and can reroute parts to avoid a disabled machine tool. About a dozen employees operate and maintain the system during the day shift, and there are even fewer people on the other two shifts. The system is designed to produce nine part types in almost any sequence desired. (It is, therefore, rather inflexible according to the current state of the art.) This system, in fact, was one of the earliest FMSs of substantial size to be designed. It was ordered in 1978 but not fully implemented until 1981.

Despite the advantages claimed for FMS, there are relatively few systems installed. Observers estimate that there are 50 such systems in Japan, 20 each in Western and Eastern Europe, and 50 in the United States. The reasons for this scarcity of application include the complexity, newness, and cost of the systems. One American manufacturer estimated that an FMS system costs $600,000 to $800,000 per machining workstation, with a minimum expenditure of $3 million to $4 million. In addition, the in-house costs of planning for installation of an FMS—a process that often takes several years—are likely to substantially increase the investment.

Automated Materials Handling Systems

Automated materials handling (AMH) systems store and move products and materials under computer control. Some AMH systems are used primarily to shuttle items to the work areas or between workstations on automated carts or conveyors, as shown in Figure 16. Automated storage and retrieval systems (AS/RS) are another form of automated materials handling, essentially comprising an automated warehouse, where parts are stored in racks and retrieved on computerized carts and lift trucks. For the purposes of this overview, this category includes only those materials-handling systems that are not classified as robots.

How AMH Systems Work

There are a wide variety of formats for automated materials handling, including conveyors, monorails, towlines, motorized carts riding on tracks, and automated carriers that follow wires embedded in the floor of the factory. Each AMH system is unique, and each is designed for the materials-han-

Figure 16
Automated Material-Handling System at Hughes Aircraft Co. Source: Hughes Aircraft Co.

dling needs of a particular factory. The common characteristics of these devices is that they are controlled by a central computer.

There are three general applications for AMH. The first is to shuttle workpieces between stations on an FMS. In this case, the AMH system operates on commands from the FMS controller. For example, when the controller receives a message that a machine tool has finished work on a certain workpiece, the controller orders the AMH system to pick up the workpiece and deliver it to the next workstation in its routing. The materials-handling portion of the FMS is one of its trickiest elements, part-transport needs tend to be logistically complicated, and the AMH system must place the part accurately and reliably for machining.

Figure 17
Chain-Driven AMH System at General Electric's Diesel Engine Facility in Erie, Penn.

Many AMH systems, such as conveyors or tow chains (shown in Fig. 17), are serial in nature—that is, there is only one path from point A to point B. This has caused FMSs to cease operating when a cart becomes stuck or a critical path becomes unusable. FMS designers have responded to this problem by designing AMH systems with backup paths and by using systems, such as the wire-guided vehicle mentioned earlier, that can be routed around disabled carts or other obstacles.

The second major application of AMH is for transporting work-in-process from one manufacturing stage to the next, within a factory. This application is similar in concept to AMH use for a flexible manufacturing system, although serving an entire factory is more complex: there is more area to cover, more potential obstacles and logistical difficulties in establishing paths for the AMH carriers, and a wider range of materials to handle. For this reason, whole-factory AMH systems are not yet widely used. However, a few years ago, General Motors agreed to purchase automatic guided vehicles from Volvo that allow automobiles to proceed independently through the plant while being assembled, as shown in Figure 18. The *robot carts* can be programmed to stop at appropriate workstations, and the cart system essentially replaces an assembly line. Volvo uses about 2000 of the carts in its own plants in Europe, and Fiat also uses such carts in Italy.

The final application for AMH is in automated storage and retrieval systems. These storage rack systems are often very tall in order to conserve space and to limit the number of automatic carrier devices needed to service the facility. In many cases, the structure housing the AS/RS is a separate building adjacent to the main factory. Design of an AS/RS depends on the size of the products stored, the volume of material to be stored, and the speed and frequency of items moving in and out of the

Figure 18
Independently Controlled AMH System Can Move Whole Cars to Required Production Workstation

system. Advocates of AS/RS cite advantages for the system as compared to non-automated systems that include reduced land needs for the plant, fewer (but more highly trained) staff, more accurate inventory records, and reduced energy use.

Applications

In theory, AMH systems can move material quickly, efficiently, and reliably, while also keeping better track of the location and quantities of the parts by use of the computer's memory, thus avoiding much paperwork. They can therefore minimize loss of parts in a factory—a common problem in materials handling.

Deere & Company, for example, uses an extensive AS/RS to store materials and inventory at one of its tractor plants. The system's computerized controller keeps track of the products stored on the shelves, and workers can order the system to retrieve parts from the shelves by typing commands at a computer terminal. After they are retrieved from the AS/RS, the parts can be automatically carried by overhead conveyors to the desired location within the plant complex.

IBM's Poughkeepsie plant has developed an AMH conveyor cart system for transporting a 65-pound computer subassembly fixture between assembly and testing stations. The manufacturing manager reports that the decision to adopt this system was prompted by logistical difficulties in keeping track of many such fixtures among a great variety of workstations, as well as by worker health problems related to transporting the fixtures manually.

AMH systems often have reliability problems in practice. According to a Deere & Company executive, for example, Deere's AS/RS was systematically reporting more engines stored on the racks than other records indicated. After long weeks of searching for the problem, the plant staff finally found the culprit: a leak in the roof was allowing water to drip past the photocell that counted the engines as they were stored. Each drip, in essence, became an engine in the computer's inventory.

Although Deere's experience is not widely applicable to AS/RSs, the notion that AMH systems present unexpected logistical and mechanical problems does seem to be generally accurate. In spite of the fact that these systems are key aspects of flexible manufacturing systems and of computer-integrated manufacturing, materials handling has long been a neglected topic in industrial research. Materials-handling system manufacturers have only recently "caught up" to other industrial systems in level of sophistication, and few companies have so far installed sophisticated AMH systems. Because of this relative lack of sophistication, materials handling for FMS and CIM, especially for a complex application, such as delivery of multiple parts to an assembly station, may be one of the biggest problems facing integrated automation.

Other CAM Equipment

There are several other kinds of CAD/CAM equipment used in manufacturing. Although they will not be addressed in detail, it is worthwhile to describe them in brief. They include:

Computer-aided Inspection and Test Equipment

For mechanical parts, the most prominent such device is the coordinate measuring machine, which is a programmable device capable of automatic and precise measurements of parts. A great variety of inspection and test equipment is also used for electronic parts. IBM's Poughkeepsie plant, mentioned above, performs the vast majority of its testing of microprocessor modules with automatic devices built in-house. In addition, robots can be used as computer-aided inspection and test devices—several two-armed, gantry-style robots are used at IBM to test the wiring for computer circuit boards. In the test, thousands of pairs of pins on the circuit board must be tested to make sure that they are correctly wired together. Each arm of the robot is equipped with an electronic needlelike probe, and by touching its probes to each pair of pins and passing an electronic signal through the probes, the robot's control computer can determine whether the circuit board's wiring is "OK."

Electronics Assembly

Increasingly, programmable equipment is used to insert components—resistors, capacitors, diodes, and so on—into printed circuit boards. One such system, shown in Figure 19, called Mini-Semler™ and manufactured by Control Automation, is capable of inserting 15,000 parts per hour.

Figure 19
Control Automation Mini-Semler™ Robot for Assembly Operations

Process Control

Programmable controllers (PCs) are being used extensively in both continuous-process and discrete-manufacturing industries. PCs are small dedicated computers that are used to control a variety of production processes. They are useful when a set of electronic or mechanical devices must be controlled in a particular logical sequence, as in a transfer line where the conveyor belt must be sequenced with other tools, or in heat treatment of metals in which the sequence of steps and temperature must be controlled very precisely. Until the late 1960s, PCs were comprised of mechanical relays, and were *hard-wired*—one had to physically rewire the device to change its function or the order of processes. Modern PCs are computerized and typically can be reprogrammed by plugging a portable computer terminal into the PC. A computerized PC is not only more easily reprogrammed than a hard-wired device, but is also capable of a wider range of functions. Modern PCs, for example, are often used not only to control production processes, but also to collect information about the process. PCs and numerical control devices for machine tools are very similar in concept—essentially, NCs are a specialized form of PC designed for controlling a machine tool.

CAD/CAM AND MANUFACTURING MANAGEMENT

Several kinds of computerized tools are becoming available to assist in management and control of a manufacturing operation. The essential common characteristic of computerized tools for management is their ability to manipulate and coordinate *data bases*—stores of accumulated information about each component of the manufacturing process. The ability to quickly and effectively get access to these data bases is an extraordinarily powerful management tool—what was a chaotic and murky manufacturing process can become much more organized, and its strengths and weaknesses grow more apparent. The following pages describe some of these tools.

Management Information Systems

Manufacturers use and store information about designs, inventory, outstanding orders, capabilities of different machines, personnel, and costs of raw materials, among other things. In even a modestly complex business operation, these data bases become so large and intricate that complex computer programs must be used to sort the data and summarize it efficiently. Management information systems (MIS) perform this function, providing reports on such topics as current status of production, inventory and demand levels, and personnel and financial information.

Before the advent of powerful computers and management information systems, some of the information now handled by MIS was simply not collected. In other cases, the collection and digestion of the information required dozens of clerks. Beyond saving labor, however, MISs bring more flexible and more widespread access to corporate information. For example, with just a few seconds of computer time, a firm's sales records can be listed by region for the sales staff, by dollar amount for the sales managers, and by product type for production staff. Perhaps most important, the goal for the MIS is that the system should be so easy to use that it can be used directly by top-level managers.

Computer-Aided Planning

Computer-aided planning systems sort the data bases for inventory and orders processing, and help factory management schedule the flow of work in the most efficient manner. Manufacturing resources planning is perhaps the best known example of computer-aided planning tools. Manufacturing resource planning, as shown in Figure 20, can be used not only to tie together and summarize the various data bases in the factory, but also to juggle orders, inventory, and work schedules, and to optimize decisions in running the factory. In some cases these systems include simulations of the factory floor to help to predict the effect of different scheduling decisions. Manufacturing resource planning systems have applicability for many types of industry in addition to metalworking.

Another kind of computer-aided planning tool is computer-aided process planning (CAPP), used by production planners to establish the optimal sequence of pro-

Figure 20
Block Diagram of a Manufacturing Resource Planning System. Reprinted with Permission of Reston Publishing Company, Prentice-Hall, from Robotics and Automated Manufacturing *by Richard Dorf.*

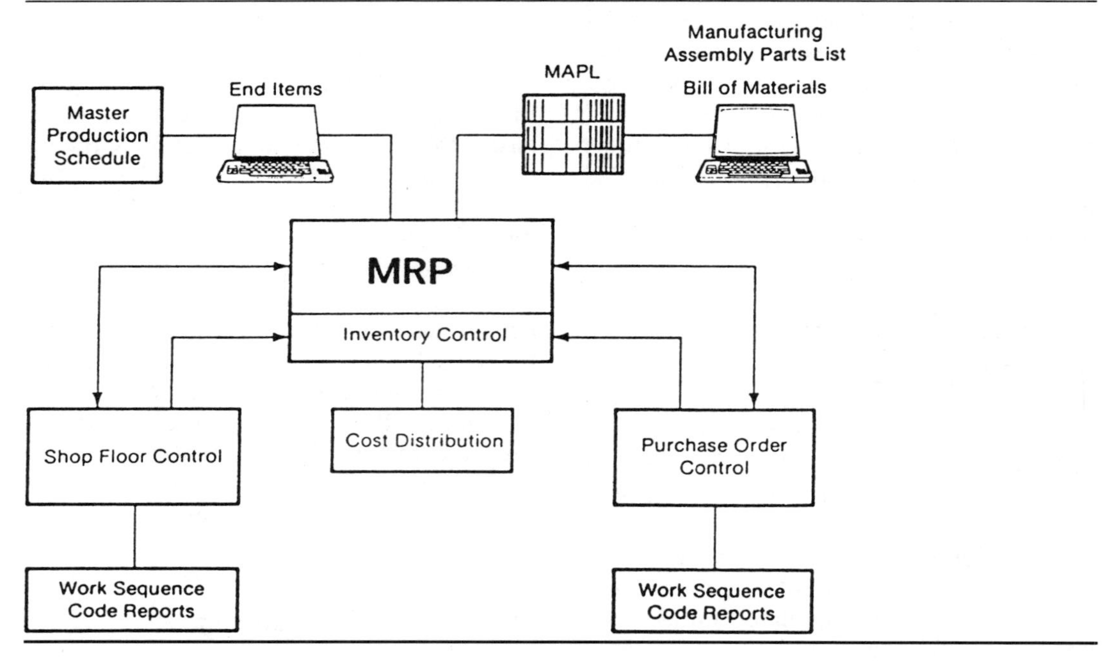

duction operations for a product. There are two primary types of CAPP systems—variant and generative.

The variant type, which represents the vast majority of such systems currently in use, relies heavily on group technology (GT). In GT, a manufacturer classifies parts produced according to various characteristics, for example, shape, size, material, presence of teeth or holes, and tolerances. In the most elaborate GT systems, each part may have a 30- to 40-digit code. GT makes it easier to systematically exploit similarities in the nature of parts produced and in machining processes to produce them. The theory is that similar parts are manufactured in similar ways. So, for example, a process planner might define a part, using GT classification techniques, as circular with interior holes, 6-inch diameter, 0.01-inch tolerance, and so forth. Then, using a group technology-based CAPP system, the planner could recall from computer memory the process plan for a part with a similar GT classification and edit that plan for the new, but similar, part.

Generative process planning systems, on the other hand, attempt to generate an ideal routing for a part based on information about the part and sophisticated rules about how such parts should be handled, together with the capabilities of machines in the plant. Unlike process plans in variant systems, generative systems produce optimal plans.

A *variant* system uses as its foundations the best guesses of an engineer about how to produce certain parts, so that the variants on that process plan may simply be variations on one engineer's bad judgment. Although *generative* CAPP may also depend on group technology principles, it approaches process planning more systematically. The principle behind such systems is that the accumulated expertise of the firm's best process planners is painstakingly recorded and stored in the computer's memory. Lockheed-Georgia, for example, developed a generative CAPP system called Genplan to create process plans for aircraft parts. Engineers assign each part a code based on its geometry, physical properties, aircraft model, and other related information. Planners can then use Genplan to develop the routing for the part, the estimated production times, and the necessary tooling. Lockheed-Georgia officials report that one planner can now do work that previously required four to eight people, and that a planner can be trained in 1 year instead of in 3 to 4 years.

Computer-Integrated Manufacturing

Computer-integrated manufacturing (CIM, pronounced "sim") involves the integration and coordination of design, manufacturing, and management using all of the preceding computer-based systems. Computer-integrated manufacturing, as diagrammed in Figure 21, is not yet a specific technology that can be purchased but is rather an approach to factory technology, organization, and management using all of the technology elements described above.

Computer-integrated manufacturing was first popularized by Joseph Harrington's book of the same name, published in 1974. Mr. Harrington recounts the history of the concept in this way:

> [CIM] came about from: 1) The realization that in many cases automation for discrete activities in manufacturing, such as design or machining, in fact often decreased the effectiveness of the entire operation—e.g., designers could conceive parts with CAD that could not be made in the factory; NC machine tools required such elaborate setup that they could not be economically programmed or used. 2) Development of large mainframe computers supported by data base management systems (DBMS) and communications capabilities with other computers. The DBMS and communications allowed functional areas to share information with one another on demand. 3) The dawning of the microcomputer age which began to allow machines in the factory to be remotely programmed, to talk to each other, and to report their activity to their ultimate source of instruction.

Although there is no quantitative measure of integration in a factory, and definitions of CIM vary widely, the concept has become a lightning rod for technology experts and manufacturers seeking to increase productivity and exploit the computer in manufacturing. For example, James Lardner, vice-president of Deere & Company, sees the current state-of-the-art manufacturing process as a series of "islands of automation"—in which machines perform essentially automated tasks,—connected by "human bridges."

The ultimate step, he argues, is to connect those islands into an integrated whole through CIM and artificial intelligence, replacing the human bridges with machines. In this essentially "unmanned factory," humans would then perform only the tasks that require creativity, primarily those of conceptual design. Lardner's vision is shared by many other prominent experts.

Experts differ, however, in their assessment of how long it might take to achieve this vision. Virtually no one believes that it is attainable in less than 10 to 15 years, while some experts would say a fully unmanned factory is at least three decades away. More important, there are other technology experts who argue that the vision may, in fact, be just a dream. For example, Bernard Roth, professor of mechanical engineering at Stanford University, argues that factories will, in reality, reach an appropriate and economical level of automation, and then the trend toward automation

Figure 21
Block Diagram of Computer-Integrated Manufacturing System

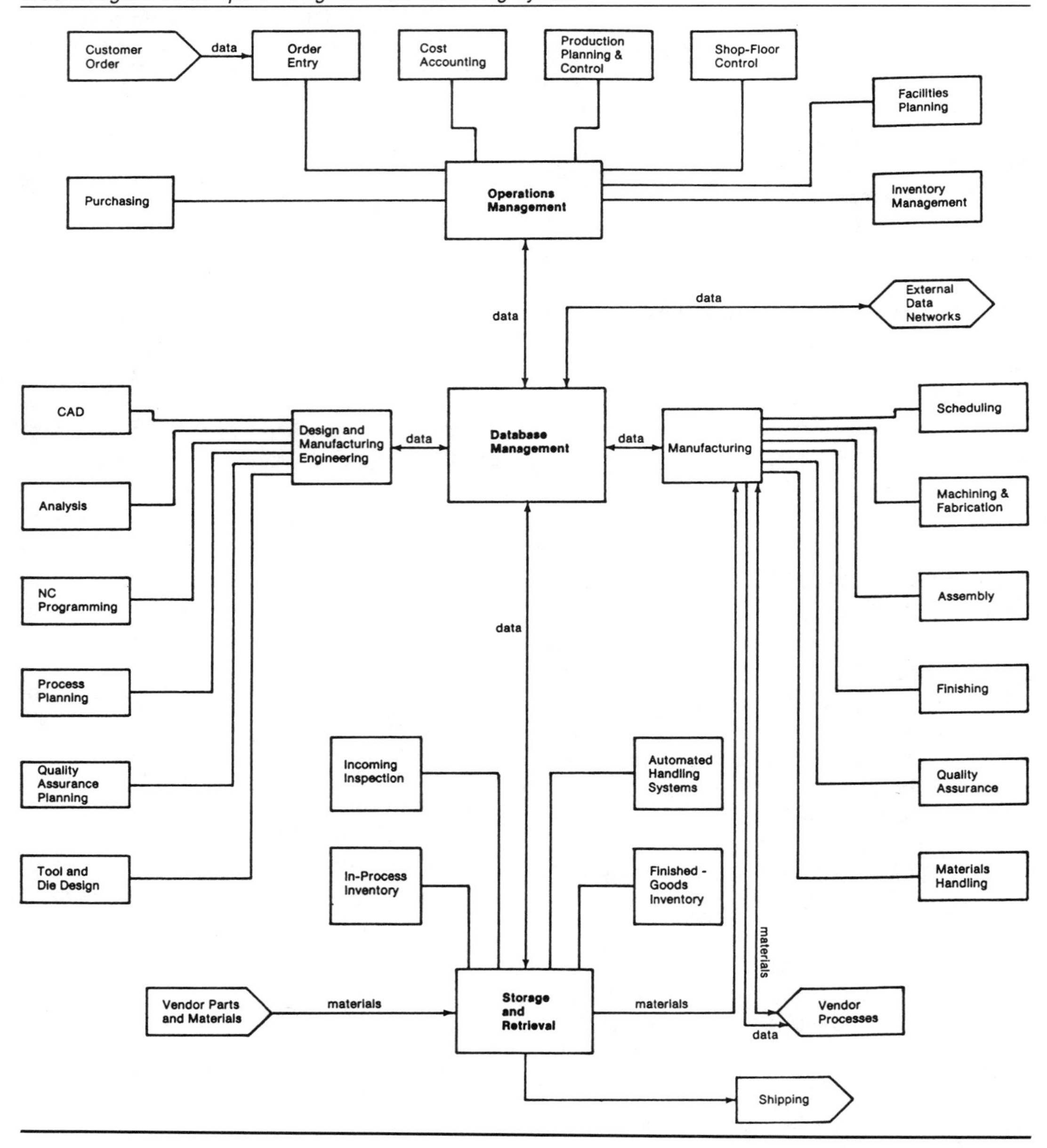

will level off. In a sense, the difference between these two views may be a difference of degree rather than kind. For many factories, the "appropriate" level of automation might indeed be very high. In others, however, a fair number of humans will remain, although the number may be significantly less than currently.

Integrated systems are often found to require more human input than expected. Indeed, as Mr. Hunt has previously stated: "There is much talk about the totally automated factory—the factory of the future—and night shifts where robots operate the factory. Whereas these situations will develop in some cases . . . many manufacturing facilities will not be fully automated. Even those that are will involve humans in system design, control, and maintenance—and the factory will operate within a corporate organization of managers and planners."

These two views do have important significance for how a manufacturer might now proceed. Many who hold the vision of the unmanned factory seem to emphasize technologies, like robotics, that can remove humans from manufacturing. Those who do not share the vision of unmanned production tend to argue that there are more practical ways to enhance productivity in manufacturing, including redesigning products for ease of fabrication and assembly.

How CIM Works

There are two different schemes for CIM. In vertically integrated manufacturing, a designer designs a product using a CAD system that then translates the design into instructions for production on CAM equipment. Management information systems and computer-aided planning systems are used to control and monitor the process. A horizontal approach to integration, on the other hand, attempts to coordinate only the manufacturing portion of the process; that is, a set of computer-aided manufacturing equipment on the factory floor is tied together and coordinated by computer instructions. A flexible manufacturing system is a good example of such horizontal integration. Vertically integrated manufacturing is what is most commonly meant by CIM, however, and many experts would consider horizontal or *shop-floor* integration to be only partial CIM. Figure 22 is a conceptual framework for CIM that illustrates the role of some of the CAD/CAM technologies at various levels of factory control.

A vertically integrated factory usually implies maximum use and coordination of all CAD/CAM technologies and can involve much more centralized control of manufacturing processes than a nonintegrated production process. Communication and shared data bases are especially important for CIM. For example, CAD systems must be able to access data from inventory on the cost of raw materials and from CAM systems on how to adapt the design to facilitate manufacture. Computer-aided manufacturing systems must be able to interpret the CAD design and establish efficient process plans. In addition, management computer tools should be able to derive up-to-date summary and performance information from both CAD and CAM data bases and be effective in helping to manage the manufacturing operation. Some parts of the above requirements are already possible, while others seem far on the horizon. Factory data bases now tend to be completely separate, with very different structures to serve different needs. In particular, the extensive communications between CAD and CAM data bases will require more sophistication in both CAD and CAM, research on how to establish such communications, and finally, major changes in traditional factory data structures in order to implement such a system.

Applications

CIM sounds like utopia to many manufacturers because it promises to solve nearly all of the problems in manufacturing; in particular, it promises to dramatically increase managerial control over the factory. Design changes are easy with extensive use of CAD; CAP and MIS systems help in scheduling; FMS and other CAM equipment cut turnaround time for manufacture, minimize production costs, and greatly increase equipment utilization; connections from CAD and CAM help create designs

Figure 22
Programmable Automation Factory Hierarchy (Simplified). Source: GCA Corp., Industrial Systems Group.

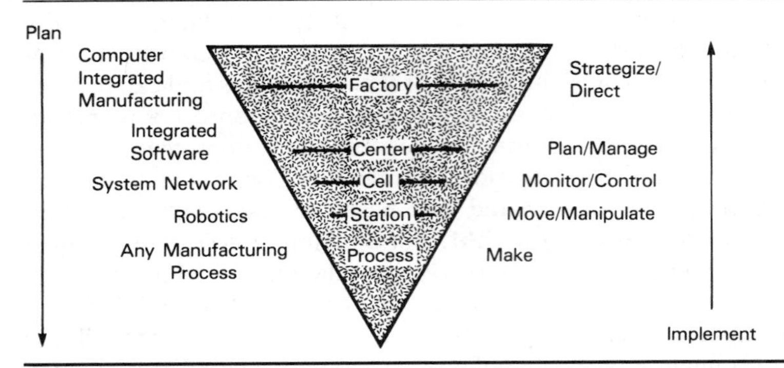

that are economical to manufacture; control and communications is excellent, with minimal paper flow; and CAM equipment minimizes time loss due to setup and materials handling.

Many of the companies that make extensive use of computers view their factories as examples of CIM, but on close examination their integration is horizontal—that is, in the manufacturing area only—or at best includes primarily manufacturing and management. General Electric, as part of its effort to become a major vendor of factory automation systems, has embarked on ambitious plans for integration at several of its factories including its Erie Locomotive Plant, its Schenectady Steam Turbine Plant, and its Charlottesville Controls Manufacturing Division.

TECHNICAL TRENDS AND BARRIERS: FUTURE APPLICATIONS

The possibilities for application of existing CIM tools are extensive and the technologies continue to develop rapidly. They depend on and share the extraordinary rate of growth in technical capabilities of computer technologies as a whole.

There are five themes in the directions for development in each of the technologies. They are:

- Increasing the power of the technologies—their speed, accuracy, reliability, and efficiency
- Increasing their versatility—the range of problems to which the technologies can be applied
- Increasing their ease of use, so that they require less operator time and training, can perform more complex operations, and can be adapted to new applications more quickly
- Increasing what is commonly called the intelligence of the systems, so that they can offer advice to the operator and respond to complex situations in the manufacturing environment
- Increasing the ease of integration of CAD/CAM devices, so that they can be comprehensively coordinated and their data bases intimately linked

Artificial Intelligence Systems for Manufacturing

Artificial intelligence is making serious progress in expanding the capabilities of factory automation.

Since the direction of manufacturing is toward more automation in the entire process, from concept planning using computer-aided design (CAD), to factory process planning and production using computer-aided manufacturing (CAM), the success of such systems implies production components that are capable of being complemented with artificial intelligence components to make them perform better.

At the present time, expert systems are best employed where there is a narrow application that can be well-defined by a source expert, who can identify rules and the criteria for applying those rules. Since the production line is by definition a well-organized and well-planned place, it provides the optimum situation for capitalizing on this new tool.

Robots currently operating under computer control are directed by software through a preplanned process that allows for no deviations from that plan. Although efficient, robots are inflexible when something not known or predicted in advance happens in their area of influence.

Newer robots rely more on sensory input to allow the tool more autonomy in its operation. Those additional forms of information retrieval are beginning to excite the artificial intelligence community into developing scenarios where a robot might perform even more freely and have even more autonomy in its work area. The implications are extremely important to the automated manufacturing process. The performance of a humanlike intellect housed in an industrial framework, impervious to the dangers of the workplace, is truly the best combination for this environment.

While artificial intelligence will play an ever-increasing role in the production robot, the more near-term applications of expert systems will be in product design specification, planning, and related chores. These parts of the production process are already being addressed by CAM, but the application of AI to the use of CAM brings the added advantage of giving the user the benefit of the experience of a resident expert that will not permit certain design mistakes to occur. It could be relatively easy to install an expert system that measures mechanical tolerances into a system while it is being created, making sure that the designer could not make the mistake of building a system that was more reliable than the components making it up. An expert system can also perform a worst-case design, given the performance bounds of the system and the overall cost limitation imposed to make the product feasible.

Imagine a tool that could provide up-front guidance to a CAD designer on the feasibility of his or her design, given a variety of variables (including size, weight, cost, reliability, environmental constraints) and then, after designing such a system, provide a parts list and process plan for making that system in the factory. This idea may seem futuristic—but the tools are already available and are used in other jobs.

Several expert systems are already in use to aid the manufacturing process. Three of these are outlined below:

- IMACS, a field prototype expert created by Digital Equipment Corporation (DEC), assists managers with capacity planning and inventory management. IMACS takes customer orders, generates build plans, and later uses those plans to monitor the implementation.
- ISIS, developed by Carnegie-Mellon, constructs factory job-shop schedules, selects the sequence of operations for the job, determines start and end times, and assigns resources to each operation. It also acts as an intelligent assistant by helping plant schedulers maintain schedules and identify decisions that are liable to result in less than optimum outcomes.
- PTRANS, developed jointly by DEC and Carnegie-Mellon, helps control the manufacture and distribution of DEC's com-

puter systems. It uses customers' orders and information about plant activities to develop a plan to assemble and test ordered systems. PTRANS monitors the progress of the technicians implementing the plan, diagnoses problems, suggests solutions, and predicts shortages or surpluses.

As sensors become more readily available and cost-effective for integration into more factory tools, especially robots, it is certain that intelligent software will be designed to allow the systems to make decisions based on those sensory inputs. At this point, the missing link is the technology to perform pattern recognition on the sensor data in a reliable and consistent manner. This area of research is now being pursued, however, and the needed technology may well become available to the manufacturing community in the near future.

IMPROVEMENTS NEEDED FOR ADVANCED MANUFACTURING TECHNOLOGY

Advanced manufacturing technology is poised to move in a number of directions in the future. The new directions are influenced, first, by the need to overcome certain current technology and management obstacles and, second, by the availability of new concepts that can be applied in practice. Future changes will affect plant management, the machines, the tools, the transport mechanisms, the control hardware and software, and the workpieces themselves.

Many of the trends in the following list are now in progress, while others are still on the horizon:

- Machine adjustable, tool-changer compatible tools
  - Cutting tools (boring bars, and others)
  - Forming tools
  - Gages
- Automatic tool loading/unloading into tool-changer magazines from the tool crib
- Automatic tool-crib operation and connection to integrated systems
- Automatic tool exchanges between machines
- Compact, large capacity, high-speed, random-access, tool-changer magazines
- Online inspection
- On-machine inspection
  - Tool-changer loading devices (e.g., smart robots)
  - Independent of tool storage, changer, and spindle
  - Machine vision inspection systems
- Adaptive error determination and compensation systems
  - Drive axes, error profile compensation
  - Alignment compensation
  - Force and weight compensation
  - Temperature compensation
  - Wear compensation
- Tighter coupling of systems
- More comprehensive and data-accessible manufacturing monitoring systems
- Improved data-driven documentation storage and retrieval
- Comprehensive production planning and management systems to handle
  - Inventory
  - Ordering
  - Batching definition
  - Scheduling
  - Routing
  - Rescheduling/rerouting contingencies
  - Reverse engineering of parts
- Fail-soft, graceful degradation of machines and systems
- Automatic generation of system initialization and startup procedures, and a "checklist" for any stoppage or failure condition
- Automatic part fixturing and defixturing; automatic part storage and retrieval
- Automatic pallet and fixture storage and retrieval, with automatic insertion into and extraction from the system
- Automatic identification and tracking of individual tools, pallets, fixtures, parts,

carts, movable elements, and others, including automatic locating during system startup

- Automated integration of more classes of manufacturing operations, including
  - Milling
  - Turning
  - Forming
  - Inspection
  - Finishing
  - Heat treating
  - Shearing
  - Assembly
  - Testing
- Built-in operation and maintenance training and information presentation via video disk
- Improvements in automated chip flushing and clearing, part cleaning, chip collection and reclaiming, and coolant reconditioning
- Improvements in automatic temperature control of parts, pallets, fixtures, machines, and coolants
- Improvements in inherent machine tool and inspection machine accuracy, as the increase in *near-net-shape* technology reduces the need for heavy machining and places the emphasis on high-precision machining
- Improved maintainability of equipment through increased use of
  - Automatic fault detection and isolation
  - Built-in diagnostic and repair aids
  - Modular construction
  - Redundant equipment configurations
- Increasingly larger systems with better integration into the factory planning and operation system (MRP, for example)
- Integrated manufacturing facilities designed for actual requirements, rather than designed around existing facility problems

PRODUCTIVITY FACTORS

Many productivity benefits are possible through the use of advanced manufacturing technology. These include:

High Capital Equipment Utilization

Typically, the throughput achieved for a set of machines in an advanced manufacturing technology system will be up to three times that achieved by the same machines in a stand-alone job-shop environment. Advanced manufacturing technology systems achieve high efficiency by having the computer schedule every part to a machine as soon as it is free—simultaneously moving the part on the automated material-handling system and downloading the appropriate computer program to the machine. In addition, the part arrives at a machine already fixtured on a pallet (this is done at a separate load station) so that the machine does not have to wait while the part is set up.

Reduced Capital Equipment Costs

The high utilization of equipment results in the need for fewer machines to carry the same workload as in conventional systems. In a job-shop situation, a reduction of 3:1 is common when replacing machining centers with advanced manufacturing technology.

Reduced Direct Labor Costs

Since machines are operated completely under computer control, machinists are not required to run each one. The only direct labor involved is that of personnel to fixture and defixture the parts at the load station, plus a system supervisor. Although the fixturing personnel in advanced manufacturing technology environments require less advanced skills than corresponding workers in conventional workplaces, labor cost reduction is somewhat offset by the need for computing and other skills that may not be required in traditional factories.

Reduced Work-in-Process Inventory and Lead Time

The reduction of work-in-process in an advanced manufacturing technology system is quite dramatic when compared to a job-shop environment. Reductions of 80 percent have been reported at some installations and may be attributed to a variety of factors that reduce the time a part waits for metal-cutting operations. Those factors include:

- Concentration of all the equipment required to produce parts into a small area
- Reduction in the number of fixtures required and the number of machines a part must travel through because processes are combined in work cells
- Efficient computer scheduling of parts batched into and within the system

Responsiveness to Changing Production Requirements

An advanced manufacturing technology system has the inherent flexibility to manufacture different products as the demands of the marketplace change or as engineering design changes are introduced. Furthermore, required spare part production can be mixed into regular runs without significantly disrupting the normal production activities.

Ability to Maintain Production

Many advanced manufacturing technology systems are designed to degrade gracefully when one or more machines fail. This is accomplished by incorporating redundant machining capability and a materials-handling system that allows failed machines to be bypassed. Thus, throughput is maintained at a reduced rate.

High Product Quality

A sometimes overlooked advantage of advanced manufacturing technology, especially when compared to machines that have not been federated into a cooperative system, is improved product quality. The high level of automation, reduction in the number of fixtures and the number of machines visited, better-designed permanent fixtures, and greater attention to part/machine alignment all result in good individual part quality and excellent consistency from one workpiece to another. Such use of advanced manufacturing technology results in greatly reduced costs of rework.

Operational Flexibility

Operational flexibility offers another increment in productivity. In some systems, the advanced manufacturing technology system runs virtually untended during the second and third shifts. This nearly "unmanned" mode of operation is currently the exception rather than the rule. It could, however, become increasingly common as better sensors and computer controls are developed to detect and handle unanticipated problems, such as tool breakages and part-flow jams. In this operational mode, inspection, fixturing, and maintenance can be performed during the first shift.

Capacity Flexibility

With correct planning for available floor space, an advanced manufacturing technology system can be designed for low production volumes initially; as demand increases, new machines can be added easily to provide the extra capacity required.

CURRENT DEVELOPMENT

The U.S. Department of Commerce International Trade Administration has recently completed a major study entitled, *A Competitive Assessment of the U.S. Flexible Manufacturing Systems Industry.** A portion of that report is adapted here to show the federal government's view of U.S. industry performance in advanced manufacturing technology development.

The first generation of advanced flexible manufacturing systems in the United States was developed at the Sundstrand Machine Tool Company in 1965. However, it was not until 1970 that the Kearney & Trecker Company built the first fully integrated system. Despite the considerable interest advanced manufacturing technology has generated, its application in the United States has spread slowly. Nine American companies have sold advanced manufacturing technology systems, and there are currently approximately 47 systems in operation in the United States and several in planning or being installed.

Suppliers

Advanced manufacturing technology systems are not off-the-shelf, turnkey operations. Each system is designed for a special purpose, and the supplier must work very closely with the customer. Few companies have the ability or are willing to undertake building such complex systems. However, many manufacturers provide equipment that can be integrated as components to the overall system.

Suppliers and potential suppliers for advanced manufacturing technology are machine tool builders, robot and material-handling manufacturers, and control manufacturers. Currently, firms in the machine tool industry provide the overall responsibility for design, development, and installation of systems.

Machine Tools

Since machine tools are a central component of advanced manufacturing technology, machine tool builders are key suppliers. Of the approximately 600 machine tool builders in the United States, most are relatively small. About two thirds employ fewer than 20 persons, and less than 1 percent employ 1000 persons or more. The 12 top producers made 85 percent of the machine tools produced in the United States in 1982. Table 6 lists the production of the top 13 machine tool builders for the years 1979 to 1983. Amca International and Oerlikon Buhrle are foreign firms that have acquired U.S. machine tool capacity (Giddings & Lewis and Motch & Merryweather, respectively).

The U.S. machine tool industry is extremely cyclical, as is demonstrated in Table 7. United States domestic shipments fell 26.7 percent from 1981 to 1982, then dropped again by 50.0 percent from 1982 to 1983 before beginning to recover in 1984. United States exports also fell 8.9 percent from 1981 to 1982 and 42.3 percent from 1982 to 1983. Exports remained depressed through 1984. This partly reflects the U.S. economic recession but also reflects decreasing worldwide competitiveness of U.S. machine tools. In 1977, the historic trend of machine tool trade surpluses was reversed. Since that time, U.S. trade in machine tools has been in deficit. Imports took 37 percent of the market in 1983 and increased their share to nearly 40 percent in 1984. Import penetration has been most severe in technologies critical to advanced manufacturing technology, such as machining centers and NC lathes.

Robotics

Over 50 firms produce robots in the United States. The International Trade Commis-

* Department of Commerce, International Trade Administration, *A Competitive Assessment of the U.S. Flexible Manufacturing Systems Industry* (Washington, D.C.: U.S. Government Printing Office, July 1985).

Table 6.
Top 13 U.S. Machine Tool Builders by Sales (millions of dollars)

Builder	1979	1980	1981	1982	1983
Cincinnati Milacron	464	563	640	515	284
Litton	160	190	200	200	267
Textron	150	150	270	210	160
Cross & Trecker	267	320	368	343	150
Ex-Cell-O	230	290	280	235	145
Oerlikon Motch	–	–	–	217	140
Giddings & Lewis	196	222	287	240	135
Bendix Corporation	285	475	400	300	130
Ingersoll Milling Machine	75	200	200	160	130
Lam Technicon	175	200	275	120	–
White Consolidated Industries	135	165	180	155	100
Esterline	–	–	–	–	98
Acme-Cleveland	200	257	252	210	–

Source: American Machinist, 1980, 1981, 1982, 1983, 1984.

sion reports that the top six firms accounted for 80 percent of U.S. shipments in 1982. In 1983 and 1984, the top six firms accounted for only 65 percent of U.S. shipments.

Major producers in the United States are listed in Table 8. These robotics companies include major corporations with an existing high-technology emphasis, such as General Electric, International Business Machines, and Westinghouse; venture capital funded firms spurred by innovation and/or the pros-

Table 7.
U.S. Machine Tool Shipments, Exports, Imports, Consumption, and Import Share of Consumption—1967 to 1984 (millions of dollars)

Year	Shipments	Exports	Imports	Domestic Consumption	Import Share
1967	$1,826	$225	$ 178	$1,780	10.0%
1968	1,723	217	164	1,670	9.8
1969	1,692	242	156	1,606	9.7
1970	1,552	292	132	1,392	9.5
1971	1,058	252	90	896	10.0
1972	1.269	238	114	1,145	10.0
1973	1,788	325	167	1,629	10.3
1974	2,166	411	271	2,026	13.4
1975	2,406	537	318	2,187	14.5
1976	2,178	515	318	1,982	16.0
1977	2,453	427	401	2,428	16.5
1978	3,143	533	715	3,325	21.5
1979	4,064	619	1,044	4,489	23.3
1980	4,812	734	1,260	5,338	23.6
1981	5,111	950	1,431	5,593	25.6
1982	3,749	575	1,218	4,392	27.7
1983	2,114	359	921	2,676	34.4
1984	2,428	383	1,333	3,377	39.5

Source: Department of Commerce, *Current Industrial Reports, Metalworking Machinery*, MQ35W, IM-146, EM-522, Washington, D.C.: U.S. Government Printing Office, 1984.

Table 8.
U.S.-Based Robot Vendors Sales (millions of dollars)

Company	1980	1981	1982	1983	1984	1985*	1986*
GMF	–	–	0.3	22.3	102.8	187.0	185.0
Cincinnati Milacron	29.0	50.0	32.0	51.0	52.5	61.0	65.0
Westinghouse/Unimation	40.0	68.0	63.0	36.0	44.5	45.0	40.0
Automatix	0.4	3.0	8.1	13.0	17.3	25.0	24.0
DeVilbiss	5.0	6.5	23.7	21.0	30.0	33.0	28.0
ASEA, Inc.	2.5	9.0	9.5	15.0	30.0	39.0	48.0
American Cimflex	–	–	–	–	—	18.0	32.0
IBM	–	–	4.5	9.0	12.0	16.0	NA
General Electric	–	–	1.8	11.0	10.0	13.0	NA
Prab Robots	5.5	8.2	12.5	13.5	11.0	12.0	NA
Intelledex	–	–	–	2.0	10.0	10.0	NA
Seiko	–	–	–	4.0	6.5	11.0	NA
GCA	–	–	1.5	6.0	13.5	34.0	NA
American Robot	–	–	–	2.8	8.0	20.0	NA
Cybotech	–	–	9.0	2.3	7.0	12.0	NA
Graco Robotics	–	–	–	5.0	9.3	20.0	NA
Advanced Robotics	1.7	0.8	6.6	3.2	6.0	NA	NA
ESAB	–	–	–	4.0	4.5	8.0	NA
Thermwood	–	1.0	2.5	2.0	4.5	NA	NA
Hobart	–	–	–	3.5	5.5	6.5	NA
Control Automation	–	–	–	0.5	3.0	NA	NA
U.S. Robots	–	–	1.0	1.7	2.5	NA	NA
Nordson	0.8	2.5	4.5	4.7	2.0	NA	NA
Mobot	0.8	0.6	1.3	1.6	1.6	NA	NA
Microbot	–	–	–	0.7	1.4	NA	NA
Precision Robots	–	–	–	0.3	1.0	NA	NA
Adept Technology	–	–	–	–	1.1	17.0	25.0
KUKA	–	–	–	–	–	30.0	25.0
Other	4.5	5.5	8.2	6.2	6.2	19.5	131.0
Total	$90.0	$155.0	$190.0	$240.0	$395.0	$595.0	$625.0

* Estimated.

Note: Material designated NA is not available or is included in other category.

Source: CIM Newsletter, Prudential-Bache Securities, April 27, 1984; June 24, 1985; and May 23, 1986.

pect of growth (e.g., Adept and American Cimflex); and established robot producers that either began in this field or entered robotics based on their machine tool/processing system expertise (e.g., Prab Robots and Cincinnati Milacron).

The automotive industry is the largest user of industrial robots in the United States. Other major users of robots are the aircraft, farm equipment, electrical equipment, and home appliance industries. It is expected that the use of robots will spread to other manufacturing industries as robots become more cost-effective.

U.S. robot shipments, exports, imports, domestic consumption, and imports share of consumption from 1979 through 1983 are listed in Table 9.

Both the quantity and value of U.S. exports have increased each year, reaching an estimated 631 units valued at $33.7 million in 1983. In early years, U.S. exports of robots exceeded U.S. robot imports. The import penetration ratio has been rising since 1980 but was still below the 1979 ratio until 1983. Shipments, exports, imports, and apparent domestic consumption have all increased fairly rapidly over the 5 years, reflecting the general increase in the application of robots in industrial settings.

United States producers' domestic shipments of robots (as opposed to total ship-

Table 9.
U.S. Robot Shipments, Exports, Imports, Consumption, and Import Share of Consumption—1979 to 1983 (millions of dollars)

Year	Shipments	Exports	Imports	Domestic Consumption	Import Share
1979	$ 28.1	$ 8.9	$ 3.8	$ 22.9	16.4%
1980	64.1	20.8	4.3	47.5	8.9
1981	113.4	23.3	10.6	100.7	10.5
1982	142.8	20.3	15.1	137.6	11.0
1983*	168.7	33.7	28.9	163.9	17.7

* Estimated.

Source: Department of Commerce, International Trade Commission, *Competitive Position of U.S. Producers of Robotics in Domestic and World Markets,* Washington, D.C.: U.S. Government Printing Office, December 1983.

ments that are reported above) increased from $19 million in 1979 to $122 million in 1982 and are estimated at $135 million in 1983. Shipments to the domestic market accounted for 85 percent of total U.S. producer shipments in 1982.

A large number of U.S. producers of robots, including most of the major vendors, are involved in agreements with major foreign robot firms. These agreements cover joint ventures, product or process licensing arrangements, marketing, distribution, and technology transfer. Because of these arrangements, dispersion of technology between countries has accelerated, and duplication of research and development (R&D) efforts has undoubtedly decreased.

CAD/CAM

There were more than 70 CAD suppliers in the United States as of February, 1984. According to a Merrill Lynch report, in 1983 five vendors of CAD accounted for 72 percent of the U.S. market, as follows: IBM (23 percent), Computervision (22 percent), Intergraph (11 percent), Calma (GE) (10 percent), and Applicon (6 percent). Other significant competitors include Auto-Trol Technology, McDonnell-Douglas (McAuto), and Gerber Scientific. The CAD/CAM market is growing and market shares have had a tendency to change fairly substantially from year to year.

The CAD installed base in 1981 and projected for 1985 and 1995 are given in the Table 10.

United States shipments of CAD/CAM turnkey systems, including net exports, were $765 million in 1981. U.S. shipments of turnkey systems are predicted to increase to $2.75 billion by 1985 and $12.0 billion by 1995.

Projected growth rates are very high, due to expectations of declines in the price of CAD/CAM systems. In current dollars, the cost of an average turnkey CAD/CAM system is expected to fall from $400,000 in 1980 to $250,000 in 1985. These costs are for relatively large systems. Smaller stand-alone workstations based on microcomputers

Table 10.
U.S. CAD Installed Base (units)

Industry Group	1981	1985	Projected 1995
Mechanical design	1,744	9,000	82,000
Electrical and electronic	1,620	3,700	51,000
Civil Engineering/ Architecture	630	2,900	37,000
Mapping/Other	568	2,400	20,000
Total installations	4,562	18,000	190,000

Source: Predicasts, *Industry Week.*

have been introduced in the $10,000 price range.

A study by Arthur D. Little projected a compound annual growth rate in CAD of about 30 percent from 1982 to 1992. The growth rate of CAM is expected to be lower, at 10 to 12 percent over this period.

The CAD/CAM market is currently dominated by U.S. turnkey suppliers, with 80 percent of world sales. It is expected that there will be more fierce competition in the future from minicomputer and mainframe computer manufacturers. In 1982, U.S. firms held nearly all of the domestic CAD/CAM market, 90 percent of the Western European market, and 70 percent of the Japanese market. Three European firms represented the most significant competition: Quest, Ferranti Cetec, and Racal-Redac. Prominent firms in Japan include Fujitsu, Hitachi, and Sharp.

CIM

Ten American companies have actually sold or produced at least one CIM system. The 10 companies with CIM sales and the companies to which the systems were sold are listed in Table 11. These 10 are not responsible for all the systems in the United States. A few have been built by foreign companies, and some have been developed by companies that have been using CNC machine tools for some time and have the in-house skills to interface their existing machines or new machines into a CIM system. The companies that have done this are listed in Table 12. The systems in operation or soon to be in operation in the United States are described in detail in the appendix.

The suppliers are all large machine tool manufacturers or end-users. They are the companies one would expect to be in the forefront of the market. They have the range of necessary skills and technology as well as the resources to develop fully integrated systems.

Table 11.
United States CIM Suppliers with Sales or Self-Built Systems

Supplier	Users
Cincinnati Milacron	FMC Corporation Vought Aero Products Cincinnati Milacron Caterpillar (Davenport) N.Y. Air Brake General Dynamics (Fort Worth) General Electric (Evandale)
Giddings & Lewis	General Electric Caterpillar (Aurora) Anderson Strathclyde McDonnell-Douglas Astronautics
Ingersoll Milling Machine	J.I. Case Ingersoll Milling (3 systems)
Kearney & Trecker	Allis Chalmers Onan Avco-Lycoming (Williamsport) Avco-Lycoming (Stratford) Hughes Aircraft Rockwell International Mercury Marine Mack Truck Warner Ishi Cummins Engine Georgetown Manufacturing Deere & Company Sundstrand Aviation
White Sundstrand	Boeing Aerospace Buick, Detroit Diesel Watervliet Arsenal Caterpillar
Oerlikon-Motch	Rockwell Motch
Acme-Cleveland	Vickers
Mazak	Caterpillar (Peoria) Mazak (Florence) (2 systems)
Dearborn	Caterpillar (Decatur)
Allen-Bradley	Allen-Bradley

Table 12.
U.S. Manufacturers with Self-Built Systems

Harris Press	1981
General Dynamics Convair (3 systems)	1982
General Electric	1984
Pratt & Whitney	1984
Westinghouse	1984
Allen-Bradley	1985

Some of the advantages reported by users integrating their own system are that the company can get the best machines and equipment to fit its needs from any source and utilize its in-house expertise at minimal additional cost. On the other hand, a company putting its own system together could underestimate the complexity of the undertaking and risk spending considerable time and money getting the system running. Companies will continue building their own systems despite the risks, particularly if, as some users perceive, suppliers are not responsive to the needs of the customer.

Some of the larger manufacturers and multinationals have worked with machine tool builders to develop systems because they are planning to become suppliers. General Electric has spent over $1 billion in the past 5 years, putting together what it calls an "across-the-board factory automation capability." It has put together its process control, electric drive, and CNC products and acquired a CAD company, a microchip maker, and several robot manufacturing licenses. Westinghouse is concentrating on developing CIM to process sheet metal, mechanical parts, and printed circuit boards. McDonnell Automation (McAuto) had a major exhibition at the 1984 International Machine Tool Show in Chicago, offering its services as a factory automation supplier. Other leading electronics companies such as Hewlett-Packard, Digital Equipment, and IBM may also look into developing integrated systems. With the substantial resources these large multinationals have, they could present a real challenge to the traditional machine tool builders in the market.

So far, the large multinationals have shown no interest in developing their own machine tools or in acquiring any machine tool companies. Instead, they have been concentrating on the ancillary products that go in CIM—particularly software. The market for ancillary products may become more lucrative than the market for machine tools. If these multinationals do become strong participants in the market, their sources of machine tools will have a direct impact on the U.S. machine tool industry.

Smaller companies with innovative computer capabilities or expertise in a narrow high-technology product line related to advanced manufacturing technology are beginning to have an effect on the market. Today, software is one of the main drawbacks of advanced manufacturing technology; companies that come up with innovative software packages and machine controls should do well in the market.

Allen-Bradley and General Electric, for example, believe that machine tool builders that now make their own CNCs will not be able to keep up with the specialist manufacturers in terms of product improvement and price reduction, and so they will start buying from specialists. However, it should be noted that the number of machine tool builders that supply their own system has held reasonably steady over the past decade.

Products such as systems monitors, terminals, and communication devices are prime targets for specialists. However, it remains to be seen whether those companies specializing in controls or software actually get into the business of putting systems together for a customer. Since this would require resources beyond the means of most of these companies, the more likely outcome will be that these smaller specialized companies will push for standards and design their equipment to be compatible with systems or sign agreements with systems builders.

Some firms are finding that the most efficient route to gaining access to additional skills and product lines is to pursue joint ventures. Joint ventures are common among companies trying to reposition themselves strategically. Examples of such ventures include Acme-Cleveland-Mitsubishi, Westinghouse-Mitsutoki, GM-Fanuc, and Cross & Trecker.

Most of these joint ventures have offered the potential for low-cost, reliable overseas manufacturing and marketing for the U.S. partner and an enhanced marketing network in this country for the foreign partner. They represent the current industry trend toward centralization. These joint ventures raise some concerns as to the effect they will have on the long-run competitiveness of machine tool manufacturing facilities located in the United States. If procuring or producing machine tools overseas and selling them in the United States becomes a widespread practice of U.S. companies, there exists a long-term danger that U.S. firms would end up more as distributor channels for foreign-built machine tools than as manufacturers in this country.

End-Users

Forty-seven advanced manufacturing technology systems have been identified to be in operation in the United States. This may seem a low number, especially since numerous articles about their application give the impression that there is an explosion of applications. There may be systems in operation that were inadvertently overlooked.

Table 13 is a list of advanced manufacturing technology users and the supplier of the system. Figure 23 shows the number of systems in operation or soon to be in operation in the United States.

From the list of users it is apparent that the farm machinery, automobile, aircraft, and locomotive industries are the pioneers of advanced manufacturing technology in the United States. These are generally large companies with the necessary capital that are under heavy pressure from foreign competition and are prime candidates for advanced manufacturing technology based on their product mix, size of part batches, families of parts, and so forth.

Table 13.
Advanced Manufacturing Technology Systems in the United States

User	Supplier
Allen-Bradley	Allen-Bradley
Vought Aero Products	Cincinnati Milacron
FMC Corporation	
Cincinnati Milacron, Plastics	
General Dynamics Convair (3 systems)	
General Dynamics (Fort Worth)	
General Electric (Evandale)	
Caterpillar (Davenport)	
New York Brake	
Caterpillar (Aurora) (2 systems)	Giddings & Lewis
General Electric (Erie)	
McDonnell Douglas	
J.I. Case (Racine)	Ingersoll Milling
Allis-Chalmers	Kearney & Trecker
Deere & Company	
Avco-Lycoming (Williamsport)	
Avco-Lycoming (Stratford)	
Rockwell-International (Newark)	
Hughes Aircraft (El Segundo)	
Mack Truck (Hagerstown)	
Onan (Minneapolis)	
Mercury Marine (Fond du Lac)	
Georgetown Manufacturing	
Cummins Engine	
Sundstrand Aviation	
Warner Ishi (Shelbyville)	
Boeing Aerospace	White Sundstrand
Boeing Aerospace	Shin Nippon Koki
Watervliet Arsenal	
Buick, Detroit Allison	
Caterpillar	
Vickers	Acme-Cleveland
Harris Press (Fort Worth)	Harris Press
Borg Warner	Comau
Buick Gear & Axle	
Caterpillar (Peoria)	Mazak
Mazak (Florence) (2 systems)	
Rockwell Motch	Oerlikon-Motch
Caterpillar (Decatur)	Dearborn
Pratt & Whitney	Pratt & Whitney
Westinghouse	Westinghouse

Foreign Penetration of the U.S Market

Of the 47 systems in operation in the United States, five were built by foreign companies, an import penetration rate of slightly less than 11 percent. Comau of Italy, Mazak

Figure 23
Advanced Manufacturing Technology Demand in the United States by Year

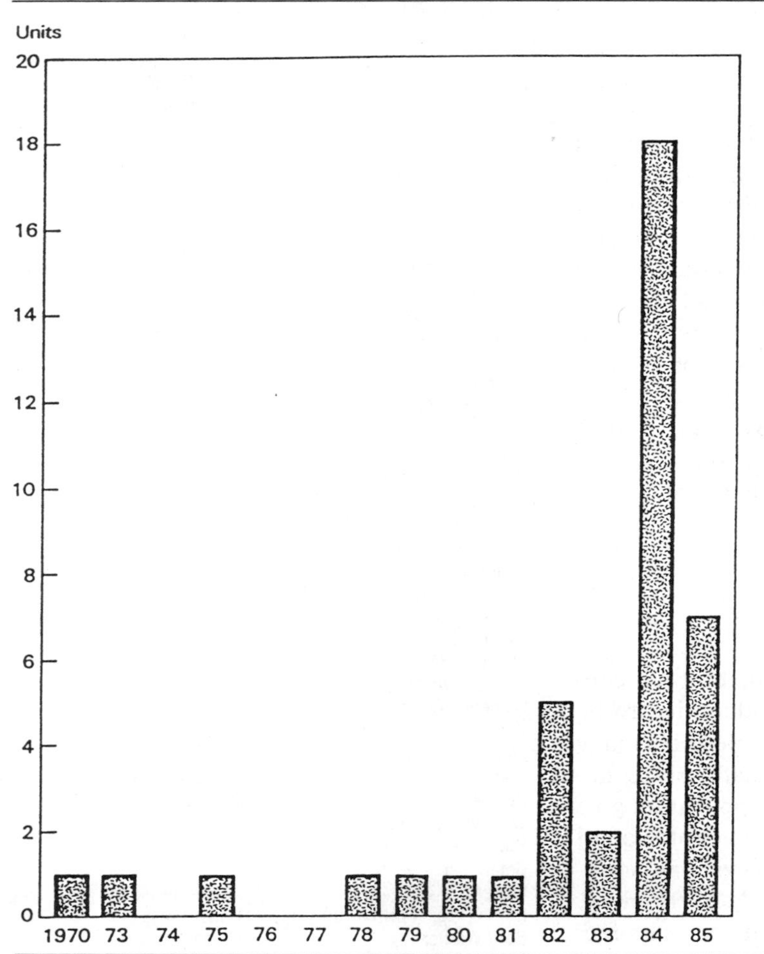

of Japan, and Shin Nippon Koki of Japan are the major foreign competitors in the U.S. market in this time. That few foreign-built systems have been sold in the United States may seem to indicate that U.S. suppliers have a firm hold on the U.S. advanced manufacturing technology market. However, this conclusion may be rather deceptive. Some foreign companies are developing considerable experience in their own countries and then moving to the United States to get into the market. At the 1984 International Machine Tool Show in Chicago, the major emphasis was on the modular approach to system development and technology. Many foreign companies offered system development, including Hitachi-Seiki, LeBlond-Makino (which had a system operating at IMTS), Toshiba, OKK, Toyoda, and SNK from Japan; and Scharmann and EMAG/UMA Corporation of West Germany.

Some of the larger foreign companies are making direct investments in the United

States to establish a presence; others are just setting up service centers. Since the supplier and the user must work very closely together for the planning, installation, and debugging stages of the project, foreign companies that do not have investments in the United States will be at a disadvantage.

Outlook for the U.S. Market

There are basically three ways in which a company can get an advanced manufacturing technology system: (1) buy the entire system from a domestic or foreign supplier who makes virtually all the machines and equipment, (2) use its existing NC machine tools or buy new machines and related equipment and develop the system itself, or (3) contract out to a company that buys the machine tools and related equipment and does the system integrating and installation.

The first method is the most common and will continue to be so over the next few years. The six suppliers that are currently leaders in the U.S. market will continue to lead, although they will be challenged by foreign suppliers and possibly U.S. machine tool builders that decide to get into the market. However, it will be difficult for other U.S. machine tool builders to get into the market unless it begins to expand more rapidly. A potential user is more apt to buy a system from an experienced supplier. The suppliers with experience should be able to underprice new machine tool builders coming into the market, since they are much farther along the learning curve and may be able to install the system more efficiently. The major competition for current U.S. suppliers for the next 3 to 5 years will come from foreign suppliers. While foreign suppliers are not yet firmly established in the U.S. market, there are strong indications of their intent to so establish themselves.

Certain large multinationals initially envisioned becoming suppliers of the complete factory of the future. Demand for this technology has failed to materialize, forcing these multinationals to revise their strategies to supply smaller systems. This has brought them into direct competition with established machine tool suppliers.

Indications are that the machine tool suppliers will continue to dominate the market for at least the next 3 to 5 years.

The second method—building your own system—is being used by some companies and will continue to be used. For companies that want to retrofit existing machines or buy new machines of their choice, building their own system, provided they have the software skills, is a viable alternative. Standardization will have offsetting effects on the use of this method. More standardization, and the willingness of suppliers to use any machine, will reduce the use of this approach. However, increased standardization will reduce the level of skills required for in-house end-user development and installation.

The third method—contracting out to a company that designs the system, buys the machines, and provides software and installation—may have the greatest potential. Increased market share will go to companies that can write the software and build the controls necessary to link the machines and equipment together. Standardization in communication between system components will promote this method, as well as those discussed above.

Growth Projections

Technology Research Corporation, after numerous discussions with end-users, suppliers, and analysts in the field, has estimated the growth in advanced manufacturing technology systems in the United States through 1990, as shown in Figure 24.

COMPETITIVE POSITION

According to the U.S. Department of Commerce, it is difficult to make a factual de-

Figure 24
Growth Projections for Advanced Manufacturing Technology Systems

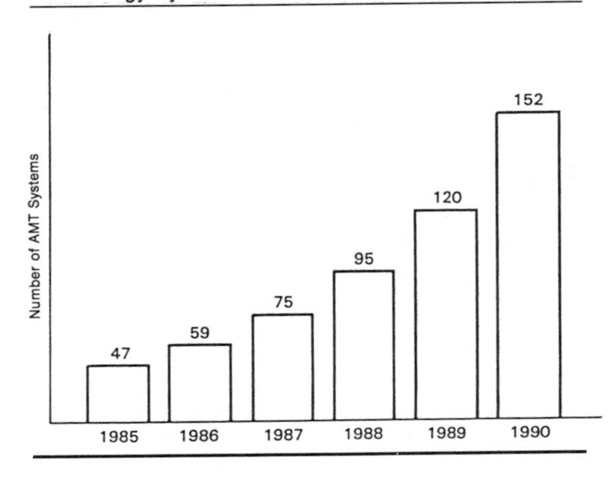

Table 14.
Number of Advanced Manufacturing Technology Systems by Country and Imports/Exports

Country	Number of Systems	Number of Imports	Number of Exports
United States	47	5	2
Japan	50	0	6
Italy	37	0	14
West Germany	25	0	2
United Kingdom	15	4	0
Sweden	14	NA	NA
The Netherlands	0	NA	NA
Switzerland	2	0	0
France	7	5	1
USSR	2	2	0
East Germany	5	1	0
Korea	1	1	NA
Taiwan	1	0	NA
China	0	0	NA
Total	206	18	25

Source: Department of Commerce, International Trade Administration, *Competitive Assessment of the U.S. Flexible Manufacturing Systems Industry*, Washington, D.C.: U.S. Government Printing Office, July 1985.

termination about the competitiveness of advanced manufacturing technology producers in one country versus those in another because (1) producers within a country can vary considerably in ability, (2) quantifiable data that forms a basis for comparison is generally proprietary, and (3) qualitative measures can be subjective. Competitiveness depends on a number of factors. Each of these factors will be considered separately followed by an overall assessment.

Trade Trends

There are no trade data for advanced manufacturing technology systems since they do not neatly fit into any SIC or other classification system. In the absence of trade data, comparisons and trends must be based on systems identified around the world. The U.S. Department of Commerce has identified more than 206 FMS systems around the world. Table 14 shows how many systems there are in each country and how many were imported into each country.

From Table 14, it is apparent that development of advanced manufacturing technology systems, at this time, is not significant. Since this is a new market, however, the trends are important. The biggest advanced manufacturing technology markets in non-Communist countries today are in Japan, the United States, Italy, West Germany, the United Kingdom, and Sweden. The United States has few foreign systems and should be expected to represent a large potential future market. It will be the target of foreign firms, particularly Japanese, Italian, and West German.

There are several possible explanations for the low current volume of development in advanced manufacturing technology systems, some of which are examined in the following discussion of competitive factors.

Prices

An important consideration, and one of the leading barriers to advanced manufacturing technology, is the price of the fully integrated system. Each system is unique so prices vary tremendously from system to system. In addition, for a true price comparison, only to considering the up-front expense is not enough. The durability and reliability must also figure into costs.

Japanese machine tool builders have been able to sell machine tools in the United States for 10 to 40 percent below U.S. producers' prices. Part of this difference is due to the strength of the dollar, but the Japanese have lower material costs and reportedly superior machine tool manufacturing facilities. Since machine tools are generally the big-ticket hardware items in advanced manufacturing technology, this cost advantage helps make the equipment made by foreign competitors the least expensive in terms of up-front costs.

Performance Features and Quality

It is difficult to determine which systems have the best performance and are the most reliable because there are so few foreign systems in the United States for comparison, and even if there were, this information is not generally disclosed candidly. Nevertheless, based on studies of the machine tool industry, the robotics industry, the CAD/CAM industry, and interviews with CIM/FMS users and suppliers, a general picture can be drawn. We will briefly examine U.S. competitiveness in each industry before moving on to advanced manufacturing technology.

Machine Tools

In the 1982 Yano survey, U.S. purchasers of both U.S.-made and foreign-made machine tools are asked to rate producers regarding the engineering of their products. Users rated U.S. producers only slightly higher than Japanese producers. When the machine tool categories were broken down into types of machine tools used, U.S. products were rated first and Japanese products second in the metalcutting categories, and vice versa in the metalforming category. According to this survey, U.S. producers have a lead in large sophisticated NC machine tools for use in production of aircraft, military equipment, and other specialized products. This is in line with a view that U.S. machine tools and CIM/FMS are most rigid and, therefore, more capable of heavy cuts. Of the European suppliers, the West Germans most nearly parallel the U.S. machine tools in construction.

The Yano report says that for such characteristics as spindle speed, maximum allowable torque, spindle motor power, and cutting efficiency (quantities of cutting chips), the United States is ahead of Europe and particularly Japan.

This may be due more to the nature of Japanese targeting than differences in technology. Japanese companies mainly manufacture small- and medium-sized machining centers for general machining; U.S. and West German companies mainly manufacture large and more powerful machine tools. The Japanese appear to hold a leadership position in designing and building advanced manufacturing technology systems for small to medium prismatic parts and for rotational parts.

Computer Hardware/Software

The electronic content in CIM/FMS, both in hardware (sensors, process controllers, CAD systems, and inspection devices) and in software, is becoming larger and more sophisticated, making it difficult for traditional machine tool builders to move into the advanced manufacturing technology business. This is one area, however, where the United States is widely recognized as the world leader. Japan lags behind the United States in this area.

Despite developments, software is considered the largest single problem area in the application of advanced manufacturing technology. The unique aspects of each system make it difficult for suppliers to use software packages from previous systems as a guide for a newly designed system.

As the scope of advanced manufacturing technology expands from machining to assembly, the importance and complexity of software will increase. This bodes well for U.S. suppliers.

Industrial Robots

Japan is further ahead in the application of industrial robots. In terms of technology, Japan, the United States and West Germany are relatively equal.

AMT/CIM/FMS Systems

Until the late 1970s, the United States had a sizable lead in advanced manufacturing technology. Today, no country stands out as being dominant in the market. Each country has its strengths and weaknesses, which tend to offset each other.

Availability

An advanced manufacturing technology system is not bought off a showroom floor, so immediate availability and delivery is not as important a factor as it is for stand-alone machine tools. The process of designing and installing a system can range from 6 months to several years. The system supplier that has the experience and can design and install a system has significant advantage over his or her competitor.

If advanced manufacturing technology systems turn into standard, turnkey operations, availability will become more important. Some companies in Japan claim to have modular systems. These are probably small FMCs. At this point, availability is a neutral issue and does not favor suppliers from any country.

Supplier/User Relationships

Suppliers and users must work closely together over an extended period covering the designing, installing, and training phases of the project. Clearly, language barriers and lack of accessibility to each other could be problems. So far, most of the systems in operation today have been built by companies in the same country; however, this trend is likely to change over the next 5 years. In order for a company to build a system in another country, it is desirable to have some form of permanent representation in that country. This is already taking place with U.S., Japanese, West German, and Italian machine tool builders and suppliers located all over the world.

Foreign companies with a subsidiary in the United States are beginning to make inroads in terms of service and responsiveness to customers. Based on discussions with U.S. users, some feel that U.S. suppliers are not as responsive to the needs of the customers, or as committed to making the system work, as are foreign suppliers. These same people expressed frustration because they feel that American equipment is as good as or better than foreign equipment.

In one domestic instance, a large foreign-built system was selected because the foreign builder offered to set up and test-run the entire system in his own plant before shipping to the customer. U.S. builders would not include this for a near-comparable price.

Marketing and Distribution

In the 1970s, FMS suppliers in the United States had a corner on the market. During this period, marketing was primarily aimed at educating potential U.S. users about the benefits of systems. While U.S. suppliers were selling the concept to potential customers, companies in other countries, many with significant government assistance, were learning how to build systems. Today, there are experienced suppliers in at least seven countries.

These suppliers have become established in their own market and are now entering foreign markets—the United States being a prime target. This competition highlights the importance to U.S. firms of aggressive marketing and establishing good distribution channels.

Effective marketing influences the perceptions users have about a product. The Japanese have been very successful in creating the perception that they are in the forefront of new advanced manufacturing technology and the application of systems. In fact, several representatives from different U.S. suppliers expressed dismay that the Japanese were getting so much attention for their advances in technology, when the technology had been developed by U.S. companies several years earlier. It does appear, however, that the systems at Mori Seiki, Iga, Yamazaki, Mino-Kamo, and Fanuc, Fugi are complex and state of the art.

Capital Availability

The cost of capital affects competitiveness in two ways. It can affect demand by raising or lowering the cost of borrowing to purchase. A customer is unlikely to use retained earnings to purchase high-priced advanced manufacturing technology systems. The cost of capital also can raise the price the supplier must charge in order to be profitable if it is necessary to borrow money to deliver the product. The domestic market is affected in both ways. Foreign markets are affected by the product price effect but not usually by the cost of borrowing to purchase.

Japanese advanced manufacturing technology suppliers, although not yet very successful in penetrating the U.S. market, have a great price advantage due to the strength of the U.S. dollar in relation to the Japanese yen and lower real interest rates. Japanese machine tool suppliers have certainly capitalized on the high value of the U.S. dollar and can be expected to press their advantage in the U.S. market.

OVERALL COMPETITIVENESS

Advanced manufacturing technology systems draw on technology from three important industries: (1) machine tools, (2) robotics and material handling, and (3) electronics. The United States is not behind, technologically, in any of these three fields, yet it is experiencing a decline in the competitiveness of its machine tool industry and is facing serious competition in the robotics industry. The computer software industry is one area in which the United States has the lead. As it turns out, computer software is a key ingredient in advanced manufacturing technology systems. This gives U.S. suppliers one big advantage. However, examination of the various factors that affect competitiveness makes it clear that no one country's industry has a clear advantage. The United States, Japan, Italy, Sweden, and West Germany are all developing a commercial capability. The U.S. industry has been responsible for virtually all the technological innovations incorporated in CIM but has been unable to translate that into a commercial advantage.

United States producers have been able to maintain their domination in the domestic market, but they have not been particularly successful at exporting the technology. On the contrary, there are signs that U.S. producers will have their hands full protecting the U.S. market from foreign penetration. It is too early to predict the fate of U.S. suppliers. They have the ability to stay competitive, and even move ahead of the competition, if they can capitalize on the U.S. edge in computer software. End-users generally agree that U.S. equipment is somewhat superior, but add that U.S. suppliers are becoming less competitive. Why?

- Service and responsiveness of suppliers is important. There is overwhelming agreement that U.S. suppliers are not as responsive to the needs of the customers, not as committed to making the system work, and are not able to deliver orders as scheduled.

- There is a general feeling that U.S. suppliers lack the hands-on experience of running a system. Most of the Japanese suppliers developed their systems for internal use before marketing the product. Some U.S. suppliers do have systems in their own factories but only installed them after much experimenting on the floors of customers.
- There is a general feeling that foreign companies are putting more money into research and development, thereby advancing technology rapidly, although not yet surpassing the United States.

United States suppliers have benefited from the tendency of American end-users to give them preference. As other countries develop experience in advanced manufacturing technology and can offer comparable equipment at equal or lower prices and equal or superior service, potential customers may turn to foreign suppliers, especially if U.S. firms do not begin to develop better relations with the end-users.

The competitiveness of U.S. industry also depends on the rapidity with which demand for advanced manufacturing technology increases in the United States. Currently, in the U.S. market, U.S. suppliers have an advantage. If demand grows fast enough to allow them to move down the learning curve more quickly than competitors (some of whom have smaller domestic market bases), the U.S. industry will be in a good position. Although the factors affecting demand are very complicated, one of them is confidence in the technology, and this factor is closely tied to confidence in U.S. suppliers.

The Definitions

A

A-size sheet
$8\frac{1}{2}$ by 11-inch engineering drawing.

abbe offset
The perpendicular distance between the displacement measuring system of a machine and the displacement to be measured.

ABEND
Acronym for ABortive END; to end a computer run before it has reached a conclusion.

aberration
Failure of a machine vision imaging device to produce exact point-to-point correspondence between an object and its image.

ablative-pit forming
An optical recording method in which the data bit is written by forming a pit, or depression, in the sensitive recording layer of the disk by removing, or ablating, surface material. This is achieved by the laser literally burning a pit in the surface. The difference in surface reflectivity caused by the pitted area is then read as a data bit in readback mode.

abnormal time
The elapsed time for any element recorded during a time study, being excessively longer or shorter than a majority or median of the elapsed time, as judged at the time of the study not to be representative for the element. It may be excluded in determining the most typical elapsed time for the element.

abort
Abnormal termination of a computer program, caused by hardware or software malfunction or operator cancellation.

absolute accuracy
The difference between a point instructed by the control system and the point actually achieved by the manipulator. Repeat accuracy is the cycle-to-cycle variation of the manipulator arm when aimed at the same point. The extent to which a machine vision system can correctly interpret an image, generally expressed as a percentage to reflect the likelihood of a correct interpretation.

absolute address
An actual location in main memory that the control unit can interpret directly. Synonymous with machine address.

absolute coordinates
Units measured from the origin point in the coordinate system (or some other fixed point), rather than expressed as relative to other objects or locations. An ordered pair or triplet of signed numbers between -1 and 1 that specifies (in two's complement arithmetic) the new location of the drawing point with respect to the origin of the unit screen. Only nonnegative absolute coordinates lie within the unit screen.

absolute dimensioning
A means of describing movement instructions as a distance from the axes. Word instructions on a numerical control system tape are noted in absolute form.

absolute language
The language in which instructions must be given to the computer. The absolute language is determined when the computer is designed.

absolute order
A display command in a program that causes the display devices to interpret the data bytes following the order as absolute data rather than as relative data.

absolute specific gravity
Ratio of weight in a vacuum of a given volume of material to weight in a vacuum of an equal volume of gas-free distilled water. Sometimes inaccurately called simply specific gravity, but it is not to be confused with specific gravity, which is based on weight measurements in air. For practical purposes, specific gravity is commonly used to refer to both measurements.

absolute system
A numerical control system in which all coordinate locations are dimensioned and programmed from a fixed or absolute zero point.

absorption
The process by which light or other electromagnetic radiation is converted into heat or other radiation, when incident on or passing through material.

absorption costing
A cost method that charges to inventory all manufacturing overhead on top of the usual direct labor and direct material costs.

acceleration time
That part of access time required to bring an auxiliary storage device, typically a tape drive, to the speed at which data can be read or written.

acceptable quality level
Usually, a small number of defective items stated as a fraction of the entire specifics or a tolerable level of failures with respect to functional and economic requirements of the end item.

acceptance criterion
A level of acceptable performance that the item under test must satisfy.

acceptance gauging
Gauging performed upon the completion of all of the operations scheduled for a part.

acceptance sampling
The extraction of a portion of a lot of material to be inspected for the purpose of determining whether the entire lot will be accepted or rejected.

acceptance test
A test for evaluating the performance, capabilities, and conformity of software or hardware to predefined specifications.

access
The manner in which files or data sets are referred to by the computer or also obtaining data from a storage device or peripheral.

access arm
The mechanical device in a disk storage unit that holds one or more reading and writing heads.

access code
A user identification or password.

access control
In a network or its components, the tasks performed by hardware, software, and administrative controls to monitor the system's operation, ensure data integrity, perform user identification, record system access and changes, and grant users access.

access methods
The technique and/or program code for moving data between main storage and input/output devices.

access points
The end points at each end of a connection.

access time (read/write)
The time interval between the instant at which data are called for from a storage device and the instant delivery is completed—that is, the read time. Also, the time interval between the instant at which data are requested to be stored and the instant at which storage is completed—that is, the write time.

accommodation
Adjustment of the focus and binocular angle of the eyes for a given viewing distance.

accounting data
Machine hours and labor hours consistent with the facilities cost-accounting system. Specific information provided by subsystems to record labor and machine hours associated with a job or activity, consistent with facility accounting requirements.

accumulation bin
A physical location in product assembly environments, used to accumulate all of the components that go into the assembly before sending the assembly order out to the assembly floor.

accumulative timing
A time-study technique utilizing two stopwatches connected so that when one is stopped, the other is simultaneously started.

accumulator
A circuit in the central processing unit of a computer that can perform arithmetical or logical operations.

accuracy
The degree to which actual position corresponds to desired or commanded position—thus, the degree of freedom from error, which is frequently confused with precision. Accuracy refers to the degree of closeness to a "correct" value; precision refers to the degree of preciseness of a measurement.

achromatic lens
A lens consisting of two or more elements that has been corrected for chromatic aberration at two selected wavelengths.

AC input module
Input/output module that conditions various alternating current input signals from user switch to the appropriate logic level for use within a processor.

ACK
Conventional name for a positive acknowledgment control character. Normally, it means that no errors were detected in the just-received data block, and transmission of the next block is welcomed. The corresponding

negative response is NAK (sometimes NACK).

acknowledgment
Output to the operator of a logical input device indicating that a trigger that has fired has been recognized and processed.

acoustic coupler
An electronic device that sends and receives digital data through a standard telephone handset. To transmit data, the digital signals are converted to audible tones that are acoustically "coupled" to a telephone handset. To receive data, the acoustically coupled audible signals are converted to digital signals.

AC output module
Module that conditions the logic levels of the processor to a proper output voltage to control a user's alternating current device.

action message
An output of a material requirements planning system that identifies the need for and the type of action to be taken to correct a current or potential material coverage problem. Examples of action messages are: RELEASE ORDER, RESCHEDULE IN, RESCHEDULE OUT, and CANCEL.

activation mechanism
The situation required to invoke a procedure, usually a match of the system state to the preconditions required to exercise a production rule.

active accommodation
Integration of sensors, control, and robot motion to achieve alteration of a robot's preprogrammed motions in response to sensed forces. Used to stop a robot when forces reach set levels, or to activate performance feedback tasks like insertions, door opening and edge tracing.

active error compensation
A means of error compensation in which the error is monitored and compensated for during actual use of the machine.

active illumination
On a robot machine vision system, illumination that can be varied automatically to extract more visual information from a scene. This can be done by turning lamps on and off, by adjusting brightness, by projecting a pattern on objects in the scene, or by changing the color of the illumination.

active inventory
Raw material, work-in-process, and finished products that will be used or sold within the budgeted period without extra cost or loss.

active medium
In laser technology, a medium in which stimulated emission of light, rather than absorption, will occur at a specific wavelength. Also known as lasing medium.

activity (time-study usage)
The number of times a given operation or occurrence is repeated during a given period, usually a year.

activity log
Administrative information provided by subsystems to record sequentially (log) all job activities accomplished by the subsystem.

activity network
A representation of two particular aspects of a project: the precedence relationship among the activities, and the duration of each activity.

activity ratio
A measurement of file activity computed as the relationship of records used to the total number of records in the file.

actual cost
An acceptable approximation of the true cost of producing a part, product, or group of parts or products, including all labor and material costs and a reasonable allocation of overhead charges.

actual cost system
A cost system that collects costs historically as they are applied to the production and allocates indirect costs based on specific costs and achieved volume.

actual time
The time taken by a worker to complete a task or an element of a task.

actuator
In robots, a motor or transducer that converts electrical, hydraulic, or pneumatic energy into motion, for example, a cylinder, a servomotor, or rotary actuator.

ADA
General-purpose computer programming language intended to be the primary language used in U.S. defense computer applications.

ADAPT
Positioning and limited contour processor system developed by IBM.

Actuator
Actuators Are Often Custom Designed for Each Manufacturing Need. (*Source:* Parker Haneflin Corp.)

adaptable
Capable of making self-directed corrections—often accomplished in robots with the aid of visual, force, or tactile sensors.

adaptive
Refers to a capability to handle a wide variation in one or more system characteristics and still perform the intended function; the control system adjustment is made automatically and responds to a changing environment.

adaptive control
A control algorithm technique for measuring performance of a process and then adjusting the process in order to obtain optimum performance. In the machine tool field, a means of automatically adjusting the feed and speed of a cutting tool from sensor feedback in order to realize the optimum cut. Sensors may measure any one or all of the variable factors involved including heat, vibration, torque, or deflection. Machine control units for which fixed speeds and feeds are not programmed. The control unit, working from feedback sensors, is able to optimize favorable situations by automatically increasing feeds and speeds. It also will reduce feeds and speeds if adverse conditions are met. This secures optimum cutter life and/or lower machining cost.

adaptive smoothing
A technique used in exponential smoothing forecasting, in which the alpha smoothing constant is periodically revised based on error analysis of previous forecasts.

addend
The number or digital quantity to be added to another (augend) to produce a result (sum).

adder
A device whose output data represent the sum of the numbers represented by its input data.

address
Identifying words on a program tape by means of a letter at the beginning of the word. Some computer operators also speak of "addressing" a computer when programming instructions are loaded into memory. A computer system location that can be referred to in a program, it defines a main memory location, a terminal, a peripheral device, a cursor location, or any other physical item in a computer system. A means of identifying information that is to be stored in a computer. In numerical control, the address usually consists of a letter noted on the tape that requires some action, the extent of which is described by numbers following the letter.

addressability
A characteristic of a device in which each storage area or location has a unique address. The address is then usable by the programmer to access the information stored at that location.

addressable point
Position on a cathode-ray tube screen that can be specified by absolute coordinates that form a grid over the display surface.

address bus
The wires in a computer that carry signals used to locate a given memory address.

address decoder
Circuitry that routes signals along an address bus to the appropriate memory cells or chips.

address format
The arrangement of the address parts of a computer instruction. The arrangement of the parts of a single address, such as those for identifying channel, module, track, etc.

address modification
Changing the address of a machine instruction by means of coded instruction.

address register
A register holding the location of memory, peripherals, or other physical system items.

address space
The complete range of addresses available to a programmer. The set of addressable points, often in a specific coordinate direction and usually limited by the data path width in a physical device (e.g., 12 bits). Some addressable points may not be visible if they fall outside the display surface boundaries.

address translation
The process of changing the address of an item of data or an instruction to the address in main storage at which it is to be loaded or relocated.

add-subtract time
The time required to perform addition or subtraction, excluding the time required to fetch the operands from storage and store the result.

adiabatic
A process condition in which there is no gain or loss of heat from the environment.

adjacency
In character recognition, a condition in which the character spacing reference lines of two consecutive characters on the same line are separated by less than a specified distance.

advanced data-communications control procedure
The proposed American National Standard for a bit-oriented synchronous data-communications protocol. Supports full- or half-duplex, point-to-point or multipoint communication. Employs a cyclic redundancy error check algorithm.

advanced manufacturing technology (AMT)
Advanced manufacturing technology tools differ from conventional automation primarily in their use of computer and communications technology. They are thus able to perform information processing as well as physical work, to be reprogrammed for a variety of tasks, and to communicate directly with other computerized devices. Advanced manufacturing technology is divided into three general categories: (1) computer-aided design; (2) computer-aided manufacturing (e.g., robots, computerized machine tools, flexible manufacturing systems); and (3) computer-aided techniques for management, such as management information systems and computer-aided planning. These systems are integrated by extensive computer-based coordination software. Because of its ability to perform a variety of tasks, advanced manufacturing technology is usually associated with batch production. However, it has been used extensively in mass production, and it could be useful in custom production as well.

advisory systems
An expert system that interacts with a person in the style of giving advice, rather than in the style of dictating commands. These systems generally have mechanisms for explaining their advice and for allowing them to interact at a comfortable detail level.

afocal
An optical system with object and image points at infinity.

agenda
A prioritized list of pending activities, usually the application of various pieces of knowledge.

aging
The chemical and/or physical changes that occur in a material after exposure to environmental conditions over a period of time. Can be natural or artificial (for testing), with specified conditions and predetermined time periods. In a metal or alloy, a change in properties that generally occurs slowly at room temperature and more rapidly at higher temperatures.

air-assist forming
A method of plastic thermoforming, using a vacuum, in which air flow or air pressure is employed to partially preform the sheet immediately prior to the final pull-down onto the mold.

air gap
In extrusion coating, the distance from the die opening to the nip formed by the pressure roll and the chill roll.

air-hardening steel
A steel containing sufficient carbon and other alloying elements to harden fully during cooling in air or other gaseous media, from a temperature above its transformation range. The term should be restricted to steels that are capable of being hardened by cooling in air in fairly large sections, about 2 inches or more in diameter.

air motor
A device that converts pneumatic pressure and flow into continuous rotary or reciprocating mechanical motion.

air ring
A circular manifold used to distribute an even flow of the cooling medium—air—onto a hollow tubular form passing through the center of the ring. In blown tubing, the air cools the tubing uniformly to provide uniform film thickness.

air-spaced doublet
A two-element compound lens where the elements are separated by an air space.

Airy disk
The image of a point source formed by diffraction-limited optics. The Airy disk appears as a bright central disk surrounded by alternately bright and dark rings. The size of the Airy disk is the diameter of the first dark ring.

algebraic language
An algorithmic language whose statements are structured to resemble algebraic expressions, for example, ALGOL, FORTRAN.

ALGOL
A programming language designed for the concise, efficient expression of arithmetic and logical processes and the control of those processes, ALGOL—an acronym for ALGOrithmic Language—resembles FORTRAN and PL/1 and, in many cases, has been replaced by Pascal.

algorithm
A prescribed set of well-defined rules or processors for the solution of a problem in a finite number of steps. Contrast with heuristic, which pertains to exploratory methods of problem-solving where solutions are discovered by evaluation of the progress made toward the final result.

aliasing
Jagged edges (*jaggies*) on lines and edges created by the limited resolution of a display or drawing program. Many graphic programs include antialiasing routines to reduce the visual impact of jagged edges.

alignment
The relative position of a scanner or light source to the target or the receiving element.

allocate
To assign a resource for use in performing a specific task.

allocated material
Material assigned to a specific future production order. Material reserved for a specific order that has been released but whose parts are not yet picked. Once allocated, these parts are not counted as available for future use.

allocation
In material requirements planning system, allocation of an item marks it as one for which a picking order has been released to the stockroom but which is not yet sent out of the stockroom. It is an "uncashed" stockroom requisition. Also, a process used to distribute material in short supply.

allowance
A time increment included in the standard time for an operation to compensate the workman for production lost due to fatigue and normally expected interruptions, such as for personal and unavoidable delays.

allowed time
The leveled time plus allowances for fatigue and delays.

alloying
The process of melting the surface of a metal and adding an alloying agent to it so that a change can occur in chemical composition near the surface.

alphabetic character
A letter or other symbol, excluding digits, used in a computer language. In COBOL, a character that is one of the 26 characters of the alphabet or a space. In FORTRAN, a character of the set A, B, C, . . . , Z.

alphabetic character set
A character set that contains letters and may contain control characters, special characters, and the space character, but not digits.

alphabetic character subset
A character subset that contains letters and may contain control characters, special characters, and the space character, but not digits.

alphabetic string
A string consisting solely of letters from the same alphabet.

alphageometric
A method of presenting graphic images, using line or line segments to compose characters and shapes.

alphamosaic
A method of presenting graphic images, using dots to make up characters and shapes. Al-

Alphanumeric Keyboard
Example of an Alphanumeric Terminal Keyboard

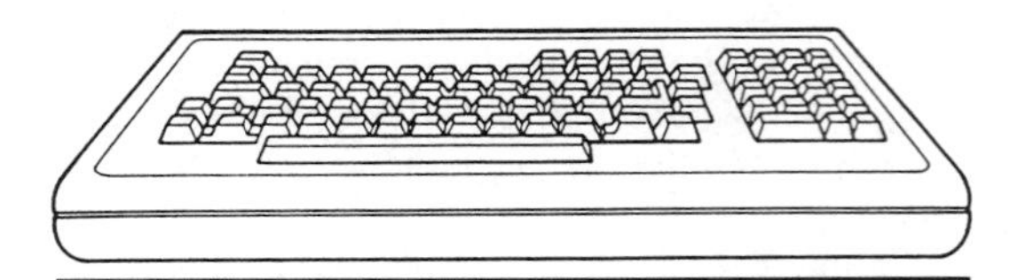

though this method can produce finer images than the competing alphageometric method, it requires more data to be transmitted for each picture.

alphanumeric
Pertaining to the characters (letters, numerals, punctuation marks, and signs) used by a computer.

alphanumeric display
A workstation device consisting of a cathode-ray tube on which text can be viewed and that is capable of showing a fixed set of letters, digits, and special characters. It allows the designer to observe entered commands and to receive messages from the system.

alphanumeric keyboard
A keyboard similar to the typewriter keyboard that allows the user to input letter (alpha) and number (numeric) instructions to the central processor.

alphanumeric text
The written form of languages, comprising alphabetic letters with or without diacritical marks, numerical digits and fractions, punctuation marks, typographical symbols, and mathematical signs as well as space and special letters, signs, and symbols.

alpha test
The stage in the research and development of a new product during which a prototype of the system is operated to ascertain that the system concept and design are functional and to identify areas that need further development and/or enhancement.

alternate-action switch
A unit that remains in a given condition until actuated by an operator into another condition.

alternate routing
An alternate method or sequence of performing an operation, a series of operations, or a complete routing. The alternate is generally used because of a machine breakdown or an excessive overload on the machine or work centers specified in the "primary" routing. An operation may be replaced by either a single alternate operation or a sequence of operations.

alternate time standard
A standard time allowed for a method of performing a task other than the established standard method.

alternate work center
A work center that can be used in case of breakdowns or overloads in the "primary" work center.

aluminizing
Forming an aluminum or aluminum alloy coating on a metal by hot dipping, hot spraying, or diffusion.

ambient light
Light that is present in the environment of a machine vision system and generated from outside sources. This light must be treated as background noise by the vision system. In a computer-aided design (CAD) environment, ambient light is the surrounding level of lighting, which needs to be much lower in a CAD area than in a traditional drafting room.

ambient temperature
A temperature within a given volume, for example, in a room or building.

ambiguity
The characteristic of an image in which more than one interpretation of the object from which the image was formed can be made. There is no ambiguity when only one interpretation is possible.

AML
Modern manipulator-oriented programming language for robot programming. A Manufacturing Language (AML) is a product of IBM.

amorphous
Not having a crystal structure; noncrystalline.

amplification
The increase in power of a laser that is generated when photons of light are reflected back and forth between mirrors at the ends of the laser cavity.

amplitude
The maximum height of an electromagnetic wave, measured from the mean value to the extreme value.

amplitude modulation
A method of information transmission, whereby the amplitude of the carrier is modulated in accordance with the amplitude of the signal waveform.

analog
The representation of a smoothly changing physical variable by another physical variable. In data transmission, the term is used in opposition to digital. In this context, analog transmission uses amplifiers—required due to attenuation of the signal and distance—that magnify the incoming signal. Digital transmission uses repeaters that generate completely new signals, based on whether the incoming signal was mostly a one or mostly a zero during the period the incoming signal was measured.

analog channel
A channel on which the information transmitted can take any value between the limits defined by the channel. Voice channels are analog channels.

analog communications
Transfer of information by means of a continuously variable quantity, such as the voltage produced by a strain gage or the air pressure in a pneumatic line.

analog computer
A computer that solves a problem by creating a physical, usually electrical, model giving continuous data simulating the behavior of the variables and their interrelationships. Used in scientific applications and in manufacturing for controlling operations.

analog control
Control involving analog signal processing devices (electronic, hydraulic, pneumatic, and others).

analog data
Data represented by a physical quantity that is considered to be continuously variable and whose magnitude is made directly proportional to the data or to a suitable function of data. Data represented in a continuous form, as contrasted with digital data represented in a discrete form.

analog device
An apparatus that measures continuous information (e.g., voltage, temperature). The measured analog signal can take an infinite number of possible values. The only limitation on resolution is the accuracy of the device.

analogic control
Pertaining to control by communication signals that are physically or geometrically isometric to the variables being controlled, usually by a human operator. A device for effecting such control.

analog interface
Used to control ancillary equipment by a signal consisting of an analog voltage. (Used in welding to control wire-feed speed).

analog-to-digital converter
An electronic device that senses a voltage signal and converts it to a corresponding digital signal (a string of 1s and 0s) for use by a digital computer system.

analog transmission
Transmission of a continuously variable signal as opposed to discretely variable signals, such as digital data. Examples of analog signals are voice calls over the telephone network and facsimile transmission.

analysis
The methodical study of a problem, and the separation of the problem into smaller related units for further detailed investigation. In computer graphics, the use of the relative position of object and light source to set shading, color, and texture of visible objects. The two principal techniques are ray tracing and boundary analysis.

AND
A logical operator that—if *X* and *Y* are two logic variables—defines the function "X and Y" according to a predetermined table. The AND operator is usually represented in electrical notation by a centered dot "·" and in FORTRAN programming by an asterisk "*" within a boolean expression.

AND gate
A logic circuit designed to compare TRUE-FALSE (or on-off or one-zero) inputs and pass a resultant TRUE signal only when all the inputs are TRUE.

android
A robot resembling a human in physical appearance.

angle of incidence
The angle between the axis of an impinging light beam and perpendicular to the specimen surface.

angle of view
The angle formed between two lines drawn from the most widely separated points in the

object plane to the center of the lens. The angle between the axis of observation and the perpendicular to the specimen surface.

angstrom unit
A wavelength measure equal to 1 ten-billionth of a meter. Ten angstrom units equal 1 nanometer.

angular dimension entity
An annotation entity designating the measurement of the angle between two geometric lines.

angular displacement
The quantitative measurement of the angular motion of a rotary axis about its intended axis of rotation.

animate
To produce perceived motion by displaying a series of slightly different images. To produce animation on a computer display, the system must be able to paint pictures rapidly enough so that the eye perceives the changes as motion (at least several dozen pictures per second). This normally requires a fast connection from memory and processor to screen and enough memory to create one image while displaying another. Specialized graphics chips are now making some animation possible even on small inexpensive computers.

anisotropic mapping
A transformation that does not need to preserve aspect ratio.

anisotropy
Physical properties of a material differing along different directions. The resultant material is said to be anisotropic. Some crystals, for example, are easier to magnetize along one axis than along another and are therefore anisotropic.

annealing
Heating to and holding at a suitable temperature and then cooling at a suitable rate, for such purposes as reducing hardness, improving machinability, facilitating cold working, producing a desired microstructure, or obtaining desired mechanical, physical, or other properties.

annotation
Process of inserting text or a special note or identification, such as a flag on a drawing, map, or diagram constructed on a CAD/CAM system. The text can be generated and positioned on the drawing using the system; also, text or symbols, not part of the geometric model, that provide information.

annualized contracts
A method of acquiring materials that helps ensure continuous supply, minimizes forward commitments, and provides the supplier with estimated future requirements.

annunciator
An indicator or action that requires a response; usually prompted by a light on the keyboard.

anode
The electrode by which electrons leave a system such as a battery, an electrolytic cell, or a vacuum tube.

answer back
A manually or automatically initiated reply message from a terminal. It usually includes the terminal address to verify that the correct terminal has been reached and that it is operational.

antecedent
With artificial intelligence, the left-hand side of a production rule. The pattern needed to make the rule applicable.

anthropomorphic
Resembling human shape or characteristics. In robotics, describes the ability of the robot arm to move in a fashion similar to the human arm.

anthropomorphic robot
A robot with all rotary joints and motions similar to a human's arm, (Also called jointed-arm robot).

antialiasing
Efforts to smooth the appearance of jagged lines (*jaggies*) created by the limited resolution of a graphic system. The most common antialiasing technique is to add extra dots (*phantom pixels*) at random points adjacent to sloping lines.

anticipated delay report
A regular report, normally issued by both manufacturing and purchasing to the material planning function, setting out jobs or purchase orders that will not be completed on time, the reasons for the delay, and when they will be completed. This is one essential ingredient of a closed-loop material requirements planning system.

anticipation inventories
Additional inventory above basic pipeline stock to cover projected trends of increasing

Anthropomorphic Robot
Work Envelope of Anthropomorphic Robot, Also Referred to as a Jointed Arm Robot

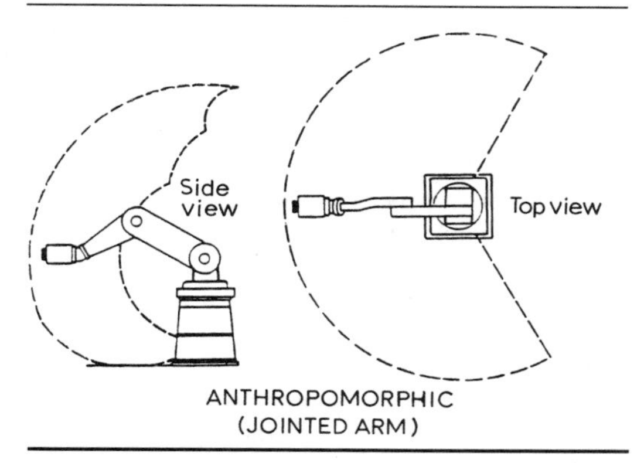

sales, planned sales promotion programs, seasonal fluctuations, plant shutdowns and vacations.

antireflection coating
A thin layer of material applied to an optical surface to reduce reflectance and thereby increase transmittance. The coating may consist of one or more layers.

antistatic agents
Methods of minimizing static electricity in plastics materials. Such agents are of two basic types: (1) chemical additives that, mixed with the compound during processing, give a reasonable degree of protection to the finished products; and (2) metallic devices that come into contact with the plastics and conduct the static to earth—such devices give complete neutralization at the time, but because they do not modify the surface of the material, it can become prone to further static during subsequent handling.

aperture
An opening that will pass light. The effective diameter of the lens that controls the amount of light reaching the image plane. One or more adjacent characters in a mask that cause retention of the corresponding characters.

APL
Acronym for A Programming Language with a syntax and character set primarily designed for mathematical applications.

append
To alter a file or program.

application
Any machine or process monitored and controlled by a computer by means of a user program.

application function
The systems function of developing programs and systems to satisfy user needs.

application generator
Software that generates an application program with information supplied by the user. Also known as a program generator.

application-oriented language
A problem-oriented language whose statements contain or resemble the terminology of the user, for example, a report program generator.

application package
A commercially available applications program. In most cases, the routines in the application packages are necessarily written in a generalized way and will need to be modified to meet each user's own specific needs.

application process
An element within a system that performs the information processing required for a specific application.

application program
A program using graphics-standard functionality to generate graphics output and/or receive graphics input.

application software
The instructions that direct the hardware to perform specific functions. Common applications include payroll, inventory control, and electronic spreadsheets.

application study
The investigation and determination of system or procedural requirements and the establishment of criteria for selection of suitable solutions.

APT
Stands for Automatically Programmed Tools, a computer-aided part programming system that consists of the input language, the APT processor, an APT postprocessor, and a computer of sufficient size to run the APT program. The APT system was initially developed for three-, four-, and five-axis milling machines, but, due to further development, is presently capable of a wide range of applications including point-to-point and turning work.

arc
In a computer map or drawing specified with a polygon model, a boundary between two regions. Each polygon is defined by its arcs, nodes marking the intersection of arcs, and a centroid that represents the total placement of the region. In welding, an arc is a sustained continuous discharge of electricity between two electrodes or between an electrode and the work.

arc brazing
Brazing with an electric arc, usually with two nonconsumable electrodes.

arc cutting
Metal cutting with an arc between an electrode and the metal itself. The terms carbon-arc cutting and metal-arc cutting refer, respectively, to the use of a carbon or metal electrode.

architecture
The basic composition and structure of a computer or other complex electromechanical device. Physical and logical structure of a computer or manufacturing process. The organizing framework imposed on knowledge applications and problem-solving activities. The knowledge-engineering principles that govern selection of appropriate frameworks for specific expert systems.

archival storage
The use of memory (magnetic tape, disks, printouts, or drums) to store data on completed designs or elements outside of main memory.

archive
A procedure for transferring information from an online storage diskette or memory area to an offline storage medium.

archive file
A file into which a structure may be archived or retrieved.

arc lamp
An electric lamp in which current passes through the ionized air between two electrodes, producing light.

arc resistance
Time required for a given electrical current to render the surface of a material conductive because of carbonization by the arc flame. A measure of resistance of the surface of an electrical insulating material to break down under electrical stress.

arc welding
Welding with an electric arc-welding electrode.

ARELEM
Acronym for ARithmetic ELEments. It is part II of the APT processor and calculates the cutter locations, based on the motion commands that were input and canonical forms of the geometry input also converted by the APT processor.

argument form
A reasoning procedure in logic.

arithmetic capability
The ability to do addition, subtraction, and (in some cases) multiplication and division with the computer processor.

arithmetic logic unit
The part of the computer processing section that does the adding, subtracting, multiplying, dividing, and/or logical tasks (comparing). A part of the central processor that performs arithmetic operations such as subtraction and logical operations such as TRUE-FALSE comparisons.

arithmetic overflow
That portion of a numeric word expressing the result of an arithmetic operation by which its word length exceeds the word length provided for the number representation.

arithmetic register
A register that holds the operands or the result of operations, such as arithmetic operations, logic operations, and shifts.

arithmetic shift
A shift applied to the representation of a number in a fixed-radix numeration system and in a fixed-point representation system, in which only the characters representing the absolute value of the numbers are moved. An arithmetic shift is usually equivalent to multiplying the number by a positive or a negative integral power of the radix, except for the effect of any rounding.

arithmetic unit
In a computing system, the section of the central processing unit (CPU) containing the circuits that do arithmetic operations and perform logical comparisons.

arm
An interconnected set of links and powered joints comprising a robot manipulator that supports and/or moves a wrist and hand or end-effector through space.

AROM
Acronym for Alterable (Read-Only Memory). A ROM that can be changed or initialized by the user.

Arc Welding
Hobart Arc-welding System with Robot Manipulation

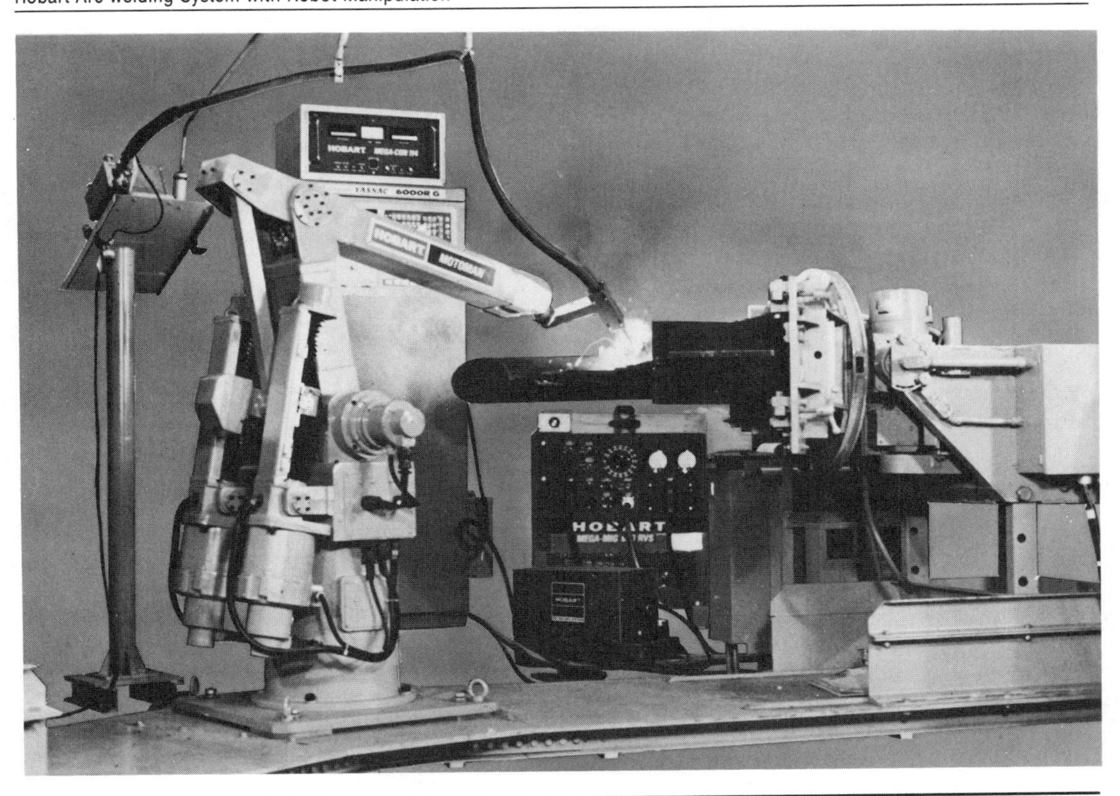

Arm
Basic Elements of Robot System Including Manipulator, End-Effector Arm, Power Drive, and Computer Controller

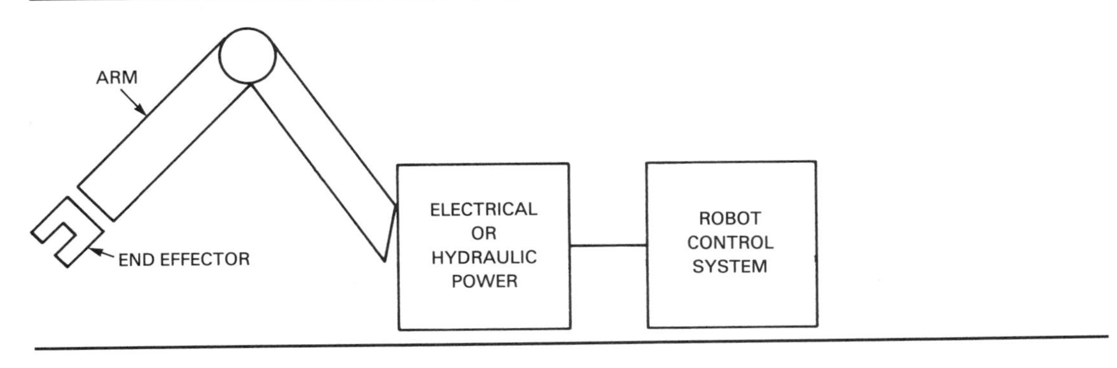

array
A named, ordered collection of data elements, all of which have identical attributes. An array has dimensions, and its individual elements are referred to by subscripts. An array can also be an ordered collection of identical structures (PL/I and FORTRAN).

arrival date
The date purchased material is due to arrive at the receiving site. Arrival date can be input, can be equal to current due date, or can be calculated from ship date plus transit time.

arrival rate
The mean number of units arriving at a service facility during a given interval of time.

articulated robots
Robots having rotary joints in several places along the arm that roughly correspond to the shoulder, elbow, and wrist in humans. They are usually mounted on a rotary base.

artificial intelligence
A subfield of computer science concerned with pursuing the possibility that a computer can be made to behave in ways that humans recognize as "intelligent." An approach to programming that emphasizes symbolic processes for representing and manipulating knowledge in solving problems. A subfield of computer science concerned with the concepts and methods of symbolic inference by a computer and the symbolic representation of the knowledge to be used in making inferences.

artwork
One of the outputs of a computer-aided design system. For example, a photo plot (in printed circuit design), a photo mask (in integrated circuit design), a pen plot, an electrostatic copy, or a positive or negative photographic transparency. Transparencies (either on glass or film) and photo masks are forms of CAD artwork that can be used directly in the manufacture of a product, such as an IC, PC board, or mechanical part.

artwork master
A highly accurate photographic representation of a circuit design generated by the system for use in the fabrication process.

as-built configuration
A description of an assembled system.

ascenders
The parts of lowercase characters that rise above the line of alphabetical data.

Artificial Intelligence
Elements of Artificial Intelligence

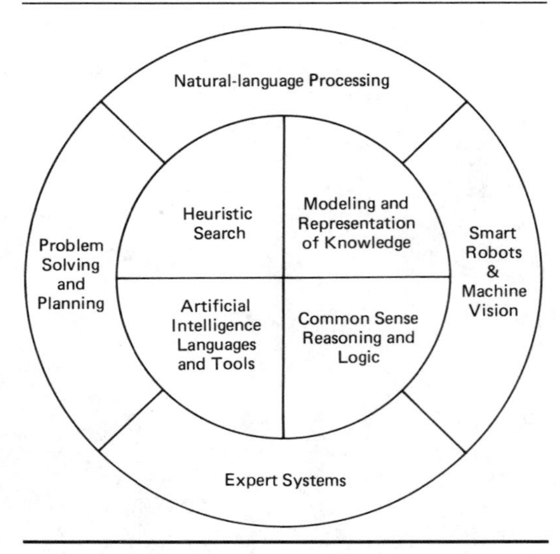

ASCII
The acronym for American Standard Code for Information Interchange, a widely used information exchange system for encoding letters, numerals, punctuation marks, and signs as binary numbers.

aspect ratio
The ratio of the width to the height of a rectangular area, such as a display surface, window, viewport, or character space. For example, an aspect ratio of 2.0 indicates an area twice as wide as it is high.

aspect source flag
A flag indicating whether a particular attribute selection is to be presented individually or bundled together with other attributes.

assembler language
A source language that includes symbolic machine-language statements in which there is a one-to-one correspondence with computer instructions. This language lies midway between high-level language and machine language but is closer to the latter. Programmers use "assemblers" to make more efficient use of the computer.

assembly
A number of basic parts or subassemblies or any combination thereof joined together to perform a specific function.

assembly language
A low-level (primitive) symbolic programming language directly translatable by an assembler into a machine (computer-executable) language. Enables programmers to write a computer program as a sequence of computer instructions, using mnemonic abbreviations for computer operation codes and names and addresses of the instructions and their operands.

assembly program
A computer program that translates a symbolic language into machine language. The symbol is generally more easily recognizable than a machine-language instruction and is, therefore, referred to as a mnemonic form.

assembly robot
A robot designed specifically for mating, fitting, or otherwise assembling various parts or components into either subassemblies or completed products. A class of generally small, lightweight, fast, accurate robots used primarily for grasping parts and mating or fitting them together.

assign
To give a new value to a variable during the running of a program.

associate
A unique field used to identify a field that can only be a parameter field of a primary field.

associative dimensioning
A computer-aided design capability that links dimension entities to geometric entities being dimensioned. This allows a dimension value to be automatically updated as the geometry changes.

associative storage
A storage arrangement in which storage locations are identified by their contents, not by names or positions. Synonymous with content-addressed storage.

associativity
Any logical linking of geometric entities (parts, components, or elements) in a computer-aided manufacturing and design data base with their nongraphic attributes (dimensions and text) or with other geometric entities. This enables the designer to retrieve, by a single command, not only a specified entity but all data associated with it. Associated data can automatically be updated on the system if the physical design changes. Also, structure entity that defines a logical link or relationship between different entities.

Assembly Robot
Assembly Robot Places Components on Printed Circuit Board

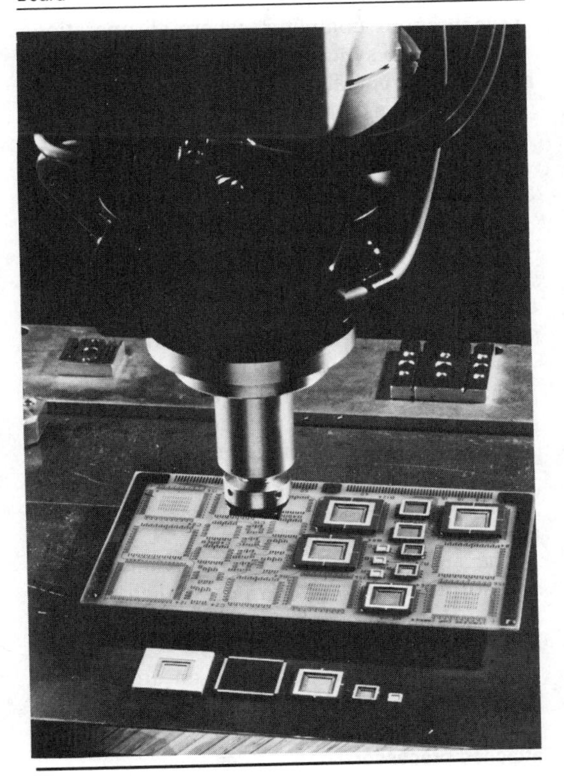

associativity definition entity
A structure entity that designates the type (link structure) and generic meaning of a relationship.

associativity instance entity
A structure entity formed by assigning specific values to the data items defining an associativity.

astigmatism
A machine vision system lens aberration in which the effective focal length of the lens is not constant around the lens' central axis. The effect is that the lens is not able to focus both arms of a cross simultaneously.

asynchronous
Not occurring at the same time.

asynchronous computer
A computer in which each operation starts as a result of a signal generated by the comple-

tion of the previous operation or by the availability of the parts of the computer required by the next event or operation. Contrast with synchronous computer, where all operations are timed to synchronize with a master clock.

asynchronous operation
Describes machine operations that are triggered successively—not by a clock, but by the completion of an operation.

asynchronous shift register
A shift register that does not require a clock. Register segments are loaded and shifted only at data entry.

asynchronous transmission
A mode of data communications transmission in which time intervals between transmitted characters may be of unequal length. Transmission is independently controlled by start and stop elements at the beginning and end of each character. Also called start-stop transmission.

atmospheric attenuation
The reduction in light energy resulting from absorption and scattering as light travels through the atmosphere.

atomic-hydrogen welding
Arc welding with heat from an arc between two or other suitable electrodes in a hydrogen atmosphere.

attached processor
A processor that communicates with a manufacturing automation protocol (MAP) mode that does not have its own end-to-end service component. An attached processor makes use of the end-to-end service component of the open-system processor with which it is attached.

attachment link
Used to connect computers and devices that do not provide full MAP functionality to MAP gateways. MAP gateways will provide required communication functions for these attached devices, allowing indirect participation in the network. The use of attachment links will be a necessity for a period of time—some computers and devices will offer only vendor-specific protocols. Eventually, MAP functionality will be implemented in most of this equipment to allow direct connection to the MAP network.

attenuation
In waveforms or signals, a loss of amplitude that can result from a distortion that alters the original form and causes a diminishing of particular portion of the signal.

attribute
A nongraphic characteristic of a part, component, or entity under design on a computer-aided design system. Examples include dimension entities associated with geometry, text with text nodes, and nodal lines with connect nodes. Changing one entity in an association can produce automatic changes by the system in the associated entity, for instance, moving one entity can cause moving or stretching of the other entity. Also, a particular property that applies to an output primitive. Transformation and viewing matrices, as well as highlighting and visibility, are also treated as attributes. Attribute also refers to information provided in specific fields within the directory entry of an entity that serve to qualify the entity definition. In addition, some properties of workstations are called workstation attributes.

attribute binding
The attachment of attributes to the output primitives they affect. The attachment occurs at structure traversal time, not when output primitive elements are inserted into a structure.

attribute definition
The process of declaring the set of possible values for an attribute.

attribute elements
Elements that describe the appearance of graphical elements.

attribute index
Attributes organized into workstation-dependent tables that are accessed via workstation-independent attributes indices.

attribute selection
The means by which the user specifies attribute values from a set of possible alternatives; the attribute or bundle of attributes so selected.

attribute specification
The processes of attribute definition and attribute selection.

audio frequencies
Frequencies that can be heard by the human ear, usually between 15 and 20,000 cycles per second.

audio-response systems
Computer data-processing systems wherein audible answers are generated in response to keyed input questions.

Automated Assembly
Example of Automated Production Workcell

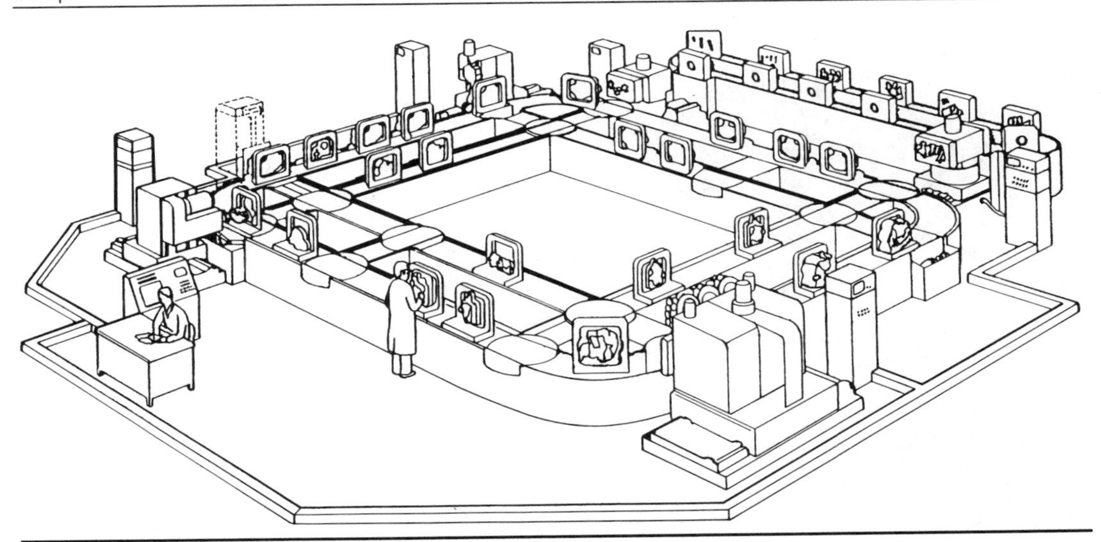

audio-response unit
A device that links a computer system to a telephone network to provide voice responses to inquiries made from telephone-type terminals.

audit trail
A clear sequence of transactions that can be used to document how a number, such as an inventory balance, was calculated.

augend
In an addition operation, a number or quantity to which other numbers or quantities (addends) are added.

authority decentralization
Placing the authority and decision-making power as close as possible to the organizational level at which the work is done.

authorized program analysis report
A request for correction of a problem caused by a defect in a current unaltered release of a program. The correction is then incorporated into subsequent releases of the program.

auto call
A machine feature that allows a transmission control unit or a station to initiate a call automatically over a switched line.

autoexecute
A utility of many computer operating systems that allows the program to automatically negotiate many of the statements necessary for the program to run.

Autohide
Computervisions's software option that allows the operator to view wire-frame geometry from a selected angle, automatically hiding all lines that would not normally be seen from that view.

auto-interactive design
A combination of design automation—where the computer executes programs or routines with no operator intervention, and computer-aided design—where the operator interacts with the computer in the design process.

automated assembly
Assembly by means of operations performed automatically by machines. A computer system may monitor the production and quality levels of the assembly operations.

automated audit
Feature whereby steps in a computer program or procedure automatically cause a review of the steps previously taken or the results previously prepared.

automated design system
Another term for a computer-aided design system.

automated guided vehicle system (AGVS)
Vehicles that interface with workstations for automatic or manual loading and unloading.

Automated Guided-Vehicle Systems (AGVS)
Kenway RoboCarrier Automated Guided-Vehicle System

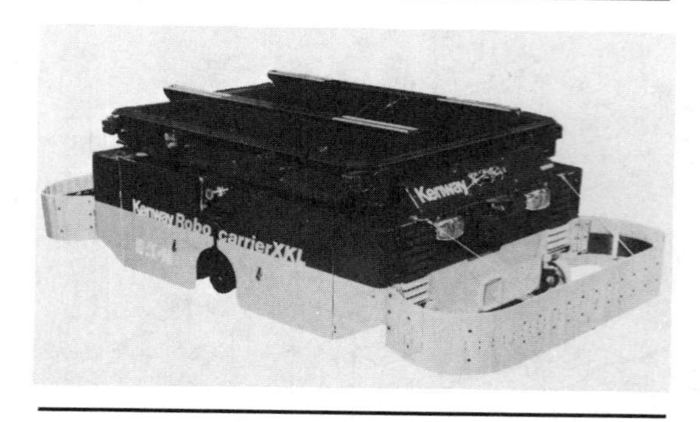

These systems have automatic guidance equipment and follow a prescribed guide path. Flexible manufacturing systems guided vehicles generally operate under computer control.

automated handling system
A system used to automatically move and store parts and raw materials throughout the manufacturing process and to integrate the flow of work pieces and tools with that process. In flexible manufacturing systems, the automated material-handling system operates under computer control.

automated process planning
The use of partial or complete computer assistance to create process plans for items in a given production family.

automated storage
A high-density rack-storage system with a retrieval system serving the rack structure for automatic loading and unloading. The storage system interfaces with automatic guided vehicle systems, car-on-track, towline or other conveyor systems for automatic storage and retrieval of loads. In flexible manufacturing systems, automated storage systems operate under computer control.

automated warehouse
A warehouse that employs automatic handling equipment to move materials from a receiving area to a bin or from a bin to an outgoing area. Instructions to the automatic handling equipment may be given either by an operator in the warehouse or by a computer.

automatic
Pertaining to a process or device that, under specified conditions, functions without intervention by a human operator.

automatic acceleration and deceleration
A feature in a computer control system enabling the machine to accelerate and decelerate smoothly. This is accomplished in some systems by denoting a GO8 or GO9 word; GO8 applying to acceleration and GO9 to deceleration.

Automatically Programmed Tools
One of the principal software languages used in computer-aided manufacturing to program numerically controlled machine tools.

automatic calling
A dialing device supplied by a common carrier that permits a business machine to dial calls automatically over communications networks.

automatic dialing unit
A device capable of generating digits automatically.

automatic dimensioning
A CAD capability that computes the dimensions in a displayed design or in a designated section and automatically places dimensions, dimensional lines, and arrowheads where required. In the case of mapping, this capability labels the linear feature with length and azimuth.

automatic message-switching center
In a communications network, location at which messages are automatically routed according to the information they contain.

automatic operation
The time during which a robot is performing its programmed tasks through continuous program execution.

automatic program
A computer program whereby a part programmer, using a relatively simple language, can prepare a manuscript covering the complete part that may then be inserted into a computer and automatically processed through to the final tape by one or several passes. Contrasts to special program, wherein the computer is used to assist in the performance of generally complicated calculations.

automatic programming
The process of using a computer to perform some stages of the work involved in preparing

a computer program. Also, an area of artificial intelligence research involved in creating software that can generate programs automatically from program specifications.

automatic rescheduling
The use of computers to automatically change due dates on orders where requirements and scheduled receipt dates do not agree. Contrasts to manual rescheduling, where messages concerning the need to reschedule open shop orders.

automatic tool change
Changing of tool (cutter, bit, and others) automatically on a machine tool by tape or computer control.

automation
The science and practice dealing with machinery and/or mechanisms that are so self-controlled and automatic that manual input is not necessary during operation. Automatically controlled operation of an apparatus, process, or system by mechanical or electronic devices that take the place of human observation, effort, and decision. The theory, art, or technique of making a process automatic, self-moving, or self-controlling.

autonomous
A system capable of independent action.

Autoplacement
A Computervision program that automatically packages integrated circuit elements and optimizes the layout of components on a printed circuit board.

autoradiography
Inspection technique in which radiation spontaneously emitted by a material is recorded photographically. The radiation is emitted by radioisotopes that are produced in or added to a material. The technique serves to locate the position of the radioactive element or compound.

auto routing and placement
In computer-aided design and manufacturing systems used for electronic design, a feature that locates circuit elements and adds the interconnections between points on the physical layout drawing. The routing and placing algorithm is usually based on a minimum total path-length calculation.

Autoroute
Computervision software that automatically determines the placement of copper on the printed circuit board to connect part pins of the same signal.

Autovision II
Vision system for robot applications. Autovision II is a product of Automatix, Billerica, Massachusetts.

auxiliary equipment
Equipment not under direct control of the central processing unit. Synonymous with ancillary equipment.

auxiliary function
An operation other than positioning or contouring performed by or associated with a machine.

auxiliary memory
The storage area that is supplementary to main memory. No manipulation of data can take place in auxiliary memory. This kind of memory is usually much slower than main memory.

auxiliary operation
An offline operation performed by equipment not under continuous control of the central processor.

auxiliary storage
Supplementary data storage other than main storage—for example, storage on magnetic tape or direct-access devices. Synonymous with external storage, secondary storage.

availability
The degree to which a system or resource is ready when needed to process data.

available machine time
The elapsed time during which a device is in operating condition, whether or not it is in use.

available material
A term usually interpreted to mean "material available for planning." Includes not only the on-hand inventory but also inventory on order.

available time
The number of hours the system is available for use. Synonymous with uptime. Consists of idle time and operating time.

available to promise
That portion of inventory or a time period's production that is still available or for sale: (beginning available inventory + scheduled receipts) − (customer orders) = available to promise.

available work
Work that is actually in a department ready to be worked on as opposed to scheduled work that may not yet be on hand.

average output power
The total laser energy per pulse times the number of pulses per second. In a pulsed laser, this determines the average power produced over time in watts (joules per second).

average sample number
The expected number of pieces that must be inspected to determine the acceptability of a lot. This is a function of the sampling plan used and the incoming quality.

average time
The arithmetical average of all the times or of all except the abnormal times, taken by a worker to complete a task or an element of a task.

avoidable delay
Any time during an assigned work period that is within the control of a worker and that he or she uses for idling or for doing things not necessary to the performance of the operation.

awareness barrier
An attachment/device that by physical and visual means warns a person of an approaching or present hazard.

awareness signal
A device that by means of audible sound or visible light warns a person of an approaching or present hazard.

axial-flow laser
A gas laser in which the lasing gas mixture is directed through the laser tube in an axial direction.

axis
A general direction of relative motion between cutting tool and workpiece. The understanding of axes in rectangular coordinates is the basic keystone to understanding numerical control. A reference line of a coordinate system, for example, the *X, Y,* or *Z* axis of the Cartesian coordinate system. A direction along which a movement of a tool or workpiece occurs. Also, rotary or translational (sliding) joint in a robot.

axis average line
The average axis of rotation of a spindle axis.

axis constraint
A mode or setting that forces any on-screen movement along the closer of the vertical or horizontal axes. It is used to make it easier to line up items precisely using on-screen positioning.

axonometric
A graphical representation that expresses three faces of an object. The faces are inclined to the plane of projection. The length, height, and depth are shown but not in perspective.

axonometric projection
A computer-aided design projection in which only one plane is used, the object being turned so that three faces show. The main axonometric positions are isometric, dimetric, and trimetric.

azimuth
Direction of a straight line to a point in a horizontal plane, expressed as the angular distance from a reference line, such as the observer's line of view.

B

back annotation
A CAD process by which data (text) is automatically extracted from a completed PC board design or wiring diagram stored on the system and used to update logic elements on the schematic created earlier in the design process. Text information can also be back-annotated into piping drawings and 3-D models created on the system.

backbone
The trunk media of a multimedia local area network separated into sections by bridges, gateways, or routers.

backbone media
Broadband coaxial cable—the recommended MAP media. Broadband cable can appear as many independent communication channels or as a multipoint link. Existing systems do not need to be altered to use broadband cable. The MAP-recommended IEEE 802 protocols include operation on broadband cable. The migration path can support existing equipment and mesh with new MAP equipment on existing cable installations.

back-chaining
A control procedure that attempts to achieve goals recursively, first by enumerating antecedents that would be sufficient for goal attainment and second by attempting to achieve or establish the antecedents themselves as goals.

back clipping plane
A plane parallel to the view plane, whose location is specified by a viewing coordinate space distance along the view plane normal from the view reference point.

back face
Rear surface plane of a three-dimensional polyhedron, normally invisible.

back focal distance
The distance from the rearmost element in a lens to the focal plane.

background
Parts of an image that will be overlaid by objects designated as foreground. Often, the background is a different color than foreground objects.

background job
A low-priority job, usually a batched or noninteractive job. Contrast with foreground job.

background program
In multiprogramming, the program with the lowest priority. Background programs execute from batched or stacked job input. Contrast with foreground program.

background processing
The execution under automatic control of lower-priority computer programs when higher-priority programs are not using the system resources. Contrast with foreground processing.

backhand welding
Welding in which the back of the principal hand (torch or electrode hand) of the welder faces the direction of travel. It has special significance in gas welding because it provides postheating.

backlash
Movement between interacting mechanical parts resulting from looseness. Free play in a power transmission system, such as a gear train, resulting in a characteristic form of hysteresis.

backlighting
The condition where the light reaching the image sensor is not reflected from the surface of the object. Often, backlighting produces a silhouette of an object being imaged. This is used when surface features on an object are not important.

backorder
An unfilled customer order or commitment. It is an immediate (or past due) demand against an item whose inventory is insufficient to satisfy the demand.

backplane
The area in a computer that accepts additional circuit boards or cards. Also known as a motherboard.

backplane clipping
If backplane clipping is enabled, only that part of the object that lies in front of and on the back clipping plane will be projected onto the CAD viewport. Clipping against the back clipping plane is controlled separately from all other clipping.

back pointer
A pointer in the parameter data section of an entity, pointing to an associativity instance of which it is a member.

back porch
That portion of a machine vision composite picture signal that lies between the trailing edge of a horizontal sync pulse and the trailing edge of the corresponding blanking pulse.

backstep sequence
A longitudinal welding sequence in which the direction of general progress is opposite to that of welding the individual increments.

backup
The provision of facilities to speed the process of restart and recovery following computer failure. Such facilities might include duplicated files of transactions, periodic dumping of core or backing up storage contents, duplicated processors, storage devices, terminals or telecommunications hardware, and the switches to effect a changeover.

backup copy
A copy of a file or data set that is kept for reference in case the original file or data set is destroyed.

backward chaining
A search technique used in production (*if then* rule) systems that begins with the action clause of a rule and works "backward" through a chain of rules in an attempt to find a verifiable set of condition clauses.

backward scheduling
Scheduling backward from the order due date by the time each operation takes, in order to arrive at an order start date.

baffle
An opaque shielding device, usually of high absorbance, used to block unwanted light rays from entering an optical system. A device to restrict or divert the passage of fluid through a pipeline or channel. In hydraulic systems the device, which often consists of a disc with a small central perforation, restricts the flow of hydraulic fluid in a high-pressure line.

bag molding
A method of applying pressure during bonding or molding in which a flexible cover, usually in connection with a rigid die or mold, exerts pressure on the material being molded, through the application of air pressure or drawing of a vacuum.

balanced line
A series of progressive related production operations with approximately equal standard times for each, arranged so that work flows at a desired steady rate from one operation to the next.

balanced loading
Loading a sequence of operations, where the capacity of the entire assembly or production line is considered and no one operation's work center is significantly overloaded or underloaded in relation to the line's capacity. For example, loading a starting department with a product mix that should not overload or underload subsequent departments.

ball plate
An artifact standard consisting of a number (more than two) of master spheres held in a stable dimension relationship to one another. Such plates are used for checking coordinate-measuring machines.

band
A unit of signaling speed equal to the number of discrete conditions or signal events per second.

bandpass
The specific range of frequencies, or wavelengths, that will be passed through a device.

bandwidth
The difference between the lower and upper limiting frequencies of a frequency band; also, the width of a band of frequencies.

bang-bang robot
A non-servo controlled, point-to-point robot that operates by "banging" into fixed stops in order to achieve the desired positions. Any robot in which motions are controlled by driving each axis against a mechanical stop.

bar chart
The graphical representation of information via rectangular columns on a **x, y** matrix.

bar code
An identification symbol where the symbol value is encoded in a sequence of high contrast bars and spaces. The relative widths of the bars and spaces contains the information.

bar code density
The number of characters that can be represented in a linear inch.

bar code label
A label that carries a bar code and is suitable to be affixed to an article.

bar code nominal size
The standard size for a bar code symbol. Most codes can be used over a range of magnifications, say from 0.80 to 1.20 nominal.

Bang-Bang Robot
Seiko Model 400. Simple Robot for Part Handling/Assembly.
Source: Seiko.

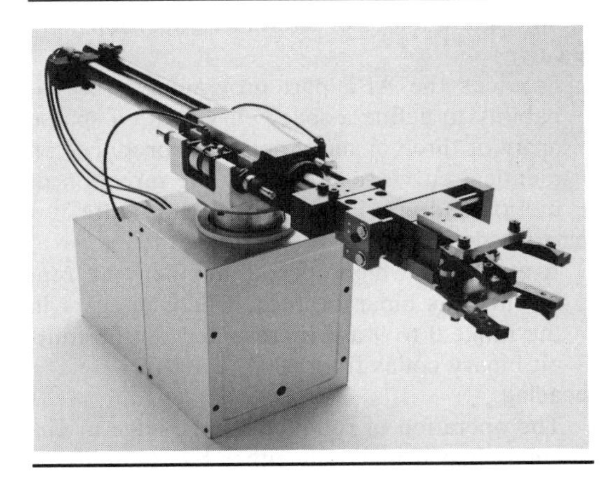

bar code reader
A device used for machine reading of bar codes. Readers may employ hand-held wands, fixed optical beams, or moving optical beams.

bar length
The bar dimension perpendicular to the bar width.

barrel distortion
An effect that makes an image appear to bulge outward on all sides like a barrel. Caused by a decrease in effective magnification as points in the image move away from the image center.

barriers
A physical means of safely separating persons from the robot-restricted work envelope.

Bar Code
Example of Bar Code Label

bar width
The thickness of a bar measured from the edge closest to the symbol start character to the trailing edge of the same bar code.

bar width reduction
Reduction of the nominal bar width dimension on film masters or printing plates to compensate for printing gain (ink spread).

base
The platform or structure to which a robot arm is attached; the end of a kinematic chain of arm links and joints opposite to that which grasps or processes external objects.

base address
A numeric value that is used as a reference in the calculation of addresses in the execution of a computer program.

baseband
Transmission of a signal at its original frequency, without modulation. Baseband signaling systems are generally limited to a single channel.

baseline fill
Coloring area of graph between data line and specified baseline.

base number
The number that is the basis for counting in a particular number system; for example, 10 is the base number for the decimal number system.

BASIC
A common algebra-like high-level, time-sharing computer programming language. It is easily symbolic-learned and used for problem-solving by engineers, scientists, and others who may not be professional programmers. The language is similar to FORTRAN II and was developed by Dartmouth College for a General Electric 225 computer system. It is now common on almost all computer systems. The BASIC acronym stands for Beginner's All-purpose Symbolic Instruction Code.

basic motion time study
A system of predetermined motion time standards. The essence of the system lies in the arbitrary definition of a basic motion as one that commences from rest and ends at rest. The system's purpose is to establish time standards for procedures that are composed of human motions, controlled only by the individual performing them, and to do so without resorting to time study.

basic telecommunications
An access method that permits read/write communications with remote devices.

batch
Group of jobs to be run on a computer in succession, without human intervention. A processing method where all input is gathered before being processed. Compare with real-time processing where input is processed as it is recorded.

batch card
A document used in the process industries to authorize and control the production of a quantity of material. Batch size usually relates to vessel size or output in a specific time period. A batch card usually contains information such as quantity and lot number of ingredients to be used, processing condition variables, pack-out instructions, and product disposition.

batched job
A job that is grouped with other jobs as part of an input stream to a computing system.

batch manufacturing
A process in which a facility produces a multitude of different parts by manufacturing them in groups, lots, or batches in which each part in the batch is identical.

batch processing
A technique in which a number of similar data or transactions are collected over a period of time and aggregated (batched) for sequential processing as a group during a machine run.

batch production
Refers to the lot size of identical parts produced in a factory. Batches range from 300 to 15,000 parts.

batch size
A quantity of items to be produced together, that is, produced without a break.

BAUDOT code
A code for the transmission of data in which five bits represent one character. It is named for Emile Baudot, a pioneer in printing telegraphy.

baud rate
A measure of speed of signal transmission or serial data flow between the CPU and the workstations it services. The term baud refers to the number of times the line condition changes per second. It can be measured in signal events (bits) per second.

Bayesian analysis
Statistical analysis incorporating uncertainty; using all available information to choose among a number of alternative decisions.

B:curve
Provides the APT part programmer the capability to define a space curve with a sparse array of three or more points. It produces an identical curve regardless of the direction of motion and provides for reflex curvature.

binary code decimal
A means by which decimal numbers are represented as binary values, where integers in the range 0 to 9 are represented by the four-bit binary codes from 0000 to 1001.

beading
The operation of rolling over the edge of circular-shaped material, either by stamping or spinning.

bead weld
A weld composed of one or more string or weave beads deposited on an unbroken surface.

beam delivery system
The use of optics, such as mirrors and lenses, arranged in such a way that a laser beam can be precisely directed to a specific location.

beam diameter
The diameter of the portion of the laser beam that contains 86 percent of the total energy of the beam.

beam divergence
The tendency of a laser beam to expand in diameter as it moves away from the source. Measured in milliradians at specified points.

beam splitting
The use of an optical device, such as a system of mirrors, to split the laser beam into two or more separate beams, so that more than one location on a part can be processed at one time, although at lower power.

bearing load
A compressive load supported by a member, usually a tube or collar, along a line where contact is made with a pin, rivet, axle, or shaft.

bearing strength
The maximum bearing load at failure divided by the effective bearing area. In a pinned or riveted joint, the effective area is calculated as the product of the diameter of the hole and the thickness of the bearing member.

benchmark
A set of standards used in testing a software or hardware product, or system, from which a measurement can be made. Benchmarks are often run on a system to verify that it performs according to specifications.

bending
The operation of forming flat materials, usually sheet metal or flat wire, into irregular shapes by the action of a punch forcing the material into a cavity or depression. Such forming is done in a punch press. When the operation involves long straight bends, the material is usually bent in a brake press.

beta test
The stage at which a new product is tested under actual usage conditions. The purpose of beta testing is to locate and correct potential problems before consumer marketing begins.

bevel gears
Mating gears having conical external shapes whose axes of rotation are nonparallel.

biaxial stress
A state of stress in which only one of the principal stresses is zero, the other two usually being in tension.

bid
An attempt to gain control over a line in order to transmit data. Usually associated with contention style of sharing a single line among several terminals.

bidirectional printing
Printing output in two directions—left to right and right to left. This is not only faster but saves wear on the printer.

bidirectional read
The ability to read data successfully whether the scanning motion is left to right or right to left.

bidirectional repeatability
The difference between the values of a measurement, when the measurement point is approached from opposite directions.

bidirectional search
An artificial intelligence search technique that combines forward chaining and backward chaining.

Big Blue
A nickname for IBM. Originally all IBM computers came in blue cabinets.

bilateral manipulator
A master-slave manipulator with symmetric force reflection, where both master and slave arms have sensors and actuators such that in any degree of freedom a positional error between the master and slave results in equal and opposing forces applied to the master and the slave arms.

billet
A solid, semifinished round or square product that has been hot-worked by forging, rolling, or extrusion. An iron or steel billet has a minimum width or thickness of $1\frac{1}{2}$ inches and the cross-sectional area varies from $2\frac{1}{4}$ to 36 square inches. For nonferrous metals, it may also refer to a casting suitable for finished or semifinished rolling or for extrusion.

bill of lading
The contractual paperwork that accompanies items shipped by common carrier. It gives description and quantity of goods and lists terms and conditions of shipment.

bill of material
A listing of all the subassemblies, parts, and raw materials that go into a parent assembly. There are a variety of formats of bill of material, such as single-level bill of material, indented bill of material, modular (planning) bill of material, transient bill of material, matrix bill of material, costed bill of material.

bill-of-material processor
The computer applications supplied by many manufacturers for maintaining, updating, and retrieving bill-of-material information on direct-access files.

bimodal
A distribution of values with two peaks.

binary
The basis for calculations in all digital computers. This two-digit numbering system consists of the digits 0 and 1, in contrast to the ten-digit decimal system.

binary coded decimal
A system of number representation in which each decimal digit is represented by four binary digits. Letters and symbols are represented by using the four numeric and the two zone channels.

binary element
A constituent element of data that takes either of two values or states. The term bit, originally

the abbreviation of the term binary digit, is misused in the sense of binary element.

binary image
A black and white image represented in memory as zeroes and ones. Images appear as silhouettes on the video display monitor.

binary interface
Used to control ancillary equipment by use of an on or off signal, usually through an input/output module.

binary notation
Any notation that uses two different characters, usually the binary digits 0 and 1, for example, the gray code. The gray code is a binary notation but not a pure binary numeration system. Fixed-radix notation where the radix is two.

binary number system
A number system using the base two, as opposed to the decimal number system, which uses the base ten. The binary system is comparable to the decimal system in using the concepts of absolute value and positional value. The difference is that the binary numbering system employs only two absolute values, 0 and 1.

binary picture
A digitized image in which the brightness of the pixels can have only two different values, such as white or black or 0 or 1.

binary search
A dichotomizing search in which, at each step of the search, the set of items is partitioned into two equal parts, some appropriate action being taken in the case of an odd number of items.

binary synchronous transmission
Data transmission in which synchronization of characters is controlled by timing signals generated at the sending and receiving stations. Contrast with asynchronous transmission.

binary system
A vision system that creates a digitized image of an object in which each pixel can have one of only two values, such as black/white or 0 or 1.

bin location file
A file that specifically identifies the physical location where each item in inventory is stored.

bin picking
The ability to remove individual parts from a bin in an oriented fashion.

bipolar
Literally, having two poles. An input signal is bipolar when one electrical voltage polarity represents a logically true input, and its opposite polarity represents a logically false input. Contrasts to unipolar, when both logical opposites are represented by the same electrical voltage polarity.

biquinary code
A notation in which a decimal digit n is represented by a pair of numerals, a being 0 or 1, b being 0, 1, 2, 3, or 4, and $(ta + b)$ being equal to n. The two digits are often represented by a series of two binary numerals.

bistable
Describes that which is able to assume one of two stable states; a flip-flop circuit is bistable.

bisynchronous
Data transmission in which synchronization of characters is controlled by timing signals generated at both the sending and receiving stations. Method of transmitting computerized data that allows for multiple error detection.

bit
In a digital system, the name of a single binary character, either a 1 or a 0. The smallest piece of data with which a digital computer can operate.

bit-aligned block transfer
Transfer of a rectangular array of pixels from one location in a bit map to another.

bit combination
An ordered set of bits (binary digits) that represents a character or a control function.

bit manipulation
The process of controlling and monitoring individual special-purpose data table bits through user-programmed instructions in order to vary application functions.

bit-mapped display
Referring to video displays and other graphic output, a system in which each possible dot of the display is controlled by a single bit of memory. Bit-mapped displays offer good resolution, fast updating, and flexibility, but they require a large amount of memory and high-speed transfers between memory and display. The bit map contains a bit for each point or dot on the screen, allowing for very fine resolution since any point on the screen can be

addressed. Movement of the scan beam is directed by software or microcode, rather than by a character generator, in order to create characters and/or graphics. Image resolution is usually referred to in pixels (picture elements).

bit plane
In computer displays, a section of memory that provides one single bit of information for each pixel (possible dot) on the screen.

bit rate
The speed at which bits are transmitted, usually expressed in bits per second.

bit-slice processor
A microprocessor designed to allow microcomputer organizations of variable word sizes, with processor units separated into two-, four-, or eight-bit slices on a single chip. These devices can be paralleled to yield an 8-, 12-, 16-, 24-, or 32-bit microcomputer, when assembled with the other necessary "overhead" components of the system.

bit storage
A single bit in any unused data table word that may be individually de-energized without directly controlling any output. However, any storage bit may be monitored as often as necessary in the user's program in order to control various outputs indirectly.

bit string
A string of binary digits (bits) in which the position of each binary digit is considered as an independent unit.

bits per inch (bpi)
The number of bits that can be stored per inch of magnetic tape. A measure of the data storage capacity of a magnetic tape.

bits per second (bps)
Transfer rate of a serial transmission.

blackboard approach
An artificial intelligence problem-solving approach, whereby the various system elements communicate with each other via a common working data storage called the blackboard.

blank
A semifinished piece of metal to be machined into a final finished form, as opposed to raw stock. A CAD command that causes a predefined entity to go temporarily blank on the CRT.

Blackboard Approach
Relationship of Elements of Artificial Intelligence Blackboard Approach

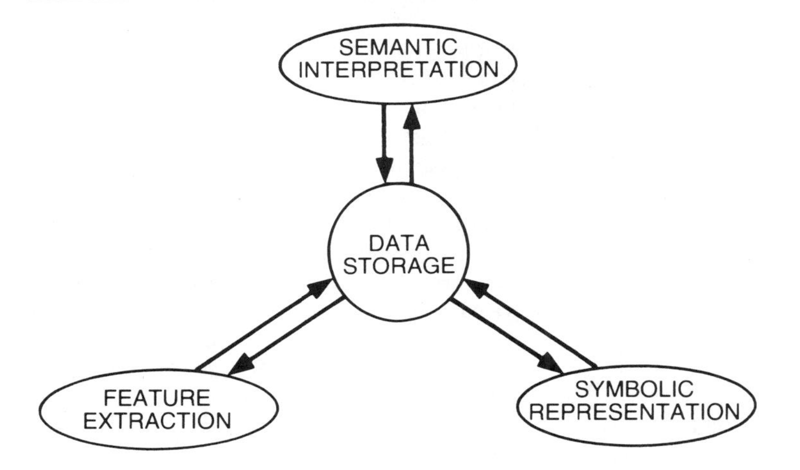

BLACKBOARD APPROACH

blank column detection
The collator function of checking for and signaling error conditions if a blank column is found in a particular data field.

blanket order
A large order that covers a long period of time (e.g., year), against which smaller periodic releases are specified. Usually used to secure a firm price or guaranteed delivery of parts or materials by reservation of the vendor's capacity.

blanket routing
A routing that lists a group of operations needed to produce a family of items. The items may have small differences in size, but they use the same sequence of operations. Specific times or tools for each individual item can be included.

blanking
The cutting of flat sheet stock to shape by striking it sharply with a punch while it is supported on a mating die. Punch presses are used. Also called die cutting. In machine vision, the suppression of the video signal for a portion of the scanning raster, usually during the retrace time.

blank status flag
A portion of the status number field of the directory entry of an entity, designating whether a data item is to be displayed on the output device.

blind search
An ordered approach in artificial intelligence that does not rely on knowledge for searching for a solution.

blinking
Refers to the flashing or pulsing of the cursor on a display screen, designed to attract the operator's attention for ease of screen manipulation.

blob
Any group of connected pixels in a binary image. A generic term including both "objects" and "holes."

blob addressing
A mechanism used to select a blob, for example, sequential addressing, *x-y* addressing, family addressing, and addressing by blob identification number.

blob labeling
A method of highlighting an addressed blob of the displayed image, for instance, by shading, cursor-marking, or alphanumeric labeling.

block
A group of machine words considered or transported as a unit. In flowcharts, each block represents a logical unit of programming. A "word" or group of "words" considered as a unit separated from other such units by an end-of-block character (EB). On a punched tape, it consists of one or more characters or rows across the tape that collectively provide sufficient information for a complete cutting operation. A unit for memory measurement. A block is equal to 256 words or 512 bytes.

block codes
Method involving the assignment of numbers in sequence by groups of various sizes other than tens, hundreds, and thousands. Instead, a block can consist of any quantity of numbers necessary to cover the items in a particular classification. In the original design of the code, a few blank numbers may be left in each block to provide for later additions.

block count readout
Display of the cumulative number of blocks that have been read from the tape. The count is triggered by the end-of-block character. Has generally been replaced by a "sequence number readout."

block delete
This CAD feature provides a means for skipping certain blocks by programming a slash (/) code immediately ahead of the block. The feature is useful when the operator desires to leave off certain cuts on a particular part configuration.

block diagram
A diagram in which a system or computer program is represented by annotated boxes and interconnecting lines to show the basic function and the functional relationship between the parts. A simplified schematic drawing.

blocking factor
The number of logical records combined into one physical record or block. If the blocking factor were four, there would then be four logical records in one physical block.

block length
A measure of the size of a block, usually specified in units such as records, words, computer words, or characters.

block sort
A sort of one or more of the most significant characters of a key to serve as a means of mak-

ing groups of workable size from a large volume of records to be sorted.

blooming
The defocusing of regions of the image where the brightness is at an excessive level.

blur circle
The image of a point source formed by an optical system at its focal point. The size of the blur circle is affected by the quality of the optical system and its focus.

Boolean algebra
A process of reasoning or a deductive system of theorems using a symbolic logic and dealing with classes, propositions, or on-off circuit elements such as AND, OR, NOT, EXCEPT, IF, THEN, and others, to permit mathematical calculations.

Boolean function
A switching function in which the number of possible values of the function and each of its independent variables is two.

Boolean logic/operation
Algebraic or symbolic logic formulas used in computer-aided design to expand design-rules checking programs and to expedite the construction of geometric figures. In a computer-aided mapping environment, Boolean operations are used either singly or in combination to identify features by their nongraphic properties (i.e., by highlighting all parcels with an area greater than 15 square kilometers, etc.).

booting
A technique for loading a program into a computer's memory in which the program's initial instructions direct the loading of the rest of the program. Usually, a few manual instructions must be entered on a keyboard or a switch implemented to initiate the process.

bootstrap program
Built-in instructions that take effect when a computer is turned on, preparing the computer for operation.

boot up
Start up of a computer system.

border area
The area of the physical CRT display screen that is outside the display area.

borrow digit
A digit generated when a difference in a digit place is arithmetically negative and is transferred for processing elsewhere. In a positional representation system, a borrow digit is transferred to the digit place with the next higher weight for processing.

boss
Protuberance on a plastic part designed to add strength, to facilitate alignment during assembly, or to provide for fastening.

bottleneck
The limiting operation in a process, that is, that operation with the lowest flow rate.

bottom up
Refers to the sequential processing by a machine vision system, beginning with the input image and terminating in an interpretation.

bottom-up control structure
An artificial intelligence problem-solving approach that employs forward reasoning from current or initial conditions. Also referred to as an event-driven or data-driven control structure.

bounce
Rapid and unwanted rebounding of a contact after closing, usually measured in either microseconds or milliseconds.

boundaries
Edge of a blob or object in a computer display. Outside limits of a process or procedure.

boundary
The line formed by the joining of two image regions, each having a different light intensity.

bounded plane
A finite region defined in a plane.

Bottom Up
Block Diagram of Hierarchical Bottom-Up Approach

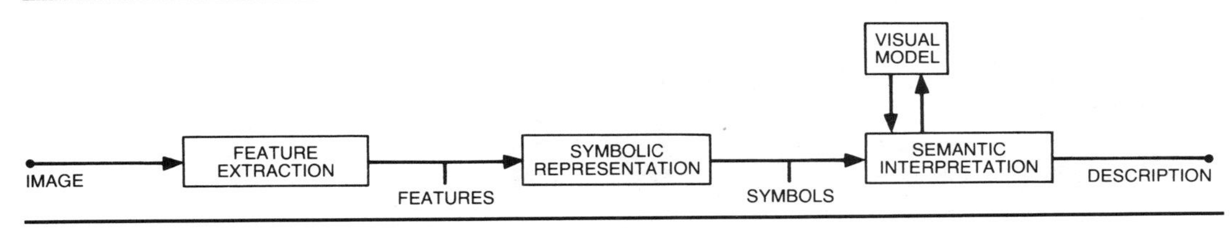

B-pocket
Provides the APT part programmer the capability to machine an enclosed area referred to as a pocket. Standard APT motion statements are used to define the periphery. The bottom may be one plane or two intersecting planes.

branch
A set of instructions that are executed between two successive branch instructions. In a data network, a route between two directly connected nodes. In the execution of a computer program, to select one from a number of alternative sets of instructions.

branching
The function of a computer program that alters the logic path, depending on some detected condition or data status. For example, the program would branch to a reorder routine when the projected available balance went negative.

branch instruction
Instructions that cause the computers to switch from one point in a program to another point, thereby controlling the sequence in which operations are performed.

branch jump
A means of departing from the sequence of the main program to another routine or sequence of operations as indicated by a branch instruction, whose execution is dependent on the conditions of the results of computer operations.

braze welding
Joining metals by following a thin layer (capillary thickness) of nonferrous filler metal into the space between them. Bonding results from the intimate contact produced by the dissolution of a small amount of base metal in the molten filler metal, without fusion of the base metal. Sometimes the filler metal is put in place as a thin solid sheet or as a clad layer, and the composite is heated as in furnace brazing. The term brazing is used when the temperature exceeds some arbitrary value, such as 800° F; the term soldering is used for temperatures lower than the arbitrary value.

breadboard
Describes the rough model for constructing another device, usually referred to as a circuit board that has a specified configuration on it.

breadth-first search
An artificial intelligence approach in which, starting with the root node, the nodes in the search tree are generated and examined level by level (before proceeding deeper). This approach is guaranteed to find an optimal solution if it exists.

break
An interruption to a transmission; usually a provision to allow a controlled terminal to interrupt the controlling computer.

break action
An implementation-dependent and workstation-dependent means for the operator to interrupt an input operation.

breakaway force
Same as static friction, although this term implies more strongly that the resistive force is not constant as the relative velocity increases.

breaker plate
A perforated plate located at the rear end of an extruder head. It often supports the screens that prevent foreign particles from entering the die.

breaking load
The load that causes fracture in a tension, compression, flexure, or torsion test.

breakpoint
A point in a computer program specified by the computer user when the program is to stop running so that the user can check it for error and change it if necessary. A member of an increasing sequence of real numbers in a CAD system that is a subsequence of the knot sequence used to specify parametric spline curves.

breathing
The opening and closing of a mold to allow gases to escape early in the molding cycle. Also called degassing.

Brewster-angle window
To achieve polarized and hence better-quality laser light. The end windows of the cavity are sometimes set at this angle, at which refracted and reflected rays of incident parallel light are mutually perpendicular. The result is that there is no reflective loss of vertically polarized light.

bridge
Transparent devices used to connect segments of a local area network whose data links are the same, or possibly different, for the purpose of extending the network or isolating these segments. Networks connected by bridges would logically appear as one network.

Bridge
Bridge Architecture

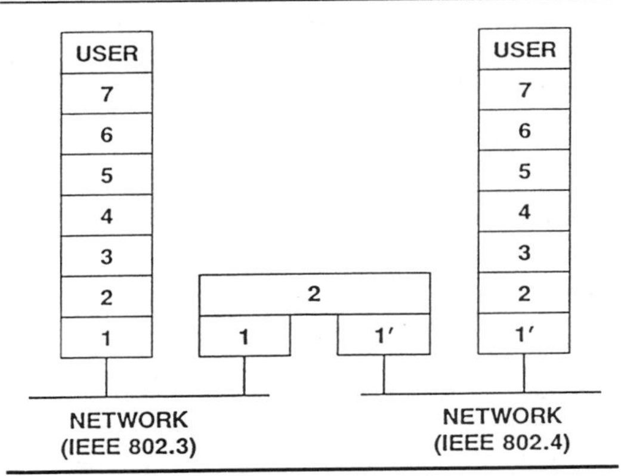

bridge configuration robot
A bridge-configuration robot is a Cartesian robot that looks like a bridge crane and in which the traveling bridge lies on elevated rails.

bridge top
Provides network expansion by connecting two physically distinct networks at the data link layer. The two networks must use a consistent addressing scheme and frame size. A bridge can extend a network beyond the design capacity of a single segment or physically isolate the segments from each other. Interconnection relies on the use of identical data link protocols across connected networks. Bridges may be used to provide connection of two identical IEEE 802 network types, such as two token bus implementations. In some circumstances, a bridge may also be used in coupling dissimilar IEEE 802 networks.

brightness
Relative amount of light reflected by a material. Measured by the reflectance of a material. The total amount of visible light per unit area. The same as luminance.

Brinell hardness test
A test for determining the hardness of a material by forcing a hard steel or carbide ball of specified diameter into it under a specified load. The result is expressed as the Brinell hardness number, which is the value obtained by dividing the applied load in kilograms by the surface area of the resulting impression in square millimeters.

broaching
The operation of cutting away the material with a succession of cutting teeth on a tool that is either pushed or pulled along the surface of the workpiece. Cutting may be applied to either holes or to the outside edges of the piece.

broadband
Of data transmission facilities, capable of handling frequencies greater than those required for high-grade voice communications, that is, greater than 300 characters per second.

broadcast
Transmission to a number of receiving locations simultaneously; normally associated with a multidrop line, where a number of terminals share the line.

brush
In a computer graphics paint program for drawing on a screen, an on-screen cursor that can move about, leaving a mark as it passes. In most paint programs, brushes are larger than pens and can have various shapes. In many programs, brush shapes can be modified for special effects.

B-size sheet
11- by 17-inch drawing.

B-spline
A sequence of parametric polynomial curves forming a smooth fit between a sequence of points in a 3-D space. The piecewise defined curve maintains a level of mathematical continuity, dependent upon the polynomial degree chosen. It is used extensively in mechanical design applications in the automotive and aerospace industries.

B-spline surface
The mathematical description of a 3-D surface that passes through a set of B-splines.

B-surf
Provides the APT part programmer the capability to machine a nonmathematically defined sculptured surface. The surface is easily defined by using standard APT motion statements in describing planar or nonplanar sections through the surface; four definition methods are available. It allows the use of ball-nose or bull-nose (flat-bottom cutter with corner radius) cutters and includes five-axis machining capability.

B-tree
Hierarchical indexing method that always maintains a balance between the number of entries in the branches of the hierarchy.

bubble forming
An optical recording method in which the laser causes the underlayer of the media to vaporize, thus pushing the upper layer to form a blister or bubble. On reading, the reflected laser light is scattered at the bubble, denoting a written bit.

bubble memories
Tiny cylinders of magnetization whose axes lie perpendicular to the plane of the single-crystal sheet that contains them. Magnetic bubbles arise when two magnetic fields are applied perpendicular to the sheet. A constant field strengthens and fattens the regions of the sheet whose magnetization lies along it. A pulsed field then breaks the strengthened regions into isolated bubbles that are free to move within the plane of the sheet. Because the presence or absence of bubbles can represent digital information, and because other external fields can manipulate this information, magnetic-bubble devices are used in data storage systems.

bubbler mold cooling
A method of cooling an injection mold in which a stream of cooling liquid flows continuously into a cooling cavity equipped with a coolant outlet normally positioned at the end opposite the inlet. Uniform cooling can be achieved in this manner.

bucketed system
An MRP system under which all time-phased data is displayed in accumulated time periods or "buckets." If the period of accumulation would be 1 week, then the system would be said to have weekly buckets.

bucketless system
An MRP system under which all time-phased data is received, stored, processed and reported by specific dates and not in weekly (or larger) time buckets, horizontal displays, or vertical displays.

buffer
Memory area in a computer or peripheral used for temporary storage of information that has just been received. The information is held in the buffer until the computer or device is ready to process it. A temporary holding area. In an automated cell or workstation, a buffer area might be used to store incoming blanks until the machine tools are ready to process them, or to store finished parts until the materials-handling system is ready to fetch them. In a computer, a buffer is an area of memory set aside to hold data that, for example, are being moved from one part of the system to another.

buffer stock
That quantity of an item of inventory that is held in stock for absorbing expected variations in usage between the time reorder action is initiated and the first part of the new order is received in stock. Inventory meant to decouple two operations so that the second operation's production rate is not dependent on the output rate of the first operation.

buffer storage
A place for storing information in a control system or computer for anticipated utilization. Information from the buffer storage section of a control system can be transferred almost instantaneously to active storage, which is that portion of the control system commanding the operation at the particular time. Buffer storage offers the ability of a control system to act immediately on stored information, rather than wait for this information to be read into the machine via the tape reader, which is relatively slow.

bug
An error or flaw in a program that renders it incapable of performing the objectives for which it was written. In 1946, Grace Hopper detected a problem with an ENIAC computer at the University of Pennsylvania. Investigation uncovered an insect lodged within the computer, causing the malfunction; she exclaimed, "There's a bug in the computer." Debugging is the process of finding and eliminating programming errors. Most programs of any complexity have to go through a *debugging* process.

build/test information
Includes detailed engineering-generated design/test data expected from a CAD data base or other source, in standard ANSI format. The data defines the part number and level, component locations, assembly/manufacturing sequence (or routing), the related product bill of material, and test data. This batch-type data

is furnished as required to update the global data base for consistency with the other information or in response to generated requests for deficient data.

bulging
The operation of expanding the metal below the opening of a drawn cup or shell.

bulk annotation
A CAD feature that enables the designer to automatically enter repetitive text or other annotation at multiple locations on a drawing or design.

bulk factor
Ratio of volume of powdered material to volume of solid piece. Also, ratio of density of solid material to apparent density of loose powder.

bulk memory
A memory device for storing a large amount of data, that is, disk, drum, or magnetic tape. It is not randomly accessible, like main memory.

bundle
A group of related attributes that may be selected and modified as a unit.

bundled
A pricing strategy in which a manufacturer includes all products—hardware, software, services, training, and so forth—in a single price.

bundle index
An index pointer for accessing a particular set of attributes in a bundle table.

bundle table
An indexed table containing a set of attributes for each index.

bundle-table entry
A single entry in a bundle table. Each entry contains one value for each attribute that applies to the corresponding type of output primitive.

bundle-table index
A pointer used to access a particular bundle-table entry. The bundle-table index itself is an attribute.

burn-in
A product test where all components are operated continuously for an extended period of time in order to eliminate early failures.

burnishing
The operation of producing smooth-finished surfaces by compressing the outer layer of the metal, either by the application of highly polished tools or by the use of steel balls in rolling contact with the surface of the piece part.

burst
In data communication, a sequence of signals counted as one unit in accordance with some specific criterion or measure.

bursting strength
A measure of the ability of materials in various forms to withstand hydrostatic pressure.

bus
A path or channel for transmitting electrical signals and data, usually between a computer and peripheral equipment.

busy hour
The peak 60 minutes during a business day when the largest volume of communications traffic is handled.

butt-fusion
A method of joining pipe, sheet, or other similar forms of a thermoplastic resin, wherein the ends of the two pieces to be joined are heated to the molten state and then rapidly pressed together to form a homogeneous bond.

button
In the Core graphics standards, an input device that supplies a signal upon command from the operator. Buttons are often combined with other devices, such as the push buttons on a mouse or light pen. They also can be simulated by activating on-screen button images with a light pen or mouse or with a keyboard.

byte
A sequence of bits operated upon as a unit and usually shorter than a computer word. The representation of a character. Often, a sequence of eight adjacent binary digits that are operated upon as a unit and that constitute the smallest addressable unit in the system.

C

C

High-level programming language that can often be used in lieu of the lower-level assembler language.

cable drive

Transmission of power from an actuator to a remote mechanism by means of a flexible cable and pulleys.

cabling diagram

A diagram showing connections and physical locations of system or unit cables and used to facilitate field installation and repair of wiring systems. Can be generated by CAD.

cache memory

A high-speed, buffer-type memory filled at medium speed from the main memory. Programs and instructions found in the cache memory can be operated at higher speeds without the necessity of loading another segment.

CAD (computer-aided design)

Describes the more demanding and elaborate preparation of complex schematics and blueprints, typically those of industry. In these applications, the operator constructs a highly detailed drawing online, using a variety of interaction devices and programming techniques. Facilities are required for replicating basic figures; achieving exact size and placement of components; making lines of specified length, width, or angle to previously defined lines; satisfying varying geometric and topological constraints among components of the drawing; and so forth. A primary difference between interactive plotting and design drafting lies in the amount of effort the operator contributes, with interactive design drafting requiring far more responsibility for the eventual result. In interactive plotting, the computation is of central importance and the drawing is typically secondary. A second difference is that design drawings tend to have structure, that is, to be hierarchies of networks or mechanical or electrical components. These components must be transformed and edited. If, in addition to nontrivial layout, the application program involves significant computation of the picture and its components, we speak of the third and most complex category, that of interactive design. In addition to a pictorial datum base, or data structure, that defines where all the picture components fit on the picture and also specifies their geometric characteristics, an application datum base is needed to describe the electrical, mechanical, and other properties of the components in a form suitable for access and manipulation by the analysis program. This data base must naturally also be editable and accessible by the interactive user.

CAD

Typical CAD Workstation. *Courtesy:* Computer Systems.

CADAM

Acronym for Computer-graphics Augmented Design And Manufacturing system developed by GE. An interactive graphics system for computer-aided design and manufacturing. The system includes a design/drafting package, together with a number of aids to design analysis.

CAD/CAM (computer-aided design/computer-aided manufacturing)

Refers to the integration of computers into the entire design-to-fabrication cycle of a product or plant.

CAE (computer-aided engineering)

Analysis of a design for basic error checking or to optimize manufacturability, performance, and economy (e.g., by comparing various possible materials or designs). Information drawn from the CAD/CAM design data

CAD/CAM

Elements of Computer-Aided Design and Manufacturing. Reprinted with Permission of McGraw-Hill Book Company, from *CAD/CAM Handbook* by Eric Teicholz.

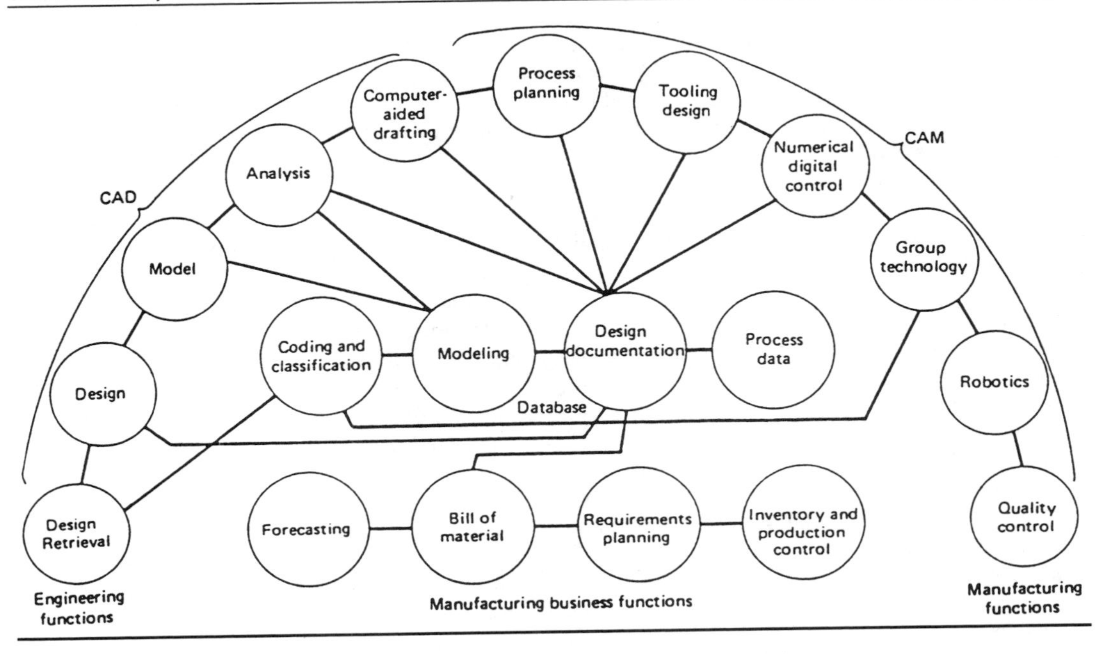

base is used to analyze the functional characteristics of a part, product, or system under design and to simulate its performance under various conditions. CAE permits the execution of complex circuit loading analyses and simulation during the circuit definition stage. CAE can be used to determine section properties, moments of inertia, shear and bending moments, weight, volume, surface area, and center of gravity. CAE can precisely determine loads, vibration, noise, and service life early in the design cycle so that components can be optimized to meet those criteria. Perhaps the most powerful CAE technique is finite element modeling.

CAI (computer-assisted instruction)
The application of computers to education. The computer monitors and controls the student's learning, adjusting its presentation based on the responses of the student.

calendar time
Refers to the passage of days or weeks as in the definition of lead time or scheduling rules in contrast with running time.

calibration
The act of determining, marking, or rectifying the capacity or scale graduations of a measuring instrument or replicating machine.

call
The action of bringing a computer program, a routine, or subroutine into effect, usually by specifying the entry conditions and jumping to an entry point.

calligraphic
Refers to displays that form images and characters out of line segments, rather than out of dots. The display system positions the writing beam or element to the end of each desired stroke and then writes the required line segment. Another term for this type of display is vector display.

calling sequence
A basic set of instructions used to begin, initialize, or transfer control to and return from a subroutine.

CAM (computer-aided manufacturing)
The use of computer and digital technology to generate manufacturing-oriented data. Data

drawn from a CAD/CAM data base can assist in or control a portion or all of a manufacturing process, including numerically controlled machines, computer-assisted parts programming, computer-assisted process planning, robotics, and programmable logic controllers. CAM can involve production programming, manufacturing engineering, industrial engineering, facilities engineering, and reliability engineering (quality control). CAM techniques can be used to produce process plans for fabricating a complete assembly, to program robots, and to coordinate plant operation.

CAM-I
Acronym for Computer-Aided Manufacturing International, a not-for-profit association of advanced manufacturing technology suppliers.

candela
The unit of luminous intensity equal to one-sixtieth the normal intensity of a 1-square-centimeter blackbody at the solidification temperature of platinum.

candlepower
Luminous intensity expressed in candelas.

canned cycle
A preset sequence of events initiated by a single command (hardware or software). Normally handled by a preparatory, or G, code. Ideal for positioning operations such as hole drilling, tapping, boring, and reaming. Canned cycles decrease programming time and length of NC tape.

canned program
A software program written to meet the expected needs of a specific application.

capability
A functional or logical feature of a solution.

capability set
The set of all capabilities relating to a given development effort.

capacity
The highest sustainable output rate that can be achieved with the current product specifications, product mix, worker effort, plant, and equipment.

capacity control
The process of measuring production output and comparing it with the capacity requirements plan, determining if the variance exceeds preestablished limits, and taking corrective action to get back on plan if the limits are exceeded.

capacity loading
Work-center loading where work will be rescheduled into other time periods if capacity is not available for it in the required time period.

capacity planning
The function of setting the limits or levels of manufacturing operations in the future, consideration being given to sales forecasts and the requirements and availability of men, machines, materials, and money. The production plan is usually in fairly broad terms and does not specify in detail each of the individual products to be made but usually specifies the amount of capacity that will be required.

capacity requirements
The projected future production capacity needs, expressed in terms of men, machines, and facilities.

CAPOSS-E
Acronym for CApacity Planning and Operation Sequence System Extended program. Designed for use whenever a large number of activities have to be allocated to limited capacity resources. It contains features that allow the overlapping and splitting of operations, the use of alternative resources, the grouping of similar activities, and the simultaneous reservation of more than one resource for any activity.

CAPP (computer-aided process planning)
A prototype software development that provides a data management framework designed to assist the functions of process planning in the manufacture of discrete parts. The system enables a process planner to automatically access standard process plan specification data in an interactive and dynamic manner.

carbide tools
Cutting tools made of tungsten carbide, titanium carbide, tantalum carbide, or combinations of these in a matrix of cobalt or nickel, having sufficient wear resistance and heat resistance to permit high machining speeds.

carbon-arc cutting
Metalcutting by melting with the heat of an arc between a carbon electrode and the base metal.

carbon-arc welding
Welding in which an arc is maintained between a nonconsumable carbon electrode and the work.

carbon dioxide laser
A gas laser that uses a mixture of CO_2, nitrogen, and helium to produce a continuous output of laser light at a wavelength of 10.6 microns.

card
The individual circuit boards that carry the necessary electronics for particular functions, for example, memory, and disk drive control.

card cage
Also referred to as card chassis. A frame for holding circuit cards in a microprocessor.

card code
The combination of punches used to represent alphabetical and numerical data on a punched card.

card field
The card column or consecutive columns used to store a particular piece of information.

card reader
A device for reading data from punched cards.

car-on-track
A conveying system consisting of a load-carrying car that is driven along a set of tracks over which the car can be programmed to stop to interface with process equipment, robots, or other material-handling systems.

carousel
A rotating work-in-process queuing system that delivers workpieces to a load/unload station that may be served by an operator or a robot.

carriage
The assembly in a printer that moves the paper past the print mechanism or vice versa.

carriage return
The carriage return on a typewriter keyboard corresponds to the end or return on a CAD alphanumeric.

carrier
A continuous frequency capable of being modulated or impressed with a signal.

carrier-sense multiple access with collision detect
A media access control scheme that uses contention techniques to allocate and control a common communication medium. *Carrier-sense* refers to the capability of any node to sense traffic on the medium. If no traffic is sensed, any node may transmit (*multiple access*). Delays due to electrical propagation of the transmitted signal may allow two nodes to begin transmitting "simultaneously," causing the transmission to collide. Each node monitors the channel to detect the possible collision of its transmission; if collisions occur, the node abandons the current transmission, then retransmits after a brief random delay.

carrier system
A measure of obtaining a number of channels over a single path by modulating each channel on a different carrier frequency and demodulating at the receiving point to restore the signals to their original form.

carrier-to-noise ratio
Related to signal-to-noise ratio and can impact the error rate of an optical recording disk. The noise component comprises substrate inconsistencies, as well as amplifier and laser noise. Typically, for 100 kHz bandwidth of carrier, noise of about 70 microvolts magnitude is read.

carry
The action of transferring a carry digit. One or more digits produced in connection with an arithmetic operation on one digit place of two or more numerals in positional notation that are forwarded to another digit place for processing there.

carry digit
A digit that is generated when a sum or a product in a digit place exceeds the largest number that can be represented in that digit place and that is transferred for processing elsewhere. In a positional representation system, a carry digit is transferred to the digit place with the next higher weight for processing.

carrying cost
The cost of having inventory. Includes cost of capital, space costs, insurance, taxes, material handling, and damaged or lost inventory. Normally expressed as a percentage of unit cost.

cart
Vehicle used to transport pallets between stations.

Cartesian coordinate robot
A robot whose manipulator arm degrees of freedom are defined primarily by Cartesian coordinates. A robot with linear motions arrayed in mutually perpendicular directions, that is, east-west, north-south, and up-down, as well as rotary motions to change orientation. A robot with motions resulting from movement along horizontal and vertical tracks, rather

Cartesian Coordinate Robot
Cartesian Coordinate Robot Work Envelope.

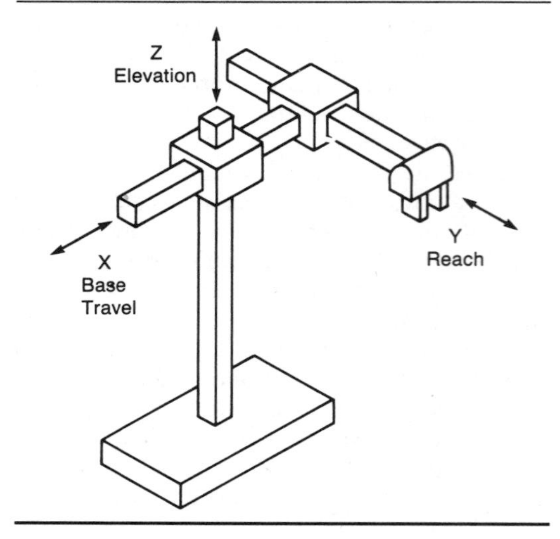

than through the use of joints—often called an orthogonal robot.

Cartesian coordinates
A set of three numbers defining the location of a point within a rectilinear coordinate system, consisting of three perpendicular axes (X, Y, Z).

cart serial number
A unique number assigned to a specific cart.

cascade control
An automatic control system in which the control units, linked in sequence, feed into one another in succession, each regulating the operation of the next line.

cascaded receiving
A method of receiving material that allows delivery quantities to be applied to the oldest open delivery record and to successive delivery records until the quantity received is exhausted or the controlling purchase order quantity is satisfied.

cascaded systems
Multistorage operations; the input to each stage is the output of a preceding stage, thereby causing interdependencies among the stages.

CASE commitment, concurrency, and recovery (CCR)
Part of the general functions provided by CASE. CCR specifies distributed synchronization, backup, recovery, and restart of work to reliably complete a distributed activity, when machines fail. It is intended for use with distributed data bases and the job transfer and manipulation (JTM) protocol.

CASPA
Acronym for Computer-Aided Sculptured Pre-APT system. CASPA supports sculptured surfaces definitions under the control of a general one-pass supervisory program. The system contains an integrated graphics package, permitting immediate plotting of sculptured geometry and CASPA-generated cutter positions from any point of view.

cassegrain optics
An optical system that uses two mirrors to fold the image path back on itself. It has the advantage of shortening the physical length of the optical system.

cast-alloy tool
A cutting tool made by casting a cobalt-base alloy that is used at machining speeds between those for high-speed steels and sintered carbides.

casting
An object at or near finished shape, obtained by solidification of a substance in a mold. Pouring molten metal into a mold to produce an object of desired shape.

catalogue
To place data sets permanently in a storage device for use when required. This technique avoids having to read in a deck of cards each time the program or data are required.

catalyst
A substance capable of changing the rate of reaction, without itself undergoing any net change.

catanet
An interconnected set of MAP/TOP networks using bridges, routers, and/or gateways.

cathode-ray tube
A special form of vacuum tube in which a focused beam of electrons is caused to strike a surface coated with a phosphor. This beam is deflected so that it traces an orthogonal presentation of two separate signals; a third independent signal may be presented as a variation of the intensity of the electron beam, and in turn, the fluorescent intensity. The principal component in a CAD display device. A CRT displays graphic representations of geo-

Cathode-Ray Tube
Cathode-Ray Tube Geometry. Reprinted with Permission of McGraw-Hill Book Company, from *CAD/CAM Handbook* by Eric Teicholz.

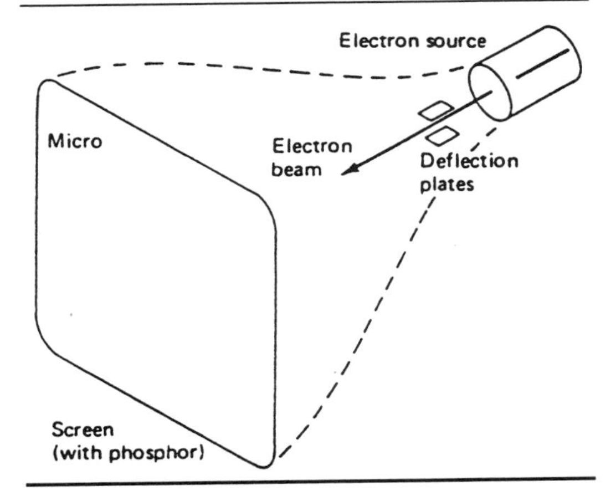

metric entities and designs and can be of various types: storage tube, raster scan, or refresh. These tubes create images by means of a controllable beam of electrons striking a screen. The term CRT is often used to denote the entire display device.

causal model
Model in which the causal relations among various actions and events are represented explicitly.

cavitation
The formation and instantaneous collapse of innumerable tiny voids or cavities within a liquid subjected to rapid and intense pressure changes. Cavitation produced by ultrasonic radiation is sometimes used to give violent localized agitation.

cavity
Depression in a mold made by casting, machining, hobbing, or by a combination of these methods. Depending on number of such depressions, molds are designated as single cavity or multicavity. In a laser the resonator, or tube, in which the lasing process occurs.

CCD camera
A solid-state television camera that uses charge-coupled device (CCD) technology.

CCITT
Acronym for Comite Consultatif Internationale de Telegraphie et Telephonie, an international consultative committee that sets international communications usages standards.

cell
A manufacturing unit consisting of two or more workstations or machines and the materials, transport mechanisms, and storage buffers that interconnect them. A computer graphic entity that is a rectangle or parallelogram. Its only aspect is color, which is specified as part of the associated output primitive.

cell array
An output primitive consisting of a two-dimensional array of cells.

cell run
A compressed representation of a contiguous group of like cells, represented by a color index and the length of the run (number of like cells).

cell run array
An output primitive consisting of a two-dimensional array of cell runs.

cemented doublet
A two-element lens, where the elements are cemented together.

center
A manufacturing unit consisting of two or more cells and the materials transport and storage buffers that interconnect them.

center drilling
Drilling a conical hole (pit) in one end of a workpiece.

centerless grinding
Grinding the outside or inside of a workpiece mounted on rollers, rather than on centers.

CCD Camera
Fairchild CCD Industrial Camera

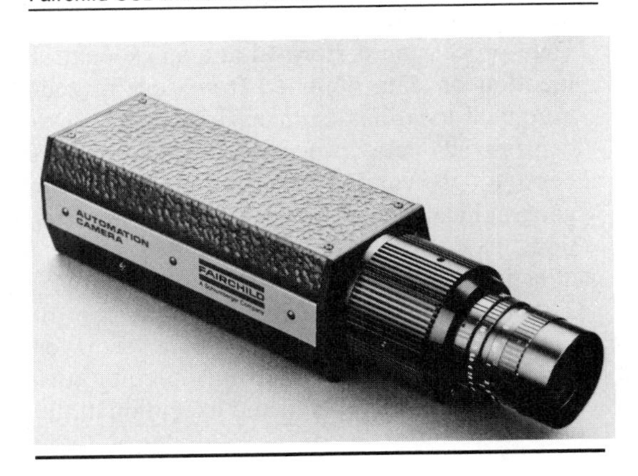

The workpiece may be in the form of a cylinder or the frustum of a cone.

centerline
The center of an NC machine tool path computed on a CAD/CAM system. Centerline data—in the form of a centerline file—ultimately controls the NC machine itself. In normal 2-D/3-D drafting, centerline means a representation of the central axis of a part or the central axes of intersecting parts.

centerline average
When measuring roundness, centerline average is functional equivalent to an arithmetic average diameter.

centerline entity
An annotation entity for representing the axis of symmetry for all symmetric views or portions of views, such as the axis of a cylinder or a cone.

center of acceleration
That point in a rigid body around which the entire mass revolves.

center of gravity
That point in a rigid body at which the entire mass of the body could be concentrated and produce the same gravity resultant as that for the body itself.

center of projection
A reference point for perspective views.

centralized computer network
A computer network configuration in which a central node provides computing power, control, or other services. Contrast to decentralized network.

centralized control
Control decisions for two or more control tasks at different locations, made at a centralized location.

centralized data processing
Data processing performed at a single central location on data obtained from several geographical locations or managerial levels. Decentralized data processing involves processing at various managerial levels or geographical points throughout the organization.

central processing unit
The arithmetic and logic unit and the control unit of a digital computer. Another term for processor. It includes the unit circuits controlling the interpretation and execution of the user-inserted program instructions stored in the computer or robot memory. The hardware part (CPU) of a computer that directs the sequence of operations, interprets the coded instructions, performs arithmetic and logical operations, and initiates the proper commands to the computer circuits for execution. Controls the computer operation as directed by the program it is executing. The brain of a CAD/CAM system that controls the retrieval, decoding, and processing of information, as well as the interpretation and execution of operating instructions—the building blocks of application and other computer programs.

centrifugal casting
A method of forming thermoplastic resins in which the granular resin is placed in a rotatable container, heated to a molten condition by the transfer of heat through the walls of the container, and rotated so that the centrifugal force induced will force the molten resin to conform to the configuration of the interior surface of the container. Used to fabricate large-diameter pipes and similar cylindrical items.

centroid
The center; in the case of a two-dimensional object the average X and Y coordinate.

Centronics interface
An eight-bit, parallel interface that has become the de facto standard for microcomputer printer connection.

certainty factor
Either a number supplied by an expert system that indicates the level of the system's confidence in its conclusion, or a number supplied by the user of an expert system that indicates the level of the user's confidence in the validity of the information being supplied to the system.

certificate of compliance
A supplier's certification to the effect that the supplies or services in question meet certain specified requirements.

certification
An authoritative endorsement of the correctness of a program, analogous to the certification of electrical equipment by the Underwriters Laboratories.

C-Gun
A spot welding tool.

chain code
A boundary representation that starts with an initial point and stores a chain of directions to successive points.

chain drive
Transmission of power from an actuator to a remote mechanism by means of a flexible chain and mating-toothed sprocket wheels.

chaining
A system of storing records in which each record belongs to a list or group of records and has a linking field for tracing the chain.

chain printer
Printer that uses a type chain or print chain as its printing mechanism.

chamfer
A beveled surface that eliminates an otherwise sharp corner. A relieved angular cutting edge at a tooth corner.

chamfer angle
The angle between a referenced surface and the bevel. On a milling cutter, the angle between a beveled surface and the axis of the cutter.

change notice
A formal notification of an alteration to a contract that will affect its terms and conditions. A means of evaluating, controlling, and possibly obtaining reimbursement for variances in the anticipated cost of the project from an established budget or control estimate.

changeover time
The time required to modify or replace an existing facility or workplace, usually including both teardown time for the existing condition and setup for the new condition.

channel
A path over which information is transmitted, generally from some input/output device to storage. With reference to magnetic or punched tape, a channel is one of the parallel tracks in which data are recorded. In data communications, a path for electrical transmission between two or more points. Also called a circuit, facility, line, link, or path. Within a computer, the device along which data flows between the input/output units of a computer and the CPU. Devices attached to the CPU communicate electronically with it via these channels. The paths parallel to the edge of punch tape in which the holes are located. The standard NC tape has eight channels; they are also known as tracks or levels.

channel capacity
The maximum bit rate that can be handled by the channel.

character
The coded symbol of a digit, letter, special symbol, or control function. The term byte is sometimes used to describe a character. One symbol of a set of elementary symbols, such as a letter of the alphabet or a decimal numeral.

character expansion factor
An attribute of text that specifies the multiplicative deviation of character aspect ratio from the defined nominal value.

character field
Means the rectangular area within which a character is displayed.

character generator
The subsystem in a display unit or printer that creates characters from the codes used to represent them.

character height
An attribute of text; the height of a capital character from baseline to cap line.

character plane
The plane on which text appears.

character printer
A device that prints a single character at a time. Contrast with line printer.

character recognition
The identification of characters by automatic means.

character set
A set of unique representations called characters, such as the 26 letters of the English alphabet or the decimal digits 0 through 9.

character spacing
An attribute of text that specifies the fraction of character height to be added between adjacent character boxes in a string.

characters per second (cps)
A measure of the speed with which an alphanumeric terminal can process data.

character subset
A selection of characters from a character set, comprising all characters that have a specified common feature. For example, in the definition of character set, digits 0 through 9 constitute a character subset.

character-up vector
An attribute of text that indicates the principal up direction of the text string.

charge-coupled device
A semiconductor device for high-density memory storage with low power consumption. Sometimes called a bucket-brigade device because of the way in which it transfers charges at prescribed intervals, resulting in a ripple process.

charge-coupled device camera
A machine vision image sensor that uses semiconductor arrays so that the electric charge at the output of one provides input stimulus to the next.

check digit
A digit added to each number in a coding system that allows for detection of errors in the recording of the code numbers. Through the use of the check digit and a predetermined mathematical formula, recording errors such as digit reversal can be noted. Synonymous with parity bit.

check plot
A pen plot generated automatically by a CAD system for visual verification and editing prior to final output generation.

checkpoint/restart facility
A facility for restarting execution of a program, at some point other than the beginning, after the program was terminated due to a program or system failure. A restart can begin at a checkpoint or from the beginning of a job step and uses checkpoint records to reinitialize the system. In teleprocessing, a facility that records the status of the teleprocessing network at designated intervals or following certain events. Following system failure, the system can be restarted and continue without loss of messages.

checksum
An error detection code that sums all 1 bits of a group of data storage location. Summing is done without carries from one column to another. The known result is stored; any variance from this result indicates data has been altered. Checksums can be prepared for any portion of logic memory, coil storage, or register content.

chip
A small piece of silicon impregnated with impurities in a pattern to form transistors, diodes, and resistors. Electrical paths are formed on it by depositing thin layers of aluminum or gold. The commonly used name for an integrated circuit. In machining, a particle of material removed from a workpiece as a result of a processing operation to alter the shape of the raw material.

chip time
Period of time in which a tool is cutting metal.

choice device
A logical input device providing a nonnegative integer defining one of a set of alternatives.

chroma
The quality of color, including both hue and saturation. White, black, and gray have no chroma—color intensity or purity of tone being a degree of freedom from gray.

chromatic aberration
An optical defect of a lens that causes different colors (different wavelengths of light) to be focused at different distances from the lens.

chromaticity coordinates
Two of three parameters commonly used in specifying color and describing color difference.

chromaticity diagram
Plot of chromaticity coordinates useful in comparing color of materials.

CID camera
A solid-state camera that uses a charge-injection imaging device (CID) to transform a light image into a digitized image. The light image focused on the CID generates minority carriers in a silicon wafer that are then trapped in potential wells under metallic electrodes held at an elevated voltage. Each electrode corresponds to one pixel of the image. To register the brightness of one pixel of the image, the voltage on the electrode that corresponds to that pixel is changed to inject the charge stored under that electrode into the substrate. This produces a current flow in the substrate that is proportional to the brightness of the image at that pixel location and is therefore capable of producing a gray-scale image. In a CID camera, pixels of the image can be read out in an arbitrary sequence. This is not possible with a CCD camera. In some CID cameras, the same image can be read out hundreds or thousands of times.

CIE diagram
A diagram developed in 1931 by the Commission Internationale de l'Eclairage (International Commission on Illumination) to show

the entire gamut of perceivable colors, expressed in chromaticity coordinates derived from tristimulus values of the spectrum under standardized viewing conditions.

circuit
A system of conductors and related electrical elements through which electrical current flows. A communication link between two or more points.

circuit board
A board on which are mounted integrated circuits in a microprocessor; also called circuit cards or cards.

circuit grade
The information-carrying capability of a circuit, in speed or type of signal. For data use, these grades are identified with certain speed ranges.

circuit switching
A method of communications, where an electrical connection between calling and called stations is established on demand for exclusive use of the circuit until the connection is released.

circular arc entity
A geometric entity; a connected portion of a circle or the entire circle.

circular interpolation
A simplified means of programming circular arcs in one plane that eliminates the necessity for segmenting the arc into calculated straight-line increments. A machine control unit with a small computer module that is able to direct a cutting tool in a complete circle or arcs when given only a few basic statements, such as coordinates of center point, radius, direction of travel, and coordinate locations of arc end points.

cirfit
A software processor that fits a series of tangential arcs to very closely approximate any curved surface in a single plane. The input is a series of rectangular or polar coordinate points that define the outline or surface of workpiece.

cladding
The process of coating a material with another material in either powder or solid form by melting the surface with a laser. The result is a strengthened or better-quality surface.

clamp
In robotics, the function of a pneumatic hand that controls grasping and releasing of an object.

clamping pressure
In injection molding and in transfer molding, the pressure that is applied to the mold to keep it closed, in opposition to the fluid pressure of the compressed molding material.

class
A group of data items pertinent to a common logical relationship in an associativity definition.

clear
To replace information in a storage unit by zero (or blank, in some machines).

cleaved-coupled laser
A tiny semiconductor laser that can be tuned electronically to transmit ultrapure light at ten or more frequencies. It is fast (up to a billion bits a second), powerful (100 miles without a regenerator) and no harder to make than conventional semiconductor lasers.

CLFILE
Output of an APT or graphics system that provides *X, Y,* and *Z* coordinates and NC information for target machine tool processing.

clip
To abbreviate or terminate the display or recording of an entity along an intersecting line, curve, or surface.

clip indicator
An indicator flag that specifies whether graphical elements are to be clipped at the limits of a clip rectangle or viewport (2-D or 3-D).

clipping
Removing parts of a design displayed on the CAD CRT that lie outside predefined bounds. The process of removing any portion of a graphical image that extends beyond a specified boundary. The effect of limiting the amplitude of a signal without limiting the gain associated with processing that signal.

clipping box
A bounding set of surfaces that abbreviate the intended display of CAD data to the portion that lies within the box.

clipping plane
A bounding plane surface that abbreviates the intended display of data to the portion that lies on one or the other side of the plane.

clipping volume
A volume in the CAD viewing coordinate system. The clipping volume is defined by the window, near and far clipping planes, and projectors of the corners of the window. Data on the planes forming the edges are considered to be within the volume.

clock
A device, usually based on a quartz crystal, that gives off regular pulses used to coordinate a computer's operations.

clock rate
The speed (frequency) at which the processor operates, as determined by the rate at which words or bits are transferred through internal logic sequences. The speed at which pulses are emitted from a clock generator. The unit of measure for a clock cycle is megahertz, or millions of cycles per second.

closed
Referring to a family of curves undergoing a stated set of graphic transformations, having the property that the operations always produce another member of the family. For example, circles are closed under the operations of translation, scaling, and rotation.

closed curve
A curve with coincident start and terminate points.

closed loop
A circuit in which the output is continuously fed back to its source for constant comparison. Also a group of indefinitely repeated computer instructions. The complete signal path in a control system represented as a group of units connected in such a manner that a signal started at any point follows a closed path and can be traced back to that point. A method of control in which feedback is used to link a controlled process back to the original command signal.

closed-loop MRP
A system built around MRP and also including the additional planning functions of production planning, master production scheduling, and capacity requirements planning. Further, once the planning phase is complete and the plans have been accepted as realistic and attainable, the execution functions come into play. These include the shop-floor control functions of input/output measurement, detailed scheduling and dispatching, plus anticipated delay reports from both the shop and vendors, purchasing follow-up and control, and so forth. The term "closed loop" implies that not only is each of these elements included in the overall system but also that there is feedback from the execution functions so that the planning can be kept valid at all times.

closed subroutine
A subroutine that can be stored at one place and can be linked to one or more calling routines.

CL tape
Abbreviation for either "centerline" or "cutter line" tape, which is the initial output of a computerized NC program giving the coordinate locations of where the cutting tool centerline will travel to machine the workpiece. The program is then postprocessed to take into account the particular features of the machine tool/control unit combination on which the program will actually be run and the part produced.

clustered operation
A set of elementary operations simultaneously performed.

coalescence
The union of particles of a dispersed phase into larger units, usually effected at temperatures below the fusion point.

coaxial gas
An inert gas that flows over the surface of a workpiece to help prevent plasma oxidation and absorption of laser energy, to blow away debris, and to control heat reaction.

COBOL
The name COBOL is derived from Common Business-Oriented Language. It was the first major attempt to produce a truly common business-oriented programming language. The COBOL character set is composed of the 26 letters of the alphabet, the numerals 0 through 9, and 12 special characters. The COBOL language consists of names to identify things; constants and literals; operations that specify some action or relationship; key words essential to the meaning of a statement; expressions consisting of names, constants, operators, or key words; statements containing a verb and an item to be acted on; sentences composed of one or more statements properly punctuated. COBOL programs are divided into four divisions: (1) the identification division is used to attach a unique identification, such as program name, program number, program version, and others, to the program; (2) the environment division is used to acquaint the processor with the computer on which the program is to be compiled and executed; (3) the data division is used to define the characters

and format of the data to be processed; and (4) the procedure division is used to describe the internal processing that is to take place.

code
A set of rules used to convert data from one representation to another. Although a variety of coding techniques are used, most codes are constructed by using numerals to represent the original data. The use of codes to express many classifications of data not only saves space but also increases efficiency by reducing the number of card columns to be processed for certain items of data.

coded character set
A set of unambiguous rules that establish a character set and the one-to-one relationship between the characters of the set and their coded representations.

code extension
Techniques for expanding the absolute character address space of a byte-oriented code into a larger virtual address space.

coder
A person whose primary duty is to write (but not design) computer programs.

code set
The complete set of representations defined by a code or by a coded character set.

code table
A set of unambiguous rules that define the mapping between received bit combinations and presentation-level characters.

coding
To prepare a set of computer instructions required to perform a given action or solve a given problem.

coding interface
An interface through which coded bit combinations are passed between communication media and receiving equipment.

cognition
An intellectual process by which knowledge is gained about perceptions or ideas.

cognitive science
The investigation of the details of the mechanics of human intelligence to determine the processes that produce that intelligence.

coherent radiation
Radiation that consists of wave trains traveling in phase with each other. Includes the properties of monochromaticity and low divergence.

cohesion
The force of attraction between the molecules (or atoms) within a single phase.

cohesive strength
The hypothetical stress in an unnotched bar causing tensile fracture without plastic deformation. The stress corresponding to the forces between atoms.

coil
In personal computer terminology, the output element of a programmed ladder sequence. This coil may be used to drive an external device internally for interlocking or other internal logic functions.

coil weld
A butt weld joining the ends of two metal sheets to make a continuous strip for coiling.

coining
The operation of forming metallic material in a shaped-cavity die by impact. All the material is restricted to the cavity, resulting in fine-line detail of the piece part.

cold boot
The first initialization of the computer, or booting just after the computer has been turned on.

cold swaging
The operation of forming or reducing a metallic material by successive blows of a pair of dies or hammers. Material flows at right angles to the pressure.

cold work
Permanent strain produced by an external force in a metal below its recrystallization temperature.

collet
A split conical sleeve that holds a tool or workpiece on the drive shaft of a lathe or other machine.

collimated
Made parallel, as rays of light.

color
In optical character recognition, the spectral appearance of the image. It is dependent on the spectral reflectance of the image, the spectral response of the observer, and the spectral composition of incident light. Property of light by which an observer may distinguish between two structure-free patches of light of the same size and shape. Neutral color qualities such as black, white, and gray—which have a zero saturation or chroma—are called ach-

romatic colors. Colors that have a finite saturation or chroma are said to be chromatic or colored. Color has three attributes: hue, lightness, and saturation.

color index
An index that serves as a pointer into a color table.

color lookup table
An ordered list of active colors (or pointers to color values) for use in a graphic display, in which each color in the list is selected from a much larger possible set. Because the lookup table is much smaller than the full-color palette, it takes fewer bits to describe each color—a factor that becomes important when displays use from thousands to millions of bits.

color map
A lookup table that is used during scan conversion of a stored digital image. Converts color map addresses into actual color values.

color map address
An ordinal number associated with each pixel in a stored digital image. Determines the address in the color map at which the actual color value of that pixel can be found.

color mass
Color, when viewed by reflected light, of a pigmented coating of such thickness that it completely obscures the background. Sometimes called overtone or masstone.

color model
The method used to describe a color to a graphics system. The color model is a specification of a three-dimensional color coordinate system and a three-dimensional subspace in the coordinate system within which each displayable color is represented by a point.

color saturation
The degree to which a color is free of white light.

color specification mode
An indicator flag that specifies whether color selection is to be direct—by specifying red, green, blue (RGB) values—or indexed—by specifying an index into a table of color values.

color subcarrier
In NTSC color format; the 3.579545 MHz carrier, whose modulation sidebands are added to the monochrome signal to convey color information.

color table
A table that maps a color index to a corresponding color.

color value
The value of the RGB (red, green, blue), HSV (hue, saturation, value), or HLS (hue, lightness, saturation) components describing a color.

column binary
Pertaining to the binary representation of data on punched cards in which adjacent positions in a column correspond to adjacent bits of data; that is, each column in a 12-row card may be used to represent 12 consecutive bits of a 36-bit word.

coma
An abberation in imaging systems that makes a very small circle appear comet-shaped at the edges of the image.

combinatorial
Refers to the rapid growth of possibilities as the search space expands. If each branch point (decision point) has an average of n branches, the search space tends to expand as n^d, as the depth of search, d, increases.

command
A signal or group of signals or pulses initiating one step in the execution of a program.

command-driven
Programs that require that the task to be performed be described in a special language with strict adherance to syntax. Compare to menu-driven.

command language
A source language consisting primarily of procedural operators, each capable of invoking a function to be executed.

comment
A string of information in a computer program meant for people to read and not for the machine to operate. Also referred to as "remarks."

COMMET
Designed as a shop-oriented system for automatic generation of NC workpiece programs and tapes, directly from blueprint data.

commodity buying
Grouping like parts or materials under one buyer's control for the procurement of all requirements to support production releases.

common application service elements
The common application service elements (CASE) protocol provides a general frame-

work and toolkit of service elements. These service elements are generic network functions used by specific application service elements (SASE) and user applications. User login and authorization are examples of CASE generic network functions. More general CASE functions cover application associations and contexts. These define the relationships between application elements that depend on what they are doing and are independent of a current connection to lower layers. These general CASE functions are intended for use in a distributed information processing environment, where computers and applications can communicate with other applications in a relevant context. To establish these relationships, the general CASE functions allow one application to set an association with a named peer application and to negotiate and agree upon the semantics (as opposed to syntax) of the information to be exchanged. Application associations and context functions allow applications to switch between different contexts and transfer information within these contexts.

common carrier
A government-regulated private company that furnishes the general public with telecommunications service facilities; for example, a telephone or telegraph company.

common field
A field that can be accessed by two or more independent routines.

common language
A coded structure that is compatible with two or more data-processing machines or families of machines, thus allowing them to communicate directly with one another.

common parts
Parts that are used in two or more products or models.

common parts bill
A type of planning bill that groups all common components for a product or family of products into one bill of material.

commonsense reasoning
Low-level reasoning based on a wealth of experience.

communicating entity
Any program, state machine, or automation of some form capable of communicating on a network.

communication
Transmission of intelligence between points of origin and reception, without alteration of sequence or structure of the information content.

communication control
In ASCII, a functional character intended to control or facilitate transmission over data networks. There are 10 control characters specified in ASCII that form the basis for character-oriented communications control procedures.

communication control program
A control program that provides the services needed to operate a communication-based information processing system.

communication line
Any medium, such as a wire or a telephone circuit, that connects a remote station with a computer for the purpose of transmitting/receiving information.

communications controllers
Dedicated computers with special processing capabilities for organizing and checking data. Information traffic to and from many remote terminals or computers, including such functions as message switching.

communications link
Any mechanism for the transmission of information; usually electrical. May be serial or parallel; synchronous or asynchronous; half-duplex or full-duplex; encrypted or clear; or point-to-point, multidrop, or broadcast. May transmit binary data or text; may use standard character codes to represent text and control information, such as the ASCII, EBCDIC, or BAUDOT (tty) codes; or may use a handshaking protocol to synchronize operations of computers or devices at opposite ends of the link such as BISYNC, HDLC, or ADCCP.

COMPACT II
A source language used in computer-aided manufacturing to program NC machine tools. COMPACT II is a registered trademark of Manufacturing Data Systems. A universal NC programming system for point-to-point and contouring applications on mills, drills, lathes (including four axes), punches, and flame cutters.

compactness
A measurement that describes the distribution of pixels within a blob, with respect to the

blob's center. A circle is the most compact blob; a line is the least compact. Circular objects have a maximum value of one and very elongated objects have a compactness approaching zero.

comparator
A device that compares two items of data and indicates the result of the comparison.

compare
To examine the relationship between two pieces of data and present the result.

compatibility
The ability of a particular hardware module or software program, code, or language to be used in a CAD/CAM system, without prior modification or special interfaces. Upward compatibility denotes the ability of a system to interface with new hardware or software modules or enhancements (i.e., the system vendor provides with each new module a reasonable means of transferring data, programs, and operator skills from the user's present system to the new enhancements). A term used to describe the degree of interchangeability of tapes or numerical control language between numerical control machine tools and their respective control systems.

compensation
Logical operations employed in a control scheme to counteract dynamic lags or otherwise to modify the transformation between measured signals and controller output to produce prompt stable response. When contouring, compensation is a displacement, always normal to the cutting path and workpiece programmed surface, to account for the difference between the actual radius or diameter of the cutting tool and the programmed dimension.

compilation time
The time during which a source language is compiled (translated) into a machine-language object program as opposed to the time during which the program is actually being run (execution time).

compile
To prepare a machine-language program from a high-level, symbolic-language program by generating more than one machine instruction for each symbolic statement, as well as performing the function of an assembler. The compiler is similar in concept to an assembler in that the programmer employs a symbolic language that is more recognizable than a machine language. A compiler goes farther than an assembler, however, in that it provides linkages to subroutines, selects the required subroutines from a library of routines, and assembles these parts into an object program. The two most common compilers are FORTRAN, which is applicable to scientific-type programming, and COBOL, which is concerned with business-type problems. Numerical control programming, because of its geometric and formular nature, generally utilizes the FORTRAN compiler. The significance of FORTRAN lies in its universal nature, as most computers will accept this language.

compiler language
A computer language, more powerful than an assembly language, that instructs a compiler in translating a source language into a machine language. The machine-language result (object) from the compiler is a translated and expanded version of the original.

complementary arc
Either of the two connected components of a closed connected curve that has been divided by two distinct points lying on the curve.

complete carry
In parallel addition, a procedure in which each of the carries is immediately transferred.

completion rate/percentage
Applied to computer-aided PC board routing, the percentage of routes automatically completed successfully by a CAD program.

complex sensors
Vision, sonar, and tactile sensors that will enable a robot to interact with the work evironment.

compliance
The quality or state of bending or deforming to stresses within the elastic limit. The amount of displacement per unit of applied force. The ability of a mechanism to flex or comply, especially when subjected to external forces.

component
A physical entity or a symbol used in CAD to denote such an entity. Depending on the application, a component might refer to an integrated circuit (IC) or part of a wiring circuit (i.e., a resistor), or a valve or elbow in a plant layout, or a substation or cable in a utility map. Also applies to a subassembly or part that goes into higher-level assemblies.

component assembly material
Includes materials used in the component-mounting operations. This may include tack or adhesive material.

composite component
A hypothetical component that contains all the features of a group of parts.

composite curve
A connected curve that is formed by concatenating two or more curve segments.

composite map
A single map created on the system from a mosaic of individual adjacent map sheets. Individual sheets are brought together and entities that fall on the neat line (geographic border of the map) are merged into a single continuous entity, whereupon the final sheet can be corrected for systematic errors.

composite routing
A routing listing a group of operations needed to produce a family of items, but that are not all used for all items. The operations used depend on the characteristics of each particular item.

composite symbol
A symbol consisting of a combination of two or more CAD symbols in a single character field, such as a diacritical mark and an alphabetic letter.

composite transformation
The matrix resulting from the concatenation, during structure traversal, of the local modeling transformation and the global modeling transformation.

composite video
In machine vision, the signal that is created by combining the picture signal (video), the vertical and horizontal syncronization signals, and the vertical and horizontal blanking signals.

compound die
Any die so designed that it performs more than one operation on a part with one stroke of the press, such as blanking and piercing, where all functions are performed simultaneously within the confines of the particular blank size being worked.

compound fixture
Fixture designed to accommodate two or more part numbers, one at a time.

compound lens
A lens made of two or more elements that may or may not be cemented together.

computational logic
A science designed to make use of computers in logic calculus.

computed path control
A control scheme wherein the path of the manipulator end point is computed to achieve a desired result in conformance to a given criterion, such as acceleration limit or a minimum time.

computer
A device capable of solving problems or manipulating data by accepting data, performing prescribed operations on the data, and supplying the results of these operations.

computer-aided design (CAD)
The use of a computer and computer graphics in the design of parts, products, and others. Some advanced CAD systems not only produce three-dimensional views of intricate parts, but also allow the designer to "test" a simulated part under different stresses, loads, and so forth.

computer-aided manufacturing (CAM)
The effective utilization of computer technology in the management, control, and operations of the manufacturing facility through either direct or indirect computer interface with the physical and human resources of the company.

computer-aided process planning (CAPP)
An application program that is interactive with CAD/CAM and assists in the development of a process/production plan for manufacturing. Process or routing planning that takes advantage of standard sequences of manufacturing operations stored in a computer. These standard sequences usually are developed for families of parts coded under the group technology concept.

computer-aided testing (CAT)
An application program that tests by modeling parts and product design and specifications through interaction with CAD/CAM.

computer architecture
The manner in which various computational elements are interconnected to achieve a computational function.

computer graphics
The input, construction, storage, retrieval, manipulation, alteration, and analysis of pictorial data. Computer graphics, in general, includes both offline input of drawings and pho-

tographs via scanners, digitizers, or pattern-recognition devices, and output of drawings on paper or (micro) film via plotters and film recorders. Among such input devices are the alphanumeric and function keyboards for typing text and activating preprogrammed subroutines, respectively, and the light pen and data tablet for identifying and entering graphic information by means of pointing and drawing. For various technological reasons, most of today's graphics concerns line drawings of two- and three-dimensional abstractions, such as electronic and mechanical circuits; structural components of buildings, cars, ships, and planes; chemical diagrams; functional plots of mathematical formulas; and flowcharts. In addition to line-drawing graphics, there is now an increase in interest in online manipulation of solid pictures with gray scale, color, and hidden line-surface representation of three-dimensional scenes.

computer-independent language (CIL)
A high-level language designed for use in any computer equipped with an appropriate compiler. CIL is relatively independent of such characteristics as word size and code representations.

computer-integrated manufacturing (CIM)
The concept of a totally automated factory in which all manufacturing processes are integrated and controlled by computer. CIM enables production planners and schedulers, shop-floor foremen, and accountants to use the same data base as product designers and engineers.

computer language
The grammar, reserved words, symbols, and techniques for providing instructions to a computer system.

computer-managed parts manufacture
Computer-aided manufacture of discrete parts, usually when a number of processing and product transport operations are coordinated by computer.

computer network
An interconnection of two or more computer systems, terminals, and communications facilities.

computer network components
Facilities that support the host computer, including the user communication interface, the communications subnetwork, and facilities for the network control function.

Computer-Integrated Manufacturing
General Electric Computer-Integrated Manufacturing System In Erie, Penn.

computer numerical control (CNC)
The use of a dedicated computer within a numerical control unit, with a capability of local data input. It may become part of the DNC system by direct link to a central computer. Using a computer to handle the numerical control of a machine tool. Such systems are more powerful and flexible than more primitive NC systems.

computer-output microfilm
A technology for generating CAD artwork. A computer-output microfilm device turns out a microfilm from data base information converted to an image on a high-resolution screen that is then photographed.

computer program
A specific set of software commands in a form acceptable to a computer, used to achieve a desired result. Often called a software program or package.

computer vision
Perception by a computer, based on visual sensory input, in which a symbolic description is developed of a scene depicted in an image. It is often a knowledge-based, expectation-guided process that uses models to interpret sensory data. Used somewhat synonymously with image understanding and scene analysis.

computer word
A sequence of bits or characters treated as a unit and capable of being stored in one computer location. Synonym: machine word.

CONAPT
A conversational version of APT processor for use with White-Sundstrand's Omnicontrol, a direct computer control of machine tools.

concatenation
A linking together of character strings to form a single character string.

concave lens
A lens with a surface that curves inward. Also known as a diverging lens.

concentrator
A device that matches a larger number of input channels with a fewer number of output channels. The input channels are usually low-speed asynchronous and the output channel(s) is high-speed synchronous. The low-speed channels may have the capability to be polled by a computer and may in turn poll terminals.

concentric
Having a common center, such as a circle or ellipse.

concept
In artificial intelligence, a descriptive schema for a class of things or a particular instance of the schema with some of its general properties specialized to characterize the specific subclass or element that instantiates the class description.

conceptual dependency
An approach to artificial intelligence natural language understanding in which sentences are translated into basic concepts, expressed as a small set of semantic primitives. An approach, related to case frames, in which sentences are translated into basic concepts, expressed in a small set of semantic primitives.

conceptual network models
In manufacturing information systems planning, definition of the major systems and how they interrelate, and determination of the data flow between and among the systems.

conceptual schema
A description of all data known to the data management system and independent of any user views (external schema) or physical storage considerations (internal schema). The conceptual time is compiled into tables and used by the external schema compiler, the internal schema compiler, and the runtime subsystems.

conceptual-to-internal runtime subsystem
A set of routines and processors within IPIP using the tables set up by the conceptual-to-internal translator to perform the transformations between conceptual schema records and internal schema records. Also processes keys by using and maintaining the files created for keyed attributes.

concurrent processing
The simultaneous processing of more than one program.

concurve
A boundary representation consisting of a chain of straight lines and arcs.

condenser
A lens used to collect and redirect light for purposes of illumination.

conditional branching
Standard branches, the simplest way to alter a robot's path program, but limited in the sense that each branch is associated with one particular input signal to the robot control.

conditional statement
A computer program step that specifies a dependence on whether certain tests of criteria are met.

conditional transfer
An instruction that may cause a departure from the sequence of instructions being followed, depending upon the result of an operation, the contents of a register, or the setting of an indicator.

conditioning
The addition of equipment to voice-grade telephone lines to provide specified minimum values of line characteristics, in ranges from C1 to C4. The common carrier will often recommend no conditioning for lines transmitting at 1200 baud, C1 conditioning for 2400 baud, C2 for 4800 baud, and C4 for speeds above 4800 baud.

conductive rubber
A material consisting of carbon granules suspended in rubber, whose electrical resistance decreases gradually as it is mechanically compressed. Used as a sensor in advanced manufacturing technology.

configuration
A particular combination of a computer, software and hardware modules, and peripherals

at a single installation and interconnected in such a way as to support certain application(s). The group of machines, devices, parts, and so forth, that make up a system.

configuration control
A means of ensuring that the product being built and shipped corresponds to the product ordered and designed.

confirming order
A purchase order issued to a vendor, listing the goods or services and terms of an order placed verbally, or otherwise, in advance of the issuance of the usual purchase document.

conflict resolution flag
A flag that instructs the graphics system to either replace a structure with a duplicate identifier, when naming conflicts occur between structures on the archive file and structures defined in the data storage, or to maintain it without change.

conflict set
The set of rules that matches some data or pattern in the global data base.

conic arc entity
A geometric entity that is a finite connected portion of an ellipse, a parabola, or a hyperbola.

conjunct
One of several subproblems. Each of the component formulas in a logical conjunction.

conjunction
The Boolean operation whose result has the Boolean value 1 if and only if each operand has the Boolean value 1.

connected curve
A curve such that, for any two points $P1$ and $P2$, one can travel from $P1$ to $P2$ without leaving the curve.

connection
The communication dialog of a pair of communication entities in the same or different processors.

connection identifier
An implementation-specific means of identifying a physical file or device or a collection of files or devices comprising a single workstation.

connectionless
Message transmission without establishing a circuit.

connectives
Operators (e.g., AND, OR) connecting statements in logic so that the truth value of the composite is determined by the truth value of the components.

connectivity
The ability of an electronic design data base to recognize connections by association of data. Connectivity facilitates design automation and CAM processing.

connectivity analysis
A technique for segmenting binary images.

connect node
In computer-aided design, an attachment point for lines or text.

connector
A termination point for a signal entering or leaving a PC board or a cabling system.

connect time
A measure of system usage by a user, usually the time interval during which the user terminal was online during a session.

Conographics
Making graphics out of curved segments, rather than from dots or straightline segments. Conographic Corporation claims the term as a trademark. Its CONO-COLOR hardware and CONO-LIB software for PC use elipses as the curve, requiring only seven values to describe each curve segment. The company claims that drawings made of curves are smoother, more attractive, and can be described using far less memory than vector drawings using lines.

consequent
The right-hand side of an artificial intelligence production rule. The result of applying a procedure.

Consight
Industrial object-recognition system, developed by General Motors, that uses special lighting to produce silhouettelike images.

consistent with the physical resolution
Positioning of display information is calculated with sufficient precision to display the information within one pixel of the true position.

console
That part of a computer used for communication between the operator or maintenance engineer and the computer.

console debugging
Debugging a program at the machine console or at a remote console by slowly stepping the machine through each instruction and observing the contents of appropriate registers and memory locations.

consortium
A group of institutions consisting of a nonprofit research institution and such other entities as a state or a subdivision thereof, an individual firm or industry association, any other nonprofit research institution, and a university and any other higher education institution.

constants
Data with a fixed value or meaning that are available for use throughout a program.

constant angular velocity
A disk recording method that records the same amount of information on each track. The data is more densely packed on tracks closer to the center.

constituent
A member of a set.

constraint
A condition or set of conditions specified by the conceptual schema author for establishing rules of integrity that must be satisfied by data within the data base, including range checks, uniqueness requirements, and subset conditions.

constraint proof
An association of a constraint with the test case used to prove the system has been validated against the constraint.

constraint propagation
A method for limiting search by requiring that certain constraints be satisfied. It can also be viewed as a mechanism for moving information between subproblems.

constraint set
The set of all constraints relating to a given development effort.

constraint test
An evaluation of the operability of a constraint.

construction plan
A strategy that establishes the procedures, tools, and methods to be used in constructing a module.

construction tool
A technique used to construct or debug a module. Some examples are: structure editors, text editors, compilers, linkers, loaders, debuggers, trace monitors, error analyzers, and subroutine libraries.

contact and coil cross-reference report
A CAD-generated report that identifies on-off page connections in multisheet diagrams and provides respective wire-source and destination data. The report gives the locations of all contacts associated with each coil adjacent to the coil symbol, underlining closed-contact references.

contact bounce
The uncontrolled making and breaking of a contact when the switch or relay contacts are closed.

contact chatter
Intermittent closure of open contacts or opening of closed contacts, resulting from contact impact.

contact sensor
A device that detects the presence of an object or measures the amount of force or torque applied by the object through physical contact with it. Contact sensing of force, torque, and touch can be usefully combined with visual sensing for many material-handling and assembly tasks. The function of contact sensors in controlling manipulation can be classified into the following basic material-handling and assembly operations: searching—detecting a part by sensitive touch sensors on the hand exterior, without moving the part; recognition—determining the identity, position, and orientation of a part, again without moving it, by sensitive touch sensors with high spatial resolution; grasping—acquiring the part by deformable, roundish fingers, with sensors mounted on their surfaces; moving—placing, joining, or inserting a part with the aid of sensors.

contact symbology
Commonly referred to as a ladder diagram, a method of expressing the user-programmed logic of the controller in relay-equivalent symbology.

contention
A condition on a communications channel or in a peripheral device when two or more stations try to transmit at the same time or access to a resource is simultaneously required by two or more users.

contents
The information in a storage location.

context
The set of circumstances or facts that define a particular situation, event, and so forth. The portion of the situation that remains the same when an operator is applied in a problem-solving situation.

continuous casting
A casting technique in which an ingot, billet, tube, or other shape is continuously solidified as the molten metal is poured, so that its length is not determined by mold dimensions.

continuous code
A bar code or symbol where the space between characters (*intercharacter gap*) is part of the code.

continuous cost system
A cost procedure used in continuous manufacturing environments, where costs are added as the manufacturing process progresses from the beginning to end.

continuous form
A supply of paper made up of numerous individual sheets separated by perforations and folded to form a pack. Sprocket holes are punched in the margins to permit automatic feed through the printer.

continuous method
The procedure of timing, used in making time studies, whereby the watch is permitted to run continuously throughout the period of study. The observer notes and records the reading of the watch at the end of each element, delay, or any other occurrence happening in the study, regardless of whether or not it has a direct bearing on the job. A method of magnetic particle testing in which the indicating medium is applied during which time the magnetizing force is present.

continuous path
In contrast to point-to-point, a robot that is controlled over the entire path traversed. An example is a spray painting robot. The end points are relatively unimportant since they are outside the dimensions of the object being painted, while the trajectory of the painting itself is crucial.

continuous path control
A control scheme whereby the inputs or commands specify every point along a desired path of motion. Continuous path control techniques can be divided into three basic categories, based on how much information about the path is used in the motor control calculations, as illustrated in the following example. The first is the conventional or servocontrol approach. This method uses no information about where the path goes in the future. The controller may have a stored representation of the path it is to follow, but for determining the drive signals to the robot's motors, all calculations are based on the past and present tracking error. This is the control design used in most of today's industrial robots and process control systems. The second approach is called preview control, also known as *feed-forward* control, since it uses some knowledge about how the path changes immediately ahead of the robot's current location, in addition to the past and present tracking error used by the servocontroller. The last category of path control is the *path planning* or *trajectory calculation* approach. Here the controller has available a complete description of the path the manipulator should follow from one point to another. Using a mathematical/physical model of the arm and its load, it precomputers an acceleration profile for every joint, predicting the nominal motor signals that should cause the arm to follow the desired path. This approach has been used in some advanced research robots to achieve highly accurate coordinated movements at high speed.

continuous path motion
A type of robot motion in which the entire path followed by the manipulator arm is programmed on a constant time base during teaching, so that every point along the path of motion is recorded for future playback.

continuous path operation
Refers to machine tool operation where the motion of the tool is controlled totally throughout the tool's movement, as opposed to point-to-point control. It involves simultaneous movement on two or more axes.

continuous production
A manufacturing process that lends itself to an endless flow of the same part or product, as in mass production.

continuous wave
A laser beam that is produced continuously, rather than as a series of pulses.

continuous weld
A weld extending continuously from one end of a joint to the other; where the joint is essentially circular, completely around the joint.

contour control system
A system that continuously controls the path of the machine (e.g., cutting tool, pen or scribe, welding head or torching head) by a coordinated simultaneous motion of two or more axes.

contour forming
The operation of reforming material sections of all types, rolled formed sections, cold-drawn or rolled shapes, extrusions, and others, normally received in straight lengths, into various contours.

contouring
The shaping that can be done to a surface by continuous path operation; involves simultaneous movement on two or more axes.

contour machining
Machining of irregular surfaces, such as those generated in tracer turning, tracer boring, and tracer milling.

contour milling
Milling of irregular surfaces.

contrast
The difference in light intensities between two regions in an image. This term is generally used to measure the difference between the lightest and darkest portion of an image.

contrast enhancement
An image-processing operation that improves the contrast of an image.

contrast ratio
Measure of hiding power or opacity. Ratio of the reflectance of a material having a black backing to its reflectance with a white backing.

contrast transfer function
A measure of the resolving capability of an imaging system. Shows the square-wave spatial frequency amplitude response of a system. The process of making a variable or system of variables conform to what is desired. A device to achieve such conformance automatically. A device by which a person may communicate his or her commands to a machine.

control board
A visual means of showing machine loading or project planning. Usually a variation of the basic Gantt chart.

control bus
The wires that carry timing and control pulses to all parts of a computer.

control character
A character whose occurrence in a particular context initiates, modifies, or stops a control operation. A control character may be recorded for use in a subsequent action. A control character is not a graphic character but may have a graphic representation in some circumstances.

control chart
Usually a large piece of graph paper used in the same manner as a control board. The control board often uses strings and pegs or movable slips of paper to represent the plan and progress, while the control chart typically would be filled in with pencil.

control dial
Graphics input device that produces a continuous range of display values.

control elements
Elements that specify delimiters, address space, clipping boundaries, initialization, error handling, and format descriptions of other elements.

control enclosure
A surrounding case designed to provide a degree of protection for equipment against a specified environment and to protect personnel against accidental contact with the enclosed equipment.

control hierarchy
A relationship of control elements, whereby the results of higher-level control elements are used to command lower-level elements. A relationship of sensory processing elements, whereby the results of lower-level elements are utilized as inputs by higher-level elements.

control key
A computer key that alters the meaning of another key; usually used to generate commands.

controlled path
A servo-driven robot with a control system that specifies or commands the location and orientation of all robot axes. This allows the robot to move in a straight line between programmed points with the added benefit of real-time velocity.

controller
The robot brain, which directs the motion of the end-effector so that it is both positioned and oriented correctly in space over time. An information-processing device whose inputs are both the desired and the measured position velocity or other pertinent variables in a process and whose outputs are drive signals to a controlling motor or actuator. (*Illus. p. 108.*)

Controller
ASEA Robot Driven by the ASEA Controller

control point
A point in definition space that appears in the numerator of the expression for a rational B-spline curve or surface. If the weights are all positive, the resulting curve or surface lies within the convex hull of the control points. Its shape resembles that of the polygon or polyhedron whose vertices are the control points. A control point is sometimes referred to as a B-spline coefficient.

control procedure
The means to control communication of information in an orderly way between stations on a data link.

control program
An operating system support program that monitors the flow of transactions in a computer system, for example, a data communication program.

control punch
The most common method used to differentiate master cards from detail cards when the decks are run through punched-card equipment. The majority of the machines are designed so that, with proper board wiring, they can recognize an 11-punch in a particular card column of a numeric field. The 11-punch, more commonly known as an x-punch, is usually used to signal the machine that a master card is passing through, as opposed to a detail card. Board wiring then causes the particular machine to initiate one set of operations for the master card and another set for the detail card.

control section
That section of the central processing unit that interprets instructions and directs the operation of all of the other units of the computer system. It has a temporary memory called a scratch pad, where contents of registers are stored.

control station
The network station that supervises control procedures such as polling, selecting, and recovery. It is also responsible for establishing order on the line in the event of contention or any other abnormal situation.

control system
Sensors, manual input and mode selection elements, interlocking and decision-making circuitry, and output elements to the operating mechanism. An administrative system that has as its primary function the collection and analysis of feedback from a given set of functions for the purpose of controlling those functions. Control may be implemented by monitoring and systematically modifying the parameters or policies used in those functions or by preparing control reports that initiate useful action with respect to significant deviations and exceptions.

control unit
Intermediary device between peripheral devices and channel. May be part of the I/O device or actual hardware. The part of a computer system that effects the retrieval of instructions in proper sequence, the interpretation of each instruction, and the application of the proper signals to the arithmetic unit and other parts of the system in accordance with this interpretation. The performance of these operations requires a vast number of "paths" over which data and instructions may be sent. Routing data over the proper paths in the circuitry, opening and closing the right "gates" at the right time, and establishing timing sequences are major functions of the control unit. All of these operations are under the control of a stored program.

convention
Standardized methodology or accepted procedure for executing a computer program. In CAD, the term denotes a standard rule or mode of execution undertaken to provide consistency. For example, a drafting convention might require all dimensions to be in metric units.

conversational
A program or a system that carries on a dialog with a terminal user, alternately accepting input and then responding to the input quickly enough for the user to maintain his or her train of thought.

conversion
The process of changing from one method of data processing to another or from one computer system to another. The so-called *controlled transition* from an old system to a new one. It involves careful planning for the various steps that have to be taken and equally careful supervision of their execution. Also applies to the representation of data.

converter
A unit that changes the representation of data from one form to another so as to make it available or acceptable to another machine, for example, from punched cards to magnetic tape.

convex lens
A lens with a surface that curves outward. Also known as a converting or convergent lens.

convolve
To superimpose a operator over a pixel area in the image, multiply corresponding points together, and sum the result.

coolant-handling system
System used to deliver, collect, move, and store cooling fluid.

cool colors
The violet, blue, green portion of the color spectrum.

coons patch
A three-dimensional surface.

coordinate
A numerical value associated with a position on an axis.

coordinated axis control
Control wherein the axes of the robot arrive at their respective end points simultaneously, giving a smooth appearance to the motion. Control wherein the motions of the axes are such that the end point moves along a prespecified type of path (line, circle, and so forth). Also called end-point control.

coordinate dial
Displays the position of the cursor, usually in *X, Y* coordinates.

coordinate dimensioning
A system of dimensioning where points are defined as a specified dimension and direction from a reference point measured with respect to defined axes.

coordinate measuring machine
A machine for measuring the shape and dimensions of small solid objects, particularly objects with a complex shape. Usually it has a small probe that is touched against or traced around the surface of the object to be measured. The machine reports the position of the tip of the probe from point-to-point in three-dimensional coordinates. The dimensions of the object can then be calculated from these coordinates. The entire operation can be automated.

coordinates
Locations on the screen, expressed as *X, Y,* and, in 3-D systems, *Z* points. Entry of data in absolute coordinates sends the cursor to a specific location on the screen. Relative coordinates move the cursor specific distances in screen measurements from its present point in the *X, Y,* and *Z* directions. Polar coordinates also move the cursor from its present position, but the directions are expressed as an angle and real-world distance (feet, inches, and so forth).

coordinate space
The total extent of defined possible positions in a computer graphics system. The size of the coordinate space depends on how many bits are allocated for the point coordinates. For example, a full 32-bit by 32-bit address defines a coordinate space of more than 4 billion points.

COPICS
Acronym for Communications-Orientated Production Information and Control System. This IBM system provides individual compatible program module support for advance function material requirements planning, bill of materials, routing data control, shop order release, facilities data control, product cost calculations, inventory planning and forecast-

ing, inventory accounting, and customer order servicing.

copious data entity
A geometric entity sometimes used as an annotation entity, containing arrays of types of real numbers to which a specific meaning has been assigned. One form number corresponds to one special meaning.

copper brazing
Brazing with copper as the filler metal.

coprocessor
An additional processor working with the main processor. It does specific tasks while the main processor executes its primary tasks. Frequently, these chips are added to speed up mathematical tasks.

copy
To reproduce data in a new location without changing the original data source, although the form the new data takes may differ from the original data source.

Core
A set of standard graphic routines and calling conventions under development by ACM/SIG-GRAPH. The Core standards allow programs to use single calls to produce output on dissimilar devices.

core image library
A library of machine-language versions of user programs that have been produced as output from link-editing. The programs in the core image library are in a format that is executable either directly or after processing by the relocating loader in the supervisor.

core loss
Active power expended in a magnetic circuit in which there is a cyclically alternative induction. Measurements are usually made with sinusoidally alternating induction.

core memory
An obsolete main memory technology. It was the computer's internal information storehouse. Core memory was fast and expensive. Information in core memory is located by "addresses." Physically, core memory is made up of tiny doughnut-shaped pieces of magnetizable material that can be in either an on or off state to represent either a binary 1 (on) or binary 0 (off).

core plane
A grid of wires upon which small iron cores are strung. A series of core planes are stacked to make up main memory.

Core primitives
Basic picture-drawing elements embodying the concepts of program portability and graphics standardization.

corner
An abrupt change in direction of a curve.

corona
In spot welding, an area sometimes surrounding the nugget at the faying surfaces, contributing slightly to overall bond strength.

correlation
The relationship between two sets of numbers, such as between two quantities such that when one changes the other changes correspondingly. If the changes are in the same direction, there is positive correlation. When changes go in opposite directions, there is negative correlation. A correspondence between attributes in an image and a reference image.

correspondence problem
In stereo machine vision imaging, the requirement to match features in one image with the same feature in the other image. The feature may be occluded in one or the other image.

corrosion
The deterioration of a metal by chemical or electromechemical reaction with its environment.

corrosion embrittlement
The severe loss of ductility of a metal resulting from corrosive attack, usually intergranular and often not visually apparent.

corrosion fatigue
Effect of the application of repeated or fluctuating stresses in a corrosive environment, characterized by shorter life than would be encountered as a result of either the repeated or fluctuating stresses alone or the corrosive environment alone.

cost center
Any department of work group where costs are gathered for accounting purposes.

cost driver
An operation that contributes disproportionately to the total cost of an enterprise.

costed bill of material
A form of bill of material that, besides providing such normal information as components, quantity per, effectivity data, and so forth, also extends the quantity per of every component in the bill by the cost of the components.

cost factors
The units of input that represent costs to the manufacturing system, for example, labor hours or purchased material.

costing units
The units of output to which costs are applied and in terms of which costs are expressed, for example, 1000s of bolts.

cost of capital
The cost of maintaining capital invested for a certain period, normally 1 year. This cost is normally expressed as a percentage and may be based upon factors such as the average expected return on alternative investments and current bank interest rate for borrowing.

counter
An electromechanical, relay-type device used to measure the number of events of some occurrence. The events to be counted may be pulses developed from mechanical operations, such as switch closures, interruptions of a light beam on a production line, or other periodic or random events. A programmable controller can eliminate hardware counters by using an equivalent software counter. The software counter can be given a present count value and will count up or down, depending on the instruction, whenever the counted event occurs. The software counter has greater flexibility than hardware counters.

counterboring
The operation of enlarging some part of a cylindrical bore or hole.

countersinking
The operation of producing a conical entrance to a bore or hole.

countouring
Controlling the path of the robot arm between successive positions or points in space.

coupler
A modem that converts digital data into signals that can be transmitted via a conventional telephone handset. At the other end of the line, another modem converts the data back to digital form.

coupling
Connecting two or more computers together at one site to share the workload and resources (e.g., disk drives, memory) and provide immediate backup for one another in case of malfunction.

coupon
A piece of metal from which a test specimen is to be prepared—often an extra piece as on a casting or forging.

covalent bond
A bond between two or more atoms resulting from the completion of shells by the sharing of electrons.

covering power
Alternative term for opacity or hiding power of paints.

CPU
Another term for processor. It includes the circuits controlling the interpretation and execution of the user-inserted program instructions stored in the computer or robot memory. The hardware part (CPU) of a computer that directs the sequence of operations, interprets the coded instructions, performs arithmetic and logical operations, and initiates the proper commands to the computer circuits for execution. The arithmetic and logic unit and the control unit of a digital computer. Controls the computer operation as directed by the program it is executing.

crash
A breakdown resulting from software or hardware malfunction.

creep
Deformation that occurs over a period of time, when a material is subjected to constant stress at constant temperature. In metals, creep usually occurs only at elevated temperatures. Creep at room temperature is more common in plastic materials. Time-dependent strain occurring under stress. The creep strain occurring at a diminishing rate is called primary creep; that occurring at a minimum and almost constant rate, secondary creep; that occurring at an accelerating rate, tertiary creep.

creep limit
The maximum stress that will cause less than a specified quantity of creep in a given time. The maximum nominal stress under which the creep strain range decreases continuously with time under constant load and at constant temperature.

creep recovery
Time-dependent strain after release of load in a creep test.

creep strength
The constant nominal stress that will cause a specified quantity of creep in a given time at

constant temperature. The constant nominal stress that will cause a specified creep rate at constant temperature.

crevice corrosion

A type of concentration-cell corrosion; corrosion of a metal that is caused by the concentration of dissolved salts, metal ions, oxygen or other gases, a such, in crevices or pockets remote from the principal fluid stream, with a resultant building-up of differential cells that ultimately causes deep pitting.

critical items

Items that have a lead time longer than the normal planning span time or items whose scarcity may impose a limit on production.

critical path method

A network technique for scheduling resources to accomplish a certain job within time constraints, preferably where time and cost estimates can be obtained with a relatively high degree of certainty.

critical path scheduling

A network planning technique used for planning and controlling elements in a project. By showing each of these elements and associated lead time, the "critical path" can be determined. The critical path identifies those elements that actually control the lead time for the project. CPM uses vector diagrams.

critical point

The temperature or pressure at which a change in crystal structure, phase, or physical properties occurs. In an equilibrium diagram, that specific value of composition, temperature, pressure, or combinations thereof, at which the phases of a heterogeneous system are in equilibrium.

critical ratio scheduling

The sequencing of jobs in the queue of a work center in accordance with their critical ratio priorities.

critical temperature

Synonymous with critical point, if the pressure is constant. The temperature above which the vapor phase cannot be condensed to liquid by an increase in pressure.

critical work center

A work center that is working close to its maximum capacity or where a bottleneck (overload) occurs. Also, a work center that processes the work of an important part of the plant or product line, or one where a breakdown would be critical, or one that consists of a machine with unique characteristics for which an alternate is not available.

cross-assembler

A computer program to translate instructions into a form suitable for running on another computer.

cross-assembler program

A program run on one computer but "built" or prepared on another computer. Small computers, especially microcomputers, generally do not have enough memory or are not equipped with the necessary peripheral devices to support many utility programs. In such a situation, another, larger computer is used to perform the assembly or compilation and the programs used are called cross-assemblers or cross-compilers. For example, a microcomputer program might be cross-assembled on a time-sharing system or a large mainframe.

cross-compiler

A program that translates instructions from a high-level language on one computer to the machine language of another computer on which the program is to be run.

cross hairs

On a cursor, a horizontal line intersected by a vertical line to indicate a point on the display whose coordinates are desired.

cross-hatching

A CAD design/drafting/editing aid for automatically filling in an outline (bounded area) with a pattern or series of symbols to highlight a particular part of the design. A series of angular parallel lines of definable width and spacing.

cross-referencing

Ties together the source and destination connections. In computer-aided design systems, a relay or network signal may need to be continued from one sheet to another; cross-referencing accomplishes this by automatically tying the appropriate text on both sheets.

CRT terminal

A terminal containing a cathode-ray tube to display program data.

crystal

A solid composed of atoms, ions, or molecules arranged in a pattern that is repetitive in three dimensions. A solid material, the atoms of which are arrayed in an orderly, repetitive manner. Ruby and YAG are crystals used as laser sources.

crystalline fracture
A fracture of a polycrystalline metal characterized by a grainy appearance.

crystallization
The separation, usually from a liquid phase on cooling, of a solid crystalline phase.

C-size sheet
The drawing for architectural size is 24 by 36 inches. The engineering size is 22 by 34 inches.

cumulative lead time
The longest length of time involved to accomplish the activity in question. For any item planned through MRP, it is found by reviewing each bill of material path below the item, and whichever path adds up to the greatest number defines cumulative material lead time.

cumulative lost time
The time when a system is not available for production. Usually referred to as downtime.

cumulative manufacturing lead time
The composite lead time, when all purchased items are assumed to be in stock.

cumulative sum
The accumulated total of all forecast errors, both positive and negative. This sum will approach zero if the forecast is unbiased.

cure
To change the physical properties of a material by a chemical reaction such as condensation, polymerization, or vulcanization; usually accomplished by the action of heat and catalysts, alone or in a combination, with or without pressure.

curing temperature
Temperature at which a cast, molded, or extruded product, a resin-impregnated reinforcing material, an adhesive, and so forth, is subjected to curing.

curing time
In the molding of thermosetting plastics, the interval of time between the instant of cessation of relative movement between the moving parts of a mold and the instant that pressure is released.

current-flow method
A method of magnetizing by passing current through a component via prods or contact heads. The current may be alternating, rectified alternating, or direct.

current position
A value that defines the current drawing location in world coordinates. The value of current position is affected by calls made to the functions that create output primitives.

Curie point
Temperature at which ferromagnetic materials can no longer be magnetized by outside forces, and at which they lose their residual magnetism (approximately 1200° F to 1600° F for many metals).

cursor
A visual tracking symbol, usually an underline or cross hairs, for indicating a location or entity selection on the CRT display. A text cursor indicates the alphanumeric input; a graphics cursor indicates the next geometric input. A cursor is guided by an electronic or light pen, joystick, keyboard, or other input device.

cursor keys
Keys that move the cursor; usually designated with arrows.

customer service ratio
A measure of delivery performance usually in the form of a percentage. In a make-to-stock company, this percentage usually represents the number of items or dollars shipped compared with the number of items or dollars on the customer's order. In a make-to-order company, it is usually some comparison of the number of jobs shipped in a given period of time, like a week, compared with the number of jobs that were supposed to be shipped in that time period.

cut and paste
A text editing function that moves text from one place to another.

cut-off
The line where the two halves of a compression mold come together; also called flash groove or pinch-off.

cutoff frequency
That frequency beyond which no appreciable energy is transmitted; usually defined as the frequency where the response is one-half the maximum response. It may refer to either an upper or lower limit of a frequency band.

cut plane
To define and intersect a plane with a 3-D object in order to derive a sectional view.

CUTS II
Computerized part-generation system written by Warner and Swasey for their SC line of turning machines.

cutter compensation
A means of manually adjusting the cutter center path on a contouring system so as to compensate for the variance in nominal cutter ra-

dius and the actual cutter radius. The net effect is to move the path of the center of the cutter closer to or away from the edge of the workpiece. Special considerations must be noted on the tape if it is anticipated that cutter compensation will be used by the operator.

cutter location data
The file produced by numerical control systems containing the cutting tool centerline at the tip of the tool, with an assumed orientation perpendicular to the *XY* plane unless otherwise indicated.

cutter location information
Describes the coordinates of the path of the center of the cutter, resulting from a basic computer program. This information is common to all machine tool system combinations and is the input to the postprocessor.

cutter offset
The distance of the cutter adjustment parallel to an axis.

cutter path
The path of a cutting tool through a part. The optimal cutter path can be defined automatically by a CAD/CAM system and formatted into a numerical control (NC) tape to guide the tool.

cutting speed
The relative velocity, usually expressed in feet per minute, between a cutting tool and the surface of the material from which it is removing stock.

cyan
The hue sensation evoked by radiations with a dominant wavelength of approximately 494 nanometers. The complement of cyan is red.

cycle
A preset sequence of events (hardware or software) initiated by a single command. A sequence of operations that is repeated regularly. The time it takes for one such sequence to occur.

cycle stealing
Taking an occasional machine cycle from a CPU's regular activities in order to control such things as an input or output operation.

cycle time
The period of time from starting one machine operation to starting another (in a pattern of continuous repetition).

cyclic redundancy check
A data transmission error detection scheme in which the check character is generated by taking the remainder after dividing all the serialized bits in a block of data by a predetermined binary number.

Cylindrical Coordinate Robot
Cylindrical Coordinate Robot Work Envelope

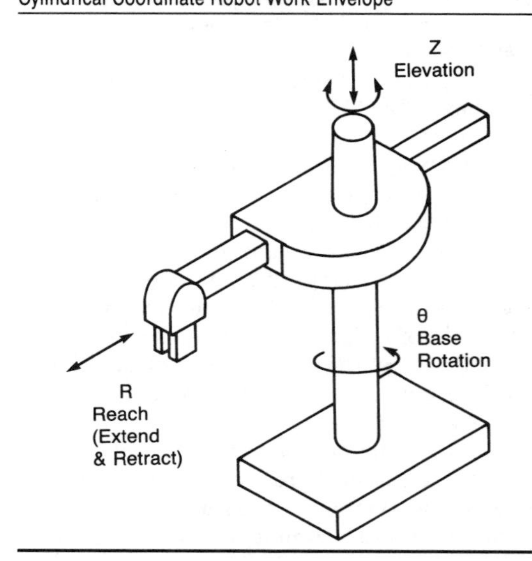

cylindrical
With robots, the manipulator is usually a vertical column with a horizontal arm mounted on a rotating base. The volume of space the end-of-arm tooling can occupy is basically a portion of a cylinder.

cylindrical coordinate robot
A robot whose manipulator arm consists of a primary vertical slide axis on a rotary axis. The vertical slide axis and a second slide are at right angles to one another in such a way that the shape traced by a point at the end of the farthest axis at full extension is that of a cylinder. A robot whose manipulator arm degrees of robot freedom are defined primarily by cylindrical coordinates.

cylindrical coordinate system
A coordinate system that defines the position of any point in terms of an angular dimension, a radial dimension, and a height from a reference plane. These three dimensions specify a point on a cylinder.

D

daisy chain
A method of propagating signals along a bus, whereby devices not requesting service respond by passing the signal on. The first device requesting the signal responds by performing an action and breaks the daisy chain signal continuity. This permits assignment of device priorities based on the electrical position of the device along the bus.

daisy wheel
An interchangeable element electronic impact printer, offering faster print speeds than a Selectric typewriter/printer and producing a fully formed character.

DAL
A user-orientated, interactive high-level computer programming language. DAL is an integral part of the Calma DDM (design, drafting, and manufacturing) CAD/CAM system.

dampeners
User input parameters to suppress the reporting of insignificant or unimportant action messages created during the computer processing of MRP.

damping
The absorption of energy, as viscous damping of mechanical energy or resistive damping of electrical energy. A property of a dynamic system that causes oscillations to die out and makes the response of the system approach a constant value.

dark current
The current generated in a photosensor when it is placed in total darkness.

darkfield
A machine vision illumination technique where the illumination is supplied at a grazing angle to the surface. Ordinarily only a negligible amount of light is reflected into the camera. Specular reflections occur off any abrupt surface irregularities and are detected in the image.

dark noise
The output of a detector that is shielded from all radiation.

dark operated
Sensor operating mode in which the output is activated when light ceases to be received from its source (or when the sensor is dark).

DART
Acronym for Data Accumulation and Retrieval of Time system. DART is a computer-based unit for generating and maintaining production time standards and standard data.

data
A general term used to denote any facts, numbers, letters, symbols, or facts that refer to or describe an object, idea, condition, situation, or other factors.

data access arrangement
Data communication equipment furnished by a common carrier, permitting attachment of privately owned data terminal and data communication equipment to the common carrier network.

data acquisition
The retrieval of data from remote sites, initiated by a central computer system. That is, retrieving data during off-hours processing from a previously mounted magnetic tape at an unattended terminal or taking periodic readings from an unattended real-time station.

data aggregate
A collection of data items within a record that is given a name and referred to as a whole.

data area
A logical grouping of data sets based on configuration control permission of the data set. Data areas related to each other in a hierarchical manner that generally follow the structure of the user's organization.

data base
A comprehensive collection of interrelated information stored on some form of mass data storage device, usually a disk. Generally consists of information organized into a number of fixed-format record types with logical links between associated records. Typically includes operating system instructions, standard parts libraries, completed designs and documentation, source code, graphic and application programs, as well as current user tasks in progress. A collection of data fundamental to an enterprise. Comprised of comprehensive files of information having predetermined structure and organization and

suitable for communication, interpretation, or processing by human or automatic means.

data base administrator

The custodian of the corporation's data—or that part of it that the system relates to. The data base administrator controls the overall structure of the data.

data base management

A set of rules about file organization and processing, generally contained in complex software, that controls the definition and access of complex, interrelated files shared by numerous application systems.

data base management system

A package of software programs to organize and control access to information stored in a multiuser system. It gives users a consistent method of entering, retrieving, and updating data in the system and prevents duplication and unauthorized access to stored information.

data base producer

A company or organization that collects and arranges the data for a data base.

data base relations

Linkages within a data base that logically bind two or more elements in the data base. For example, a nodal line is related to its terminal connection nodes because they all belong to the same electrical net.

data bus

The wires in a computer that carry data to and from memory locations.

data cartridge

A small, self-contained reel of magnetic tape used to store data.

data center

An abbreviated term applied to a computer-equipped central location. The center processes data and converts it to a desired form, such as reports or other types of management information records.

data channels

Highly specialized central processing units that manage and control one or more I/O control units.

data class

A category of logically related information, for example, customer, vendor, customer orders, parts inventory, and appropriations.

data code

In data communication, rules and conventions according to which the signals representing data should be formed, transmitted, received, and processed.

data collection

The act of bringing data from one or more points to a central point.

data communication equipment

The equipment that provides the functions required to establish, maintain, and terminate a connection. The signal conversion and coding required for communication between data terminal equipment and data circuit. The data communication equipment may or may not be an integral part of a computer, for example, a modem.

data communications

The transmission and reception of data, often including operations such as coding, decoding, and validation. Much data communication is carried over ordinary telephone lines, but often it requires specially conditioned leased lines where, in effect, several telephone lines are linked "side by side" to provide the required wide carrier bandwidths that carry a heavy and broad flow of information traffic. This is in contrast to voice-grade communication for which narrower carrier bandwidths are sufficient. The transmission of data (usually digital) from one point (such as a CAD/CAM workstation or CPU) to another point via communication channels such as telephone lines.

data compression

A technique that saves storage space by eliminating gaps, empty fields, redundancies, or unnecessary data to shorten the length of records or blocks. For data transmission, a byte string of data is transmitted as a count plus a string value.

data concentration

Collection of data at an intermediate point from several low- and medium-speed lines for retransmission across high-speed lines.

data conversion

The process of changing data from one form of representation to another.

data definition

The means by which data is described through its physical and logical characteristics.

data definition language

Language used by the data base administrator to describe the content of the conceptual, external, and internal schemes.

data division
One of the four main component parts of a COBOL program. The data division describes the files to be used in the program and the records contained within the files. It also describes any internal working-storage records that will be needed.

data element
The smallest unit of data stored in some medium to which a reference or name may be assigned.

data entry
Data entered by an operator from a single data device, such as a card or badge reader, numeric keyboard, or rotary switch.

data extract
Capability to obtain printed reports from the data base.

data file
A collection of data records usually organized on a logical basis. For example, all the records pertaining to one month's transactions would comprise a monthly data file. Data files are currently being replaced in information system design by a data base.

data flowchart
A flowchart that represents the path of data in the solving of a problem and that defines the major phases of the processing as well as the various data media used.

data hierarchy
A data structure consisting of sets and subsets such that every subset of a set is of lower rank than the data of the set.

data independence
Often cited as being one of the main attributes of a data base. The term implies that the data and the application programs that use them are independent so that either may be changed without changing the other.

data integrity
A performance measure based on the rate of undetected errors.

data interface
An interface between software modules or devices comprising one or more packets containing commands and data, as contrasted with a subroutine call interface.

data item
The smallest unit of named data. It may consist of any number of bits or bytes. A data item is often referred to as a field or data element.

data link
The communication line, related controls, and interface for the transmission of data between two or more computer systems. Can include modems, telephone lines, or dedicated transmission media such as cable or optical fiber. Linkage of a teletypewriter with a remote computer, using public telephone lines. Generally on a time-shared basis, wherein a number of teletype units at different locations share the computer on an effective simultaneous basis.

data link layer
Provides for and manages the transmission of individual frames of data. The MAP data link layer may also detect and take action to correct errors in the physical layer. In the IEEE 802, the data link layer is divided into two sublayers, namely, the media access control sublayer and logical link control sublayer. The media access control sublayer manages access of the physical media by the data frames. The logical link control sublayer performs error checking, addressing, and the other functions necessary to ensure accurate data transmission between nodes. It is important to note that the operation of the logical link control sublayer is independent of the particular media access method used.

data logging
Recording data about events in the time sequence in which they happen.

data management
A major function of operating systems that involves organizing, cataloging, locating, storing, retrieving, and maintaining data.

data management system
Assigns responsibility for data input and integrity, within the organization, to establish and maintain the data bases. Also provides necessary procedures and programs to collect, organize, and maintain the data required by the information systems.

data manipulation language
The interface between the application program and the data base management system. Referred to as the data manipulation language, it is embedded in a host language such as COBOL. It is desirable that it should have a syntax compatible with the host language because the application program has host language and data manipulation language state-

ments intimately mixed. In fact, it should appear to the programmer as though he or she is using a single language. There should be no enter or exit requirements from one language to the other.

data processing
The use of computers to gather, manipulate, summarize, and report information that flows through an organization.

data processing system
A network of machine components capable of accepting information, processing it according to a plan, and producing the desired results.

data protection
Measures to safeguard data from undesired occurrences that intentionally or unintentionally lead to modification, destruction, or disclosure of data.

data rate
The rate at which a channel carries data, measured in bits per second (bps).

data reduction
The massaging of raw data into a more comprehensible form, for example, curve smoothing.

data set
The major unit of data storage and retrieval in the operating system, consisting of a collection of data in one of several prescribed arrangements and described by control information to which the system has access.

data set organization
In data organization and storage, the arrangement of information in a data set. For example, sequential organization or partitioned organization.

data sharing
The ability of users or computer processes at several nodes to access data at a single node.

data storage
The preservation of data in various data media for direct use by the system. A conceptually centralized collection of structures. Structure editing and manipulation functions act on the contents of the data storage.

data structure
The form in which data is stored in a computer.

data tablet
A CAD/CAM input device that allows the designer to communicate with the system by placing an electronic pen or stylus on the tablet surface. There is a direct correspondence between positions on the tablet and addressable points on the display surface of the CRT. Typically used for indicating positions on the CRT, for digitizing input of drawings, or for menu selection.

Data Tablet
Data Tablet with Design Pen

data terminal equipment
The equipment comprising both a data source and a data link with respect to a communications network. Typical DTEs are computer systems and/or terminals. DTEs interface to data communication equipment to access data circuits in a network.

data transmission
The sending of data from one part of a system to another part.

data units
Messages passed over connections.

dating routine
A set of instructions that computes or stores the date (such as the current day's date).

dead band
The range through which an input can be varied to the servo portion, without initiating response at the machine tool. Generally, the narrower the dead band, the better the response of a machine tool system combination. Analogous to mechanical backlash. A range within which a nonzero input causes no output.

deadlock
Unresolved contention for the use of a resource.

deadly embrace
A state of a system in which it is logically impossible for the activity to continue. A deadly

embrace may result, for example, when the existence of a response is not recognized.

dead zone
A range within which a nonzero input causes no output.

deblocking
Separating blocked data into individual logical records.

debugging
Process of detecting, locating, and correcting mistakes in hardware or software programs. Synonymous with troubleshooting.

decarburization
The loss of carbon from the surface of a ferrous alloy, as a result of heating in a medium that reacts with the carbon at the surface.

decentralized network
A computer network where some of the network control functions are distributed over several network nodes.

decimal digit
In decimal notations, or in the decimal numeration system, one of the digits 0 through 9.

decimal hour stopwatch
A two-handed timing device whose movement may be started or stopped manually and whose large outer dial is divided into 100 spaces, each of which represents 0.0001 hour.

decimal minute stopwatch
A two-handed timing device whose movement may be started and stopped manually and whose large outer dial is divided into 100 spaces, each of which represents 0.01 minute.

decimal numeration
The fixed-radix numeration system that uses the decimal digits and the radix 10 and in which the lowest integral weight is 1.

decision table
A table that combines contingencies to be considered in the description of a problem, along with the action to be taken. Decision tables are sometimes used instead of flowcharts to describe and document problems.

decision trees
An approach toward identifying risks and probabilities in a problem situation involving uncertainty, or chance events by sketching in the form of a "tree" various courses that might be undertaken. The expected value of each alternative is the sum of the various possible outcomes weighted by their probability of occurring.

deck
A collection of punched cards; commonly, a complete set of cards that have been punched for a specific purpose.

deckle rod
A small rod or similar device inserted at each end of the extrusion coating die that is used to adjust the length of the die opening.

declarative knowledge
Facts about objects, events, and situations.

declarative statement
A source program statement that specifies the format, size, and nature of data.

DECnet
Trademark for Digital Equipment Corporation's communications network architecture that permits interconnection of multiple DEC computers.

decode
To apply a code so as to reverse some previous encoding. In machine operation, to translate data and/or instructions to determine exactly how and where signals are to be sent. To translate from coded language into an easily recognizable language. The process of translating from a language that a computer or machine control unit can understand to one that a human can comprehend. The reverse process is called "encoding."

decoder logic
The electronic package that receives the signals from the scanner, performs the algorithm to interpret the signals into meaningful data, and provides the interface to other devices.

decoding
To operate on a single input to extract the message and present it in usable form at its destination.

dedicated
Used with a single terminal by one user. Said of a microprocessor generally used for one type of work. A dedicated system stands alone and does not hook in to any larger computer to complete work.

dedicated line
A line permanently assigned to specific data terminals not part of a switched network.

deduction
A process of reasoning in which the conclusion follows from the premises given.

deductive capability
Conclusions drawn from knowledge, rules, and general principles.

deep structure
The underlying formal canonical syntactic structure represented in artificial intelligence, associated with a sentence, that indicates the sense of the verbs and includes subjects and objects that may be implied but are missing from the original sentence.

default
The predetermined value of a parameter required in a CAD/CAM task or operation. It is automatically supplied by the system whenever that value (i.e., text, height, or grid size) is not specified. The system automatically returns to one function type. Normally, it is the most common one. The default for line generation, for example, would be a solid line.

default attribute value
The attribute value inherited by a root structure.

default selection
A CAD/CAM feature that allows a user to preselect certain parameters for a product under design. These default parameters are then used each time a command is given. The designer can override them by selecting a different parameter when entering the command.

default value
A value to be used when the actual value is unknown. A value that is used until a more valid one is found.

deferral mode
Deferral-state value that specifies when additions to the structure must be reflected in the displayed image.

deferral state
A control function that allows a workstation to delay, for a certain period of time, actions requested by the application program.

deferred addressing
A method of addressing in which one indirect address is replaced by another to which it refers a predetermined number of times or until the process is terminated by an indicator.

definition
The fidelity of an image to the original scene.

definition level
The graphics-display level (or layer) on which one or more entities have been defined.

definition matrix
The matrix that transforms the coordinates represented in the definition space into the coordinates represented in the model space.

definition space
A local Cartesian coordinate system chosen to represent a geometric entity for the purpose of mathematical simplicity.

definition space scale
A scale factor applied within an entity definition space.

deflection
Alternation of voltages at the CRT yoke in order to move electron beam across the monitor surface.

degaussing
Process of neutralizing magnetic buildup in a system.

degenerate primitive
An output primitive whose definitions reduce the dimensionality of the primitive. For example, if the vertices of a polygon are all collinear, the polygon is degenerate.

degradation factor
A measure of the loss in performance that results from reconfiguration of a data-processing system, for example, a slowdown in runtime due to a reduction in the number of central processing units.

degree of disorder
Robots cannot operate in a disorderly environment. Parts to be handled or worked on must be in a known place and have a known orientation. For a simple robot, this must always be the same position and attitude. For a more complex robot, parts might be presented in an array; however, the overall position and orientation of the array must always be the same. On a conveyor, part position and orientation must be the same and conveyor speed must be known. Sensor-equipped robots (vision, touch) can tolerate some degree of disorder; however, there are definite limitations to the adaptability of such robots today. A machine vision system, for example, enables a robot to locate a part on a conveyor belt and to orient its hand to properly grasp the part. It will not, however, enable a robot to remove a part incorrectly oriented from a bin of parts or from a group of overlapping parts on a conveyor belt. A touch sensor enables a robot to find the top part on a stack. It does not, however, direct the robot to the same place on each part if the stack is not uniform or is not always in the same position relative to the robot.

delay
A period during which conditions (except those that intentionally change the physical or chemical characteristics of an object) do not permit or require immediate performance of the next planned action. The time between input and output of a pulse or other signal that undergoes normal distortion.

delay allowance
A time increment included in a time standard to allow for contingencies and minor delays beyond the control of the worker. A separate credit (in time or money) to compensate the worker or incentive for a specific instance of delay not covered by the piece rate or standard.

delay distortion
Distortion resulting from nonuniform speed of transmission of the various frequency components of a signal through a transmission medium.

delegation
To grant or confer responsibility and commensurate authority from one executive or organizational unit to another, in order to accomplish a particular assignment, often used in the sense of a superior in the organization delegating to subordinates.

delete
To remove data from context; to remove a record from a file of records or a file from a library of files.

delimiter
In data communications, a character that separates and organizes elements of data.

delinquent order
A line item on the customer open order that has an original schedule ship date prior to the current date.

delivery cycle
The time from the receipt of the customer order to the time of the shipment of the product or of the supplying of the service.

delivery instruction
A statement of conditions the system must meet prior to delivery.

delivery policy
The company's goal for the time to ship product after receipt of a customer's order. The policy is sometimes stated as "our quoted delivery time."

delivery schedule
The required or agreed time or rate of delivery of goods or services purchased for a future period.

delivery system
The hardware system on which a program is used. The delivery system may be the same as the development system; however, the delivery system often is less sophisticated and less expensive than the development system.

Delphi method
A forecasting approach where the opinions of experts are combined in a series of questionnaires. The results of each questionnaire are used to design the next questionnaire, so that convergence of the expert opinion is obtained.

demagnetization
Reduction in degree of residual magnetism in ferromagnetic materials to an acceptable level.

demand
A need for a particular product or component. The demand could come from any number of sources, that is, customer order, forecast, interplant, branch warehouse, service part, or to manufacturing the next-higher level. At the finished goods level, *demand data* is usually different from *sales data* because demand does not necessarily result in sales, that is, if there is no stock there will be no sale.

demand filter
A standard that is set to monitor individual sales data in forecasting models. Usually set to be tripped when the demand for a period differs from the forecast by more than some number of mean absolute deviation.

demand management
The function of recognizing and managing all of the demands for products to ensure that the master scheduler is aware of them. It encompasses the activities of forecasting, order entry, order promising, branch warehouse requirements, interplant orders, and service parts requirements.

density
A measure of the complexity of an electronic design. For example, IC density can be measured by the number of gates or transistors per unit area or by the number of square inches per component. In magnetic tape storage capacity, high density might be 1600 bits/inch,

and low, 800 bits/inch. A measure of the light transmitting or reflecting properties of an area; it is expressed by the logarithm of the ratio of incident to transmitted or reflected light flux. The number of bits in a single linear track measured per unit of length of the recording medium.

department
An organizational unit established to operate in, and be responsible for, a specified activity or a physical or a functional area.

department overhead rate
The overhead rate applied to jobs passing through a department.

departmental stocks
Stock informally held in reserve in a production department. This action is taken as a protection from being out of stock in the stockroom or for convenience; however, it results in increased inventory investment and possible degradation of the accuracy of the inventory records.

dependency
A relation between the antecedents and corresponding consequents produced as a result of applying an inferential rule. Dependencies provide a record of the manner in which decisions are derived from prior data and decisions.

dependent demand
Demand to or derived from the demand for other items or end products. Such demands are therefore calculated and need not and should not be forecast. A given inventory item may have both dependent and independent demand at any given time.

depletion
A lessening of the value of an asset due to a decrease in the quantity available.

depreciable value
The difference between the first cost of an asset to the current owner and the net recoverable value at the time of its disposal. The recorded or book value of an asset at any time.

depreciation
The actual decline in the value of an asset due to exhaustion, wear and tear, and obsolescence.

depreciation base
The actual or adjusted initial cost of an asset to which a depreciation rate is applied in computing depreciation cost or expense.

depreciation expense
A periodic accounting charge or operating cost arising from the systematic writing off of assets in the accounting records.

depth-first search
In artificial intelligence, a search that proceeds from the root node to one of the successor nodes and then to one of that node's successor nodes, and so forth, until a solution is reached or the search is forced to backtrack.

depth of field
The in-focus range of an imaging system. It is measured from the distance behind an object to the distance in front of the object within which objects appear to be in focus.

depth of focus
The range of distance from lens to image plane for which the image formed by the lens appears to be in focus.

depth perception
The ability to perceive differences in distance from an observer to one point relative to the distance to another point.

derivative control
Control scheme whereby the actuator drive signal is proportional to the time derivative of the difference between the input and the measured actual output.

description
A symbolic representation of the relevant information, for example, a list of statistical features of a region.

description table
A defined data structure containing static (fixed) information. Application programs can obtain description-table contents through inquiry.

descriptor
Data that designates a record, allowing it to be called, classified, sorted, and so forth.

descriptor elements
Describe the metafile's functional content, format, default conditions, identification, and characteristics.

design
A desired arrangement of things and/or activities selected from any source.

design anomaly
A feature of a solution structure indicating undesirable partitioning of capabilities with regard to applicable technology disciplines.

design automation
The technique of using a computer to automate portions of the design process and reduce human intervention.

design goal
A stated aim that the design structure is to achieve.

design goal set
The set of all design goals for a given development effort.

design file
Collection of information in a CAD data base that relates to a single design project and can be directly accessed to a separate file.

design limitation
A constraint on the way in which a solution can be partitioned.

design optimization
A process that uses a computer to determine the best graphic design to meet such criteria as fuel efficiency, cost of production, and ease of maintenance. In CAD, algorithms may be applied to rapidly evaluate many possible design alternatives in a comparatively short time, for example, to optimize PC board component placement in terms of minimal total length of all interconnections.

design-related parts
Parts that have similar forms (e.g., parts of rotation, prismatic parts).

design rules checking
A CAD software program that automatically examines a displayed PC or IC design or layout for any violations of user-selected design rules and manufacturing tolerances. Usually associated with electrical applications.

design structure
A framework of system elements, including their relationships, that the solution is to possess.

destructive read
Reading that erases the data in the source location.

detail card or deck
A card containing changeable information, as opposed to a master card or deck. Information from a master card is often transferred into a detail card at machine speed rather than by rekeypunching.

detail drawing
The drawing of a single part design containing all the dimensions, annotations, and so forth necessary to give a definition complete enough for manufacturing and inspection.

detail file
A temporary reference file, usually containing current data to be processed against a master file at a later date.

detailing
The process of adding the necessary information to create a detail drawing.

detail printing
Printing information from each punched card passing through the machine. A method of printing in which the accounting machine prints one line per card.

detector
A device that converts incident radiation into another form such as electrical, visual, or chemical. A device that alters an electrical flow in response to incident radiation.

detent
A piece of a mechanism that, when disengaged, releases the operating power, or by which the action is prevented or checked; a catch, as in a lock; a pawl.

deterministic model
A mathematical model that, given a set of input data, produces a single output or a single set of output. An example would be an equation for computing the optimum level of inventory for a given product; such a model generates a single number that, given the assumptions of the model, is considered to be the correct answer.

developable surface
A surface that can be unrolled onto a plane.

development time
The time used for debugging new routines or hardware. Considered part of the operating time.

deviation
The difference, usually the absolute difference, between a number and the mean of a set of numbers or between a forecast value and the actual datum.

device
A system hardware module external to the CPU and designed to perform a specific function (i.e., a CRT, plotter, printer, hard-copy unit, etc.).

device coordinates
The coordinates used by workstation hardware. The coordinate units may be in fractions

between 0 and 1; in meters, for devices such as a plotter; or in an arbitrary subrange of integers, for devices such as raster CRTs.

device-dependent display file
A file containing device coordinate information, used by all but the least interactive graphics systems for maintaining a refresh display and for modifying individual segments of the picture; may be storage buffer in the host computer's memory or in the display device.

device driver
The device-dependent part of a graphics implementation that supports a physical device. The device driver generates device-dependent output and handles device-dependent interaction. Software package that maps normalized device coordinates to physical device (screen) coordinates and handles device-dependent interaction; sometimes referred to as the interface software.

device independence
The ability to request I/O operations without regard for the characteristics of specific types of input/output devices.

device-independent display file
An optional file containing device-independent commands, used primarily for postprocessing graphic data on a variety of graphic output devices; sometimes referred to as a pseudodisplay file.

device or screen coordinates
Device-dependent coordinates that are typically either in integer raster units or fractions between 0 and 1.

device space
The space defined by the addressable points of a display device.

diagnostic
Pertaining to the detection, discovery, and further isolation of an equipment malfunction or a processing error.

diagnostic program
A user-oriented test program to help isolate hardware malfunctions in the computer or robot and the application equipment.

diagnostic routine
A test program used to detect and identify hardware malfunctions in the computer and its associated I/O equipment. A specific computer routine designed to locate an error or malfunction in the program. A *printout* can be made available that will describe the location of the error and prescribe its correction.

diagnostic study
A brief investigation or cursory methods study of an operation, process, group, or individual to discover causes of operational difficulties or problems for which more detailed remedial studies may be feasible. An appropriate work measurement technique may be used to evaluate alternatives or to locate major areas requiring improvement.

diagram
A drawing made up of a series of symbols. These are usually not drawn to scale.

dial up
To initiate station-to-station communications with a computer via a dial telephone, usually from a workstation to a computer.

dial-up line
A communications circuit that is established by a switched circuit connection.

diameter dimension entity
An annotation entity designating the measurement of a diameter of a circular arc.

diamond tool
A diamond, shaped or formed to the contour of a single-pointed cutting tool, for use in the precision machining of nonferrous or nonmetallic materials.

dibit
A pair of bits treated as one information element. In modulation schemes with more than two states, multiple bits are represented by each state. The term dibit originally was used for the AT&T 201 series of modems that used four phase states to encode information; each state could therefore represent two bits.

dichroic mirror
A semitransparent mirror that selectively reflects some wavelengths more than others and so transmits selectively.

dichromatic vision
A form of defective color vision in which all colors can be matched by a mixture of only two stimuli. The spectrum is seen as two regions of different hues, plus a colorless band.

die
Tool that imparts shape to material primarily because of the shape of the tool itself. Examples are blanking dies, cutting dies, drawing dies, forging dies, punching dies, and threading dies. The part or parts making up the confining form into which a powder is pressed.

die blades
Deformable member(s) attached to a die body that determine the slot opening and that are adjusted to produce uniform thickness across the film or sheet produced.

die block
The tool steel block into which the desired impressions are machined and from which forgings are produced.

die body
The stationary or fixed part of a die.

die casting
A casting made in a die. A casting process where molten metal is forced under high pressure into the cavity of a metal mold.

die gap
The distance between the metal faces forming the die opening.

die welding
Forge welding between dies.

difference reduction
An approach to problem-solving that tries to solve a problem by iteratively applying operators that will reduce the difference between the current state and the goal state.

differential piecework
A wage incentive plan that employs two or more piece rates. One piece rate is paid if the expected output is not attained. A higher piece rate is paid if the expected piece rate is attained and exceeded.

differential positioning
The position difference obtained by providing pulses of compressed air to the air motor in opposite directions, resulting in more accurate positioning.

differentiator
A device whose output function is proportional to the derivative of the input function with respect to one or more variables; for example, a resistance-capacitance network used to select the leading and trailing edges of a pulse signal.

diffraction grating
A substrate with a series of very closely spaced lines etched in its surface. The surface may be transparent or reflective. Light falling on the grating is dispersed into a series of spectra.

diffraction limited
An optical system of such quality that its performance is limited only by the effects of diffraction.

diffuse
A process where incident light is redirected over a range of angles (scattered) while being reflected from or transmitted through a material.

diffuse light transmission factor
Ratio of transmitted to incident light for translucent reinforced plastic building panels. Property is an arbitrary index of comparison and is not related directly to luminous transmittance.

diffuse luminous transmittance
Ratio of light scattered by a material to light incident on it.

diffuse transmittance
Ratio of light scattered by material to light incident on it. The ratio depends on spectral distribution of incident energy. The analogous spectral ratio is diffuse luminous transmittance.

diffusion value
Index of scattering or diffusion of light by a material compared to a theoretically perfect light-scattering material (rated as 1.0).

digital
Refers to information and values expressed in discrete terms. In a digital computer, such terms are generated by a combination of binary on/off or positive/negative signals. This is the opposite of analog, where a fluctuating signal strength determines the fluctuations of values. The representation of data as discrete points. The process of digitizing an image converts analog light into an array of digital elements, each with a discrete value.

digital communications
Transfer of information by means of a sequence of signals called bits, each of which can have one of two different values. The signals may, for example, take the form of two different voltage levels on a wire or the presence or absence of light in a fiber-optic light guide. Can be made arbitrarily insensitive to external disturbances by means of error control procedures.

digital computer
A computer that processes information represented by combinations of discrete or discontinuous data, as compared with an analog computer for continuous data. More specifically, it is a device for performing sequences of arithmetic and logical operations, not only

on data but its own program. Still more specifically, it is a stored program digital computer capable of performing sequences of internally stored instructions, as opposed to calculators, such as card-programmed calculators, on which the sequence is impressed manually.

digital data
Data represented by digits, perhaps with special characters and the space character.

digital data acquisition system
Used in plant floor automation to collect digitized data, occasionally from limit switches, but most often from information inserted in workstations by means of punched cards and readable tags, as well as information inserted by the operator, using dials or a keyboard.

digital data communications message protocol
A uniform discipline for the transmission of data between stations in a point-to-point or multipoint data communications system. DDCMP supports parallel, serial asynchronous, and serial synchronous physical data transfer methods. The DDCMP protocol is byte-oriented, using a header field to define the length of the data field as well as to provide control functions; in this manner, data cannot be misinterpreted as control information.

digital device
An electronic device that processes electrical signals that have only two states, such as on or off, high or low voltages, or "positive" or "negative" voltages (i.e., programmable controller). In electronics, *digital* normally means binary or two-state.

digital image analysis
A multistage process that leads to the "understanding" of a digital image, the recognition of certain attributes in given objects in the image. Stages of the image analysis process may be image digitizing, image preprocessing, feature extraction, and pattern recognition.

digital readout
A display device that shows values in discrete numbers, such as the display on a pocket calculator.

digitization
The process of converting an analog video image into digital brightness values (gray scale) that are assigned to each pixel in the digitized image. Analogous to a computerized snapshot.

digitize
To use optical sensing of physical surfaces (lines, points, and curves) to produce Cartesian coordinate approximations of these surfaces. To turn a drawing or diagram into a set of coordinate points. In most cases, this is done by manually tracing out the image on a digitizer pad that transmits position information to the computer. To convert a measurement into a number. A substantial amount of data that reaches the computer starts as physical quantities, pressures, temperatures, rates of flow, and so forth, and is converted to a voltage and then to a number. This is also called analog to digital (A to D) conversion.

digitized image
A representation of an image as an array of pixels having different brightness values. This is necessary before an analysis can be performed on the image.

digitizer
A term commonly used to describe a graphics tablet. A computer peripheral device that sends position information to the computer, either on command from the user (point digitizing) or at regular intervals (continuous digitizing). Digitizers come in tabletop models (digitizer pads) through large stand-alone units. Some offer translucent backlit surfaces and resolutions of as much as 1000 lines per inch. In most designs, the user traces the object to be digitized with a pen stylus or puck.

digitizing
The method by which data is entered on a graphics tablet. A stylus or puck is "touched down" at a particular location. This touching down is called digitizing.

dimension
The maximum number and order of a series of related items.

dimensionality
The number of coordinates (two or three) needed to specify a point in coordinate space.

dimensional stability
Ability of a material to retain the precise shape in which it was molded, fabricated, or cast.

dimpling
The operation of forming semispherical-shaped impressions in flat material.

dinking
The operation of blanking a piece out of a soft flat material by the use of sharp beveled cut-

ting edges arranged to produce a part of a specific configuration. Dinking is similar to steel rule die.

dip brazing
Brazing by immersion in a molten salt or metal bath. Where a metal bath is employed, it may provide the filler metal.

direct access
Retrieval or storage of data in the system by reference to its location on a tape, disk, or cartridge, without the need for processing on a CPU.

direct-access storage device
A basic type of storage medium that allows information to be accessed by positioning the medium or accessing mechanism directly to the information required, thus permitting direct addressing of data locations. The time required for such access is independent of the location of the data most recently accessed. Synonymous with random access. File organizations can be sequential, direct, or indexed sequential. Contrast with serial or sequential access.

direct-chill casting
A method for the continuous production of ingots or billets for sheet or extrusion by pouring the metal into a short mold. The base of the mold is a platform that is gradually lowered as the metal solidifies, the shell of solidified metal acting as a retainer for the liquid metal below the wall of the mold. The ingot is usually cooled by the impingement of water directly on the mold or on the walls of the solid metal as it is lowered. The length of the ingot is limited by the depth to which the platform can be lowered; therefore, it is often called semicontinuous casting.

direct color
A color selection scheme in which color values are specified directly, without requiring an intermediate mapping via a color table.

direct curve
A curve with an associated direction derived from the start and terminate points.

direct-deduct inventory transaction processing
A method of doing bookkeeping that decreases the book inventory of an item as material is issued from stock and increases the book inventory as material is received into stock. The key concept here is that the book record is updated coincident with the movement of material out of or into stock. As a result, the book record is a representation of what is physically in stock.

direct delivery
The consignment of goods directly from the vendor to the buyer. Frequently used where a third party acts as intermediary agent between vendor and buyer.

direct digital controller
A special-purpose machine that replaces an analog setpoint controller, such as a flow-rate or temperature controller. Such controllers compare a measured value to a set value and compute a correction signal. The desirability of a digital implementation is indicated by the need for accuracy, for a means to change setpoints dynamically, and for the ability to modify algorithms easily. Since the cost of even a special-purpose digital machine would be greater than that of the analog device, all designs for DDCs have been based on the capability of one processor to handle at least 16 and as many as 1000 control loops.

directed curve
A curve with an associated direction derived from the start and terminate points.

directed graph
A knowledge representation structure consisting of nodes (representing, for example, objects) and directed connecting arcs (labeled edges, representing, for example, relations).

directional solidification
The solidification of molten metal in a casting in such a manner that feed metal is always available for that portion that is just solidifying.

direct labor standard
A specified output or a time allowance established for a direct labor operation.

direct material
All material that enters into and becomes part of the finished product, the cost of which can be identified with and assessed against a particular part, product, or group of parts or products accurately and without undue effort and expense.

direct memory access
A method of transferring blocks of data directly between a peripheral device and system memory, without the need for CPU intervention. This powerful input/output technique significantly increases the data transfer rate, hence system efficiency.

direct numerical control
Refers to the operation of a number of machines directly from a single relatively large computer. Unlike CNC, in which a computer, usually a minicomputer or microcomputer, replaces hard-wired electronic components, a DNC system normally supplements conventional electronics and automatically regulates the operation of a number of machines. There can be as many as several hundred machines involved in a DNC system. One of the advantages of a DNC system is overall factory control. Another definition for DNC that has evolved is Distributed Numerical Control. In this case, tape image data are sent directly from a main computer to a CNC control system. After the complete program has been stored in the CNC system, the link between the main computer and the CNC system is cut and the CNC system operates on its own stored program.

directory
A named space on the disk or other mass storage device in which are stored the names of files and some summary information about them.

directory entry section
The section of an IGES file, consisting of fixed field data items for an index and attribute list of all entities in the file.

directory services
Local area network system function to provide required addressing, given the global name.

direct planning
Day-to-day production planning and control, involving "how much" and "when" to make the order quantity calculated on a discount that the vendor will give if a fairly large minimum quantity is ordered.

direct quenching
Quenching carburized parts directly from the carburizing operation.

direct read after write
An error detection method that ensures that the laser wrote the last data correctly before proceeding on to the next data block. The drawback to this method is that the disk must make a full rotation to read the data before it can write again.

direct read during write
A refinement of DRAW technology. The data is read as it is written. Thus, if an error is detected, the recording head can skip to the next good sector. The advantage to this error detection method is speed, because it does not require a full disk rotation, as does the DRAW method. The disadvantage is that it requires more sophisticated and expensive light-path technology than that used on DRAW devices.

directrix
The curve entity used in the definition of a tabulated cylinder entity.

direct statement
An instruction in a program language that is not numbered and therefore is executed by the computer immediately. A direct statement cannot be stored or executed again during a program.

direct view storage tube
One of the most widely used graphics display devices. DVST generates a long-lasting, flicker-free image with high resolution and no refreshing. It handles an almost unlimited amount of data. However, display dynamics are limited since DVSTs do not permit selective erase. The image is not as bright as with refresh or raster.

dirty power
Vague term used to describe electric power that is sufficiently adulterated with harmonic distortion and other impurities to cause electricity-using machinery to operate less than optimally.

disable
The capability to disconnect a logic line coil or a discrete input from its normal control and force it on or off.

disassemble
The basic element denoting the removal of a part of a unit or assembly.

disbursement list
A printed list containing the identity and quantity of parts and assemblies to be withdrawn from stock and dispatched to the starting point of manufacture or assembly on a specific day.

discontinuity
Any interruption in the normal physical structure or configuration of a part such as cracks, laps, seams, inclusions, or porosity.

discount
An allowance or deduction granted by the seller to the buyer, usually when certain stipulated conditions are met by the buyer, which reduces the cost of the goods purchased. A

quantity discount is an allowance determined by the quantity or value of purchase. A cash discount is an allowance extended to encourage payment of invoice on or before a stated date that is earlier than the net date. A trade discount is a deduction from an established price for items or services, often varying in percentage with volume of transactions, made by the seller to those engaged in certain businesses and allowed irrespective of the time when payment is made.

discounted cash flow

A method of investment analysis in which future cash flows are converted, or discounted, to their value at present time. The rate of return for an investment is that rate at which the present value of all related cash flow equals zero.

discrete

Pertaining to distinct elements or to representation by means of distinct elements, such as characters. References that can be either on or off; can be input, output, or internal references.

discrete code

A bar code or symbol where the spaces between characters (*intercharacter gap*) are not part of the code.

discrete components

Components with a single functional capability per package; for example, transistors and diodes.

discrete part manufacturing

A manufacturing process that produces discrete parts in comparatively small lots, or batches, of from 1 to perhaps 50,000. The discrete-parts industry represents about 35 percent of all U.S. companies and about 75 percent of all U.S. trade in manufactured goods. These are mostly small companies—about 87 percent of these firms employ fewer than 50 workers.

discrete word recognizer

Computer control device, commanded by human voice, for entry of data into a computerized system containing discrete word structure and formatted data fields; nonintelligent word recognizers are typically limited by small fixed vocabulary.

discrimination

The degree to which a machine vision system is capable of sensing differences in light intensity between two regions.

discrimination instruction

One of the class of instructions that comprise branch instructions and conditional jump instructions.

disengage

The basic element employed to break the contact between one object and another.

disk

A secondary memory device in which information is stored on one or both sides of a magnetically sensitive rotating disk. The disk is rotated by a disk drive and information is retrieved/stored by means of one or more read/write heads mounted on movable or fixed arms. Disks are hard or floppy. Disk memory has a much faster access time than magnetic tape but is slower than semiconductor memory.

disk check

A test of numerical control contouring accuracy performed by recording the difference between a nominal tool position and a master disk of known size and roundness during circular contouring.

disk drive

The mechanism that rotates a storage disk and reads or records data. A typical unit consists of a floppy disk drive, power supply (100–125 v AC, 60 Hz) cooling fan disk buffer and address select electronics. It is capable of storing more than 300,000 words on flexible disk.

diskette

A flexible or floppy disk. Magnetic-coated mylar disk enclosed in a protecting envelope. A standard diskette is $5\frac{1}{4}$-inches in diameter, with a capacity of approximately 75 text pages (4000 characters per single-spaced page).

disk memory

A nonprogrammable, bulk-storage, random-access memory consisting of a magnetizable coating on one or both sides of a rotating thin circular plate.

disk memory unit

A secondary memory device often used in real-time control systems. A disk memory unit consists of one or more disks stacked on top of one another. The disks are rotated continuously at a uniform speed. Information is stored on the magnetic disks in concentric tracks.

disk operating system

The operating system that manages the disk; also known as DOS.

disk pack
A removable direct-access storage media containing magnetic disks on which data is stored. Disk packs are mounted on a disk storage drive.

disk sector
A subdivision of a track on a magnetic disk; usually, the smallest unit of storage operated on by the disk drive at one time.

disk storage
Information recording on continuously rotating magnetic platters. Handles huge amounts of storage online. Storage is random access, meaning the recording arms move quickly to any address on any track on any disk to read or write information.

dispatching
The selecting and sequencing of available jobs to be run at individual workstations and the assignment of these jobs to workers.

dispatching rule
The logic used to assign priorities to jobs at a work center.

dispersion
Finely divided particles of a material in suspension in another substance.

display
The representation of data in visible form, for instance, in a cathode-ray tube or with lights or indicators on the console of a computer. A CAD/CAM workstation device for rapidly presenting a graphic image so that the designer can react to it, making changes interactively in real time. Usually refers to a CRT.

display area
The rectangular part of the physical display screen in which information is visibly displayed. The display area does not include the border area.

display change mode
The deferral-state value that specifies whether implicit regeneration will be suppressed or allowed.

display console
In computer graphics, a console of at least one display device (CRT) and usually one or more input devices such as an alphanumeric keyboard, functions key, tablet, joystick, control ball, or light pen.

display device
An output device that stores results from computer operations and translates them into graphic, numerical, or literal symbols to be seen by the computer user. A graphics device (e.g., refresh display, storage tube display, plotter) on which pictures can be represented.

display element
A basic graphic element that can be used to construct a display image.

display file
The list of graphics primitives and parameters sent to the system or hardware to create an image.

display image
A collection of display elements or segments that are represented together at one time on display surface.

display parameters
Data that control the appearance of graphical entities, for example, line fonting.

display processing unit
Hardware that fetches the device-dependent display commands from the device-dependent display file and interprets them as motions of the electron beam of the CRT.

display space
A portion of the device space or display surface corresponding to the area available for displaying images. Display space is also used to refer to the working space of an input device, such as a digitizer.

display surface
The physical area on a display device on which a graphics picture is generated, such as a display screen or a hard-copy medium (also called view surface).

display symbol
A method for graphically representing certain entities (plane, point, section) for identification purposes.

dispose
An element of a total operation that involves the laying aside and releasing or otherwise getting rid of a part, assembly, tool, or other object during or at the end of the operation.

distal
Away from the base, toward the end-effector of the robot arm.

distance/perimeter guard
A fixed barrier or fence designed to prevent normal access to a dangerous prohibited area. The guard used in robotics may take the form of a fixed barrier or fence designed to a height

so as to prevent normal access to a danger area, or a fixed tunnel that prevents access to a danger point by reason of the relationship of the opening dimensions of the guard to the length of tunnel.

distortion

An undesired change in the shape of an image or waveform from the original object or signal. Any change from the original waveform or signal. Normally, it refers to nonpredictable changes that interfere with interpretation of the result.

distributed computing

Computing performed within a network of distributed computing facilities. The processors for this type of system usually function with control distributed in time and space throughout the network. Associated with the distributed process are distributed storage facilities.

distributed control

A control technique whereby portions of a single control process are located in two or more places.

distributed function

In data communication, the use of programmable terminals, controllers, and other devices to perform operations that were previously done by the processing unit, such as managing data links, controlling devices, and formation of data.

distributed network

A network configuration in which all node pairs are connected either directly or through redundant paths through intermediate nodes.

distributed numerical control

The use of a computer for distributing part program data via communication lines to a plurality of remote NC machine tools.

distributed processing

Refers to a computer system that employs a number of different hardware processors, each designed to perform a different subtask on behalf of an overall program or process. Ordinarily, each task would be required to queue up for a single processor to perform all its needed operations. But in a distributed processing system, each task queues up for the specific processor required to perform its needs. Since all processors run simultaneously, the queue wait period is often reduced, yielding better overall performance in a multitask environment. Distributed-intelligence systems differ from multiprocessing systems in the way that tasks are handled. Although both systems use multiple processors, the tasks assigned to a distributed system remain fixed. By contrast, in a multiprocessing environment, a continuous stream of assignments is fed to a single node and allowed to be distributed according to complex resource allocation algorithms across the entire network.

distributed systems

Refers to various arrangements of computers within an organization in which the organization's computer complex has many separate computing facilities, all working in a cooperative manner, rather than the conventional single computer at a single location.

distribution by value

A method of analyzing a line or products, usually in the form of a plot of the cumulative frequency distribution of the annual dollar sales of each item in the product line. This distribution is useful in the estimation of required cycle stocks and safety stocks in an inventory.

distribution of forecast errors

Tabulation of the forecast errors according to the frequency of occurrence of each error value. The errors in forecasting are, in many cases, normally distributed even when the observed data do not come from a normal distribution.

distribution requirements planning

The function of determining the needs to replenish inventory at branch warehouses. Frequently, a time-phased order-point approach at the branch warehouse level is "exploded" via MRP logic to become gross requirements on the supplying source. In the case of multilevel distribution networks, this explosion process can continue down through the various levels of master warehouse, factory warehouse, and so forth, and become input to the master production schedule. Demand on the supplying source is recognized as dependent; standard MRP logic applies.

distribution resource planning

The extension of distribution requirements planning into the planning of the key resources contained in a distribution system: warehouse space, manpower, money, freight cars, trucks, and so forth.

dither
To place dots in an area of an image to soften an edge or visually smooth a jagged line.

dithering
A technique used to overcome breakaway friction by periodic excitation of the part to be moved.

diverging beam
A beam of light that is optically controlled so the light is diverged from the source.

division of work
The separation of tasks into less complex subtasks. This may be to use simpler skills or to make use of special skills.

DO
The basic computer command element that accomplishes in full or in part the purpose of the operation. It includes the basic elements use and assemble and may sometimes be expressed in terms of other basic elements.

document
A medium and the data recorded on it for human use, for example, a report sheet or a book. By extension, any record that has permanence and that can be read by man or machine.

documentation
The process of collecting and organizing documents or the information recorded in documents. Usually refers to the development of material specifying units, operations, and outputs to a computer program.

documentation aids
Materials that help automate the documentation process, including flowcharts, programs, and others.

documentation design
A set of specifications used to construct user documentation for the system.

document reader
An OCR device that scans one to five lines of data in fixed locations on a document at a single pass. Generally, no rescanning of a portion of the document is possible, one direction of the scan being provided by movement of the form past the reading head.

domain
A universal set of values from which the actual values appearing in a column of a table are drawn. The sphere of concern. The task world. A set of allowable inputs. The problem area of an expert system.

domain expert
A person with expertise in the field of the expert system being developed. The domain expert works closely with the knowledge engineer to capture the expert's knowledge in a knowledge base.

dominant wavelength
The wavelength of a light of a single frequency that matches a given color when combined with an appropriate amount of a reference "white" light.

dopant
An impurity (most often boron or phosphorous) that is added to silicon for the purpose of enhancing certain electrical properties.

dot
The smallest visible point that can be displayed on a display surface.

dot graphics
Graphic images that are made up of tiny dots, rather than out of lines and curves. Almost all dot graphics are raster graphics, where the dots used are selected from among those created by drawing a repeating set of adjacent lines across the screen or paper.

dot matrix
A method of display character generation in which each character is formed by a grid or matrix pattern of dots. A group of closely spaced dots with a printed pattern that looks like the shape of the desired character.

dot-matrix plotter
A CAD peripheral device for generating graphic plots. Consists of a combination of wire nibs, spaced 100 to 200 styli per inch, which place dots where needed to generate a drawing. Accuracy and resolution are not as great as with pen plotters. Also known as electrostatic plotter.

double break
A contact arrangement where the moving switch element bridges across two fixed contacts so that the circuit is broken in two places simultaneously.

double density
Term describing the storage of information on a diskette such that the capacity is twice that of a standard diskette. This is accomplished by either doubling the number of tracks per inch or doubling the serial bit density or a combination of both.

double precision
A data format used in CAD/CAM systems. Typically refers to a 64-bit floating-point data format, where the high-order bit is used as a sign, the next 8 bits for the exponent, and the remaining 55 bits for the mantissa. This provides a precision of approximately 16 significant figures. Single-precision floating-point formats have half the significant figures and constitute 32 bits.

double rail logic
Pertaining to self-timing asynchronous circuits in which each logic variable is represented by two electrical lines that together can take three meaningful states—zero, one, and undecided.

double-shot molding
A means of turning out two-color parts in thermoplastic materials by successive molding operations.

double-sided diskette
A type of diskette that utilizes both of its sides for the storage of information. A double-sided diskette can be loaded into a floppy disk drive with a dual read/write head assembly or used on a standard single-head drive, taken out, flipped and reinserted for read/record operations on both sides.

doublet
A two-element compound lens.

double word
A continuous sequence of bits or characters that comprises two computer words and is capable of being addressed as a unit.

dowel pin
Used for maintaining alignment between two or more parts of a mold.

downtime
The period during which a production line, computer, communications line, or other device is malfunctioning or not operating correctly because of mechanical or electronic failure, as opposed to available time, idle time, or standby time.

drag
On a graphics display, to move an image or part using a cursor, tablet, pen, or other positioning device. Dragging is used to position items by trial and error and to replicate image parts for repetitive use.

draft
The angle or taper on the surface of a punch or die, or the parts made with them, that facilitates the removal of the work. The change in cross-section in rolling or wire-drawing. Taper put on the surfaces of a pattern so that it can be withdrawn successfully from the mold.

drape forming
Method of forming thermoplastic sheet in which the sheet is clamped into a movable frame, heated, and draped over high points of a male mold. Vacuum is then pulled to complete the forming operation.

drawback
A refund or customs duties paid on material imported and later exported.

draw-down ratio
The ratio of the thickness of the die opening to the final thickness of the product.

drawing
The traditional graphic hard-copy representation of a design used to communicate the design during all stages from conception through manufacturing. The formation of cylindrical or otherwise shaped parts from flat stock by the action of a punch pushing the metal into a cavity having the same shape as the punch.

drawing entity
A structure entity that specifies the projection(s) of a model onto a plane, with any required annotation and/or dimension.

drawing number
Number assigned to an engineering graphic representation of a specific item; representation may be computer based.

drawing point
A logical indicator of the position at which the next geometric graphic primitive will commence execution. This is not normally marked by a drawing-point symbol.

drift
The tendency of a system's response to gradually move away from the desired response. A flat piece or steel of tapering width used to remove taper shank drills and other tools from their holders. A tapered rod used to force mismated holes in line for riveting or bolting.

drift test
A type of test used to detect environmental thermal effects on a machine. The test consists of continuously recording the output of a multiaxis displacement sensor placed in the spindle (or tool position) of a machine reading

against a sample part over a reasonable period of time.

drilling
The operation of producing cylindrically shaped holes, usually a cone shape at the bottom.

drive power
The source or means of supplying energy to the robot actuators to produce motion.

drop
A connection for a terminal unit on a transmission line.

drop delivery
A method of introducing an object to the workplace by gravity. A method whereby a chute or container is so placed that, when work on a part in question is finished, it will fall or drop into a chute or container or onto a conveyor with little or no "transport" by the worker. The laying aside of a part by releasing it so that it falls or moves away from the work area, either through the force of gravity or by mechanical or other means.

drop in/out
An error in the storage into or in the retrieval from a magnetic storage device, revealed by the reading of a binary character not previously recorded. Drop in/out is usually caused by defects in or the presence of particles in the magnetic surface layer.

drop shipment
A distribution arrangement in which the seller serves as a selling agent by collecting orders but does not maintain inventory. The orders are sent to the manufacturer, who ships directly to the customer.

dross
The scum that forms on the surface of molten metals largely because of oxidation but sometimes because of the rising of impurities to the surface.

drum memory
A storage device generally used for auxiliary memory. The unit resembles a drum and information is stored magnetically on its surface.

drum plotter
An electromechanical pen plotter that draws an image on paper or film mounted on a rotating drum. In the CAD peripheral device, a combination of plotting-head movement and drum.

Drum Plotter
Drum Plotter for Plotting Data and Making Engineering Drawings

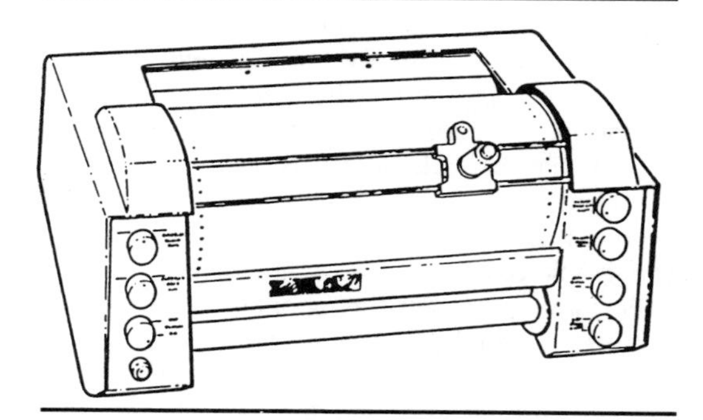

drum sequencer
A mechanical programming device that can be used to operate limit switches or valves to control a robot.

drum storage
Direct-access storage that records data magnetically on a rotating cylinder. A type of addressable auxiliary storage associated with some computers. Drum storage is almost never used today.

dry sand mold
A mold shaped of wet sand that is then dried.

dry method
Magnetic-particle inspection in which particles employed are in dry powder form. In magnetic-particle inspection, a method in which a dry powder is used to detect magnetic-leakage fields.

dual display CAD
The use of two CRTs. One screen is used for drawing preparation, the other for instructions and prompting.

dual inline package (DIP)
The most common integrated circuit package having dual, or parallel, rows of pins at 0.1-inch intervals.

ductile crack propagation
Slow crack propagation that is accompanied by noticeable plastic deformation and requires energy to be supplied from outside the body.

ductility
The extent to which a solid material can be drawn into a thinner cross section. The extent to which a material can sustain plastic deformation without rupture. Elongation and re-

duction of area are common indexes of ductility.

due date
The calendar data at which an operation or part or order is to be completed.

due date rule
A dispatching rule that directs the sequencing of jobs by the earliest due date.

dumb terminal
A device with a keyboard for inputting data and a display screen for the output of data but lacking local processing capability. A dumb terminal provides remote access to a computer but cannot itself be programmed.

dummy argument
Temporary storage that is created automatically to hold the value of an argument that is a constant, an operational expression, a variable whose attributes differ from those specified for the corresponding parameter in a known declaration, or is enclosed in parentheses.

dump
To transfer all of the information contained in a record into another storage medium. For example, a disk record could be dumped onto tape. Usually, however, dump refers to copying from an internal storage device to an external storage device for a specific purpose such as to allow other use of the storage, as a safeguard against faults or errors or in connection with debugging. Also referred to as core dump, tape dump, or disk dump.

duodecimal
Characterized by a selection choice or condition that has 12 possible different values or states.

duplex
Simultaneous two-way independent transmission in both directions. Also referred to as full duplex.

duplex fixture
A fixture designed to simultaneously hold two parts of the same type (number).

duty cycle
The fraction of time during which a device or system will be active or at full power.

dwell
A time delay that is programmed on the tape.

dynamic accuracy
Degree of conformance to the true value when relevant variables are changing with time. Degree to which actual motion corresponds to desired or commanded motion.

dynamic allocation
A software feature that allows resources (such as CPU time, file storage space, or disk transfer time) to be allocated on a request basis. For example, if a user on the system needs extra CPU time and there is unused CPU time available, it will be allocated to that user. After the resource has been used, it is returned to the pool. Dynamic allocation provides greater flexibility to use resources more efficiently.

dynamic dump
Dumping performed during the execution of a computer program, usually under control of that computer program.

dynamic error
An error that varies in time. Dynamic errors are those that change in time at a rate exceeding 1 hertz.

dynamic memory
A type of semiconductor memory in which the presence or absence of an electrical charge represents the two states of a storage element. Without refresh, the data represented by the electrical charge would be lost.

dynamic motion
Simulation of movement using CAD software so that the designer can see on the CRT screen 3-D representations of the parts in a piece of machinery as they interact dynamically. Thus, any collection of interference problems are revealed at a glance.

dynamic programming
A mathematical programming technique involving sequential multistage problem situations. At each stage, a decision must be made among alternatives.

dynamic RAM
Random-access memory that cannot be retained without continuous electrical regeneration. Faster, denser, and more expensive than static RAM.

dynamic range
The difference, usually expressed as a ratio, of the maximum acceptable signal level and the minimum acceptable signal level.

dynamics
The capability of a CAD system to zoom, scroll, and rotate.

dynamic storage
A device storing data in a manner that permits the data to move or vary with time such that the specified data are not always available for recovery. Magnetic drum and disk storage are dynamic nonvolatile storage. An acoustic delay line is a dynamic volatile storage device.

dynamic threshold
A threshold that varies with time. It is controlled either by current requirements or by local or global image parameters.

dynamic tool display
A CAD/CAM feature for graphically displaying a figure representing an NC cutting tool. The figure is dynamically moved along an NC tool path displayed on the CRT in order to verify and simulate the cutting procedure. Rotation provides the motion.

dynamic tracking
The location of an object is previewed by flashing it as it is moved into position. When the right position is reached, the object is placed. Works well with simple symbols but the response slows for complex objects.

E

earned hours
The time in standard hours credited to a worker or group of workers as a result of their completion of a given task or group of tasks.

echo
The immediate notification of the current value or measure of a logical input device provided to the operator, for example, by displaying the command or function just completed or a design entity being worked on.

echo check
An error control technique wherein the receiving terminal or computer returns the original message to the sender to verify that the message was received correctly.

echoplex
An echo check applied to network terminals operating in two-way simultaneous mode.

echo type
A parameter of device initialization that selects the echoing technique for a particular logical input device.

eddy current loss
The part of core loss that is due to current circulation in magnetic materials as a result of electromotive forces induced by varying induction.

eddy currents
Currents caused to flow in an electrical conductor by time or space variation, or both, of an applied magnetic field.

eddy current testing
Nondestructive testing method in which eddy-current flow is induced in the test object. Changes in the flow caused by variations in the object are reflected into a nearby coil or coils for subsequent analysis by suitable instrumentation and techniques.

edge
The apparent termination of an object's image. For a curved object such as a sphere, an edge represents the perimeter of the image formed by a series of lines from the observer and tangent to the surface of the object. An edge represents the outer boundary of an image. A distinguishable change in pixel values between two regions. Edges correspond to changes in brightness from a discontinuity in surface orientation, surface reflection, or illumination. The edge of an object can be defined by thresholding, that is, defining all pixels above a "threshold" value as the object and all at that value or below as background.

edge-based stereo
A stereographic technique based on matching edges in two or more views of the same scene taken from different positions.

edge detection
Any of several techniques to identify edges in an image.

edge following
A segmentation algorithm for isolating a region in an image by following the perimeter of its edge.

edge operators
Templates for finding edges in images.

edge vertex
A method of geometric modeling in which a 2- or 3-D object is represented by curve segments connected to points or vertices of the object. A higher level of topological information can be contained in such a model than is implied by a *wire-frame* terminology.

edit
To rearrange information. Editing may involve the deletion of unwanted data, the selection of pertinent data, and the insertion of symbols. To change an image or part. If the image was created by a paint program that draws on screen, editing can be done directly on the screen. If the image is the result of presenting the items in a display list, the list must be altered.

editor
A software tool to aid in modifying a software program.

editor program
A computer program designed to perform such functions as the rearrangement, modification, and deletion of data in accordance with prescribed rules.

edit run
A computer run to validate that the data is within allowable parameters and/or to perform other edit functions.

effective address
The contents of the address part of an effective instruction. The address that is derived

by applying any specified indexing or indirect addressing rules to the specified address and that is actually used to identify the current operation. The address used by a computer to execute an instruction. Due to a modification in the instruction, this address will differ from the original address in storage.

effective date
The date on which a component or an operation is introduced or severed from a bill of material or an assembly process, as the case may be. The effective dates are used in the explosion process to create demands for the correct material or assembly labor. Normally, bill-of-material systems provide for an effectivity *start date* and *stop date*, signifying the start or stop of a particular relationship. Effectivity control may also be by serial number rather than date.

effector
A robot actuator, motor, or driven mechanical device.

efficiency
The relationship between the planned labor requirements for a task and the actual labor time charged to the task. The ratio of standard performance time to actual performance time, usually expressed as a percentage.

effort
The evidence of the will to work as manifested by a worker performing an operation. The sum total of the mental absorption and physical participation that may be required by a worker on a given operation.

effort rating
That part of any performance rating technique concerned with evaluating the extent or degree to which the will to work is exhibited by a worker.

ejector pin
A pin or thin plate that is driven into a mold cavity from the rear as the mold opens, forcing out the finished piece.

ELAN
A conversational, live, interactive graphics programming system for lathes, mills, machining centers, punch presses, and other NC machine tools.

elapsed time
The actual time taken by a worker to complete a task, an operation, or an element of an operation. The total time interval from the beginning to the end of a time study.

elastic deformation
The part of the deformation of an object under load that is recoverable when the load is removed.

elasticity
The ability of a material to return to its original configuration when the load causing deformation is removed.

elastic limit
The maximum stress to which a material may be subjected without any permanent strain remaining upon complete release of stress.

elbow
The joint that connects the robot's upper arm and forearm.

electrically alterable ROM
A type of memory that combines the characteristics of RAM and ROM. It is nonvolatile (like ROM) but can be written into by the processor (like RAM). The EAROM, however, has a substantially longer writing time (currently about 2 microseconds versus 400 nanoseconds), as well as a limited number of writes (about 1,000,000), before the chip can no longer be reprogrammed.

electrical-optical isolator
A device that couples input to output, using a light source and detector in the same package. It is used to provide electrical isolation between input circuitry and output circuitry.

electrical schematic
A diagram of the logical arrangement of hardware in an electrical circuit/system, using conventional component symbols. Can be constructed interactively by CAD.

electric-beam pattern generator
An offline device using data postprocessed by a CAD/CAM system to produce a reticle from which the IC mask is later created. It exposes the IC reticle, using an electron beam in a raster-scan procedure.

electrode
In arc welding, a current-carrying rod that supports the arc between the rod and work or between two rods, as in twin carbon-arc welding. It may or may not furnish filler metal. In resistance welding, a part of a resistance welding machine through which current and, in most cases, pressure are applied directly to the work. The electrode may be in the form of a rotating wheel, rotating roll, bar, cylinder, plate, clamp, chuck, or modification thereof. An electrical conductor for leading current into or out of a medium.

electroformed molds
A mold made by electroplating metal on the reverse pattern on the cavity. Molten steel may be then sprayed on the back of the mold to increase its strength.

electroforming
Making parts by electrodeposition on a removable form.

electrogalvanizing
The electroplating of zinc upon iron or steel.

electroless plating
Immersion plating where a chemical reducing agent changes metal ions to metal.

electroluminescence
Similar to phosphorescence or fluorescence, except that the emission of light is stimulated by an electrical signal.

electrolysis
Chemical change resulting from the passage of an electric current through an electrolyte.

electrolyte
An ionic conductor. A liquid, most often a solution, that will conduct an electric current.

electrolytic cell
An assembly, consisting of a vessel, electrodes, and an electrolyte, in which electrolysis can be carried out.

electromagnetic radiation
Radiation that is emitted from vibrating charged particles and travels outward at the speed of light in the form of a wave.

electron gun
Hardware at rear of a television set or monitor that focuses an electron beam on the display screen, causing phosphors to emit light.

electronic mail
The electronic transmission of letters, messages, and memos from one computer to another.

electronic pen
Another term for a light pen; a pen-shaped device used to indicate position on a video display.

electronics
The science or use of electron-flow devices, such as vacuum tubes and transistors, with no moving parts.

electro-optical imaging sensors
Until recently, the most commonly used "eyes" for industrial robots and visual inspection. Standard television cameras, using vidicons, plumbicons, and silicon target vidicons, they interface with a computer and provide inexpensive and easily available imaging sensors. These cameras scan a scene, measure the reflected light intensities at a raster of approximately 320 × 240 pixels (picture elements), convert these intensity values to analog electrical signals, and feed this stream of information serially into a computer, all within 1/60 second.

electrosensitive printer
A nonimpact printer that employs electrically charged dots to develop specially coated paper.

electrostatic
Related to electrical phenomena that result from the storage of a static charge, as opposed to the flow of current through a circuit. Such forces can be used to bend an electron beam inside a cathode-ray tube.

electrostatic plotter
Device that produces drawings from a computer in the form of tiny dots. Not as accurate as pen plotters but much faster.

electrostatic printer
A nonimpact printer that employs electrically charged dots to attract ink that is then embedded onto the paper by heat and pressure.

element
In computer-aided design, the lowest-level design entity having an identifiable logical, electrical, or mechanical function.

elementary diagram
A wiring diagram of an electrical system in which all devices are drawn between vertical lines that represent power sources. Contains components, logic elements, wire nets, and text. Can be constructed interactively on a CAD system. Also called a wiring elementary or ladder diagram.

elementary operation
Unit of work on a workpiece that, for purposes of scheduling, is normally not subdivided, for example, drilling a single hole, rough milling in a single surface, or gauging a bore.

elementary operation time
The period of time required for the completion of an elementary operation; does not include tool movement time.

element breakdown
The subdivisions of an operation, each of which is composed of a distinct, describable and measurable sequence of one or several fundamental motions or machine or process activities.

element time
The term used to indicate either the actual, observed, selected, normal, or standard time to perform an element of an operation.

element type
A structure element's identifying classification, such as polygon, label, application data, or line width scale factor.

elevation
Direction of a straight line to a point in a vertical plane, expressed as the angular distance from a reference line, such as the observer's line of view.

elongation
A machine vision shape factor (blob feature) measurement. Equal to the length of the major axis divided by the length of the minor axis. Squares and circles have minimum of one; elongated objects have a value greater than one. In tensile testing, the increase in gage length, measured after fracture of the specimen within the gage length, usually expressed as a percentage of the original gage length.

embed
To write a computer language on top of (embedded in) another computer language.

embossing
The production operation of placing raised patterns on the surface of materials by use of dies or plates forced onto the material to be embossed.

embrittlement
Reduction in the normal ductility of a metal due to a physical or chemical change.

emergency close
An emergency procedure to terminate the graphics-generation process while saving as much graphical information as possible. Can be called by either the application program or by an error handler.

emergency stop
A method using hardware-based components that overrides all other robot controls and removes drive power from the robot actuators and brings all moving parts to a stop.

emissivity
The ratio of the radiance emitted by a source to the radiance emitted by a blackbody radiator at the same temperature and wavelength.

empirical data
Data originating in or based on observation or experience.

emulation
The use of programming techniques and special machine features to permit a computing system to execute programs written for another system. This form of imitation is done primarily by means of software. Emulation is generally used to minimize the impact of conversion from one computer system to another and is used to continue the use of production programs—as opposed to *simulation*, which is used to study the operational characteristics of another (possibly theoretical) system.

emulator
A combination of programming techniques and special machine features that permits a computing system to execute programs written for another system.

enable
To restore a suppressed feature, as an interrupt.

encapsulating
Enclosing an article (usually an electronic component or the like) in a closed envelope of plastic by immersing the object in a casting resin and allowing the resin to polymerize or, if hot, to cool.

enclosure
A surrounding case designed to provide a degree of protection for equipment against a specified environment and to protect personnel against accidental contact with the enclosed equipment.

encode
To apply a set of rules specifying the manner in which data may be represented, so that a subsequent decoding is possible. To translate from an easily recognizable language into a coded language.

encoder
A transducer used to convert angular or linear position or velocity into electrical signals. A device used to convert one form of information into another. A feedback device that generates pulses as it rotates. The source of the pulses is often an interrupted light beam. Encoders are becoming popular with CNC systems because of their digital quality that can be readily compared with the pulses generated by the CNC system. The robot system uses an incremental optical encoder to provide position feedback for each joint. Velocity data

is computed from the encoder signals and used as an additional feedback signal in order to assure servostability.

encoder accuracy
The maximum positional difference between the input to an encoder and the position indicated by its output. Includes both deviation from theoretical code transition positions and quantizing uncertainty caused by converting from a scale having an infinite number of points to a digital representation containing a finite number of points.

encoder ambiguity
Inherent encoder error, caused by multiple bit changes at code transition positions, that is eliminated by various scanning techniques.

encoding
Inscribing or imprinting magnetic ink character recognition characters on checks, deposits, and other documents to be processed by a reader; or the introduction of data on a medium such as a magnetic strip on plastic cards.

encryption
The conversion of data into code form for security purposes during data communications. The data is reconverted at the receiving end.

end-around borrow
The action of transferring a borrow digit from the most significant digit place to the least significant digit place.

end-around carry
The action of transferring a carry digit from the most significant digit place to the least significant digit place. An end-around carry may be necessary when adding two negative numbers that are represented by their diminished radix complements.

end-effector
A tool or gripping mechanism attached to the "wrist" of a robot to accomplish some task. While gripping mechanisms can be thought of as robotic "hands," end-effectors also include such single-purpose attachments as paint guns, drills, and arc-welders.

end-medium character
A control character used to identify the physical end of the data medium, the end of the used portion of the medium, or the end of the wanted portion of the data recorded on the medium.

end-of-address code
One or more control characters transmitted on a line to indicate the end of nontext characters, for example, addressing characters.

end-of-axis control
Controlling the delivery of tooling through a path or to a point by driving each axis of a robot in sequence. The joints arrive at their preprogrammed positions in a given axis before the next joint sequence is actuated.

end-of-block character
A character punched on the tape that denotes the end of a block of data.

end of file
The point at which a quantity of data is complete.

endogeneous variables
Variables whose values are determined by relationships included within the model.

end-point control
Any control scheme in which only the motion of the manipulator end point may be controlled and the computer can control the actuators at the various degrees of freedom to achieve the desired result.

end-point rigidity
The resistance of the robot hand, tool, or end point of a manipulator arm to motion under applied force.

end system
Also called *peer* system (meaning that any end system is equally capable of communicating with any other end system), an OSI end system contains an implementation of protocols for each of the OSI model's seven layers. A TOP end system is an OSI end system, using TOP-specified protocols. (*Illus. p. 142.*)

engineering change
A revision to a parts list, bill of materials, or drawing made and authorized by the engineering department and usually identified by a control number. Frequently classified as "safety change," "functional change," "cost revision," and so forth.

engineering change control
Estimating the cost and effect of an engineering change, the parts to be replaced or reworked, the loads on production facilities and vendors, and so forth. The term also implies the ability to maintain consistency of data: related changes on parent and component items, item data, product structures, requirements,

End System
Matching of Communication Layers. *Source*: Boeing Company.

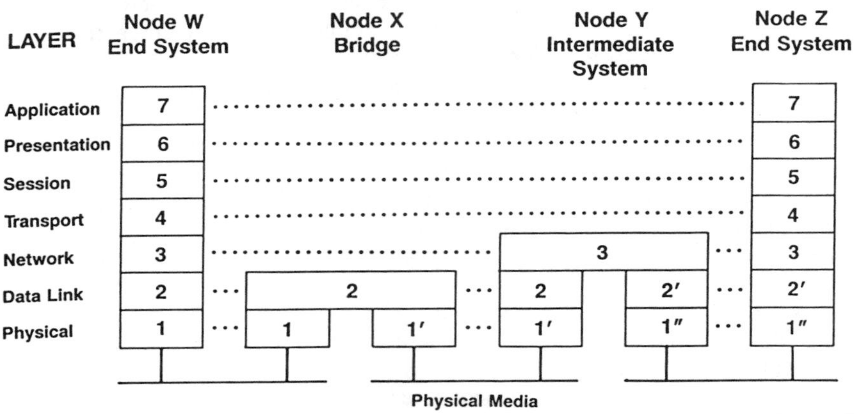

drawings, routines, tools, jigs and fixtures, gages, test instructions, parts catalogs, maintenance instructions, commercial documentation, and others.

engineering units
Units of measure as applied to a process variable, for example, PSI, degrees F, and so forth.

engineering workstation
A computer graphics system designed for use in electrical or mechanical engineering. Generally, it features programs for the appropriate type of design, such as circuit design aids or printed circuit board layout programs, along with graphics-display hardware and software. These systems both raise the productivity of engineers and speed project completion.

enhanced pulsing
Electronic modulation of a laser beam to produce a very high peak power at the initial stage of a pulse.

enhancements
Software or hardware improvements, additions, or updates to a CAD/CAM system.

enquiry character
A transmission control character used as a request for a response from the station with which the connection has been set up; the response may include station identification, the type of equipment in service, and the status of the remote station.

enter key
A special function key on a terminal keyboard, used to transmit a line of data on a display screen to a computer. The enter key is pressed after the message is complete.

entity
A geometric primitive—the fundamental building block used in constructing a design or drawing, that is, arc, circle, line, text, point, spline, figure, or nodal line; or a group of primitives processed as an identifiable unit. Thus, a square may be defined as a discrete entity consisting of four primitives (vectors), although each side of the square could be defined as an entity in its own right. In CAD, a pictorial element treated as a basic building block out of which drawings such as an arc, plane, or cube can be constructed. Entities may be assigned characteristics, replicated, and combined into more complex objects.

entity label
A one- to eight-character identifier for an entity. This term may implicitly include the entity subscript, providing for additional characters.

entity subscript
A one- to eight-digit unsigned integer associated with the entity label. The label and subscript specify a unique instance of an entity within an array of entities.

entity type number
An integer used to specify the kind of entity. For example, circular arcs have an entity type number of 100.

entity use flag
A portion of the status number field of the directory entry of an entity to designate whether the entity is used as geometry, annotation, structure, logical, or other. For example, a circle used as part of a point dimension would have an entity use flag that designates annotation.

entry point
The address of the label of the first instruction executed upon entering a computer program, a routine, or a subroutine.

enumeration type
A discrete type whose values are represented by enumeration literals that are given explicitly in the type declaration. These enumeration literals are either identifiers or character literals.

envelope
The set of points representing the maximum extent or reach of the robot hand or working tool in all directions. The work envelope can be reduced or restricted by limiting devices that establish limits that will not be exceeded in the event of any foreseeable failure of the robot or its controls. The maximum distance that the robot can travel after the limit device is actuated will be considered the basis for defining the restricted (or reduced) work envelope. The smallest cylinder, regular prism, or rectangular prism that can enclose the part on a CAD display.

environment
The mode of operation of a computer.

environmental stress cracking
The susceptibility of a thermoplastic article to crack or craze formation under the influence of certain chemicals and stress.

equalization
In data communications, a compensation for the increase of attenuation with frequency. Its purpose is to produce a flat frequency response.

equivalent
Has the same truth value (in logic).

equivalent binary digit factor
The average number of binary digits required to express one radix digit in a nonbinary numeration system. For example, approximately three and one third times the number of decimal digits is required to express a decimal numeral as a binary numeral.

erase
To obliterate information from a storage medium by clearing or overwriting. To replace all the binary digits in a storage device by binary 0s. To remove data from a magnetic surface or other memory unit.

eraser
In a graphics program that allows drawing on screen, a special pen or brush deletes images over which it is moved. In some systems, the eraser leaves behind a selected color (usually default to black), while in others it restores the color and image that was displayed at that spot before the erased element was added.

ergonomics
The study of the interaction between people and machines.

error
The difference between the actual response of a machine to a command issued according to the accepted protocol of that machine's operation and the response to that command anticipated by that protocol. This definition is intended to apply only to machine tools and measuring machines. A difference between a computed value and the theoretically correct value.

error avoidance
A means of error reduction in which the source of the error or its coupling mechanism is eliminated.

error burst
In data communication, a sequence of signals containing one or more errors but counted as only one unit in accordance with some specific criterion or measure.

error compensation
A means of error reduction whereby the effect of the error is canceled.

error control
An arrangement that detects the presence of errors. In some systems, refinements are added that will correct the detected errors, either by operations on the received data or by retransmission from the source.

error control procedure
The inclusion of redundant information in a message, for example, parity bits, checksums,

cyclic redundancy check characters, Hamming codes, and fire to permit the detection of errors that arise from noise or other disturbances in the transmission medium. May involve retransmission of messages until they are received correctly.

error correcting code
A code in which each acceptable expression conforms to specific rules of construction that also define one or more equivalent nonacceptable expressions, so that if certain errors occur in an acceptable expression, the result will be one of its equivalents and thus the error can be corrected.

error detecting and correcting
A system employing an error detecting code and so arranged that a signal detected as being in error automatically initiates a request for retransmission.

error detecting code
A code with specific rules of construction designed so that if certain errors occur the presence of the errors is detected. Synonymous with self-checking code. Such codes require more transmission data than necessary to convey the fundamental information.

error logging
Recording information on the occurrence of an error condition.

error map
The representation of the amount of error compensation as a function of the input variables.

error of direction
An error of motion of a machine, caused by nonorthogonality or nonparallelism of axes.

error rate
The ratio of the amount of data incorrectly received to the total amount of data transmitted.

error reaction
A predefined action taken after error detection.

error reduction
The elimination or reduction of machine errors by any means.

error reporting
The communication of the error condition to the application program.

error signal
The difference between desired response and actual response.

error state list
A conceptually defined data structure that contains information on the current error condition.

escalation
An amount or percent by which a contract price may be adjusted if specified contingencies occur, such as changes in the vendor's raw material or labor costs.

escape
A function used to access implementation-dependent or device-dependent features not concerned with the generation of graphical output.

escape character
A code extension character used, in some cases, with one or more succeeding characters to indicate by some convention or agreement that the coded representations following the character or the group of characters are to be interpreted according to a different code or according to a different coded character set. A character that signifies that the next character is a member of a different character set.

escape data record
A data structure containing detailed information about a particular escape function.

escape elements
Elements that describe device- or system-dependent elements used to construct a picture but are not otherwise standardized.

estimated time
Estimate of the time required for an operation, usually the standard time for the operation and is expressed as time for one piece, or time for 100 pieces, and so forth. It is used as a basis for capacity requirements, planning, and cost estimating.

Ethernet
Local area network for sending messages between computers by way of a single coaxial cable that snakes through all of the computers to be connected.

eutectic
An isothermal reversible reaction in which a liquid solution is converted into two or more intimately mixed solids on cooling, the number of solids formed being the same as the number of components in the system. An alloy having the composition indicated by the eutectic point on an equilibrium diagram. An alloy structure of intermixed solid constituents formed by an eutectic reaction.

eutectic melting
Melting of localized microareas whose composition corresponds to that of the eutectic in the system.

evaluation function
In artificial intelligence, a function (usually heuristic) used to evaluate the merit of the various paths emanating from a node in a search tree.

even parity
The condition that occurs when the sum of the number of ones in a binary word is always even.

event driven
In artificial intelligence, a forward-chaining problem-solving approach based on the current problem status.

event input
One of three input modes. In event input mode, asynchronous input is placed on the event queue when a trigger fires.

event queue
A time-ordered collection of event input items.

event report
A logical input value saved on an event queue or input queue.

evolutionary development
The practice of iteratively designing, implementing, evaluating, and refining computer applications, especially characteristic of the process of building expert systems.

EXAPT
A numeric control processor that selects the work sequence and tool and determines the cutting data such as feed rates, spindle speeds, depths of cut, and so on. The processor is based on APT.

exception
An abnormal condition, such as an I/O error encountered in processing data set or a file.

exception reports
Reports that list or flag only those items that exceed a specified range of acceptable values.

excess-three code
The binary coded decimal notation in which a decimal digit n is represented by the binary numeral that represents $n + 3$.

excitation purity
On a CIE chromaticity diagram, the ratio of the sample-color and spectrum-color distances from the reference "white" color, as measured along the dominant-wavelength vector. For nonspectrum colors, the denominator is the distance to the purple boundary.

excitation rms
The rms alternating current required to produce a specified induction in a material. For inductions of less than 10 kilogausses, rms excitation (Hz) is usually expressed in oersteds; for high inductions, in amp-turns/inch.

exclusive-OR
Logical operation between two binary digits, yielding a result of 1 if one and only one of the digits has value 1 and yielding a result of 0 otherwise.

execute file
A text file containing a sequence of computer commands that can be run repeatedly on the system by issuing a single command. Allows the execution of a complicated process without having to reissue each comand.

execution time
The total time required for the execution of one specific operation. For a computer system, the time at which an object program actually performs the instructions coded in the procedure division, using the actual data provided. For the arithmetic and logic unit portion of the CPU, the time during which an instruction is decoded and performed.

executive
A collection of routines considered part of the programmable controller/computer itself. They are usually stored in a nonvolatile memory. They are the coordinating, directing, or modifying routines that control the operations of other routines or programs. They also schedule, allocate, and control system resources rather than process data to produce results. Synonymous with supervisory routines.

executive control program
A main system program designed to establish priorities and to process and control other programs.

executive routine
A segment of the operating system that controls the execution of other portions of the operating system. Synonymous with supervisory routine.

exerciser programs
Training programs that allow the user to gain familiarity with the system and the software

by using the various capabilities of the particular software. The programs are provided with on-screen documentation and are designed to minimize the need for previous training.

existing system
A system currently under operation and maintenance. A system is a collection of elements so arranged as to address some problems. Systems are composed of lesser aggregates of elements called subsystems. The smallest elements identified as being within a system are called components.

exit
An instruction in a computer program, in a routine, or in a subroutine after the execution of which control is no longer exercised by that computer program, that routine, or that subroutine.

exogeneous variable
A variable whose values are determined by consideration outside the model in question.

exoskeleton
An articulated mechanism whose joints correspond to those of a human arm, and which, when attached to the arm of a human operator, will move in correspondence to his or her arm. Exoskeleton devices are sometimes instrumented and used for master-slave control of manipulators.

expansion card
Circuit board or card that can be inserted into the motherboard of the computer to increase its capacities.

expansion slot
The space into which additional cards are inserted into a motherboard.

expectation-driven
Processing approaches that proceed by trying to confirm models, situations, states, or concepts anticipated by the system.

expectation-driven reasoning
A control procedure that employs current data and decisions to formulate hypotheses about yet unobserved events and to allocate resources to activities that confirm, disconfirm, or monitor the expected events.

expected completion quantity
The planned quantity of a manufacturing order after expected scrap (cf. scrap factor).

expected demand
The quantity expected to be withdrawn from stock during the lead time when usage is at the forecasted rate. Consequently, it is identified with the forecast of demand during a lead time.

expected value
The average value that would be observed in taking an action an infinite number of times. The expected value of an action is calculated by multiplying the outcome of the action by the probability of achieving that outcome.

expediting
The ''rushing'' or ''chasing'' of production orders that are needed in less than the normal lead time.

expertise
The set of capabilities that underlines the high performance of human experts, including extensive domain knowledge, heuristic rules that simplify and improve approaches to problem-solving, metaknowledge and metacognition, and compiled forms of behavior that afford great economy in skilled performance.

expert system
A computer program that uses knowledge and reasoning techniques to solve problems normally requiring the abilities of human experts. A computer program containing knowledge about objects, events, situations, and courses of action that emulates the reasoning processes of human experts in a particular domain. The components of an expert system are the knowledge base, inference engine, and user interface. Types of expert systems include rule-based systems and model-based systems.

explanation
Motivating, justifying, or rationalizing an action by presenting antecedent considerations such as goals, laws, or heuristic rules that affected or determined the desirability of the action.

explanation facility
The component of an expert system that can explain the system's reasoning and justify its conclusions.

explode
Break apart a symbol or object into its basic components so the symbol may be changed.

explosion
An extension of a bill of materials into the total of each of the components required to manufacture an assembly or subassembly quantity.

Expert System
Basic Block Diagram of Key Elements of Expert System

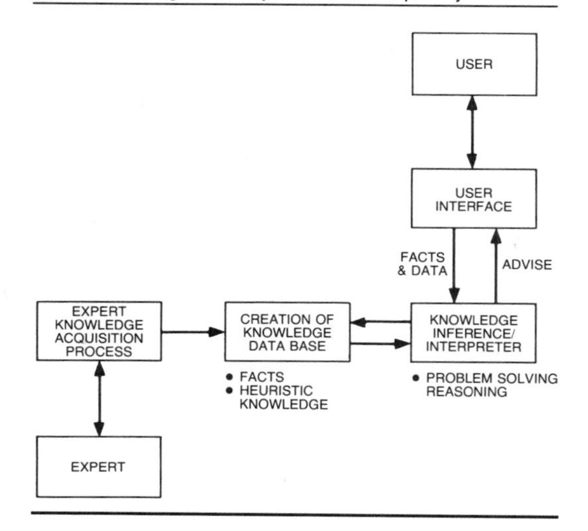

exponent
In a floating-point representation, the numeral that denotes the power to which the implicit floating-point base is raised before being multiplied by the fixed-point part to determine the real number represented.

exponential smoothing
A weighted, moving-average technique in which past observations are geometrically discounted according to their age. The heaviest weight is assigned to the most recent data. The smoothing is termed *exponential* because data points are weighted in accordance with an exponential function of their age.

Extended Binary Coded Decimal Interchange Code
A specific code using eight bits to represent a character; often abbreviated as EBCDIC.

extension
A linear motion in the direction of travel of the sliding motion mechanism or an equivalent linear motion produced by two or more angular displacements of a linkage mechanism. A set of letters (not more than three) used for indication of a file type.

extensometer
An instrument for measuring changes caused by stress in a linear dimension of a body.

extent
A collection of physical records that are contiguous in secondary storage. The number of records in an extent depends on the physical volume and the user's request for space allocation. Associated records are not necessarily stored contiguously; this depends on the storage organization.

external elements
Elements that communicate information not directly related to the generation of a graphical image.

external function testing
The verification of the external system functions as stated in the external specifications.

external interface
A juncture, point of contact, or point of communication between a part of a system and the environment of the system.

external memory
A storage device, such as punched cards, punched tape, or magnetic tape, that is external to the computer.

external sensor
A feedback device that is outside the inherent makeup of a robot system or a device used to effect the actions of a robot system that are used to source a signal independent of the robot's internal design.

external sort
A sort that requires the use of auxiliary storage because the set of items to be sorted cannot be held in the available internal storage at one time.

external storage
The storage of data on a device such as magnetic tape that is not an integral part of a computer but is in a form prescribed for use by a computer.

extrapolation
Estimating the future value of some data series based on past observations. Statistical forecasting represents a common example.

extreme infrared
Radiation with a wavelength greater than 15 microns.

extrinsic forecast
A forecast based on a correlated leading indicator, such as estimating engine sales based on auto production estimates. Extrinsic forecasts tend to be more useful for large aggregations, such as total company sales, than for individual product sales.

extrusion
A method of entering a three-dimensional object into a computer graphics system. In this

method, a two-dimensional drawing is first entered using a digitizer or other input device. Elevations are assigned to each closed surface. The system then extends the surfaces into the third dimension, seemingly extruding material using the two-dimensional image as a die. In a more sophisticated system, the program then removes hidden lines and fills the surfaces. Conversion of a billet into lengths of uniform cross section by forcing the plastic metal through a die orifice of the desired cross-sectional outline. In *direct extrusion*, the die and ram are at opposite ends of the billet, and the product and ram travel in the same direction. In *indirect extrusion* (rare), the die is at the ram end of the billet, and the product travels through and in the opposite direction to the hollow ram. A *stepped extrusion* is a single product with one or more abrupt cross-section changes and is obtained by interrupting the extrustion by die changes. *Impact extrusion* (cold extrusion) is the process or resultant product of a punch striking an unheated slug in a confining die. The metal flow may be either between the punch and die or through another opening. *Hot extrusion* is similar to cold extrusion, except that a preheated slug is used and the pressure application is slower.

extrusion billet
A cast or wrought metal slug used for extrusion.

extrusion defect
A defect of flow in extruded products, caused by the oxidized outer surface of the billet flowing into the center of the extrusion. It normally occurs in the last 10 to 20 percent of the extruded bar.

extrusion ingot
A solid or hollow cylindrical casting used for extruding into rods, bars, shapes, or tubes.

eyepiece
Also called an ocular. A lens system used between the final real image and the eye in an optical system. The eyepiece often serves as a magnifier.

EZAPT
An APT-based NC processor language that is implemented on a microcomputer system with two disk drives. The processor can be used to program all NC machine tools with two-axis circular, three-axis linear, and four-axis position (rotary tables and so on) capabilities.

F

fabrication
A term used to distinguish production operations for components, as opposed to assembly operations. Processing of natural or synthetic materials for desired modification of shape and properties.

fabrication level
The lowest production level. The only components found are parts (as opposed to assemblies or subassemblies). These parts are either procured from outside sources or fabricated within the manufacturing organization.

fabrication order
A manufacturing order to a component-making department, authorizing it to produce component parts.

FABRIPOINT V
Processor for CNC sheet metal punching machines. Includes geometric, macro, and random pattern definitions, with full editing and verification capabilities, and optional plotter.

facilities control
Management of the development or acquisition of facilities to support information systems. Facilities control requires a function or activity responsible for information systems facilities planning, sequencing of facilities development or acquisition that is related to information systems development requirements, facility costs included in the funding and accounting for major systems, and operational procedures for facilities such as security and emergency procedures.

facility
A machine or piece of equipment that is necessary for the performance of a production operation on a product item.

facsimile (FAX)
A system of telecommunication to transmit images for reproduction on hard copy. The original image is scanned and converted to an electrical signal, and the electrical signal is subsequently converted to a replica of the original image at the receiving terminal.

facsimile transmission
An electronic means for transferring an image (precise reproduction) from one place to another.

fact
A proposition or datum whose validity is accepted.

factor comparison
A type of job evaluation plan in which relative values for each of a specified number of factors of a job are established by direct comparison with the values established for these same factors on selected key jobs.

factory
A manufacturing unit consisting of two or more centers and the materials transport, storage buffers, and communications that interconnect them.

fadeometer
An apparatus for determining the resistance of resins and other materials to fading. This apparatus accelerates the fading by subjecting the article to high-intensity ultraviolet rays of approximately the same wavelength as those found in sunlight.

fail safe
Failure of a device without danger to personnel or damage to product or plant facilities.

fail soft
Failure in performance of some component part of a system without immediate major interruption or failure of performance of the system as a whole and/or sacrifice in quality of the product.

failure analysis activity
A program of systematic collection and analysis of data, the detection and selection of significant deviations or variations from established limits, isolation of the cause, analysis of the defect, and, finally, the recommendation for corrective action with follow-up.

fair day's work
The amount of work that can be produced during a working day by a qualified individual with average skill who follows a prescribed method, works under specified conditions, and exerts average effort.

falldown
Performance that is less than standard.

false add
To form a partial sum, that is, to add without carries.

families
Convenient groupings of related orders or similar parts.

families of computers
Series of CPUs allegedly of the same logical design but of different speeds and configura-

tion rules. This is intended to enable the user to start with a slow, less expensive CPU and grow to a fast one as his or her workload builds up, without having to change the rest of the computer.

family contracts
Grouping of families of similar parts together on one purchase order to obtain pricing advantages and a continuous supply of material.

family mold (injection)
A multicavity mold wherein each of the cavities forms one of the component parts of the assembled finished object.

family of parts
A collection of previously designed parts with similar geometric characteristics (i.e., line, circle, ellipse) but differing in physical measurement (i.e., height, width, length, angle). When the designer preselects the desired parameters, a special CAD program creates the new part automatically, with significant time savings.

FAPT
An interactive APT processor with graphic display that provides instant program verification. It simultaneously provides a readout of the commands programmed and generates 2-D contours or simulated 3-D surfaces for visual program check.

far
Another name for the back clipping plane, the imaginary surface that limits how far back objects are shown in computer graphics showing three-dimensional objects. Still another name for this plane is yon.

far distance
The distance from the back clipping plane to the view reference point.

far infrared
Radiation with a wavelength between 6 and 15 microns.

fatigue
The phenomenon leading to fracture under repeated or fluctuating stresses having a maximum value less than the tensile strength of the material. Fatigue fractures are progressive, beginning as minute cracks that grow under the action of the fluctuating stress. Permanent structural change that occurs in a material subjected to fluctuating stress and strain. In general, fatigue failure can occur with stress levels below the elastic limit.

fatigue life
The number of cycles of stress that can be sustained prior to failure for a stated test condition.

fatigue limit
The maximum stress below which a material can presumably endure an infinite number of stress cycles. If the stress is not completely reversed, the value of the mean stress, the minimum stress, or the stress ratio should be stated.

fatigue ratio
The ratio of the fatigue limit for cycles of reversed flexural stress to the tensile strength.

fatigue strength
The maximum stress that can be sustained for a specified number of cycles without failure, the stress completely reversed within each cycle unless otherwise stated. The magnitude of fluctuating stress required to cause failure in a fatigue test specimen after a specified number of loading cycles.

fatigue-strength reduction factor
The ratio of the fatigue strength of a member or specimen with no stress concentration to the fatigue strength with stress concentration. The fatigue-strength reduction factor has no meaning unless the geometry, size, and material of the member or specimen and stress range are stated.

fault
A condition that causes any physical component of a system to fail to perform in its normal fashion.

fault diagnosis
Determining the trouble source in an electromechanical system.

fault tolerance
The ability of a program or system to operate properly, even if faults occur.

fault tree analysis
A logical approach to identify the areas end system that are most critical to uninterrupted and safe operation.

faying surface
The surface of a piece of metal (or a member) in contact with another to which it is to be joined.

feature
An optical characteristic of an image that describes it in such a way that an interpretation can be made of the object that formed the

image. Features may include position, geometric characteristics, or light-intensity distribution over the image. Any characteristics descriptive of an image or a region in an image. Simple image data attributes, such as pixel amplitudes and edge-point locations, as well as more elaborate image patterns such as blob elongation, compactness, size, and centroid.

feature extraction
Determining image features by applying feature detectors.

feature vector
A set of features of an object (such as area, number of holes, etc.) that can be used for its identification.

feedback
The return of part of the output of a machine, process, or system to the computer as input for another phase, especially for self-correcting or control purposes. Actual performance can thus be compared with planned performance.

feedback control
A type of system control obtained when a portion of the output signal is operated upon and fed back to the input in order to obtain a desired effect. A guidance technique used by robots to bring the end-effector to a programmed point.

feedback control loop
A closed transmission path that includes an active transducer and consists of a forward path, a feedback path, and one or more mixing points arranged to maintain a prescribed relationship between the loop input and output signals.

feedback data
Data describing the result of a previous decision or action and used to determine actual status and deviation from a plan so as to initiate corrective action.

feedback devices
Installed to sense the positions of the various links and joints and transmit this information to the controller. These feedback devices may be simply limit switches actuated by the robot's arm or position-measuring devices such as encoders, potentiometers, or resolvers and/or tachometers to measure speed. Depending on the devices used, the feedback data are either digital or analog.

feedback loop
The components and processes involved in correcting or controlling a system by using part of the output as input.

feedback sensing
A mechanism through which information from sensing devices is fed back into the robot's control unit. The information is utilized in the subsequent direction of the robot's motion.

feed function
The relative travel motion between the cutting tool and the workpiece. The movement is usually expressed in inches per minute (ipm) but may also be expressed as inches per revolution (ipr).

feedhold
A method to arrest machine slide travel manually, without losing tape data.

feeding
The process of placing or removing material within or from the point of operation.

feedrate number
A coded number read from the tape that describes the feedrate function. The coding method may differ among systems. The feedrate number is denoted as the *F word* on the tape and usually follows the *dimension words*.

feedrate override
A manual function, usually a rotary dial, that can override the programmed feedrate. The range can be from a few percent to over 100 percent of the programmed feedrate and is usually infinitely variable.

ferrite
A solid solution of one or more elements in body-centered cubic iron. Unless otherwise designated (for instance, as chromium ferrite), the solute is generally assumed to be carbon.

ferromagnetic
A term applied to materials that can be magnetized or strongly attracted by a magnetic field.

ferromagnetics
In computer technology, the science that deals with the storage of information and the logical control of pulse sequences through the utilization of the magnetic polarization properties of materials.

fetch
In machine operation, to access data and/or instructions from memory and bring it to the CPU to be operated on.

fettling
The removal (automatic or manual) of flash, sprues, parting lines, and so forth from castings by grinding, milling, nibbling, and so on.

fiber-optic image cable
A bundle of optical fibers that can transmit a real image through its length, without distortion.

fiber optics
A communication technique where information is transmitted in the form of light over a transparent fiber material, such as a strand of glass. Advantages include noise-free communication not susceptible to electromagnetic interference. A light source for machine vision systems by which light is transmitted through a long flexible fiber of transparent material through a series of internal reflections.

field
A group of related characters treated as a unit in computer operations. A set of one or more columns of a punched card consistently used to record similar information. Each field consists of one or more consecutive columns that are reserved for punching specific types of data. The length of a field is determined by the maximum length of the particular type of data to be recorded on it. Also referred to as a reserved area in a record that serves a similar function in all records of that group; the data contained in two or more core positions and treated as a unit. One of two equal parts into which a television frame is divided in an interlaced system of scanning.

field developed program
A licensed program product that performs a specific user application. It may interact with other program products, or it may be a standalone program. Designed originally for the specific needs of an existing computer system installation.

field ID primary
Identifies a field that describes the primary function of the field group. It is the first field in time order of a functional field group.

field lens
A lens used to effect the transfer of an image formed by an imaging system.

field of view
The maximum angle of view that can be seen through a camera. As an object moves closer to the camera, the angle does not change, but a smaller portion of the object is seen by the camera. The region of the scene that is imaged by an imaging system.

field or branch warehouse
A distribution point between the plant and the customer.

fifth-generation project
The research project in which the Japanese are investigating parallel processing and other advanced computing techniques in an attempt to develop a *fifth-generation* of computer systems.

figure
A symbol or a part that may contain primitive entities, other figures, nongraphic properties, and associations.

file
A collection of related information in the system that may be accessed by a unique name. May be stored on a disk, tape, or other mass-storage media.

file maintenance
The activity of keeping a file up-to-date by adding, changing, or deleting data.

file organization
The view of the data as perceived by the application programmers. The storage method that determines the method by which file contents can be accessed.

file protection
A means of protecting the disk or tape from being erased or written over. Labels can be pasted onto floppy disks and plastic rings can be inserted on tape reels. In logical file protection, software protects files on disk or tape.

file purging
Erasing the contents of old files to make room for new ones.

fill area
An output primitive consisting of a polygon (closed boundary) that may be hollow or may be filled with a uniform color, a pattern, or a hatch style.

fill-area bundle table
A table associating specific values for all workstation-dependent aspects of a fill-area primitive with a fill-area bundle index. The table contains entries for interior style, style index, and color index.

filled surface
The transition surface that blends together two surfaces, for example, an airplane wing and the plane's body.

fillet
Rounded corner or arc that blends together two intersecting curves or lines. May be generated automatically by a CAD system. A radius (curvature) imparted to inside meeting surfaces. A concave corner piece used on foundry patterns.

fillet weld
A weld, approximately triangular in cross section, joining two surfaces essentially at right angles to each other in a lap, tee, or corner joint.

fill speed
How fast a graphics system can update images by making the dots in a bounded area a new color. The fill-speed figure is usually given in pixels per second (pixels/sec) or million pixels per second (mps).

filter
A device or process that selectively transmits frequencies. In optics, it is a material that selectively absorbs or reflects wavelengths of light. Device to suppress interference that would appear as noise.

filtering
A computer-aided mapping capability that employs one of numerous algorithms to delete selected points used to define a linear entity. Filtering removes unnecessary points from the data base and simplifies maps being converted to a smaller scale.

final assembly schedule
Also referred to as the *finishing schedule,* as it may include other operations than simply the final operation. It is a schedule of end items, either to replenish finished-goods inventory or to finish the product for a make-to-order product. For make-to-order products, it is prepared after receipt of a customer order and constrained by the availability of material and capacity, and it schedules the operations that are required to complete the product from the level where it is stocked (or master scheduled) to the end item level where it is stocked (or master scheduled) to the end item level.

final character
The last character of an ESCAPE sequence.

finished-goods inventories
Inventory of end items available for customer sales. It can include spare parts.

finished-products inventories
Those inventories on which all manufacturing operations, including final test, have been completed. These may be either finished parts, like renewal parts, or finished products that have been authorized for transfer to the finished stock account. These products are now available for shipment to the customer, either as end items or repair parts.

finishing lead time
The time that is necessary to finish manufacturing a product after receipt of a customer order. The finishing lead time should be equal to or less than the company's goal for shipping its product after receipt of a customer order.

finite capacity loading
Loading a factor or work center only to its capacity. This process automatically schedules lower-priority items into the next available time period if the present time period's capacity is fully utilized.

finite element
A small part of a structure defined by the connection of nodes, material, and physical properties.

finite element analysis
A method used for determining the structural integrity of a mechanical part or physical construction under design by mathematical simulation of the part and its loading conditions.

finite element modeling
The creation on the system of a mathematical model representing a mechanical part or physical construction under design. The model, used for input to a finite element analysis (FEA) program, is built by first subdividing the design model into smaller and simpler elements such as rectangles, triangles, bricks, or wedges that are interconnected. The finite element model is comprised of all its subdivisions or elements, and its attributes (such as material and thickness), as well as its boundary conditions and loads (including mechanical loadings, temperature effects, and materials fatigue).

finite elements
Finite elements represent the simple subdivision of a complex structure in a finite element model.

finite loading
A computer technique that involves automatic shop priority revision in order to level the load

operation by operation. Conceptually, the term means putting no more work into a factory than the factory can be expected to execute.

firm planned order
A planned order that can be frozen in quantity and time. The computer is not allowed to automatically change it; this is the responsibility of the planner in charge of the item that is being planned. This technique can aid planners working with MRP systems to respond to material and capacity problems by firming up selected planned orders. Additionally, firm planned orders are the normal method of stating the master production schedule.

firmware
Computer programs, instructions, or functions implemented in user-modifiable hardware. Such programs or instructions, stored permanently in programmable read-only memories, constitute a fundamental part of system hardware. The advantage is that a frequently used program or routine can be invoked by a single command instead of multiple commands, as in a software program.

first-ended, first-out
A queuing scheme whereby messages on a destination queue are sent to the destination on a first-ended, first-out basis within priority groups. That is, higher-priority messages are sent before lower-priority messages; when two messages on a queue have equal priority, the one whose final segment arrived at the queue earliest is sent first.

first generation
In the NC field, the period of technology associated with vacuum tubes, relays, and stepping switches. In the computer field, the technology using vacuum tubes, offline storage on drum or disk, and programming in machine language.

first-level code
A telegraph code that utilizes five impulses for describing a character. Start and stop elements may be added for asynchronous transmission.

first-level message
Under time-sharing option, a diagnostic message that identifies a general condition. More specific information is issued in a second-level message if the test is followed by a "+".

first order predicate logic
A popular form of logic used by the AI community for representing knowledge and performing logical inference. First order predicate logic permits assertions to be made about variables in a proposition.

first-piece time
The time required to produce the first of a number of identical units, including all necessary setup and makeready time.

fixed coordinate system
A coordinate system fixed in time and space.

fixed cost
An expenditure that does not vary with the production volume, for example, rent, property tax, salaries of certain personnel.

fixed disk
A nonremovable disk.

fixed expense
Expenditures that remain constant with respect to time, regardless of the volume of production, such as taxes, insurance, certain administrative salaries, and the like.

fixed guard
A barrier, not readily removable, to entry of personnel into potentially dangerous areas.

fixed-head disk
A disk memory unit on which a separate read/write head is provided for each track of each disk surface. A typical disk has about 800 tracks per side. Accessing time on a fixed-head disk is typically about 10 milliseconds and much faster than access time for a moving-head disk, which is typically 25 to 50 milliseconds.

fixed-length commands
Computer commands having the same length.

fixed-length record
A record having the same length as all other records with which it is logically or physically associated. Contrast with variable-length record.

fixed order quantity
A lot-sizing technique in MRP that will always cause a planned order to be generated for a predetermined fixed quantity (or multiples thereof if net requirements for the period exceed the fixed order quantity).

fixed-partition memory management
A memory management technique in which main memory is subdivided into a number of fixed-length partitions.

fixed-period order
An ordering method where the duration of the time period for ordering is fixed and the quantity needed or ordered over this time period can vary.

fixed point
Refers to an integer number system in which the position of the radix is fixed.

fixed-point arithmetic
A method of calculation in which the computer does not consider the location of the decimal point. Fixed-point arithmetic uses the storage-to-accumulator or the accumulator-to-accumulator concepts that were used by most scientific-oriented second-generation computers. Fixed-point arithmetic operations are performed using fixed-length binary data fields. The results of either the storage-to-accumulator or the accumulator-to-accumulator operation will be stored in either one or two general registers. Addition, subtraction, multiplication, division, and comparison operations take one operand from a register and another from either a register or storage and return the result to the general register.

fixed-point part
In a floating-point representation, the number that is multiplied by the exponential implicit floating-point base to determine the real number represented.

fixed-point representation
A number system in which the position of the decimal point is fixed with respect to one end of the string of numerals, according to some convention.

fixed-radix numeration system
A radix numeration system in which all the digit places, except perhaps the one with the highest weight, have the same radix. The weights of successive digit places are successive integral powers of a single radix, each multiplied by the same factor. Negative integral powers of the radix are used in the representation of fractions.

fixed sequential format
A numerical control format where each word in the format is identified by its position. A means of identifying a word by its location in the block. Words must be presented in a specific order and all possible words preceding the last desired word must be present in the block. An address code is therefore not necessary; however, more characters are generally required per block than with the word address format. The fixed sequential format system has steadily been losing favor to the preferred word address format and may now be considered obsolete.

fixed-stop robot
A robot with stop-point control but no trajectory control. That is, each of its axes has a fixed limit at each end of its stroke and cannot stop except at one or the other of these limits. Such a robot with *N* degrees of freedom can therefore stop at no more than two *N* locations (where location includes position and orientation). Some controllers do offer the capability of program selection of one of several mechanical stops to be used. Often, very good repeatability can be obtained with a fixed-stop robot. Also called a nonservo robot.

fixed-variable budget
A fixed and variable budget segregates those costs that are fixed from those that vary with production volume, and measure fixed costs against a fixed target and variable costs against a target based on the actual production volume. There are normally two ways of classifying costs into fixed and variable categories. The first method is to analyze each account and classify it as either fixed or variable. The second method is to use a statistical technique. This method determines, for each individual type of manufacturing overhead, a fixed portion and variable rate.

fixed word length
Pertaining to a storage device in which the capacity for digits or characters in each unit of data is a fixed length, as opposed to a variable length.

fixture
A device that holds a workpiece in position in a machine tool for machining. The workpiece must be held in a precise position, with no room for slippage, so if the shape is at all complex, a special fixture is usually built to hold the piece. Device used to hold a part such that its reference axes are in a defined orientation with respect to the reference axes of a tool; may or may not be an integral part of a pallet. One challenge in the automation of small-batch manufacturing is the development of

flexible fixturing systems that can adapt to a wide variety of differently sized and shaped workpieces.

fixture serial number
A unique number assigned to a specific fixture.

flag bit
A bit that specifically indicates condition or status to be met by arithmetic operations, for example, carry, overflow, zero, sign, parity. An indicator bit. A single bit (segment) of a memory location, used to detect and remember the occurrence of some event.

flag note entity
An annotation entity that takes label information and formats it so that the text is circumscribed by a flag symbol.

flame annealing
Annealing in which the heat is applied directly by a flame.

flame cutting
The operation of severing ferrous metals by rapid oxidation from a jet of pure oxygen directed at a point heated to the fusion point.

flame-retardant resin
A resin that is compounded with certain chemicals to reduce or eliminate its tendency to burn. For polyethylene and similar resins, chemicals such as antimony trioxide and chlorinated paraffins are used.

flame spraying
Method of applying a plastic coating in which finely powdered fragments of the plastic, together with suitable fluxes, are projected through a cone of flame onto a surface.

flame straightening
Correcting distortion in metal structure by localized heating with a gas flame.

flame treating
A method of rendering inert thermoplastic objects receptive to inks, lacquers, paints, adhesives, and so forth, in which the object is bathed in an open flame to promote oxidation of the surface of the article.

flange focal distance
The distance from the lens mounting flange to the image plane.

flanging
Essentially the same as beading or curling, except the portion of metal rolled over is left in a flat form. The operation is only performed on parts with trimmed edges, since the normal drawing operation will always begin with a flat piece forming the flange, which is later trimmed to size.

flashlamp
A tube filled with krypton or xenon used to produce stimulated emission in a solid-state laser. Often used in the form of a helical coil wrapped around the laser tube.

flash mold
A mold designed to permit excess molding material to escape during closing.

flash point
Temperature at which vapor of a material will ignite when exposed to a flame in a specially designed testing apparatus.

flash radiography
High-speed radiography in which exposure times are sufficiently short to give an unblurred photograph of moving objects, such as fired projectiles or high-speed machinery.

flashtube
A gas discharge tube used for generating very brief, very bright flashes of light, for example, a strobe light.

flash welding
A resistance butt-welding process in which the weld is produced over the entire abutting surface by pressure and heat, the heat being produced by electric arcs between the members.

flatbed plotter
A CAD/CAM peripheral device that draws an image on paper, glass, or film mounted on a flat table. The plotting head provides all the motion.

Flatbed Plotter
Illustration of Flatbed Plotter

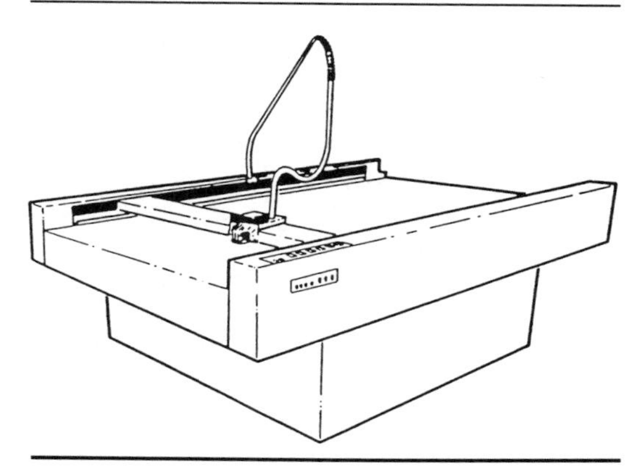

flatness of field
Degree in which the image appears flat; a measure of how well a plane in the object is imaged as a plane.

flat pattern generation
A CAD/CAM capability for automatically unfolding a 3-D design of a sheet metal part into its corresponding flat pattern design. Calculations for material bending and stretching are performed automatically for any specified material. The reverse flat pattern generation package automatically folds a flat pattern design into its 3-D version. Flat pattern generation eliminates major bottlenecks for sheet metal fabricators.

flat screen
A thin video display screen, as opposed to the boxy cathode-ray tubes (CRTs) predominantly in use.

flexible
Manufacturing systems or facilities easily adapted to produce new products; easily redirected or reprogrammed. Pliable or capable of bending. In robot mechanisms, this may be due to joints, links, or transmission elements. Flexibility allows the end point of the robot to sag or deflect under load and to vibrate as a result of acceleration or deceleration. Multipurpose; adaptable; capable of being redirected, trained, or used for new purposes.

flexible automation
Refers to the multitask, multipurpose, adaptable, reprogrammable capability of advanced manufacturing technology systems.

Flexible
Illustration of Flexible Assembly Line. *Courtesy:* General Motors.

Flexible Machining Center
Illustration of Flexible Machining Center. *Courtesy:* General Motors.

flexible budgets
Refer to fixed-variable budgets and step budgets.

flexible machining center
Usually a multirobot system that comprises CNC turning or grinding machines, with robots loading and unloading parts that are conveyed into and through the system.

flexible manufacturing
Production with machines capable of making a different product without retooling or any similar changeover. Flexible manufacturing is usually carried out with numerically controlled machine tools, robots, and conveyors under the control of a central computer.

flexible manufacturing cell
Best suited for applications where a high-variety, low-volume family of parts are manufactured. The FMC can take a number of con-

figurations, but it generally has more than one machine tool with some form of pallet-changing equipment, such as an industrial robot or other specialized material-handling device. In most cases, the grouping of machines is small and often utilizes a common pallet or part-fixturing device for the specific part requirements. The FMC generally has a fixed process, and parts flow sequentially between operations. The cell lacks central computer control with real-time routing, load balancing, and production scheduling logic.

flexible manufacturing system (FMS)
Includes at least three elements: (1) a number of workstations, (2) an automated material-handling system, and (3) system supervisory computer control. The FMS is typically designed to run for long periods, with little or no operator attention. FMS fills the need for machining in a batch environment where equipment dedicated to high-volume production, such as transfer lines, are cost-prohibitive. FMS, unlike the transfer line, can react quickly to product and design changes. Automatic tool changing, in-process inspection, parts washing, automated storage and retrieval systems (AS/RS), and other computer-aided manufacturing (CAM) technologies are often included in the FMS. Central computer control over real-time routing, load balancing, and production scheduling distinguish FMS from FMC.

flexible manufacturing system simulation
A computer simulation used to help in the design or analysis of an FMS, for example, used to determine the relationships between system parameters such as production rate, number of modules, module types, and so forth.

flexible molds
Molds made of rubber or elastomeric plastics used for casting plastics. They can be stretched to remove cured pieces with undercuts.

flexible manufacturing technology
Currently falls into three basic categories: stand-alone machines, the flexible manufacturing cell (FMC), and the flexible manufacturing system (FMS). The use of these three

Flexible Manufacturing System
Block Diagram of a Flexible Manufacturing System. Reprinted with Permission of Reston Publishing Company, Prentice-Hall, from *Robotics and Automated Manufacturing* by Richard Dorf.

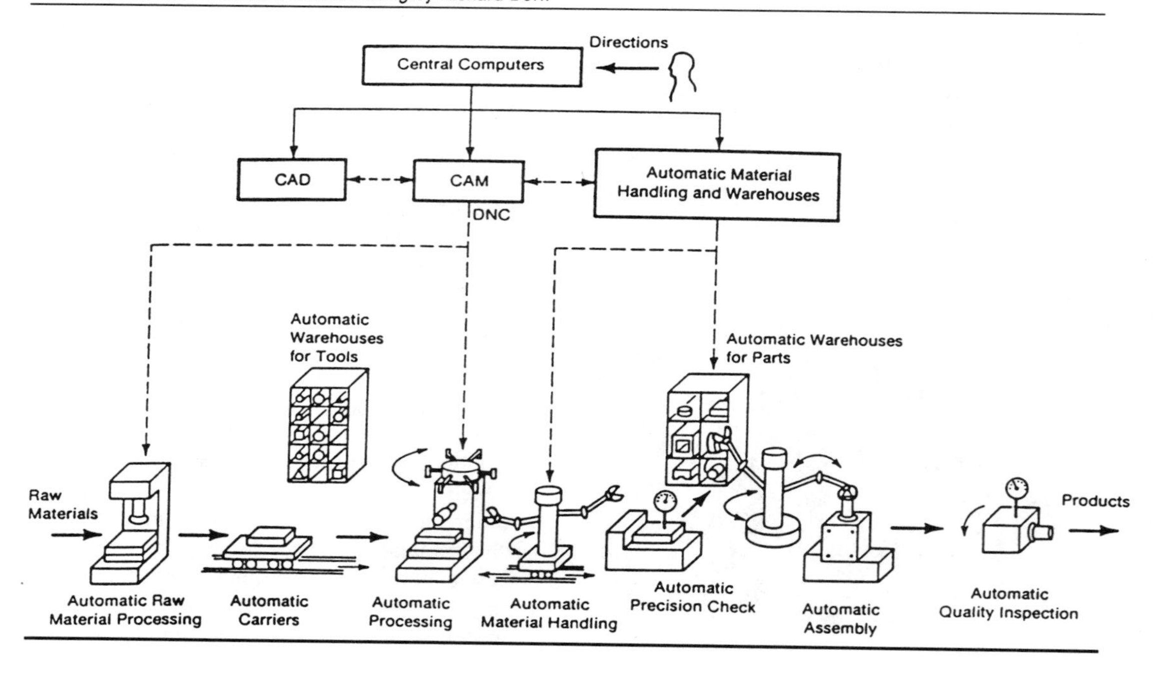

specific levels of flexible manufacturing technology is not intended to imply these are the only configurations in use in industry. However, the term FMS has been used to identify everything from the machining center to the unattended flexible manufacturing system. The earlier FMSs were less flexible in scope and limited in control process and material handling. As more experience with current technology is gained and advances continue, a fourth category, advanced manufacturing facility (AMF), is getting closer, albeit slowly, to a reality. AMF represents the full development of all aspects of computer-integrated manufacturing (CIM), a term often used synonymously with computer-aided manufacturing (CAM).

flexible printed circuit
An arrangement of printed circuit and components utilizing flexible base materials with or without flexible cover layers.

flexibility
The ability of a robot or other advanced manufacturing technology to perform a variety of different tasks.

flexion
Orientation or motion toward a position where the joint angle between two connected bodies is small.

flexivity
Index of change in flexure with temperature of thermostat bimetals. Change of curvature of longitudinal centerline of specimen per unit temperature change per unit thickness.

flicker
An undesired visual effect on a CRT when the refresh rate is low.

flip-flop
A bistable device that may assume a given stable state, depending upon the pulse history of one or more input points, and having one or more output points. The device is capable of storing a bit of information, controlling gates, and so forth.

float
In manufacturing, refers to work-in-process. In CPM, it refers to the extra time before an activity becomes critical.

floating point
Refers to a number system in which the radix varies according to the power of the number base and the value of the mantissa or coefficient. A method of representing a numeric value that contains a decimal point, that is, not necessarily a whole number.

floating-point arithmetic
Floating-point arithmetic saves programming time and makes possible the solutions of complex problems that would otherwise be almost impossible. The basic idea of floating-point arithmetic is that each quantity is represented as a combination of two items: a numerical fraction and a power of 16 by which the fraction is multiplied to get the number represented. Floating-point operations are performed with one operand from a floating-point register or storage, with the result placed in a floating-point register.

floating-point base
In a floating-point representation system, the implicit fixed-position integer base, greater than unit, that is raised to the power explicitly denoted by the exponent in the floating-point representation and then multiplied by the fixed-point part to determine the real number represented.

floating-point processor
A processor utilizing floating-point arithmetic.

floating-point representation
A number representation system in which each number, as represented by a pair of numerals, equals one of those numerals times a power of an implicit fixed-position integer base, where the power is equal to the implicit base raised to the exponent represented by the other numeral.

floating-point system
The more common format for the way a program keeps track of where objects are located. It is a bit more accurate but slower than its alternative, the integer system. Floating point and integer refer to the type of numbers the program uses to store coordinates.

floating zero
A characteristic of a numerical control system that allows the zero reference point (target point) to be established readily by manual adjustment at any position over the full travel of the machine tool. The control system retains no information on the location of any previously established zero. One advantage of a floating zero system is the ability to utilize negative as well as positive coordinate values, thus reducing the part programming time re-

quired on left- and right-hand parts and parts that are symmetrical about centerlines.

floor stock

Inventory issued to the plant in excess of immediate requirements.

floor-to-floor time

The total time elapsed for picking up a part, loading it into a machine, carrying out operations, and unloading it (back to the floor, bin, pallet). Generally applies to batch production.

floppy disk

A flexible disk with a magnetic film on both sides. Digital information is stored on one or both sides of the floppy disks in concentric circles or tracks. The term *floppy* is used because the disk is soft and bends easily. The disk is usually contained in a rigid protective envelope.

flow analysis

Detailed examination of the progressive travel, either of personnel or material, from place to place and/or from operation to operation.

flow brazing

Brazing by pouring molten filler metal over a joint.

flowchart

A systems analysis or programming tool to graphically present a procedure in which symbols are used to designate the logic of how a problem is solved. A flowchart represents the path of data through a problem solution. It defines the major phases of the processing, as well as the various data media used. Flowcharts also enable the designer to conceptualize the procedure necessary and to visualize each step and item on a program. A completed flowchart is often a necessity to the achievement of accurate final code. A program is coded by writing down the successive steps that will cause the computer to perform the necessary logical operation for solving the problem as presented by the flowchart.

flow control

A production control system that is based primarily on setting production rates and feeding work to meet these planned rates. Flow control has its most successful application in repetitive production.

flow diagram

A representation of the location of activities or operations and the flow of materials between activities on a pictorial layout of a process. Usually used with a flow process chart.

flowline

The direction taken either by personnel or material as they progress through the manufacturing or processing sequence of events. The path along which personnel or material travel in progressing through the plant.

flowline system

A system of manufacturing that orients itself to grouping machines of various functions to service the particular needs of machining a part.

flow path

The route taken and/or space occupied by the personnel, material, subassembly, or assembly as these progress through the manufacturing process.

flow shop

A shop in which machines and operators handle a standard, usually uninterrupted, material flow. The operators tend to perform the same operations for each production run. A flow shop is often referred to as a mass-production shop, or is said to have a continuous manufacturing layout. The shop layout (arrangement of machines, benches, assembly lines, and so on) is designed to facilitate a product "flow." The process industries (chemicals, oil, paint, and others) are extreme examples of flow shops. Each product, though variable in material specifications, uses the same flow pattern through the shop. Production is set at a given rate, and the products are generally manufactured in bulk.

fluerics

The area within the field of fluidics in which components and systems perform functions such as sensing, logic, amplification, and control without the use of mechanical parts.

fluidics

That branch of science and technology concerned with sensing, control, information processing, and actuation functions performed through the use of fluid dynamic phenomena.

fluorescence

The emission of characteristic electromagnetic radiation by a substance, as a result of the absorption of electromagnetic or corpuscular radiation having a greater unit energy than that of the fluorescent radiation. It occurs only as long as the stimulus responsible for it is maintained.

fluorescent liquid penetrant
Highly penetrating liquid used in performance of liquid penetrant testing and characterized by its ability to fluoresce under black light.

fluorescent magnetic-particle inspection
A magnetic-particle inspection process employing a powdered ferromagnetic inspection medium coated with material that fluoresces when activated by light of suitable wavelength.

fluoroscopy
An inspection procedure in which the radiographic image of the subject is viewed on a fluorescent screen, normally limited to low-density materials or thin section of metals because of the low light output of the fluorescent screen at safe levels of radiation.

flute
As applied to drills, reamers, and taps, the channels or grooves formed in the body of the tool to provide cutting edges and to permit passage of cutting fluid and chips. As applied to milling cutters and hobs, the chip space between the back of one tooth and face of the following tooth.

flux
In metal refining, a material used to remove undesirable substances, such as sand, ash, or dirt, as a molten mixture. It is also used as a protective covering for certain molten-metal baths. Lime or limestone is generally used to remove sand, as in iron smelting; sand, to remove iron oxide in copper refining. In brazing, cutting, soldering, or welding, material used to prevent the formation of, or to dissolve and facilitate removal of, oxides and other undesirable substances. Also, the amount of light energy per unit area reaching a surface.

flux leakage
Magnetic lines of force that leave and enter the surface of a part due to a discontinuity that forms poles at the surface of the part.

FMILL
The name of a preprocessor program that generates an arbitrary three-dimensional surface that is defined by a sparse array of points comprising the surface.

f-number
The ratio of the focal length of a lens system to the diameter of the entrance pupil.

focal length
The distance from a lens' principal point to the corresponding focal point. Also referred to as the equivalent focal length and the effective focal length.

focal plane
A plane through the focal point at right angles to the principal axis of a lens.

focal point
The point at which light rays refracted by a lens meet. The focal point of a focused laser beam is the point of highest energy concentration. The point at which a lens will focus parallel incident light.

following error
The distance lag between the actual position and command position in a contouring machine at any specific time.

font
A character set in a particular style and size of type, including all alpha characters, numerics, punctuation marks, and special symbols.

font change character
A control character that selects and makes effective a change in the specific shape or size of the graphics for a set of graphemes, the character set remaining unchanged.

font characteristic
An integer that is used to identify a text font. Positive font-characteristic numbers indicate an IGES-defined text font. Negative numbers are interpreted as a text font definition entity.

footcandle
A unit of illumination equal to the illumination that occurs when uniformly distributed luminous flux impinges on an area at a rate of 1 lumen per square foot.

footing
Adding fields of information vertically.

footlambert
A unit of luminance equal to 1 candela per square foot or to the uniform luminance at a perfectly diffusing surface emitting or reflecting light at the rate of 1 lumen per square foot. A lumen per square foot is a unit of incident light, and a footlambert is a unit of emitted or reflected light. For a perfectly reflecting and perfectly diffusing surface (no absorption of light), the number of lumens per square foot is equal to the number of footlamberts.

footprint
The area and shape of the floor space required by the robot and its controller.

force
A push button that can be used to change the state of a disable reference. The reference will be changed to on or on to off every time this push button is depressed.

force feedback
A sensing technique using electrical or hydraulic signals to control a robot end-effector.

force sensor
A sensor capable of measuring the forces and torques exerted by a robot at its wrist. Such sensors usually contain six or more independent sets of strain gages, plus amplifiers. Computer processing (analog or digital) converts the strain readings into three orthogonal torque readings in an arbitrary coordinate system. When mounted in the work surface, rather than the robot's wrist, such a sensor is often called a pedestal sensor. The force and torque acting on each point of a manipulator can be sensed directly. If the joint is driven by an electric DC motor, then sensing is done by measuring the armature current; if the joint is driven by a hydraulic motor, then sensing is done by measuring the back pressure. Sensing joint forces directly has the advantage of not requiring a separate force sensor. However, the force (or torque) between the hand and its environment is not measured directly. Thus, the accuracy and resolution of this measurement are adversely affected by the variability in the inertia of the arm and its load and by the nonuniform friction of the individual joints.

forearm
That portion of a jointed arm that is connected to the wrist.

forecast
A prediction of future events, using an algorithm that references historical data.

forecast error
The difference between actual demand and forecast demand, typically stated as an absolute value.

forecast horizon
The end of the period of time into the future for which a forecast is prepared.

forecast period
The time unit for which forecasts are prepared, such as monthly, weekly, or quarterly.

foreground
In multiprogramming, refers to the environment in which high-priority programs are executed.

foreground job
A high-priority job, usually a real-time job. A teleprocessing or graphic display job that has an indefinite running time, during which communication is established with one or more users at local or remote terminals.

foreground processing
High-priority processing, usually for real-time activities, automatically given precedence, by means of interrupts, over lower-priority "background" processing.

foreign element
An interruption that is not a regular occurrence in the work cycle or operation, and one for which no provision was made in the normal sequence of elements of a time study.

foreign exchange (FX) line
A line offered by a common carrier in which the user is assigned a telephone number belonging to a remote location to minimize long-distance charges.

forge welding
Welding hot metal by pressure or blows only.

forging
The operation of forming relatively thick sections of metallic material into a desired shape. Material may be worked cold or hot, between shaped forms or shaped rolls. In the case of roll dies, the material is squeezed into shape. In the case of shaped dies, the material is formed by a succession of blows, forcing the material into the shaped cavity, or the material may be squeezed into the shaped cavity by a steady pressure of the moving form on the metal.

formal symptom
A symptom that has been subjected to validation procedures and formally identified as such.

FORMAT
A contraction meaning the FORM of MATerial, designating the predetermined arrangement of characters of data for input/output.

form feed
A format-effector character that causes the print or display position to move to the next predetermined first line on the next form, the next page, or the equivalent.

form feed key
A printer control key that advances the paper in the printer to the top of the next page.

forming
Making a change, with the exception of shearing or blanking, in the shape or contour of a metal part without intentionally altering the thickness.

form number
An integer that is used to further define a type of entity when there are several interpretations of an entity type. For example, the form number of the conic arc entity indicates whether the curve is an ellipse, hyperbola, parabola, or unspecified. The form number is also used to supply information in an entity's directory entry for decoding the parameters in its parameter data entry.

formulation
A listing of all the components, including equipment and/or manpower resources, that are used to produce a parent product. Also shown is the quantity of each component required to make one unit of the parent product.

for/next loop
A looping instruction in BASIC, used to repeat segments a number of times without rewriting the program. Instructions are placed between the FOR and NEXT commands.

FORTH
A flexible programming language for control applications, used widely in industrial and now in personal computers. FORTH builds the language around the application, enabling added words to become commands in the language.

FORTRAN
Stands for FORmula TRANslation and is probably the most popular computer program for scientifically oriented problems. Because of its universal nature, statements in this language will be acceptable to practically all scientific-type computers. FORTRAN is machine-independent and the programmer, therefore, need not know the details of how the computer operates. FORTRAN is used in handling the APT system.

FORTRAN-S
Computervision's version of FORTRAN. High-level programming language, used primarily for scientific or engineering applications.

forward-chaining
An artificial intelligence control procedure that produces new decisions recursively by affirming the consequent propositions associated within an inferential rule with antecedent conditions that are currently believed. As new affirmed propositions change the current set of beliefs, additional rules are applied recursively.

forward scheduling
Scheduling a process from a start date forward by operation time to arrive at a finish or due date.

foundry
A commercial establishment or building where metal castings are produced.

four-address
Pertaining to an instruction format containing four address parts.

Fourier processing
A mathematical method of representing an image as a series of superimposed sinusoidal functions.

Fourier transform
A technique to convert data from the space or time domain to the frequency domain. It represents data as a series of sinusoidal waves.

four-plus-one address
Pertaining to an instruction that contains four operand addresses and the address of the next instruction to be executed.

four-wire channel
Provision of two-wire pairs (or logical equivalent) for simultaneous two-way transmission.

four-wire circuits
Indicates the capability of the switching system to accommodate connections to special four-wire circuits.

fractals
A type of mathematical description that features noninteger dimensions. They are widely used in sophisticated computer graphics, especially for describing natural, irregular objects such as clouds or landscape surfaces.

fracture stress
The maximum principal true stress at fracture; usually refers to unnotched tensile specimens. The (hypothetical) true stress that will cause fracture without further deformation at any given strain.

fracturing
The division of IC graphics by CAD into simple trapezoidal or rectangular areas for pattern-generation purposes.

frame
The total area, occupied by a television image, which is scanned while the video signal is not blanked. A data structure for representing stereotyped objects or situations. A frame has slots to be filled for objects and relations appropriate to the situation. A data structure for grouping information on a whole situation, complex object, or series of events. A knowledge representation scheme that associates one or more features with an object in terms of various slots and particular slot values. Similar to property list, schema, unit, and record.

frame buffer
An electronic device capable of storing a digitized image in a digital memory for later readout and processing. A separate memory or area of main memory used for storing a complete video image, used to allow the computer to build up an image in steps and then quickly switch to display it on screen. Without a buffer, the construction process would be visible and probably very distracting.

frame grabber
A device that interfaces with a camera and, on command, samples the video, converts the sample to a digital number, and stores the number in a computer's memory.

frame rate
The number of complete images presented per second in a video display system.

frames
An artificial intelligence technique for representing knowledge that stores a list of an object's typical attributes together with the object. Each attribute is stored in a separate slot.

framing bits
In data transmission, noninformation-carrying bits used to make possible the separation of characters in a bit stream. Synonymous with synch bits.

free fit
Various clearance fits for assembly by hand and free rotation of parts.

free machining
Pertains to the machining characteristics of an alloy to which an ingredient has been introduced to give small broken chips, lower power consumption, better surface finish, and longer tool life; among such additions are sulfur or lead to steel, lead to brass, lead and bismuth to aluminum, sulfur or selenium to stainless steel.

freezing point
Temperature at which a material solidifies on cooling from molten state under equilibrium conditions.

frequency
The number of times a given event occurs within a specified period. It most commonly refers to the number of pulses per second occurring in various electronics devices. The standard unit of measure is Hz, for cycles per second.

frequency division multiplexing
Division of the available transmission frequency range into narrower bands, each of which is used for a separate channel.

frequency modulation
A method of information transmission whereby the frequency of the carrier wave is modulated to correspond to changes in the signal wave.

frequency response
The output of a system with a periodic input. Frequency response may be defined in terms of the Fourier coefficients or the gain and phase at each multiple of the period. The characterization of system output to a continuous spectral input, according to a continuous plot of gain and phase as a function of frequency.

fresnel lens
A lens composed of discrete concentric circles.

fretting (fretting corrosion)
Action that results in surface damage, especially in a corrosive environment, when there is relative motion between solid surfaces in contact under pressure.

friction
The resistive forces resulting from two bodies sliding relative to one another.

friction feed
A paper-feeding mechanism on printers that employs rollers that hold the paper against the platen, much like a typewriter.

from-to
A pair of points between which an electrical or piping connection is made. Hard-copy

from-to lists can be generated automatically on a CAD/CAM system. In computer-aided piping, from-to reports are generated automatically to describe the origins and destinations of pipeline.

front clipping plane
A plane parallel to the view plane which is specified by a distance of viewing coordinate space along the view plane normal from the view reference point. When front clipping is enabled, portions of objects in front of the front clipping plane are discarded.

front end processor
A dedicated communications computer at the "front end" of a host computer. It may perform line control, message handling, code conversion, error control, and applications functions such as control and operation of special-purpose terminals.

front lighting
The use of illumination in front of an object so that surface features can be observed. Illumination arranged so that light reaching the image sensor is reflected off the objects in the scene.

front plane clipping
Front plane clipping is defined as clipping against that part of the object that lies within and on the view volume, behind and on the front clipping plane. Front plane clipping is independent of back plane clipping.

front porch
That portion of a composite video signal that lies between the leading edge of the horizontal blanking pulse and the leading edge of the corresponding sync pulse.

f-stop
The setting of a mechanism that varies the effective f-number of a lens.

full adder
A combinational circuit that has three inputs that are an augend D, and addend E, and a carry digit transferred from another digit place F, and two outputs that are a sum without carry T and a new carry digit R, and in which the outputs are related to the inputs.

full annealing
Annealing a ferrous alloy by austenitizing and then cooling slowly through the transformation range.

full-duplex
Pertains to a communications circuit that permits two-way simultaneous transmission. Contrasts with half-duplex, which permits alternate or one-way-at-a-time communications.

full graphics
Another name for all points addressable or dot graphics. Unlike character graphics (which are built up out of letter-sized images), in full graphics mode, each dot can be controlled.

full pegging
Refers to a process where pegging occurs all the way from the start level to the end item (parent) of the part or the customer order being traced.

full subtractor
A combinational circuit that has three inputs that are a minuend L, a subtrahend J, and a borrow digit K transferred from another digit place, and two outputs that are a difference W and a new borrow digit X, and in which the outputs are related to the inputs.

full text data base
This type of source data base contains complete textual records of primary sources. These sources include newspapers, specifications, court decisions, journals, and others.

full-travel
A switch travel that has both a pronounced pretravel and overtravel.

fully connected network
A network in which each node is directly connected with every other node.

function
A mathematical entity whose value, that is, the value of the dependent variable, depends in a specified manner on the values of one or more independent variables, not more than one value of the dependent variable corresponding to each permissible combination of values from the respective ranges of the independent variables. A specific purpose of an entity or its characteristic action.

functional application
The generic task or function performed in an application.

functional design
The specification of interrelationships between the parts of a system.

functional diagram
A diagram of the functional relationships between the parts of a system.

functionality
Refers to a set of system capabilities in terms of the functions they provide.

functional-level information systems
A MIS design that normally parallels organizational boundaries. Its vertical orientation may restrict system modularity, particularly when data needs to be shared across functional organization lines.

functional system
A way of orienting the manufacturing process to groupings of machine tools by the same function to facilitate materials handling to take advantage of computer control.

functional systems design
The development and definition of the business functions to be accomplished by a computer system, that is, the work of preparing a statement of the data input, data manipulation, and information output of proposed computer system in common business terms that can be reviewed, understood, and approved by a user organization. This statement, after approval, provides the basis for the computer systems design.

function board
A keyboard that allows for graphic data entry of various functions; for example, pressing a LINE program button will allow the input of a line onto the CRT.

function keys
Separate keys on a keyboard that do not produce characters but execute commands. There are two kinds: (1) fixed or (2) programmable (soft keys).

function of network manager
Actions actually performed by the MAP network manager.

function table
Two or more sets of data so arranged that an entry in one set selects one or more entries in the remaining sets—for example, a tabulation of the values of a function for a set of values of the variable; a dictionary.

fundamental
The lowest tone in a complex sound. The others are called harmonics.

fuse
In plastisol molding, to heat the plastisol to the temperature at which it becomes a single homogeneous phase. In this sense, cure is the same as fuse.

fusing
An operation that causes separate blobs to fuse into one larger blob.

fusion welding
Used to join metals by melting and fusing them at high temperatures, using an electric arc or combustible gases.

fuzzy set
A generalization of set theory that allows for various degrees of set membership, rather than all or none.

G

gage
A mechanical device of high precision, used for checking a part for dimensional conformance to specifications.

gain
The amount of increase in a signal as it passes through a control system or a control element. The gain in a control system would refer to its sensitivity and its ability to raise the power of a signal to a required output.

galling
Developing a condition on the rubbing surface of one or both mating parts, where excessive friction between high spots results in localized welding with subsequent spalling and a further roughening of the surface.

gamma
A numerical expression of the transfer function of a system. It is the exponent of that power law that is used to approximate the relationship of the output magnitude versus input magnitude.

gamma correction
A modification of system response to provide for linear transfer characteristic from the input to output. The correction factor for a vidicon is the reciprocal of the 2.2 root.

gangpunching
The automatic punching of data read from a master card into the following detail cards.

gantry robot
A bridgelike frame along which a suspended robot moves. A gantry creates a much larger work envelope than the robot would have if it were pedestal mounted.

Gantt chart
The earliest and best-known type of control chart designed to show graphically the relationship between planned performance and actual performance. Named after its originator, Henry L. Gantt, scientific management pioneer. Used especially for machine loading, where one horizontal line is used to represent load against capacity, or for following job progress where one horizontal line represents the production schedule and another parallel line represents the actual progress of the job against the schedule in time.

Gantry Robot
GMF Model G-200 Gantry Robot. *Courtesy:* GMF Robotics Corp.

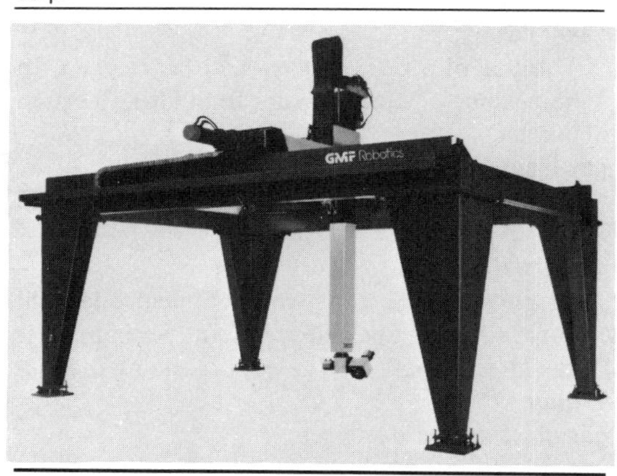

Gantt task and bonus plan
A wage incentive plan that originally provided a low guaranteed rate but offered a strong inducement to meet or better standard performance by paying a step bonus when standard performance was reached and the equivalent of a piece rate thereafter.

gap
Between two entities on a computer-aided design, the length of the shortest line segment that can be drawn from the boundary of one entity to the other without intersecting the boundary of the other. CAD/CAM design-rules checking programs can automatically perform gap checks.

gap character
A character that is included in a computer word for technical reasons but does not represent data.

garbage
Unwanted or meaningless information being stored in a file or used in a process.

garbage collection
A technique for recycling computer memory cells no longer in use.

gas holes
Round or elongated, smooth-edged dark spots that may occur individually, in clusters, or dis-

tributed throughout a casting section. Gas holes are usually caused by trapped air or mold gases.

gas jet assist
The use of a coaxial gas, such as oxygen, to help achieve high powers for cutting certain metals.

gas laser
A laser that uses a gas or a gas mixture as the active medium.

gas panel
Display device consisting of electrode grid encased in a panel filled with an ionizing gas; gas glows where energized, producing the display.

gas-shielded arc welding
Arc welding in which the arc and molten metal are shielded from the atmosphere by a stream of gas, such as argon, helium, argon-hydrogen mixtures, or carbon dioxide.

gate
A circuit having one output and one or a number of inputs. It is designed so that the output is energized when certain input conditions are met. There are three types of circuit gates: (1) AND gates, (2) OR gates, and (3) NOR gates.

gate array
An integrated circuit characterized by a rectangular array of logic sites. These sites consist of identical collections of diffused or implanted transistors, diodes, and resistors. Surrounding the array are the input/output circuits for off-chips connections.

gate array design
A shortcut method of creating an IC simply by customizing the interconnections to a partially fabricated IC having predefined logic sites.

gateway
Connects different network architectures by performing protocol conversion and may use all seven layers of the OSI mode. The architecture of a gateway will depend on the architecture of the proprietary network to which it is connected. Since a gateway connects networks with different address structures, it will have a different network address on each attached network. Gateways are normally used for connecting OSI networks such as TOP and MAP networks to non-OSI networks.

gateway work center
The work center where the first operation or actual direct labor first occurs on a part.

gauging
The process of measuring a dimension or group of dimensions according to some standard.

Gaussian
The normal, or bell-shaped, curve that represents the normal distribution of a large number of possible events. The Gaussian curve represents the shape of the profile of a laser-drilled hole.

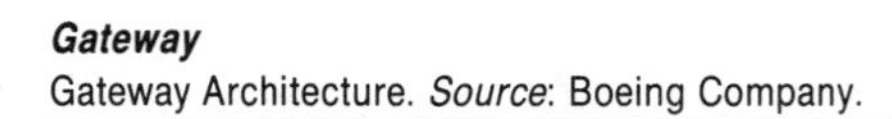

Gateway
Gateway Architecture. *Source*: Boeing Company.

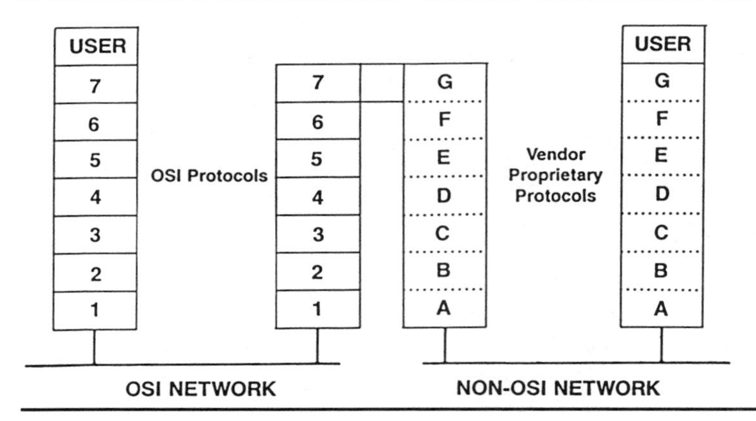

Gaussian filtering
A convolution procedure in which the weighting of pixels in the template fall off with distance, according to a Gaussian distribution.

gel point
Stage at which a liquid begins to exhibit pseudoelastic properties. Gel-point stage may be conveniently observed from inflection point on a viscosity-time plot.

generalized cone
A volumetric model defined by a space curve, called the spine or axis, and a planar cross section normal to the axis. A *sweeping rule* describes how the cross section changes along the axis.

generalized drawing primitive
An output primitive used to access implementation-dependent and workstation-dependent geometrical capabilities, such as curve drawing.

generalized ribbon
A planar region approximated by a medial line (axis) and the perpendicular distances to the boundary. The 2-D version of a generalized cone.

general label entity
An annotation entity consisting of a general note with one or more associated leaders.

general note entity
An annotation that consists of text that is to be displayed in some specific size and at some specific location and orientation.

general processor
The program containing the basic intelligence for NC work that is stored in computer memory before individual workpiece programs can be processed.

general-purpose computer
A computer designed to solve a large variety of problems, for example, a stored-program computer that may be adapted to any of a very large class of applications.

general-purpose vision system
A vision system that is universally applicable. A system that is based on generic rather than specific knowledge. A system that can deal with unfamiliar or unexpected input.

generate
To produce a program from skeletal coding, under the control of parameters.

generate and test
A common form of state space search based on reasoning by elimination. The system generates possible solutions, and the tester prunes those solutions that fail to meet appropriate criteria.

generating function
Pertaining to a given series of functions or constants, a mathematical function that, when represented by an infinite series, has those functions or constants as coefficients in the series.

generative process planning
A process planning function that synthesizes a plan for a part by reasoning from first principles.

generator
A computer program that constructs other programs to perform a particular type of operation—for example, a report program generator.

generatrix
The defining curve that is to be swept to generate a tabulated cylinder or revolved to generate a surface of revolution.

GENESIS
A two-axis contouring and simultaneous third-axis linear capability processor language for all types of NC/CNC maching tools. GENESIS features many macros that provide automatic cycles that minimize the programmer's time and eliminate repetitive calculations.

genlock
A capability to syncronize the internal sync generator of some part of a camera system to an external source.

geometric
Having to do with the shape information (points, curves, surfaces, and volumes) necessary to represent some object.

geometric configuration
The group of features that describe the shape of an image, such as area, centroid, perimeter, dimensions, and so on.

geometric design rules
User-defined design constraints aimed at ensuring a functional, reliable product.

geometric distortion
Any aberration that causes the reproduced image to be geometrically dissimilar to the perspective plane projection of the original scene.

geometric error
A difference between the true position of a machine carriage and its position as measured by the machine feedback element while the

machine is in a quasistatic state. Position is used here to denote either linear or angular position.

geometric graphic primitive
A locally stored picture-drawing algorithm that can be called by means of a specified code and associated operand.

geometric modeling
Mathematically specifying a part or product by its geometric form or properties.

geometric processing
The process of taking measurements that are characteristic for the geometry of certain objects in an image. Examples are: area (size), coordinates of center of gravity, perimeter, number and location of holes.

geometry modifier
A function used to alter a drawing on the CRT screen. For example, shortening the length of a line is considered a geometry modifier or manipulator.

GETURN
General Electric Company's Information Services Division offers GETURN for the programming of lathe parts. It was developed with the cooperation of the TNO Metaalinstituut of the Netherlands and has been operational in Europe under the name MITURN since 1970.

g function
A preparatory code on a program tape to indicate a special function. The "g" is actually a lowercase letter when shown on a printout.

GKS
Stands for Graphics Kernal System, an international standard for computer graphics presentation devices. The standard provides a logical interface between graphics programs and the characteristics and capabilities of actual devices. The standard is better defined but less comprehensive than the competing Core standard. The GKS standard adopted in the United States (by the American National Standards Institute, or ANSI, a voluntary standards body) allows for a "minimal" GKS, with less features and better suited for lower resolution devices, such as the PC. The International Standards Organization (ISO) requires exact compliance.

GKS metafile
Any type of sequential file that can be written or read by Graphics Kernal System and used for long-term storage (and transmittal and transferral) of graphical information.

glass laser
A solid-state laser in which the active medium is a glass rod doped with rare-earth atoms.

glassy transition
Change in an amorphous polymer or amorphous regions of partially crystalline polymer from viscous or rubbery condition to a hard and relatively brittle one.

glitch
A hardware malfunction, as opposed to a software error, which is called a bug.

global
Attribute of an operation that is uniformly applied to an entire image.

global address
A unique address for each communicating entity among all communicating entities that are interfaced.

global data base
Complete data base describing the specific problem, its status, and that of the solution process.

global logical data organization
Concerned with the overall organization for the data base from which multiple file organizations may be derived. It is a logical view of the data, entirely independent of the physical storage organization. It is described in a data description language that is part of the data base management software.

global method
A method based on nonlocal aspects, for example, region splitting by thresholding based on an image histogram.

global modeling transformation
The composite modeling transformation of the parent structure. Composed with the local modeling transformation during structure transversal to form the current composite modeling transformation.

global name
A unique name for a specific communicating entity among all communicating entities, regardless of its global address or location.

global operation
Transformation of the gray-scale value of the picture elements, according to the gray-scale values of all elements of the picture.

global section
The section of an IGES file consisting of general information describing the file, the file

generator (preprocessor), and information needed by the file reader (postprocessor).

gloss
Ratio of light reflected from surface of a material to light incident on surface when angles of incidence and reflectance are equal numerically but opposite in sign.

goal-driven
A problem-solving approach that works backward from the goal.

goal regression
A technique for constructing a plan by solving one conjunctive subgoal at a time, checking to see that each solution does not interfere with the other subgoals that have already been achieved. If interferences occur, the offending subgoal is moved to an earlier noninterfering point in the sequence of subgoal accomplishments.

GOTO
A branch instruction in high-level language.

GPUNCH
GPUNCH is designed to instruct users step-by-step in the programming process for turret punch presses. The overall system is structured on the assumption that tool setup in the turret is generally standard; however, a command by way of a remote terminal on the shop floor will cause the program to wait for changes.

graceful degradation
Decline in performance of some component part of a system, without immediate and significant decline in performance of the system as a whole and/or decline in the quality of the product.

graceful failure
Failure in performance of some component part of a system, without immediate major interruption or failure of performance of the system as a whole and/or sacrifice in quality of the product.

gradient
The rate of change of pixel intensities over a small local interval. A vector indicating the change of gray-scale values in a certain neighborhood of a pixel.

gradient space
A coordinate system in which p and q are the rates of change in depth (gray value) of the surface of an object in the scene along the x and y directions (the coordinates in the image plane).

gradient vector
The orientation and magnitude of the rate of change in intensity at a point in the image.

grain
An individual crystal in a polycrystalline metal or alloy.

grain growth
An increase in the size of grains in polycrystalline metal, usually effected during heating at elevated temperatures. The increase may be gradual or abrupt, resulting in either uniform or nonuniform grains after growth has ceased. A mixture of nonuniform grains is sometimes termed duplexed. Abnormal grain growth (exaggerated grain growth) implies the formation of excessively large grains, uniform or nonuniform.

grain size
For metals, a measure of the areas or volumes of grains in a polycrystalline material, usually expressed as an average when the individual sizes are fairly uniform. Grain sizes are reported in terms of number of grains per unit area or volume, average diameter, or as a grain-size number derived from area measurements.

grammar
In artificial intelligence, a scheme for specifying the sentences allowed in a language, indicating the syntactic rules for combining words into well-formed phrases and clauses.

granular fracture
A type of irregular surface produced when metal is broken, characterized by a rough grainlike appearance as differentiated from a smooth silky, or fibrous, type. It can be subclassified into transgranular and intergranular forms. This type of fracture is frequently called crystalline fracture, but the inference that the metal has crystallized is not justified.

granulation
The production of coarse metal particles by pouring the molten metal through a screen into water or by agitating the molten metal violently during its solidification.

graph
An image representation in which nodes represent regions and arcs between nodes represent properties of and relations between these regions.

graphic
A symbol produced by a process such as handwriting, drawing, or printing.

graphical elements
Elements that describe images.

graphical numerical control
A part programming system using interactive graphics, mills, lathes, flame cutters, drills, and large machining centers. Graphical numerical control provides effective tape generation by providing graphic displays of the part, the tool path, and the tools themselves.

graphic character
One of the printing characters (those that actually produce an image by themselves) plus the space character. In the ASCII character set, these are the 95 characters that come after the 32 control characters. Normally, graphic character sets include the capital Roman letters, decimal digits, and common symbols. To this may be added lowercase letters and special symbols, plus non-Roman alphabets.

graphic character repertoire
The list of graphic characters defined in the standard, including accented letters and characters obtained by the composition of two or more graphic symbols.

graphic display station
The unit that displays the image or drawing. The most popular graphic display stations are CRTs.

graphic library
A collection of standard, often-used symbols, components, shapes, or parts stored in the CAD data base as templates or building blocks to speed up future design work on the system. Generally, an organization of files under a common library name.

graphic primitives
The set of graphic operations that are done by the system or hardware and that can be called upon by a graphics program or graphic routines in other programs.

graphics data
The collection of data required to produce a picture.

graphics device
A device (e.g., refresh display, storage tube display or plotter) on which display images can be represented.

graphics terminal
A CRT terminal capable of displaying user-programmed graphics.

graphic system
A system that collects, uses, and presents information in pictorial form.

graphic tablet
An input device having a flat surface on which the work is done. A stylus or a puck is used for the graphical data entry. Information is transmitted to the CRT by means of an electrically controlled grid beneath the tablet's surface. A CAD/CAM input device that enables graphic and location instructions to be entered into the system, using an electronic pen on the tablet.

GRAPL
Acronym for Graphic Interactive Programming Language. GRAPL is an APT syntax language that allows for the definition of geometry, drafting entities, and interactive sequences coupled with a FORTRAN-like processing capability.

Grasp
The basic element employed when the predominant purpose is to secure sufficient control of one or more objects with the fingers or the hand to permit the performance of the next required basic element.

gravity feed
A method of supplying materials into a machine or to a workstation by the force of gravity.

gray code
A binary code in which sequential numbers are represented by binary expressions, each of which differs from the preceding expression in one place only.

gray level
A quantized measurement of image irradiance (brightness) or other pixel property.

gray scale
The range of gray levels used in the image. Also denotes a true portrayal of the image, that is, black appears as black and white appears as white.

gray-scale image
An image consisting of an array of pixels that can have more than two values. Typically, up to 16 levels are possible for each pixel.

gray-scale picture
A digitized image in which the brightness of the pixels can have more than two values, typically 128 or 256. It requires more storage space and much more sophisticated image processing than a binary image but offers potential for improved visual sensing. A video picture with many shades of brightness. In CAD

Gray-Scale Image
GMF True Gray-Scale Vision System in Assembly Operation.
Courtesy: GMF Robotics Corp.

systems with a monochromatic display, variations in brightness level (gray scale) are employed to enhance the contrast among various design elements. This feature is very useful in helping the designer discriminate among complex entities on different layers, displayed concurrently on the CRT.

grid
A network of uniformly spaced points or crosshatch optionally displayed on the CRT and used for exactly locating and digitizing a position, inputting components to assist in the creation of a design layout, or constructing precise angles. For example, the coordinate data supplied by digitizers is automatically calculated by the CPU from the closest grid point. The grid determines the minimum accuracy with which design entities are described or connected. In the mapping environment, a grid is used to describe the distribution network of utility resources. A regular pattern of intersecting lines (or dots set along imaginary lines), usually at right angles, that defines intersection points and divides the screen or drawing space into a number of small regions. The grid may be used merely as a visual aid during drawing, or it may become a part of the image. The set of (ui, vj) where ui and vj are the breakpoints on the u and v coordinates respectively used to specify a parametric spline or rational B-spline surface. The term grid is also applied to the projected image on the spline surface.

grinding
The operation of separating material from the surface of a solid by means of abrasion. When restricted to flat materials, the term is more generally known as surface grinding. Abrasion may be by means of an abrasive wheel or belt. When performed by a belt, it is more commonly known as abrasive belt grinding.

gripper
An actuator, gripper, or mechanical device attached to the wrist of a manipulator by which objects can be grasped or otherwise acted upon. (*Illus. p. 174.*)

grit size
Nominal size of abrasive particles in a grinding wheel corresponding to the number of openings per linear inch in a screen through which the particles can just pass. Sometimes called grain size.

groove
Geometric feature that results in a deviation of the silhouette of a part from the silhouette of its envelope on the side (not the corner as with a step or chamfer) of the silhouette of the envelope. A combination of a groove and a step or chamfer will be considered as a groove if the step or chamfer is too small.

ground plane
A conductor layer or portion of a conductor layer (usually a continuous sheet of metal with suitable clearances) used as a common reference point for circuit returns, shieldings, or heat sinking.

group
Cluster of primitives that can be modified individually.

group associativity
A predefined associativity for forming any collection of entities.

group classification code
A part of material classification technique that provides for designation of characteristics by successively lower order groups of code. Classification may denote, for example, function, type of material, size, and shape.

Gripper
Typical Robot Grippers

group indication
Printed information identifying a group of data.

groupmark
A mark that identifies the beginning or the end of a set of data that may include blocks, characters, or other items.

group processing
Usually, a setup whereby several users jointly use, rent, or own data-processing equipment. In some instances, a group of concerns form a separate data-processing organization to make it economically feasible to utilize EDP techniques.

group technology
A system for coding parts, based on similarities in geometrical shape or other characteristics of the parts. The grouping of parts into families, based on similarities in their production so that the parts of a particular family could then be processed together. The means of coding parts on the basis of similarities of the parts. The grouping of parts into production families, based on similarities in their production so that the parts of a particular "family" could then be processed together. The grouping of diverse machines together to produce a particular family of parts.

group timing technique
A work measurement procedure for multiple activities that enables one observer using a stopwatch to make a detailed element time study on from 2 to 15 men or machines at the same time.

G set
One of the four sets, G0, G1, G2, or G3. Each set consists of 94 or 96 character positions arranged in six columns of 16.

G-set repertory
The collection of available code sets that may be designated as one of the G sets.

guard
A physical means of separating persons from danger from robots or hazardous machines.

guide pins
Devices that maintain proper alignment of the force plug and cavity as the mold closes.

H

hacker
Person devoted to intricate computer programming, particularly that programming done for its own sake. A good hacker is an expert programmer. A bad hacker is a poor programmer.

half-adder
A combinational circuit that has two inputs, *A* and *B*, and two outputs, one being a sum without carry *S* and the other being a carry *C*, and in which the outputs are related to the inputs.

half-adjust
To round by one-half of the maximum value of the number base of the counter.

half-bridge robot
A half-bridge robot is a Cartesian robot in which there is a north-south axis and an up-down axis but no east-west axis.

half-duplex
A circuit designed for transmission alternately in either direction but not both directions simultaneously. Contrast with duplex.

half effect
Deflection by a magnetic field of an electric current traveling through a thin film. Force experienced by the current is perpendicular to both magnetic field and direction of current flow.

half-subtractor
A combinational circuit that has two inputs that are a minuend *G* and a subtrahend *H* and two outputs that are a difference *U* and a borrow digit *V*, and in which the outputs are related to the inputs.

halfword
A continuous sequence of bits or characters that comprise one-half a computer word; it is capable of being addressed as a unit.

halo
The appearance of a black border around exceptionally bright objects in an image.

hand
A clamp or gripper used to grasp objects that is attached to the end of the manipulator arm of an industrial robot.

handle point
Where the cursor grabs on to an object or group to move it.

handler
A program with the sole function of controlling a particular input, output, or storage device, a file, or the interruption facility.

handling time
The time required to perform the manual portion of an operation. The time required to move materials or parts to or from a workstation.

hand mold
A mold taken out of the press after each shot for part removal.

handshaking
A preliminary exchange of predetermined signals performed by modems and/or terminals and computers to verify that communication has been established and can proceed.

hard automation
A production technique where equipment has been specifically engineered for a unique manufacturing sequence. "Hard" automation implies programming with hardware in contrast to "soft" automation that uses software or computer programming.

hard copy
Machine output in a permanent visually readable form, for example, printed reports, listings, documents, and summaries. The term has gained greater significance compared to magnetic records that cannot be read by humans and require computer processing for conversion to printed records or reports or CRT displays that are transient.

hard disk
Rigid random-access, high-capacity magnetic storage medium. Disks may be removable (cartridges), providing offline archival storage, or nonremovable. Capacities range from 5Mb to well over 300Mb (250 to 750,000 pages) per disk (calculated at 4000 characters per single-spaced page).

hardener
An alloy, rich in one or more alloying elements, added to a melt to permit closer composition control than possible by addition of pure metals or to introduce refractory elements not readily alloyed with the base metal.

hardening
The process of heating metal and then allowing it to cool through self-quenching so that the metal becomes stronger.

hardness
Resistance of metal to plastic deformation, usually by indentation. However, the term may also refer to stiffness or temper or to resistance to scratching, abrasion, or cutting. Indentation hardness may be measured by various hardness tests, such as Brinell, Rockwell, and Vickers. For grinding wheels, same as grade. A measure of the resistance of a material to localized plastic deformation.

hard sectored
A term used to describe a particular diskette format and a way of recording information on the diskette. Hard-sectored diskettes employ a single index hole placed between any two of 32 equidistant sector holes. The index hole is used to designate the beginning of the disk; the sector holes designate the location of the information on each disk. Since hard-sectored diskettes are not preformatted, they have more potential storage capacity than the soft-sectored variety, employing up to 300K bytes out of a possible 400K for text storage.

hardware
The mechanical, electrical and electronic, pneumatic, hydraulic devices that compose a computer, controller, robot, or other advanced manufacturing technology systems.

hardware alphanumerics
An approach to machine vision, based on a series of ordered processing levels in which the degree of abstraction increases from the image level to the interpretation level.

hardware compensation
A specific example of error compensation in which the map of the errors is stored in a mechanical device.

hardware debugging
Process of finding and fixing malfunctioning electronic equipment, particularly digital equipment.

hardware independence
Computer software not dependent or constrained by a specific make of computer.

hard-wired
A processing system employing wired circuitry to implement system functions. Such equipment is generally cheaper than software programmed systems; it is also less flexible.

hard-wired link
A technique of physically connecting two systems by fixed circuit interconnections using digital signals. Hard-wired memories in early robots employed a wire matrix to register voltages from feedback potentiometers.

harmonic distortion
Nonlinear distortion of a system or transducer, characterized by the appearance in the output of harmonics other than the fundamental component when the input wave is sinusoidal.

harmonic smoothing
An approach to forecasting based upon fitting some set of sine and cosine functions to the historical pattern of a time series.

hash total
The sum of numbers in a specified field of a record or batch of records used for checking or control purposes. The total may be insignificant except for audit purposes, as in the case of part numbers or customer numbers. An arithmetic total of data used for checking the accuracy of one or more corresponding fields of a file such as job numbers, invoicing serial numbers, and so forth, that ordinarily would not be summed, to see that all transactions have been processed. When the hash totals agree, the data are considered verified.

hatch
One possible representation of the interior of a polygon primitive. The interior of the image is filled with a pattern of parallel and/or crossing hatch lines.

hatch table
A workstation-dependent table of possible hatch values.

hazard
A condition or changing set of circumstances that presents a potential for injury, illness, or property damage. The potential or inherent characteristics of an activity, condition, or circumstance that can produce adverse or harmful consequences.

hazardous motion
Unintended or unexpected robot motion that may cause injury.

haze
A measure of the extent to which light is diffused in passing through a transparent material. Percentage of transmitted light that, in passing through the material, deviates from the incident beam by forward scattering.

head
A device, usually a small electromagnet on a storage medium such as magnetic tape or a magnetic drum, that reads, records, or erases information on that medium. The block assembly and perforating or reading fingers used for punching or reading holes in paper tape.

head changer
A module that can automatically interchange multiple-spindle heads so that a variety of hole patterns requiring drilling and/or taping can be produced.

head crash
The physical collision of the read/write head and the recording surface of magnetic media. It usually results in the destruction of data.

header
The initial portion of a message containing any information, control codes, and so on that are not part of the text (e.g., routine, priority, message type, destination addressee, and time of origination).

header card
A prepunched record of the basic information pertaining to a specific individual or firm that is used to create automatically the upper portion of a document.

header data
Specific data stored in the data bank consisting of data set name, version, owner ID, creation date and time, security, retention information, processing histories, and so on.

header record
A record containing common, constant, or identifying information for a group of records that follow.

head gap
The distance between the read/write head and the surface of the magnetic medium.

heat-affected zone
The portion of a material adjacent to the point of a laser beam/metal interaction in which heat from the laser beam affects the material.

heating chamber
In injection molding, that part of the machine in which the cold feed is reduced to hot melt. Also called heating cylinder.

heat of fusion
Amount of heat per unit weight absorbed by a material in melting.

heat treatment
Heating and cooling a solid metal or alloy in such a way as to obtain desired conditions or properties. Heating for the sole purpose of hot-working is excluded from the meaning of this definition.

helium neon-laser
The type of laser most commonly used in bar code scanners. Because the laser beam is bright red, bars must not be printed with red ink since they would be indistinguishable from the background.

help
Frequently, a command whose response is a menu displaying all available program choices from which the operator can select the next action to be performed.

hertz
A unit of frequency equal to 1 cycle per second; used to measure the repetition rate of a laser pulse.

heterarchical approach
An image interpretation control structure in which no processing stage is in sole command but in which each stage can control other stages to its needs as required.

heterogeneous network
A network of different host computers, such as those of different manufacturers.

heuristic
Pertaining to artificial intelligence exploratory methods of problem-solving in which solutions are discovered by evaluation of the progress made toward the final result. Experiential, judgmental knowledge; the knowledge underlying "expertise"; rules of thumb, rules of good guessing, that usually achieve desired results but do not guarantee them. A form of problem-solving where the results or rules have been determined by rule of thumb or intuition instead of by optimization.

heuristic program
A set of instructions that imitates the behavior of human operations (i.e., response modification based on previous, current, and anticipated conditions that are not preplanned).

heuristic search techniques
Graph searching methods that use heuristic knowledge about the domain to help focus the search. They operate by generating and testing intermediate states along potential solution paths.

hexadecimal numbering system
A numbering system using the equivalent of the decimal number 16 as a base. Most third-

generation computers operate on a principle that utilizes the hexadecimal system, as it provides high utilization or computer storage and an expanded set of characters for representing data. In base 16 (hexadecimal), 16 symbols are required. Because only a single character is allowed for each absolute value, the hexadecimal system uses the 10 symbols of the decimal system for the values 0 through 9, and the first six letters of the alphabet to represent values 10 through 15 (A through F, respectively). The positional significance of hexadecimal symbols is based on the progression of powers of 16. The highest number that can be represented in the units position is 15.

hidden lines
Line segments that would ordinarily be obscured from view in 3-D display of a solid object because they are behind other items in the display. On a CAD system with 3-D capabilities, hidden lines can be displayed or removed as the user specifies.

hidden surface removal
The process of eliminating the representation of surfaces that would be obscured by opaque foreground surfaces if photographed by a hypothetical three-dimensional camera.

hiding power
Ability of paint to obscure surface. Also called opacity. Usually measured by contrast ratio.

hierarchical approach
An approach to machine vision, based on a series of ordered processing levels in which the degree of abstraction increases from the image level to the interpretation level.

hierarchical control
A computer control scheme in which the data processing necessary to accomplish a task is split into discrete levels, with the outputs of higher levels being used as input commands for lower levels. Upper levels of the hierarchy split complex tasks into subtasks, and each subtask is similarly split up by a lower element in the hierarchy. Such systems tend to be fast and efficient, because they can be designed so that decisions are made no higher in the architecture than necessary.

hierarchical network
A computer network in which processing and control functions are performed at several levels by computers specially suited in capability for the functions performed.

hierarchical planning
A planning approach in which first a high-level plan is formulated considering only the important (or major) aspects. Then the major steps of the plan are refined into more detailed subplans.

hierarchy
A tree structure consisting of a root and one or more dependents. In general, the root may have any number of dependents, each of which may have any number of lower-level dependents, and so on, to any number of levels.

higher levels
The interpretative processing stages in machine vision, such as those involving object recognition and scene description, as opposed to the lower levels corresponding to the image and descriptive stages.

higher-order languages
A computer language, such as FORTRAN or LISP, requiring fewer statements than machine language and usually substantially easier to use and read.

high-level language
Programming language that generates machine codes from problem or function-oriented statements. ALGOL, FORTRAN, Pascal, and BASIC are four commonly used high-level languages. A single functional statement may translate into a series of instructions or subroutines in machine language, in contrast to a low-level assembly language in which statements translate on a one-for-one basis.

highlighting
A device-dependent way of emphasizing a segment by modifying its visual attributes. An attribute indicating whether the images of subsequent output primitive encountered during structure traversal are to be distinguished in some workstation-dependent manner. In simple terms, blinking of graphic characters.

highlights
The maximum brightness of the image, which occurs in regions of highest illumination.

high-order position
The leftmost position of a number or word.

histogram
Relative frequency of the gray-scale value distribution in a digital image. The histogram of an image provides very important basic infor-

Hierarchy
Hierarchical Structure

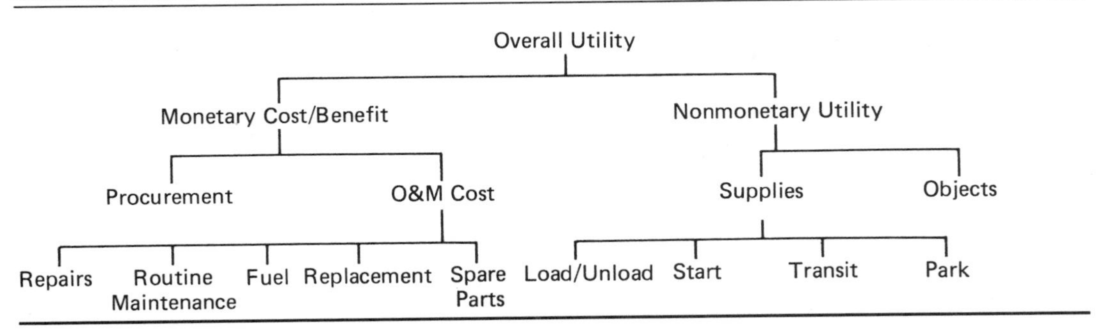

mation about the image. Frequency counts of the occurrence of each intensity (gray level) in an image.

hit
In graphics, to reverse an area in which an image is not drawn. Conceptually, it is similar to clipping a drawing at the edges except, in this case, the edge marks an inscribed area.

hobbing
The operation of producing teeth on the outside edge of the workpiece. Tool and workpiece move with respect to each other.

hold
A stopping of all movement of the robot during its sequence in which some power is maintained on the robot. For example, on hydraulically driven robots, power is shut off to the servovalves but is present on the main electrical and hydraulic systems.

holding time
The length of time a communications channel or facility is in use for each transmission. Includes both message time and operating time.

hole size
Diameter of the hole in a reference block, which determines the area of the hole bottom.

Hollerith code
An alphanumeric punched-card code invented by Dr. Herman Hollerith in 1889, in which the top three positions in a column are called *zone* punches (12, 11, and 0, or Y, X, and O, from the top downward), and are combined with the remaining punches, or digit punches (1 through 9) to represent alphabetic, numeric, and special characters.

hollow
One possible representation of the interior of a polygon-type primitive. The image is the boundary line only (the interior itself is empty).

hologram
A film image created and viewed with the help of a laser beam. The hologram records a window in a three-dimensional scene. By moving one's head, one can change the point of view of the three-dimensional scene.

homeostatic system
Any cybernetic system so arranged as to maintain one particular state or to maintain equilibrium among its component parts.

home position
A fixed location in the basic coordinate axes of the machine tool. Usually, the point at which everything is fully retracted for tool change and so on.

homogeneous network
A network of similar computers, such as those of one model by the same manufacturer.

horizontal bar code
A bar code or symbol presented in such a manner that its overall length dimension is parallel to the horizon. The bars are presented in an array that look like a "picket fence."

horizontal blanking
Blanking of the picture during the period of horizontal retrace.

horizontal display
In MRP, a method of displaying output from the system where requirements, scheduled receipts, projected balance, and so forth are dis-

played horizontally, that is, across the page. Horizontal displays are difficult to use in conjunction with bucketless systems.

horizontal machining center
Primarily for working on cube-shaped parts. Using a cutting tool rotating on a horizontal axis, it can drill holes, mill out pockets, cut grooves, or plane metal from a surface.

horizontal retrace
The return of the scanning (beam) from one side to the other of the scanning raster after the completion of one scan line.

Horn clause
A set of statements joined by logical ANDs. Used in PROLOG.

host
One central place where the data resides.

host computer
The primary or controlling computer in a multicomputer network. Large-scale host computers typically are equipped with mass memory and a variety of peripheral devices, including magnetic tape, line printer, card readers, and possibly hard-copy devices. Host computers may be used to support, with their own memory and processing capabilities, not only graphics programs running on a CAD/CAM system, but also related engineering analysis. The primary or controlling computer in a multiple computer network operation. This computer normally provides high-level services such as computation, data base access, or special programs or programming languages for other computers in the network. A computer used to prepare programs for use on another computer or on another data-processing system, for example, a computer used to compile, link edit, or test programs to be used on another system.

host interface
The interface between a communications processor or network and a host computer.

host satellite system
A CAD/CAM system configuration characterized by a graphic workstation with its own computer (typically holding the display file) but connected to another, usually larger, computer for more extensive computation or data manipulation. The computer local to the display is a satellite to the larger host computer, and the two constitute a host satellite system.

hot gas welding
A technique of joining thermoplastic materials (usually sheet) whereby the materials are softened by a jet of hot air from a welding torch and joined together at the softened points. Generally, a thin rod of the same material is used to fill and consolidate the gap.

hot runner mold
A mold in which the runners are insulated from the chilled cavities and are kept hot. Parting line is at gate of cavity, and the runners are in separate plate(s), so they are not, as is the case usually, ejected with the piece.

hot swaging
Essentially the same as cold swaging, except that the metal is heated and the arrangement of the dies causes them to be carried in a slow rotary motion. This operation is also known as rotary swaging.

hot upsetting
Essentially the same as cold beading, except that the material is heated in order to aid ductility in forming.

hough transform
A global parallel method for finding straight or curved lines, in which all points on a particular curve map into a single location in the transform space.

housekeeping
Operations or routines that do not contribute directly to the solution of the problem but do contribute directly to the operation of the computer.

housekeeping routine
That part of a program usually performed only at the beginning of machine operations that establishes the initial conditions for instruction addresses, accumulator setting, switch setting, and so on.

housing
A surrounding case designed to provide a degree of protection for equipment against a specified environment and to protect personnel against accidental contact with the enclosed equipment.

HSV
A hue (color), saturation (purity), value (brightness) color model.

hue
The dominant wavelength of light representing the color of an object. It is the redness, blueness, greenness, and so forth of an object.

Hueckel operator
A method for finding edges in an image by fitting an intensity surface to the neighborhood of each pixel and selecting surface gradients above a chosen threshold value.

human factors
The field of effort and body of knowledge devoted to the adaptation and design of equipment for efficient and advantageous use by people considering physiological, psychological, and training factors.

human interface
One subsystem of an expert system (or any computing system) with which the human user deals routinely. It aims to be as "natural" as possible, employing language as close as possible to ordinary language (or the stylized language of a given field), and understanding and displaying images, all at speeds that are comfortable and natural for humans.

hybrid circuit
A circuit constructed of several independently fabricated, interconnected ICs and embodying various component manufacturing technologies, such as monolithic IC, thin films, thick films, and discrete components. Hybrid circuit design has been automated, in several of its stages, through CAD/CAM.

hybrid computer
A data-processing device using both analog and discrete data representation.

hydraulic motor
An actuator consisting of interconnected valves or pistons or vanes that converts high-pressure hydraulic or pneumatic fluid into mechanical shaft translation or rotation.

hydroforming
The operation of drawing sheet metal using a shaped punch forced against the material and a rubber pad backing the material, while hydraulic pressure on the opposite side is applied.

hydrogen brazing
Brazing in a hydrogen atmosphere, usually in a furnace.

hydrogen embrittlement
A condition of low ductility in metals, resulting from the absorption of hydrogen.

hysteresis
Lag in response of an instrument or process when a force acting on it is abruptly changed. A component of bidirectional repeatability caused by other mechanisms than drivetrain clearance.

hysteresis loop
Curve showing relationship between magnetizing force and magnetic induction in a material in a cyclically magnetized condition. For each value and direction of magnetizing force, there are two values of induction: (1) when the magnetizing force is increasing and (2) when the magnetizing force is decreasing. Result is actually two smooth curves joined at ends to form a loop.

I

iconic
Imagelike.

IDEF
Acronym for ICAM DEFinition method (a modeling technique for analyzing systems).

$IDEF_0$
The acronym for ICAM DEFinition method, version 0 (also called "Structured Analysis and Design Technique" and available commercially under that name from SofTech, the developer). A technique for modeling the functions (activities) of a system.

$IDEF_1$
Stands for ICAM DEFinition method, version 1 (also called the information modeling technique). It is used to create a model of the information required to accomplish the functions defined using $IDEF_0$.

$IDEF_2$
Acronym for ICAM DEFinition method, version 2 (also called the dynamic modeling technique). Enables the analyst to create a graphic model of the system that relates system operation to time.

identification
The process of providing personal, equipment, or organizational characteristics or codes to gain access to computer resources. The process of identifying an object through reading symbols.

identification division
One of the four main component parts of a COBOL program. The identification division identifies the source program and the object program and, in addition, may include such documentation as the author's name, the installation, where written, date written, and so forth.

identifier
A symbol whose purpose is to identify or name data in a programming language.

identity
Two propositions (in logic) that have the same truth value.

identity element
A logic element that performs an identity operation.

identity operation
The Boolean operation, the result of which has the Boolean value 1 if and only if all the operands have the same Boolean value. An identity operation on two operands is an equivalence operation.

idle time
Lost production time on a machine due to setup, maintenance, or waiting for labor, parts, or tooling.

IDSS
Acronym for ICAM Decision Support System. IDSS is a computer-based simulation capability for exercising the dynamic model ($IDEF_2$), to permit evaluation of alternate design solution to improvement concepts, using quantitative measures of performance.

IEEE
Institute of Electrical and Electronic Engineers.

IEEE802
Activity to establish standards for LAN. The 802 standard has sections as follows:
802.1 – higher-layer interface
802.2 – local link control
802.3 – CSMA/CD
802.4 – token bus
802.5 – token ring
802.6 – metropolitan area network

if-and-only-if element
A logic element that performs the Boolean operation of equivalence.

if-then element
A logic element that performs the Boolean operation of implication.

if-then-else statement
A program statement used in high-level languages, indicating alternative responses to an initial action. If the initial statement is true, then the next expression is executed; if the initial statement is not true, then the else, or alternative, expression is executed.

IHS
Acronym for Intensity, Hue, and Saturation; a form of representing a color image. Also called luminance, hue, and saturation.

illegal character
A character or combination of bits that is not valid according to some criterion, for example, with respect to a specified alphabet, a character that is not a member.

illuminance
Luminous flux incident on a surface; luminous incidence.

illumination
The use of a light source to generate a light intensity distribution, based upon the way in which light is reflected from an object's surface.

ILP
Acronym for Ingersoll Lathe Program. ILP is a general two-axis NC "lathe" language for matching small lot sizes and/or short cycle times.

ILS-11
Acronym for Interactive Layout System. ILS-11 is a software processor designed to facilitate the generation of cams for cam automatics and the workpiece programming of camless automatics.

image
A two-dimensional representation of an object or a scene formed by creating a pattern from the light received from the scene. A particular view of one or more objects or parts of objects as presented on a view or display surface. A projection of a scene into a plane. Usually represented as an array of brightness values.

2-D image
An image $f(x, y)$ where f represents a brightness value at the coordinates x and y. This image does not in itself contain any range information.

2.5-D image
An image where the distance (range) to one or a small number of points has been determined by independent or auxiliary means. It also includes the situation where a single average range value to the scene is determined.

2.75-D image
Defines a scheme where range is determined for a significant number of points in the scene but not for all of the pixels. It may be helpful to think of this as a range grid of low resolution, overlaid on a high 2-D image resolution. It approaches the operation of the human eye.

3-D image
Described by the term *dense range map*, a condition where the range to every pixel is determined.

image burn
A desensitizing of an image pickup tube target, where the light has been intense.

image enhancement
The process of enhancing the quality of the appearance of an image. Image enhancement operations may be noise filtering, contrast sharpening, edge enhancing, and so forth.

image intensifier
A device coupled by fiber optics to an image sensor to increase sensitivity. Can be single or dual stage.

image plane
The plane where the desired scene is imaged. A pixel memory area one bit deep. Four planes are used on the Octek IAP to store a 16-gray-level image with a resolution of 320 × 240 pixels (76,800 pixels).

image preprocessing
A computational step prior to the feature extraction step in an image analysis procedure. Preprocessing may serve the purpose of enhancing the features to be extracted or adjusting the image in other ways to certain conditions set forth by the subsequent feature extraction procedure.

image processing
A wide variety of techniques for processing pictorial information by computer. The information to be processed is usually input to the computer by sampling and analog-to-digital conversion of video signals obtained from some type of two-dimensional scanner device. Initially, the information is in the form of a large array in which each element is a number representing the brightness (and perhaps color) or a small region in the scanned image. A digitized image array is sometimes called a digital picture, its elements are called "points," "picture elements," or "pixels." The values of these elements are typically six-bit or eight-bit integers. They usually represent brightness (or gray color).

image processor
A device such as a microprocessor that converts an image to digital form and then further enhances the image to prepare it in a form suitable for computer analysis.

image subtraction
An image-creating operation that creates each pixel of a new image by subtracting corresponding pixels of other images.

image transformation
A segment attribute, applied after the viewing transformation, that causes the image defined by a segment to appear translated, rotated, and/or scaled on the view surface, often referred to as geometric transformation. A retained segment dynamic attribute that permits the image defined by a segment to appear at varying sizes, orientation, and/or positions on the view surface.

image understanding
Employs geometric modeling and the artificial intelligence techniques of knowledge representation and cognitive processing to develop scene interpretations from image data. Image understanding has dealt extensively with 3-D objects. Image understanding usually operates not on an image but on a symbolic representation of it. Image understanding is somewhat synonymous with computer vision and scene analysis.

immediate access storage
A storage device whose access time is negligible in comparison with another's operating time.

immediate address
The contents of an address part that contains the value of an operand, rather than an address.

immediate instruction
An instruction that contains within itself an operand for the operation specified, rather than an address of the operand.

impact extruding
The operation of producing cup-shaped parts by a single blow against a confined slug of cold metal until the desired length and wall thickness is reached.

impact printer
A printer that forms characters by physically striking a ribbon and paper.

impact resistance
Resistance to fracture under shock force.

impact strength
The energy required to fracture a specimen subjected to shock loading, as in an impact test.

impedance
Acoustical impedance is the complex ratio of the sound pressure to the product of the sound velocity and the area at a given surface. It is frequently approximated by only the product of the density and velocity. Electrical impedance is the complex property of an electrical circuit, or the components of a circuit, that opposes the flow of an alternating current.

implementable design
An interpretation of the design specifications that are used as a basis for constructing components.

implementation
The act of finishing or installing a program or a system.

implementation dependent
Some aspect of a graphics system that cannot be completely standardized but will vary from implementation to implementation. Implementations must document their interpretations of implementation-dependent values and actions.

implementation mandatory
A property that must be realized identically in all implementations of a standard.

implicator
The dyadic Boolean operation, the result of which has the Boolean value 0 if and only if the first operand has the Boolean value 0 and the second has the Boolean value 1.

implicit regeneration
An immediate or deferred regeneration of the picture on a display surface that can occur when changes to the picture definition invalidate the displayed image but a regeneration has not been explicitly requested by the application program.

implies
A connective in logic that indicates that if the first statement is true, the statement following is also true.

implosion
The aggregation of lower-level detailed data into a summary report; the reverse of explosion; commonly refers to the process of generating a where-used report.

impregnation
The treatment of porous castings with a sealing medium to stop pressure leaks. The process of filling the pores of a sintered compact, usually with a liquid such as a lubricant. The process of mixing particles of a nonmetallic substance in a matrix of metal powder, as in diamond-impregnated tools.

impulse sealing
A heat-sealing technique in which a pulse of intense thermal energy is applied to the sealing area for a very short time, followed immediately by cooling. It is usually accomplished by using an RF-heated metal bar that is cored for water cooling or is of such a mass that it will cool rapidly at ambient temperatures.

impurity
An active medium used in solid-state lasers by being embedded in the crystal rod.

incident light
The light that falls directly on an object.

inclusions
Isolated, irregular, or elongated variations of magnetic particles occurring singly, in a linear distribution, or scattered randomly in feathery streaks.

incomplete fusion
Fusion that is less than complete. Failure of weld metal to fuse completely with base metal or preceding leads.

incremental analysis
A method of economic analysis in which the cost of a single additional unit is compared to its revenue. When the net contribution of an additional unit is zero, total contribution is maximized.

incremental computer
A computer in which incremental representation of data is mainly used. A special-purpose computer that is specifically designed to process changes in the variables as well as the absolute value of the variables.

incremental cost
Cost added in the process of finishing a part or assembly, assembling a group of parts or adding part(s) or assembly(s) to a higher-level assembly. If the cost of the components of a given assembly equals $5.00 and the additional cost of assembling the components is $1.00, then the incremental assembly cost is $1.00, while the total cost of the finished assembly is $6.00. Additional cost incurred as a result of a decision selecting a different method of procuring a part, achieving a goal, fulfilling a requirement, and so on.

incremental digitizing
The sending of selected points to a computer graphics system representing edges, intersections, or other significant points, rather than sending position information at regular time intervals.

incremental dimension word
A word defining a dimension or movement with respect to the preceding point in a sequence of points. Most contouring systems utilize incremental dimensioning. Most CNC systems can operate in either an absolute or incremental mode.

incremental integrator
A digital device modified so that the output signal is maximum negative, zero, or maximum positive when the value of the input is negative, zero, or positive.

incremental system
Programming whereby each coordinate location is given in terms of distance and direction along rectangular axes from the previous position and not from a fixed zero location.

indented bill of materials
A product structure type of bill-of-materials format where the highest-level subassemblies are shown closest to the margin, all components going into these subassemblies are shown indented to the right of the margin, and subsequent levels of components are indented farther to the right. A complex product might have ten or more levels of components and there would, therefore, be 10 or more indents on the bill of materials.

independent demand
Demand for a finished good or a component unrelated to the demand for other items. Demand for finished goods, parts required for destructive testing, and service parts requirements are some examples of independent demand.

index
In computer programming, a subscript or integer value that identifies the position of an item of data with respect to some other item of data. An integer used to specify the location of information within a table or program.

index address
An address that is modified by the content of an index register prior to or during the execution of a computer instruction.

indexed color
A color selection scheme in which the color index is used to retrieve color values from a color table.

indexed sequential file
A file in which records are organized sequentially, with indexes that permit quick access to undivided records as well as rapid sequential processing.

indexed sequential organization
A file organization used on direct-access storage devices in which records are arranged in logical sequence by key. Indexes to these keys permit direct access to individual records.

index hole
A hole in a floppy disk that indicates the start of the first sector.

indexing
A method of address modification that is performed automatically by the data-processing system.

index of refraction
A property of a medium that measures the degree that light bends in when passing between the medium and a vacuum. Also called refractive index. Ratio or velocity of light in vacuum to its velocity in the material. Ratio of the sine of the angle of incidence of light to the sine of the angle of refraction.

index register
A register whose content may be added to or subtracted from the operand address prior to or during the execution of an instruction.

index word
A storage position or register, the content of which may be used to modify automatically the effective address of any given instruction.

indications
Eddy-current signals caused by any change in the uniformity of a tube.

indicators
Internal switches that are turned on or off, depending on the results of arithmetical or logical comparisons.

indirect address
An address that designates the storage location of an item of data to be treated as the address of an operand but not necessarily as its direct address.

indirect cost
Cost not directly incurred by a particular job or operation. Certain utility costs, such as plant heating, are often indirect. An indirect cost can either be a fixed cost or a variable cost.

indirect file
A collection of keyboard commands to be executed sequentially, grouped together as a file.

indirect labor
Labor that does not add to the value of a product but that must be performed to support its manufacture. May not be readily identifiable with a specific product or service.

indirect material
Material consumed in the process of production or manufacture that does not become a part of the finished product or cannot be readily identified with or charged to a particular part, product, or group of parts or products.

individual
A nonvariable element (or atom) in logic that cannot be broken down further.

induction motor
An alternating current motor in which torque is produced by the reaction between a varying or rotating magnetic field that is generated in stationary field magnets and the current that is induced in the coils or circuits of the rotor.

inductive sensors
The class of proximity switch using an RF field, typically employing one half of a ferrite core, whose coil is part of an oscillator circuit. When a metallic object enters this field, at some point the object will absorb enough energy from the field to cause the oscillator to stop oscillating. It is this difference between oscillating or not oscillating that is detected as the difference between an object present or not present.

INDUCTOSYN
Trademark for Farrand Controls resolver, in which an output signal is produced by inductive coupling between metallic patterns in two glass members separated by a small air space. Produced in both rotary and linear configurations.

industrial robot
A reprogrammable, multifunctional manipulator designed to move material, parts, tools, or specialized devices through variable programmed motions for the performance of a variety of tasks. (*Illus. p. 187.*)

industrial robot components
The three principal components of an industrial robot are: (1) one or more arms, usually situated on a fixed base, that can move in several directions; (2) a manipulator, the working tool of the robot—the ''hand'' that holds the tool or the part to be worked; and (3) a controller that gives detailed movement instructions.

industrial robot system
A system that includes industrial robots, the end-effectors, any equipment devices and sensors required for the robot to perform its tasks, and includes communication interfaces for sequencing or monitoring the robot.

inert-gas shielded-arc cutting
Metalcutting with the heat of an arc in an inert gas, such as argon or helium.

Industrial Robot
CYRO™ Model 750 Industrial Robot. *Courtesy:* Advanced Robotics Corp.

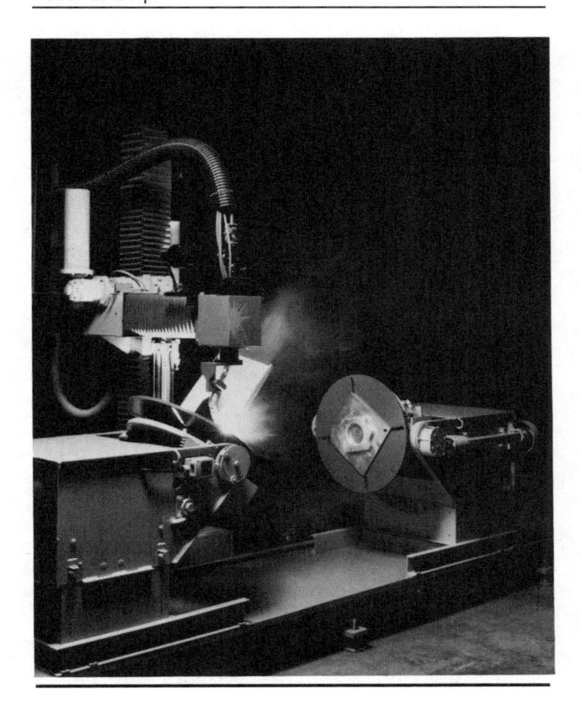

inert-gas shielded-arc welding
Arc welding in an inert gas, such as argon or helium.

inertia
The tendency of a mass at rest to remain at rest and of a mass in motion to remain in motion. The Newtonian property of a physical mass that a force is required to change the velocity, proportional to the mass and the time rate of change of velocity.

infer
To derive by reasoning. To conclude or judge from premises or evidence.

inference
The process of reaching a conclusion based on an initial set of propositions, the truths of which are known or assumed.

inference engine
The component of an expert system that controls its operation by selecting the rules to use, accessing and executing those rules, and determining when an acceptable solution has been found. This component sometimes is called the *control structure* or the *rule interpreter*.

infinite capacity
Loading a plant or work center without regard to the capacity of that location. Used to show where overloads exist so that they can be corrected.

infix notation
A method of forming mathematical expression, governed by rules of operator precedence and using parentheses in which the operators are dispersed among the operands, each operator indicating the operation to be performed on the operands or the intermediate results adjacent to it.

in-floor towline
A conveying system that consists of a load-carrying cart towed by a chain running in a track imbedded in the factory floor. Carts may be stopped, accumulated, and switched off of the chain automatically and may interface with equipment for automatic load transfer.

information
The meaning derived from data that has been arranged and displayed in such a way that it can be related to that which is previously known.

information path
The functional route by which information is transferred in a one-way direction from point to point.

information retrieval system
A complete application for cataloging vast amounts of stored data so that any part or all of this data can be called out at any time.

Inference
Inference Chain for Inferring the Spill of Material

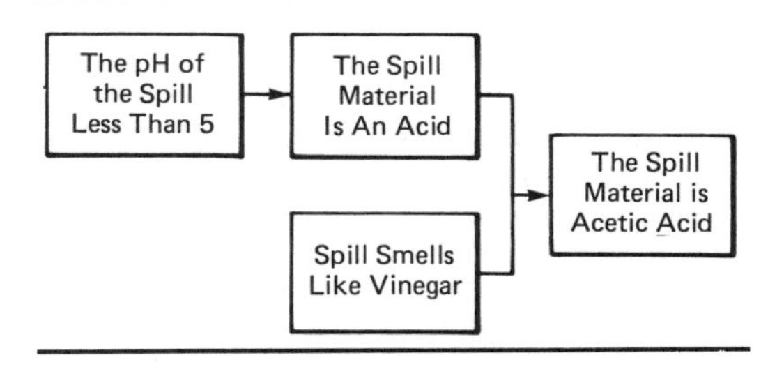

information separator
Any control character used to delimit like units of data in a hierarchic arrangement of data.

information storage
A complex system in which information is placed on file in such a manner that future retrieval may be accomplished in a variety of ways. The methods and procedures for recovering specific information from stored data.

information system
A logical group of subsystems and data required to support the information needs of one or more business processes.

information systems independence
An information system defined so that data is independent of the organizational structure of the business.

information systems network
A network of multiple operational-level information systems and one management-oriented information system (centered around planning, control, and measurement processes). The network retrieves data from data bases and synthesizes that data into meaningful information to support the organization.

information systems plan
A plan for managing an information systems network implementation, including definition of the major actions, schedules, and resources required. Must be integrated with the business plan and should be developed from the point of view and with the active participation of top management.

information theory
The branch of learning concerned with the likelihood of communication of messages subject to transmission failure, distortion, and noise.

infrared
Part of the electromagnetic spectrum between the visible light range and the radar range. Radiant heat is in this range, and infrared heaters are much used in sheet thermoforming. The region of the electromagnetic spectrum adjacent to the visible portion of the spectrum but having longer wavelengths from 0.75 to 1000 microns.

infrared reflectance
Measure of infrared brightness. Two direct optical methods are used. In each, a standard beam of light is thrown on a specimen normal to the surface and reflectance is measured from the same position, using filters to select specified wavelengths for viewing in the infrared range.

infrasound
Sounds whose frequencies lie below the range of human hearing.

inherent weakness failure
Failures attributable to weakness inherent in the item itself when subjected to stresses within the stated capabilities of that item.

inherited error
The error in the value of quantities that serve as the initial conditions at the beginning of a step in a step-by-step calculation.

inhibitor
A substance that slows down a chemical reaction. Inhibitors are sometimes used in certain types of monomers and resins to prolong storage life.

initial graphics exchange specification (IGES)
An interim CAD/CAM data base specification until the American National Standards Institute develops its own specification. IGES attempts to standardize the communication of drawing and geometric product information between computer systems.

initialize
A programming term that refers to the act of establishing fixed values in certain areas of memory. Generally, the term applies to all the housekeeping that must be completed before the main part of the program can be executed. Set counters, switches, or addresses to starting values at prescribed points in the execution of a program, particularly for reexecution of a sequence of code.

initial program loader
The utility routine that loads the initial part of a computer program, such as an operating system or other computer program, so that the computer program can then proceed under its own control.

initiator
A substance that speeds up the polymerization of a monomer and becomes a component part of the chain.

injection molding cycle
The complete time cycle of operation utilized in injection molding of an object including injection, die close, and die open time.

injection pressure
The pressure on the face of the injection ram at which molding material is injected into a mold. It is usually expressed in psi.

injection ram
The ram that applies pressure to the plunger in the process of injection molding or transfer molding.

inking
The process of drawing on the screen by using a program that connects on screen the points corresponding to the sampled positions of a digitizing device. To users, it seems as if they are directly controlling an on-screen drawing pen by moving the digitizer.

ink jet printer
A nonimpact printing technique that uses droplets of ink to form copy images. The print head, as it moves across the surface of the copy paper, shoots a stream of tiny electrostatically charged ink drops at the page, placing them precisely to form individual print characters.

in-line procedures
In COBOL, the set of statements that constitutes the main or controlling flow of the computer program and that excludes statements executed under control of the asynchronous control system.

in-line processing
The processing of data in random order, without preliminary editing or sorting.

in-process gauging
A type of active error compensation in which the workpiece is used as a master. Gauging performed on partially finished parts while they are still in the manufacturing process.

in-process inventory
Product in various stages of completion throughout the factory, including raw material that has been released for initial processing and completely processed material awaiting final inspection and acceptance as finished product or shipment to a customer.

input
The data to be processed. Also the transfer of data to be processed from keyboard or an external storage device to an internal storage device.

input class
A set of input devices that are logically equivalent with respect to their function. Typical input classes are locator, stroke, valuator, choice, pick, and string.

input data record
A data structure containing detailed information about the measure and echo type of a logical input device. Provided by the application program when the device is initialized.

input devices
A variety of devices (such as data tablets or keyboard devices) that allow the user to communicate with the CAD/CAM system, for example, to pick a function from many presented, to enter text and/or numerical data, to modify the picture shown on the CRT, or to construct the desired design. Devices such as limit switches, pressure switches, push buttons, and so on, that supply data to a robot controller. These discrete inputs are two types: (1) those with common return and (2) those with individual returns. Other inputs include analog devices and digital encoders.

input media
Data sources for computers, such as punched cards, punched tape, or encoded documents. The word input used alone often includes the medium or media. Any process that transfers data from an external source to an internal storage is designated as input.

input mode
One of three possible means of requesting and obtaining data from a logical input device: request, sample, or event.

input/output
Pertaining to either input or output signals or both. A general term for the equipment used to communicate with a computer. The data involved in such communication. The media carrying the data for input/output.

input/output analysis
A matrix that provides a quantitative framework for the description of an economic unit. Basic input/output analysis is a unique set of input/output ratios for each production and distribution process. If the ratios of inputs per unit of output are known for all production processes, and if the total production of each end product of the economy (or of that section being studied) is known, it is possible to compute precisely the production levels required at every intermediate stage to supply the total sum of end products. Further, it is possible to determine the effect at every point in the pro-

duction process of a specified change in the volume and mix of end products.

input/output control
Measuring a work center's (or device's) performance by comparing the rate at which work is coming into the work center against the rate the work is leaving the work center.

input/output controller
A functional unit in a data-processing system that controls one or more units of peripheral equipment.

input/output control system
A set of programs used in an operating system. It handles all input and output work such as opening files, closing files, backspacing tape, moving tape forward when a bad spot is encountered, and so forth.

input/output port
An outlet on a computer circuit board for attaching input or output devices, such as keyboards or printers.

inquiry
A request for information from storage; or a machine statement to initiate a search of library documents.

inquiry station
A terminal where inquiries can be entered directly into the computer. The inquiry terminal can be geographically remote from the computer or at the computer console and usually includes a typewriter keyboard.

inscribing
To read the data recorded on a document and write the same data on the same document but in such a form that the document becomes suitable for automatic reading by a character reader.

insert
To create and place entities, figures, or information on a CRT or into an emerging design on the display.

inspection by attributes
Inspection where the unit of product is classified simply as defective or nondefective, with respect to a single requirement or set of requirements.

inspection by variables
Inspection where a specified quality characteristic on a unit of product is measured on a continuous scale.

installation
The execution of the steps or measures necessary to introduce a technique, procedure, or proposed course of action into an organization and to get it functioning properly. A technique procedure or equipment arrangement that is being set up or used by a company.

installation testing
The validation of each particular installation of the system with the intent of pointing out any errors made while installing the system.

installation time
Time spent in installing, testing, and accepting equipment and/or programs.

instance
A particular occurrence of some item or relationship. Several instances may reference the same item.

instancing
A method of defining an object to appear once in a data base and replicating it (without copying) multiple times in what may be different positions, sizes, orientations, and other attributes as inherited by the instance.

instantiation
In artificial intelligence, an object that fits the general description of some class or, specifically, a pending process that associates specific data objects with the parameters of a general procedure.

instruction
A statement to the computer that specifies an operation to be performed by the system and the values or locations of all operands. An instruction is usually made up of an operation code and one or more operands.

instruction address
The address of an instruction word.

instruction address register
A register from which the address of the next instruction is derived.

instruction card
Written information supplied to a worker that specifies method, machines, and when appropriate, their speeds, feeds, and depth of cut, tools, fixtures, specification limits, and the like to be used in performing a task.

instruction control unit
In a central processing unit, the part that receives instructions in proper sequence, interprets each instruction, and applies the proper signals to the arithmetic and logic unit and other parts in accordance with the interpretation.

instruction cycle
That part of a machine cycle during which a computer instruction is transferred from a

specified primary storage location to the instruction register in the control unit, where the instruction is decoded before being executed.

instruction register
The register that stores the current instruction governing a computer operation. A counter that indicates the instruction currently being executed.

instruction repertory
The set of instructions that a computer data-processing system is capable of performing.

instruction set
The set of the instructions of a computer, of a programming language, or of the programming languages in a programming system.

instruction time
The portion of a central processing unit cycle during which an instruction is fetched and decoded.

instruction word
A word that represents an instruction.

insulation resistivity
A measure of the ability of a material to resist flow of electric current. Insulation resistance per unit volume. This property makes no distinction between volume resistivity and surface resistivity, even where no such distinction is made in the test. However, it is more common to use the term volume resistivity.

integer
A numeric data item or literal that does not include any character positions to the right of the decimal point, actual or assumed, that is, a whole number.

integer-based
A computer-aided design (CAD) system that keeps coordinates as integers. Using integers allows much faster mathematical calculations but does not represent numbers as precisely as using real numbers.

integer programming
In operations research, a class of procedures for locating the maximum or minimum of a function subject to constraints, where some or all variables must have integer values.

integer system
The less common way a CAD program keeps track of where objects are located.

integral control
A control scheme whereby the signal that drives the actuator equals the time integral of the error signal.

integrated adapter
An integral part of a central processing unit that provides for the direct connection of a particular type of device and uses neither a control unit nor the standard I/O interface.

integrated circuit
An electronic circuit, all of whose components are formed on a single piece of semiconductor material, usually silicon.

integrated-circuit diode-matrix
An integrated circuit containing a matrix of diodes that may be individually open-circuited or short-circuited to represent a program.

integrated computer-aided manufacturing (ICAM)
A program and development plan (sponsored by the U.S. Air Force in cooperation with the aerospace industry) for producing systematically related modules of machining for flexible manufacturing and control.

integrated data processing
A system that treats all data-processing requirements as a whole to reduce or eliminate duplicate recording or processing while accomplishing a sequence of data-processing steps or a number of related data-processing sequences.

integrated system
A CAD/CAM system that integrates the entire product development cycle—analysis, design, and fabrication—so that all processes flow smoothly from concept to production.

integration
The sharing of data or information among subsystems and systems.

integration testing
The verification of the interfaces among system parts (modules, components, and subsystems).

integrator
Any device that integrates a signal over a period of time. Unit in a computer that performs the mathematical operation of integration, usually with reference to time. A device whose output is proportional to the integral of the input variable with respect to time.

integrity
Preservation of data or programs for their intended purpose.

intelligence
Of a robot, the ability to perform artificial intelligence functions that are normally associated with human intelligence such as reasoning, planning, problem-solving, perception,

pattern recognition, cognition, understanding, and learning.

intelligent assistant
An artificial intelligence computer program (usually an expert system) that aids a person in the performance of a task.

intelligent computer-assisted instruction (ICAI)
An area of artificial intelligence research that attempts to create educational programs that can analyze a student's learning pattern and modify their teaching techniques accordingly.

intelligent robot
A robot that can be programmed to make performance choices contingent on sensory inputs.

intelligent terminal
A terminal with some logical capability; a remote device that is capable of performing some editing or other function upon input or output data. The intelligent terminal has flexible design for simplified user interface including custom keyboards, modularity to meet a variety of user requirements including control of other terminals, and buffering capability to simplify the communications interface and the impact on host computer software.

intelligent voice terminal
Intelligent terminal operated by human voice; software resident in terminal is user-programmable. Best for applications suiting an intelligent terminal but where hands-free data entry is cost-advantageous.

intensity
The relative brightness of an image or portion of an image.

Interact IV
Computervision's trade name for an automated drafting table used to plot and/or digitize drawings.

interactive
Pertaining to an application in which each entry elicits a response, as in an inquiry system. An interactive system may also be conversational, implying continuous dialog between the user and the system. Processing of data on a two-way basis and with human intervention, providing redirection of processing in a predetermined manner.

interactive device
Any graphics device that supports both graphics output and input.

interactive environment
A computational system in which the user interacts (dialogues) with the system (in real time) during the process of developing or running a computer program.

interactive graphics
Capability to perform graphics operations directly on the computer, with immediate feedback.

interactive graphics system
A CAD/CAM system in which the workstations are used interactively for computer-aided design and/or drafting, as well as for CAM, all under full operator control, and possibly also for text processing, generation of charts and graphs, or computer-aided engineering. The designer (operator) can intervene to enter data and direct the course of any program, receiving immediate visual feedback by means of the CRT. Bilateral communication is provided between the system and the user(s). Often used synonymously with CAD.

interactive mode
A method of operation that allows online man/machine communication. Commonly used to enter data and to direct the course of a program.

interactive operation
Online operation where there is a give-and-take between person and machine. Also called *conversational* mode. User presents problem to computer, gets results, asks for variation or amplification of results, gets immediate answer, and so on.

interactive plotting
Type of graphic application in which the display console is used simply to "browse" through output of computational processes, typically in the form of a graph prepared by the host computer. Little interaction is required except for the real-time display of successive frames on operator command and simple *menu selection* (i.e., by means of labeled function key or light pen identification of a command name displayed in a control area on the screen) to direct the browsing or further computation. Such applications as computer-assisted instruction and command and control operations fit into this category of primarily predefined pictures. It describes probably three fourths of all graphics application programs.

interactive system
A system in which bilateral communication is established between a computer or operating program and a user.

interblock gap
An area on a data medium used to indicate the end of a block or record. Same as block gap.

interchange point
A location where interface signals are transmitted.

intercharacter gap
The space between two adjacent bar code characters.

interdependence
The extent to which the behavior of the elements of a system affect each other individually and collectively.

interface
Those connections of one system whereby the system is matched to another system that is distinctly different in nature. A shared boundary between system elements defined by common physical interconnection characteristics, signal characteristics, and meanings of interchanged signals. A boundary between the robot and machines, transfer lines, or parts outside of its immediate environment.

interface characteristic proof
An association of a particular interface characteristic and the test case that verifies system compliance.

interface data unit
This unit of data transferred across the *N* service access point of an *N* entity and an (N + 1) entity in a single interaction. Each interface data unit contains interface control information and may also contain all or part of an *N* service data unit. The unit of information transferred across the service access point between an entity and an entity in the layer above it or below it in a single interaction. The total unit of information transferred across the service access point.

interfacing
A shared boundary. Except in a few applications, most robots need to communicate and interact with the outside world. This can take the form of simple on-off signals generated by electrical or pneumatic contacts or of more complex electronic signals. Inputs are the lines over which the robot accepts signals from the outside world, and outputs are the lines over which it sends signals to external equipment.

interference
Any undesired electrical signal induced into a conductor by electrostatic or electromagnetic means.

interference checking
A CAD/CAM capability that enables plant or mechanical designers to examine automatically a 3-D data model, pinpointing interferences between pipes, equipment, structures, or machinery. A computer analysis produces a summary of interferences within selected tolerances.

interference time
The period of time during which one or more machines are not operating because the worker or workers assigned to operate them are busy operating other machines to which they are assigned or are performing necessary duties related to operating such other machines.

intergranular corrosion
Corrosion occurring preferentially at grain boundaries.

interior
Referring to a style of dimensioning, in which the value of an interval is shown, with arrows leading out from the value to two guidelines.

interlace
The process of scanning or displaying an image by alternating between two sets of scan lines to reduce flicker in the display.

interlaced scanning
A scanning process in which the distance from center to center of successively scanned lines is two or more times the normal line spacing, and in which the adjacent lines belong to different fields.

interleaved bar code
A bar code in which characters are paired together using bars to represent the first character and spaces to represent the second.

interleaving
A multiprogramming technique in which segments of one program are inserted into other programming so that during delays in the processing of one of the programs, the other program can be processed.

interlock
A safety device designed to ensure that a piece of apparatus will not operate until certain precautions have been taken.

Intermediate System
Intermediate System Architecture. *Source*: Boeing Company.

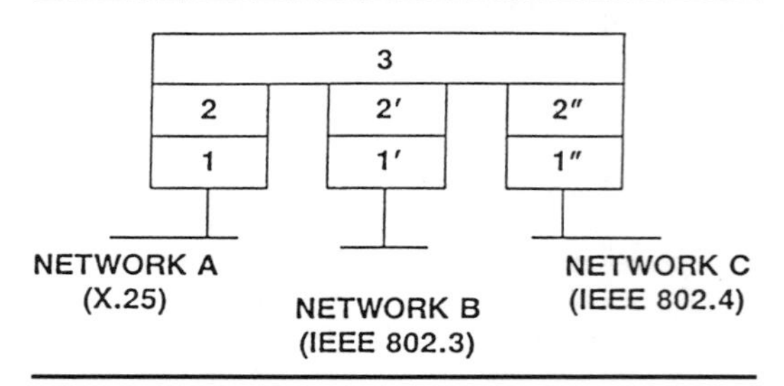

intermediate annealing
Annealing wrought metals at one or more stages during manufacture and before final treatment.

intermediate character
Any character that occurs between an ESCAPE character and the final character in an ESCAPE sequence.

intermediate system
In the Boeing Technical and Office Protocol (TOP), an intermediate system (router) interconnects two or more physically distinct networks by implementing an OSI network layer three protocol. This device has one common network address known to all networks attached to it. No restrictions are placed on either the interconnected data link or physical layer protocols. In this configuration, the intermediate system provides path selection, message relaying, hop-by-hop enhancement of network services, and alternate routing based upon destination network layer addresses and status of connected networks.

intermittent element
An element, essential to an operation, that does not occur at regular intervals, compared to the regular or cyclic elements.

intermittent production
A production system in which the productive units are organized according to function. The jobs pass through the functional departments in lots, and each lot may have a different routing.

intermittent weld
A weld in which the continuity is broken by recurring unwelded spaces.

internal element
Any element of an operation performed by a worker while the machine he or she controls is operating automatically. A short-duration task performed by one hand while the other hand is performing a more time-consuming element.

internal runtime subsystem
A set of routines and processors that use the tables set up by the conceptual-to-internal translator to actually perform the data transfers between conceptual schema records and internal schema records. Also processes keys by using and maintaining the files created for keyed attributes.

internal schema
A tabular description of how the data corresponding to a conceptual schema physically resides on the mass-storage devices. It describes to the data management system what data fields are to be treated as keys, how large the physical records are, as well as other details needed by the runtime system to store and retrieve data from the devices.

internal schema compiler
A translator that produces an internal schema object consisting of tables containing essential data items from the internal schema source. These tables are stored in the metadata base by the data inventory management subsystem. The compiler checks syntax and performs validation checks with conceptual schema information in the metadata base.

internal sensor
A feedback device in the robot manipulator arm that provides data to the controller on the position of the arm.

internal sort
A sorting technique that creates sequences of records or keys. Usually, an internal sort is a prelude to a merge phase in which the sequences created are reduced to one by an external merge.

internal storage
Addressable main storage memory directly controlled by the central processing unit of a digital computer. Types of internal storage are core storage, monolithic integrated circuits, and thin-film memory rods.

International Organization for Standardization
An organization established to promote the development of standards to facilitate the in-

ternational exchange of goods and services and to develop mutual cooperation in areas of intellectual, scientific, technological, and economic activity.

interplant demand
Material to be shipped to another plant or division within the corporation. Although it is not a customer order, it is usually handled by the master production scheduling system in a similar manner.

interpolation
The means in NC by which curved sections are approximated by a series of straight lines or parabolic segments.

interpret
To translate into or restate in human language. To print at the top of a punched card the information punched on it, using a machine called an interpreter.

interpretation
Establishing a correspondence between the scene and a set of models. Assigning names to objects in a scene.

interpretation-guided segmentation
Using models to help guide image segmentation, by the process of extending partial matches.

interpreter
A program that translates and executes each source-language expression before translating and executing the next one. A routine that decodes instructions and produces a machine-language routine to be executed at a later time.

interpretive language
Allows a source program to be translated by an interpreter for use in a computer. The interpreter translates the interpretive-language source program into machine code, and instead of producing an object program, lets the program be immediately operated on by the computer. Interpretive language programs are used for solving difficult problems and for running short programs.

interrecord gap
An area on a data medium used to indicate the end of a block or record. Same as record gap.

interrogate
Retrieve information from computer files by use of predefined inquiries or unstructured queries handled by a high-level retrieval language.

interrupt
A break in the normal flow of a system or program occurring in such a way that the flow can be resumed from that point at a later time. Interrupts are initiated by signals of two types: signals originating within the computer system with the outside world (e.g., an operator or a physical process); or signals originating exterior to the computer systems to synchronize the operation of the computer system with the outside world (e.g., an operator or a physical process).

intersection
The point at which two lines meet. For example, the crossing of two centerlines is the intersection that defines the center of a circle.

in transit lead time
The time lag between the date of shipment (at the supplier's shipping point) and the date of receipt (at the customer's dock). Normally, a customer's orders specify the date by which goods should be at his or her dock. Consequently, this date should be offset by in transit lead time for establishing a ship date for the supplier.

intrinsic characteristics
In machine vision, properties inherent to the object, such as surface reflectance, orientation, incident illumination, and range.

intrinsic images
In machine vision, a set of arrays in registration with the image array. Each array corresponds to a particular intrinsic characteristic.

in-use
The code sets or attributes that will be used to interpret or be applied to subsequently received commands.

inventory
All the materials, parts, supplies, tools, and in-process or finished products recorded on the books by an organization and kept in its storerooms, warehouses, or plants.

inventory accounting
The administrative or bookkeeping aspect of inventory management. It covers the entry, auditing, control, and processing of inventory transactions, including transaction history and audit trails, as well as the gathering of transaction data from remote locations, physical control over stock, and inventory counting procedures.

inventory control
The activities and techniques of maintaining stock of items at desired levels, whether they be raw materials, work-in-process, or finished products.

inventory file
A file containing the net quantity of all items normally maintained in inventory.

inventory management
Management of the inventories, with the primary objectives of determining items that should be ordered and in what quantity; the timing of order release and order due dates; changes in the quantity called for; and the rescheduling of orders already planned. Its two broad areas are inventory accounting, which is the administrative aspect, and inventory planning and control, which consists of planning procedures and techniques that lead to inventory order action.

inventory policy
A definite statement of the philosophy of management on inventories.

inventory turnover
The number of times that an inventory "turns over" or cycles during the year. One way to compute inventory turnover is to divide the average inventory level into the annual cost of sales. For example, if average inventory were 3 million dollars and cost of sales were 21 million dollars, the inventory would be considered to "turn" seven times per year.

inventory valuation
The value of the inventory at either its cost or its market value. Because inventory value can change with time, some recognition must be taken of the age distribution of inventory. Therefore, the cost value of inventory, under accounting practice, is usually computed on a first-in-first-out (FIFO) or last-in-first-out (LIFO) basis or a standard cost system to establish the cost of goods sold.

inventory write-off
A deduction of inventory dollars from the financial statement because the inventory is no longer salable or because of shrinkage, that is, the value of the physical inventory is less than its book value.

inverted file
In information retrieval, a method of organizing a cross-index file in which a keyword identifies a record. The items, numbers, or documents pertinent to that keyword are indicated.

inverter
A logic element that receives a single input and changes it to its opposite state.

investment casting
Casting metal into a mold produced by surrounding (investing) an expendable pattern with a refractory slurry that sets at room temperature after which the wax, plastic, or frozen mercury pattern is removed through the use of heat. Also called precision casting or lost-wax process. A casting made by the process.

investment compound
A mixture of a graded refractory filler, a binder, and a liquid vehicle, used to make molds for investment casting.

invoke
Causes a designated code set to be represented by the bit combinations in the in-use table.

I/O address
A unique number assigned to each channel of an I/O module. The I/O address is determined by the location of the I/O module when plugged into the rail. The address number is used when programming, monitoring, or editing the ladder program element associated with the specific I/O channel.

I/O allocation table
A portion that controls how input and output data is interpreted relative to its channel number and address index position. Due to its function, I/O allocation also is called the traffic cop.

I/O bound
I/O bound refers to programs with a large number of I/O (input/output) operations that result in much wasted CPU Time.

I/O channel
A single input or output circuit of an I/O module. Each user input or output device is connected to an I/O channel. There are two channels on each input and output module. Each channel on a module is identical.

I/O electrical isolation
Separation of the field wiring circuits from the logic level circuits of the PC, typically done with optical isolation.

I/O forcing
The process of overriding the true state of an input or output. This function is normally used

as a debugging tool during system (PC) start-up.

I/O isolation
Separation of the field wiring circuits from the logic level circuits of the controller, typically done with optical isolation.

I/O module
The printed circuit board that normally is the termination for field wiring.

I/O rail
A mounting unit that provides plug-in sockets for eight I/O modules. The rail also contains circuitry for communication with the processor.

I/O scan
The time required for the PC processor to monitor all inputs and control all outputs. The I/O scan repeats continuously.

IPAD
The IPAD graphics software package that includes the IPAD graphics primitives, the interface software (device drivers) and the higher-level graphics-related software.

IPAD graphics primitives
IPAD standard lower-level graphics primitives, based on work done by SIGGRAPH.

IPIP precompilers
Programs that process application programs containing embedded data-manipulation commands that are translated into source-language communication areas and subroutine calls for communication with the IPIP runtime subsystems.

iris
An adjustable aperture built into a camera lens to permit control of the amount of light passing through the lens.

ironing
The operation of reducing the walls of drawn shells to assure uniform thickness.

iron loss
Alternate term for specific core loss.

irradiance
The brightness of a point in the scene.

island of automation
Stand-alone automation products (robots, CAD/CAM systems, NC machines) without the computer and system interface integration required for a cohesive system.

island of isolation
Machines without intelligence that in the factory cannot be integrated and controlled by a central computer.

ISO
Acronym for International Standards Organization.

isolated I/O module
A module that has each input or output electrically isolated from every other input or output on that module. Each input or output has a separate return wire.

isometric view
A drawing in which the horizontal lines of an object are drawn at an angle to the horizontal and all verticals are projected at an angle from the base. In plant design, pipes are drawn in isometric form for fabrication purposes and to facilitate coding for stress analysis. Such isometrics are normally presented schematically with unsealed pipe lengths and equal-size fittings. Isometrics can be generated automatically by a CAD system.

isomorphic representation
A representation in which there is a one-to-one correspondence between the scene and its representation (e.g., an image or a map).

ISO Reference Model for Open System Interconnect
The ISO draft proposal DP7498, a international standard for network architectures that defines a seven-layer model, specifying services and protocols for each layer.

isothermal annealing
Austenitizing a ferrous alloy and then cooling it to and holding it at a temperature at which austenite transforms to a relatively soft ferrite carbide aggregate.

isothermal transformation
A change in phase at any constant temperature.

isotropic mapping
A transformation that preserves aspect ratio.

isotropy
Quality of having identical properties in all directions.

issue cycle
The time required to generate a requisition for material, to pull the material from an inventory location, and to move it to its destination.

item
Any unique manufactured or purchased part or assembly, that is, end product, assembly, subassembly, component, or raw material.

item master file
Typically, this computer file contains identifying and descriptivve data, control values

(dead times, lot sizes, and so forth) and may contain data on inventory status, requirements, and planned orders. There is normally one record in this file for each stock-keeping unit. Item master records are linked together by product structure records, thus defining the bill of material.

iterate

To repeatedly execute a series of steps in a computer program or routine, for example, in the optimization of a design on the system. When numerous iterations are required, a CAD/CAM system can significantly speed-up the search process to find optimal solutions to problems such as component placement on PC boards, automatic dimensioning, and bulk annotation. In numerical analysis, CAD/CAM accelerates the process of converging to a solution by making successive approximations.

izod test

A pendulum type of single-blow impact test in which the specimen, usually notched, is fixed at one end and broken by a falling pendulum. The energy absorbed, as measured by the subsequent rise of the pendulum, is a measure of impact strength or notch toughness.

J

jaggies
A colloquial term for the jagged edges formed on raster-scan displays when displaying diagonal lines.

jet molding
Processing technique characterized by the fact that most of the heat is applied as the material passes through a nozzle or jet, rather than in a heating cylinder as is done in conventional processes.

jet spinning
For most purposes, similar to melt spinning. Hot-gas jet spinning uses a directed blast or jet of hot gas to "pull" molten polymer from a die lip and extend it into fine fibers.

jetting
Turbulent flow of resin from an undersize gate or thin section into a thicker mold section, as opposed to laminar flow of material progressing radially from a gate to the extremities of the cavity.

jig
Tool for holding component parts of an assembly during the manufacturing process or for holding other tools. A device that holds and locates a workpiece but also guides, controls, and limits one or more cutting tools. Also called a fixture.

jig boring
Boring with a single-point tool where the work is positioned upon a table that can be located so as to bring any desired part of the work under the tool. Thus, holes can be accurately spaced. This type of boring can be done on milling machines or *jig borers*.

jitter
Instability of a signal in either amplitude or phase or both.

job
A unit of work for the computing system from the standpoint of installation accounting and/or operating system control. A job consists of one or more steps or programs. Usually includes all necessary computer programs, linkages, files, and instructions to the operating system.

job analysis
Determination of the characteristics of a job through detailed observation and evaluation of the activities, facilities required, conditions of work, and the qualifications needed in a worker.

job assignment
Assignment of an employee to a machine or team, or of a job to a machine, employee, or team. Assignment, made by the manager, is based on information supplied by the computer through the terminal.

job breakdown
A description of a task in terms of its elements.

job class
A group of jobs or positions having approximately the same relative worth as determined by a job evaluation plan. A group of jobs or positions with duties and responsibilities so similar that individuals with approximately equivalent education, experience, skills, and the like are required for their satisfactory performance.

job classification
The grouping of jobs on the basis of the nature of the functions performed or level of pay, or on the basis of job evaluation, historic groupings, collective bargaining, or arbitrary determination.

job comparison scale
A listing of job factors and the points or money assigned to key jobs under each factor.

job content
The duties, functions, and responsibilities that constitute a given job.

job control
An operating system program that is called into storage to prepare each job or job step to be run. Some of its functions are to assign I/O devices to symbolic names, set switches for program use, log (or print) job control statements, and fetch the first phase of each job step.

job control language
A programming language used to code job control statements. These statements supply information to the operating system and the operators about the program, for example, name of user, how much memory is required, estimated runtime, priority, tapes required, other programs, and so on. The JCL for modern operating systems is often quite complex,

and there are probably nearly as many user-prepared jobs that fail to execute due to JCL errors as failures due to compiler language errors.

job control statement
A statement in a program used to identify the job or describe its requirements to the operating system.

job dispatching
Instructions given to an employee concerning the next job he or she is to perform.

job lot
A relatively small number of a specific type of part or product that is produced at one time.

job lot layout
An arrangement of machines, equipment, and facilities specially set up or arranged to handle job lot production.

job order costing
A costing system in which costs are collected to specific jobs. This system can be used with either actual or standard costs in the manufacturing of distinguishable units or lots of products.

job shop
A discrete-parts manufacturing facility characterized by a mix of products of relatively low-volume production in batch lots. A batch-oriented production shop capable of processing many different types of work, where each job typically flows through the shop in a different sequence and number of operations.

job status
The stage of activity toward a defined task or responsibility at any given time. May be a measurement against scheduled requirements of an overall task or plan.

job step
A unit of work for a computing system, presented as a request for execution of an explicitly identified program and a description of resources it requires.

job stream
The stack of jobs that are to be inputted, awaiting initiation and processing. Sometimes called input job stream. It follows the same principle as batching jobs.

job transfer and manipulation protocol
Job transfer and manipulation (JTM) is a protocol for implementing distributed batch jobs. JTM specifies the way users should specify division of a job, as well as where and when the pieces are to be processed and delivered. JTM is geared not to interact with the user.

joggle
An offset in a flat plane consisting of two parallel bends in opposite directions by the same angle.

joint
A rotational or translational degree of freedom in a manipulator system.

jointed arm robot
A robot whose arm consists of two links connected by "elbow" and "shoulder" joints to provide three rotational motions. This robot most closely resembles the movement of the human arm.

joint interpolated motion
A method of coordinating the movement of the joints, so that all joints arrive at the desired

Jointed-Arm Robot
Jointed-Arm Robot Work Envelope

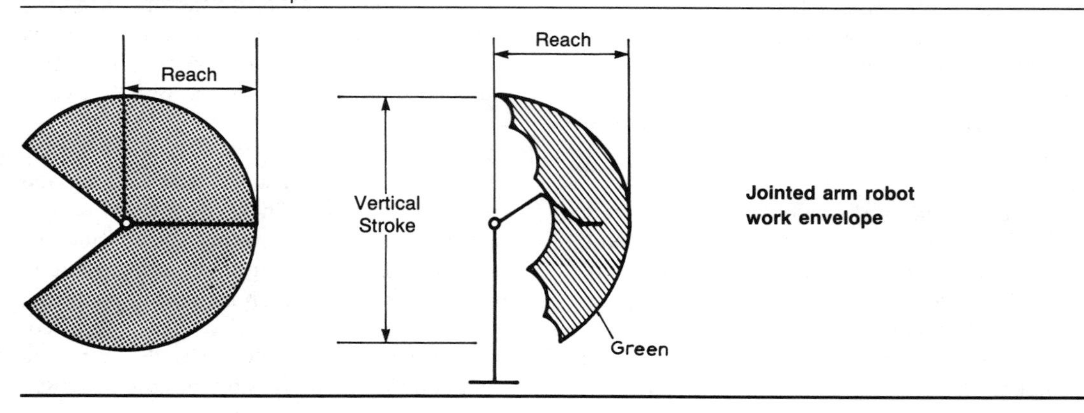

location simultaneously. This method of servocontrol produces a predictable path regardless of speed and results in the fastest cycle time for a particular move.

joint space

The space defined by a vector whose components are the angular or translational displacement of each joint of a multi-degree-of-freedom linkage relative to a reference displacement for each such joint.

Josephson junction

An experimental class of integrated circuit designed to operate at extremely high speeds (roughly 1 billionth of a second per operation) and at temperatures only a few degrees above absolute zero (−459.7° F).

joule

In pulsed lasers, a measurement of energy per pulse. The rate at which energy is generated is a measure of output power, with 1 watt equal to a rate of 1 joule per second.

joystick

An input device that directly controls the cursor. The stick is moved in the same direction as the user wishes the cursor to move on the screen. A CAD data-entry device employing a hand-controlled lever to manually enter the coordinates of various points on a design being digitized into the system.

Julian calendar

The calendar that is used in data processing. The dates are represented by five-digit numbers; the first two digits pertain to the year and the last three (001 through 365 or 366) to the day of the year.

jump

Change in sequence of the execution of the program instruction, altering the program counter (i.e., a branch). A departure from sequence in executing instructions in a computer.

jumper

A short length of conductor used to make a connection between terminals, around a break in a circuit, or around an instrument.

kan ban
Japanese term for just-in-time scheduling—a system that dramatically reduces work-in-process inventory by delivery of raw materials, parts, and subassemblies to production in small batches as they are needed.

Karnaugh map
A rectangular diagram of a logic function of variables, drawn with overlapping rectangles representing a unique combination of the logic variables, such that an intersection is shown for all combinations.

keep-out
On the Computervision system, defines the boundaries of a printed circuit board and the areas inside of the board that are not to be used for routing.

keep-out areas
User-specified areas on a printed circuit board layout where components or circuit paths must not be located and that therefore must be avoided by CAD/CAM automatic placement and routing programs.

kerf width
The width of the cut made by a laser or other cutting tool in a material.

kernal
An array of numbers that are designed to be used by an operator on repetitive sections of an image to generate one pixel of a new image.

key
One or more characters within a set of data that contain information about the set, including the identification.

keyboard
Similar to a typewriter. The difference is that the information is translated by a device that transforms each word into a series of electrical signals that open and close the switches in the computer in order to store information and instruction.

keyboard terminal
A terminal through which data can be entered to a data-processing system by means of a typewriterlike keyboard.

key data-entry devices
Keyboard-equipped devices used to prepare data so that the computer can accept it, including keypunches (card punches) and the newer key-to-tape and key-to-disk units.

keyhole
The rapidly produced, deep hole that is made by a laser in deep-penetration welding.

keypunch
A keyboard-actuated device that punches holes in cards.

keyword
One of the significant and informative words in a title or document that describe the content of that document.

keyword-in-context index
An index that lists available programs in alphabetical order by each keyword in the title. A keyword-in-context index is prepared by highlighting each keyword of the title in the context of words on either side of it and aligning the keywords of all titles alphabetically in a vertical column.

kill
To eliminate or erase. Frequently, a control character in a word-processing program meaning to drop or purge a line of text or a blank line.

kilobyte (K or K byte)
Often used as a measure of memory capacity; 1024 bytes (1024 being one K, or two to the tenth power).

kinematic error
Kinematic errors of coordination between the various moving bodies that make up a machine that is, deviations from ideal.

kinematic mount
A mount that mechanically constrains an object by the minimum number of constraints necessary to prevent undesired motion.

kinematics
A computer-aided engineering (CAE) process for plotting or animating the motion of parts in a machine or a structure under design on the system. CAE simulation programs allow the motion of mechanisms to be studied for interference, acceleration, and force determinations while still in the design stage.

kit
The components of an assembly that have been pulled from stock and readied for movement to the assembly area.

Knowledge Acquisition

Typical Knowledge Acquisition Process for Building an Expert System

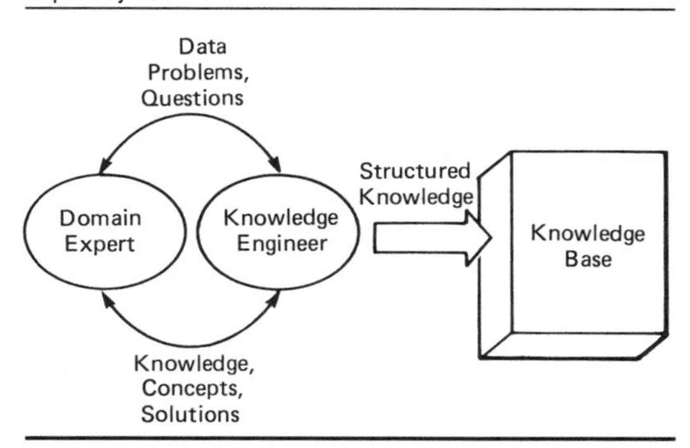

kitting
The process of removing components of an assembly from the stockroom and sending them to the assembly floor as a kit of parts. This action may take place automatically, whenever a full set of parts is available, and/or it may be done only upon authorization by a designated person.

knoop hardness
Microhardness determined from the resistance of metal to indentation by a pyramidal diamond indenter, having edge angles of 172°, 30′, and 130°, making a rhombohedral impression with one long and one short diagonal.

knot sequence
A nondecreasing sequence of real numbers used to specify parametric spline curves.

knowledge
In artificial intelligence, facts, beliefs, and heuristic rules.

knowledge acquisition
The extraction and formulation of knowledge derived from extant sources, especially from experts.

knowledge base
The repository of knowledge in a computer system. Facts, assumptions, beliefs, and heuristics; "expertise"; methods of dealing with the data base to achieve desired results, such as a diagnosis or an interpretation or a solution to a problem.

knowledge base management
Management of a knowledge base in terms of storing, accessing, and reasoning with the knowledge.

knowledge base management system
One of three subsystems in an expert system. This subsystem "manages" the knowledge base by automatically organizing, controlling, propagating, and updating stored knowledge. It initiates searches for knowledge relevant to the line of reasoning upon which the inference subsystem is working. The inference subsystem is one of the other two subsystems in an expert system; the third is the human interface subsystem with which the end-user communicates.

knowledge engineer
An artificial intelligence specialist responsible for the technical side of developing an expert system. The knowledge engineer works closely with the domain expert to capture the expert's knowledge and place it in a computer system's knowledge base.

knowledge engineering
The use of artificial intelligence techniques and a base of information or knowledge (facts, rules, and procedures) about a specific activity to control systems automatically. This type of system is called a *knowledge-based system*. The discipline that addresses the task of building expert systems; the tools and methods that support the development of an expert system.

knowledge information processing systems
The new "fifth" generation of computers that the Japanese propose to build that will have symbolic inference capabilities, coupled with very large knowledge bases, and superb human interfaces, all combined with high processing speeds, so that the machines will greatly amplify human intellectual capabilities.

knowledge representation
The techniques used to store information in a knowledge base. Knowledge representation techniques include logic, semantic networks, production rules, frames, and scripts.

knowledge source
Generally, a body of domain knowledge relevant to a specific problem. In particular, a codification made applicable for an expert system.

knurling
Impressing a design into a metallic surface, usually by means of small hard rollers that carry the corresponding design on their surfaces.

L

label
An ordered set of characters used to symbolically identify an instruction, a program, a quantity, or a data area. A structure element consisting of an identifier unique to the structure that can be used as a marker to facilitate structure editing.

label display associativity
A predefined associativity that is used by those entities that have one or more possible displays for their entity label. Entities requiring this associativity will have pointers in their directory entry to a label display associativity instance entity.

labeling
The process of assigning different numbers to the picture elements of different blobs in a binary image.

labor cost
That part of the cost of goods, services, and the like attributable to wages. It commonly refers to direct workers, but may include indirect workers as well.

labor loading
The process of applying expected labor requirements against the capacity for that labor.

labor productivity
The rate of output of a worker or group of workers per unit of time, usually compared with an established standard or expected rate of output.

laborsaving ratio
The relation of the unit labor cost of an improved method to the unit labor cost of another method.

labor ticket
A form used to record the application of labor to specific jobs or production operations.

ladder diagram
An electrical engineering technique to illustrate schematically functions in an electrical circuit (relays, switches, timers, etc.) by diagraming them in a vertical sequence resembling a ladder.

ladder elements
The symbolic circuit components used to create a ladder diagram program sequence. Ladder program elements are: (1) normally closed contact, (2) normally open contact, (3) coil, (4) branch tie, (5) horizontal line, and (6) horizontal space.

ladder matrix
A rectangular array of ladder elements, determined by the allowable number of series contacts that can be programmed across a row and the number of paralleled branches that can be programmed to form a single program sequence.

lag
The tendency of the dynamic response of a passive physical system to be delayed. The phase difference between input and response sinusoids. Any time parameters that characterize the delay of a response relative to an input. The difference between the actual position of a machine and its commanded position during motion. The persistance of image information in an image sensor for two or more scan periods after excitation (light) is taken away.

LAMA-25
A multiaxis contouring processor with line, circle, and arc capability for use with CNC machine tools.

lambert
A unit of luminance equal to 1 + candela per square centimeter, and therefore equal to the uniform luminance of a perfectly diffusing surface emitting or reflecting light at the rate of 1 lumen per square centimeter.

lambertian surface
A diffusing surface having the property that the intensity of light emanating in a given direction from any small surface component is proportional to the cosine of the angle. The brightness of a lambertian surface is constant, regardless of the viewing angle.

lamination factor
Measure of effective volume of laminated structure that is composed of strips of magnetic material. Ratio of volume of structure as calculated from weight and density of strips to measured solid volume of structure under pressure.

lancing
The operation of piercing a material in a shape that will leave the portion cut affixed to the part.

language
A defined group of representative characters or symbols, combined with specific rules necessary for their interpretation. The rules enable an assembler or compiler to translate the characters into forms (such as digits) meaningful to a machine, a system, or a process. A set of rules and conventions used to convey information.

language binding
The binding of a functional specification to the syntax of a particular programming language.

language processor
A computer program that performs such functions as translating, interpreting, and other tasks required for processing a specified programming language.

language translator
A general term for any assembler, compiler, or other routine that accepts statements in one language and produces equivalent machine-language instructions.

lapping
The operation of removing from a surface the undulations, roughness, and tool marks left by previous operations.

large interactive surface
An automated drafting table used as an input device to digitize large drawings into storage or working registers.

large-scale computer
A computer with the highest operating characteristics. Large-scale computers provide complex and powerful programmable logic to attach complex problems that require highly centralized computing power. Examples: CDC 7600, Cray Amdahl 470, ILLIAC IV, and others. Some operate at speeds of 100 million instructions per second (mips).

large-scale integration
A classification for a scale of complexity of an integrated electron circuit chip. Integrated circuits (ICs) containing 100 or more gate equivalents or other circuitry of similar complexity. Other classes are medium-scale integration (MSI) and small-scale integration (SSI).

laser
Acronym for Light Amplification by Stimulated Emission of Radiation. A device that produces a coherent monochromatic beam of light. A device that transmits an extremely narrow and coherent beam of electromagnetic energy in the visible light spectrum. In metalworking, a laser is a multipurpose noncontact processing tool.

laser-assisted machining
The use of a laser beam to heat and soften a metal workpiece just ahead of a cutting tool to make the workpiece easier to machine.

laser cavity
The laser resonator or tube in which the lasing process occurs.

laser disk
An analog or digital storage medium, written and read by laser. Also called video disk.

laser head
The enclosure containing the active medium, resonator cavity, and all other laser accessories and components except the power supply.

laser machining center
A laser combined with a machine tool or workpiece handling/positioning system, under the control of a computer, to perform a variety of processing tasks.

laser pump
The flashlamp used to excite, or ''pump,'' the atoms of the lasing medium.

laser rod
The solid material in which a lasing medium, or impurity, is embedded in a solid-state laser; generally it is a crystal.

laser scanner
An optical bar code reading device using low-energy laser light beam as its source of illumination. An optical scanning device that uses the intense monochromatic light beam given off by a laser as its source of illumination.

lasing
The process of generating a laser beam through stimulated emission of radiation.

last-in-first-out (LIFO)
A system of inventory control that designates material be disbursed in the reverse order as received.

latch
A discrete reference that can be utilized to remember the status of a logic line coil during a power failure, such that when power is restored, the line can be returned to the condition (on or off) it held prior to the power failure if properly programmed. Also referred to as delayed outputs, due to their unique timing.

latching relay
A relay constructed so that it maintains a given position by mechanical means, until released mechanically or electrically.

latency
The time between an address interpretation and the start of the actual transfer. Latency includes the delay associated with access to storage devices.

latent heat
Thermal energy absorbed or released when a substance undergoes a phase change. Heat that must be applied to material to effect change in state without change in temperature. For example, latent heat of fusion of ice water is 80 cal/gm.

lay
The direction of the predominant surface pattern that remains after cutting, grinding, lapping, or other processing.

layer
A logical concept used to distinguish subdivided group(s) of data within a given drawing. May be thought of as a series of transparencies (overlayed) in any order, yet having no depth. Operator may specify display elements (layer) to be visible. In CAD, one of the image planes built up to make a composite picture. Layering allows separation by function so that portions of the image can be examined separately or ones changed without affecting others.

layer discrimination
The process of selectively assigning colors to a layer, or highlighting entities by means of gray levels, to graphically distinguish among data on different layers displayed on a CRT.

layering
A method of logically organizing data in a CAD/CAM data base. Functionally different classes of data (i.e., various graphic/geometric entities) are segregated on separate layers, each of which can be displayed individually or in any desired combination. Layering helps the designer distinguish among different kinds of data in creating a complex product, such as a multilayered PC board or IC.

layers
User-defined logical subdivisions of data in a CAD/CAM data base that may be viewed on the CRT individually or overlaid and viewed in groups.

layout
A visual representation of the physical components and mechanical and electrical arrangement of a part, product, or plant. A layout can be constructed, displayed, and plotted in hard copy on a CAD system. The specific geometric design of an IC chip in terms of regions of diffusion, polysilicon metalization, and so forth. The IC layout is a specific implementation of the functionality described in a schematic design and can be created interactively on a CAD system.

leader
A line that can be generated on the CAD system, leading from a displayed note or dimension and ending in an arrowhead touching the part to which attention is directed.

leader entity
An annotation entity, also referred to as an arrow, that consists of an arrowhead and one or more line segments. In the case of an angular-dimension entity, the line segment is replaced by a circular arc segment. In general, these entities are used in connection with other annotation entities to link text with some location.

lead error
The linear error generated by a screw of incorrect pitch.

leading zeros
Zeros that have no significance in the value of an arithmetic integer; all zeros to the left of the first significant integer digit of a number.

leading-zero suppression
Feature of a control system, whereby the zeros preceding the first nonzero integer of a number need not be described on the tape. Contrasted to trailing-zero suppression.

lead screw
A precision machine screw that, when turned, drives a sliding nut or mating part in translation.

lead-screw compensation
Automatic compensation for measured errors in each lead screw. Can be accomplished by cam compensation or by an error table stored in the memory of the CNC system. Can improve positioning accuracy.

lead-through
Programming or teaching by physically guiding the robot through the desired actions. The speed of the robot is increased when programming is complete.

lead-through programming
A means of teaching a robot by leading it through the operating sequence with a control console or a hand-held control box.

lead time
A span of time required to perform an activity. In a production and inventory control context, the activity in question is normally the procurement of materials and/or products, either from an outside supplier or from one's own given lead time, queue time, move or transportation time, receiving and inspection time.

lead time offset
A term used in MRP where a planned order receipt in one time period will require the release of that order in some earlier time period, based on the lead time for the item. The difference between the due date and the release date is the lead-time offset.

leaf
A terminal node in a tree representation.

leak testing
Technique of liquid-penetrant testing, wherein penetrant is applied to one side of a material and observation is made on the opposite side to ascertain the presence of voids extending throughout the material.

learning
The process of improving performance by changing knowledge or control.

learning control
A control scheme whereby experience is automatically used to change control parameters or algorithms.

learning curve
A calculation made in, planning technique based on the premise that workers will be able to produce a new product more quickly after they get used to making it. Learning curve theory states that in some production situations, each time that cumulative production volume doubles the cost necessary to produce the items decreases by some percentage due to the workers' greater production experience. A graph plotting the course of learning in which the vertical axis plots a measure of proficiency while the horizontal axis represents some measure of practice or experience. Also known as experience curve, a quantitative model relating the factors of time, productivity improvement, and cost applied to prediction of future costs under conditions of production volume increases.

leased line
A communication channel leased for exclusive use from a common carrier and frequently referred to as a private line.

least commitment
A technique for coordinating decision making with the availability of information so that problem-solving decisions are not made arbitrarily or prematurely but are postponed until there is enough information.

least significant character
The extreme right-hand character in a group of characters. Contrast to most significant character.

least squares method
A method of smoothing or curve fitting that selects the fitted curve so as to minimize the sum of squares of deviations from the given points. A method of curve fitting that selects a line of best fit through a plot of data so as to minimize the sum of squares of the deviations of the given points from the line.

least total cost
A dynamic lot-sizing technique that calculates the order quantity by comparing the carrying cost and the setup (or ordering) costs for various lot sizes and selects the lot where these are not nearly equal.

least unit cost
A dynamic lot-sizing technique that adds ordering cost and inventory carrying costs for each trial lot size and divides by the number of units in the lot size, picking the lot size with the lowest unit cost.

LED display
An illuminated visual readout composed of light-emitting diode (LED) alphanumeric character segments.

left-handed coordinate system
A three-dimensional coordinate system in which, if the positive X axis (thumb) points to the right, and the positive Y axis (index finger) points up, then the positive Z axis (coming out of the palm of the left hand) points away from the viewer.

left justified
A field of numbers (decimal, binary, and so forth) that exists in a memory cell, location, or register, possessing no zeros to its left.

lens
A transparent optical component consisting of one or more pieces of optical glass (elements) with curved surfaces (usually spherical) that converge or diverge transmitted light rays.

lens speed
The ability of a lens to transmit light; represented as the ratio of the focal length to the diameter of the lens.

lens system
Two or more lenses so arranged as to act in conjunction with one another.

LETS
LETS is a conversational computer graphics software system for the preparation of layout, cost estimation, tool design and tooling plan, and setup instructions preparation implemented on an Apple II plus 48K desktop computer.

letter-quality printer
A printer that generates output that is suitable for high-quality business correspondence. Term implies that output quality matches that of a standard office typewriter.

level
Every part or assembly in product structure is assigned a level code signifying the relative level in which that part or assembly is used within that product structure. Normally, the end items are assigned level "0" and the components/subassemblies going into its level "1" and so on. MRP explosion process starts from level "0" and proceeds downward one level at a time. In data management structures, the degree of subordination in a hierarchy. An entity attribute that defines a graphic display level to be associated with the entity.

leveled time
The average time, adjusted to account for differences in skill, effort, conditions, and consistency between workers and the factors surrounding an operation.

leveling
A method of performance rating in which the causes for the observed performance, considered to be skill, efforts, conditions, and consistency are evaluated.

level loading
Arranging the load on a work center or factory so that the loads per time period are approximately equal.

level of automation
The degree to which a process has been made automatic. Relevant to the level of automation are questions of automatic failure recovery, the variety of situations that will be automatically handled, and the situation under which manual intervention or action by humans is required.

level of service
A measure of the demand that is routinely satisfied by inventory, for example, the percentage of orders filled from stock; the percentage of dollar demand filled from stock.

library
A collection of organization information. In electronic data processing, a program library is a collection of available computer programs and routines. The libraries used by an organization are known as its data bank.

library programs
A software collection of standard routines and subroutines by which problems and parts of problems may be solved on a given computer.

library routine
A proven routine that is stored in a program library.

light
Electromagnetic radiation. Visible light, light detectable by the eye, has wavelengths in the range of 400 to 750 nanometers.

light button
An on-screen image of a button or choice that is activated with a light pen, mouse, or other pointing device. Light buttons can be called up quickly or removed by the program and can be selected without the user looking away from the screen.

light-emitting diode (LED)
A semiconductor diode, generally made from gallium arsenide, that can serve as a near-infrared light source when voltage is applied continuously or in pulses. LEDs have extremely long lifetimes when properly operated; being solid state, they are very resistant to shock and vibration.

lightness
The attribute of color that permits any color to be classified as equivalent to one of a series of grays ranging between black and white. The term *shade* is often used to describe differences in lightness. Lightness difference is measured as part of the color-difference test.

light operated
Condition in which the control operates when the light beam is uninterrupted.

light pen
In computers, the input device that moves the cursor on the screen and sets points when the end of the pen is touched against the screen. A hand-held scanning wand that is used as a contact bar code reader held in the hand. A tool for CRT terminal operators that causes the computer to change or modify the display

Light Pen
Light Pen, Used to Create Graphics on a CAD System

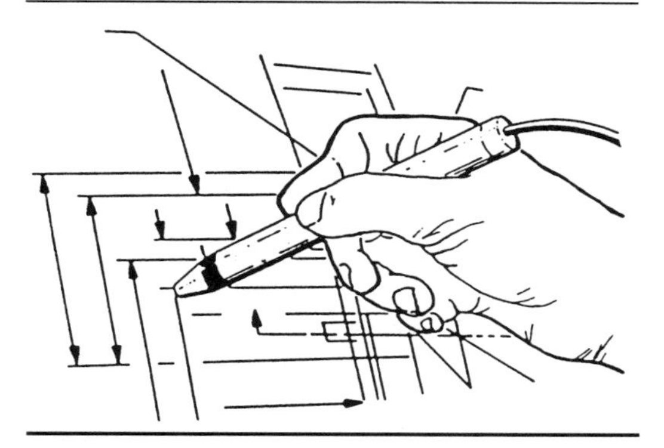

on the cathode-ray tube. The pen's response to light from the display is transmitted to the computer, which, in turn, relates the computer action to the section of the image being displayed. In this way, the operator can delete or add text, maintain tighter control over the program, and choose alternative courses of action.

light pen detect
Light pen contact with a program-detectable display element; the system returns identifying data for the detected element.

lightwave communications
A term coined to identify the use of light as an information carrier. The term is used in place of optical communications to avoid confusion with visual information and image transmission. Fiber-optic cables (light guides) are a direct replacement for conventional coaxial cables and wire pairs. The glass-based transmission facilities occupy far less physical volume for an equivalent transmission capacity.

limit-detecting hardware
A device for stopping robot motion independently from control logic.

limited continuous word/speech
Voice recognition capability for certain sets of words uttered without pause (typically digits, such as part numbers or postal zip codes), that can be trained into user-programmable voice equipment designed with proper recognition-processing algorithms.

limited-degree-of-freedom robot
A robot able to position and orient its end-effector in fewer than six degrees of freedom.

limited sequence robot
A simple or nonservo type of robot. Movement is controlled by a series of limit or stop switches. Also called bang-bang robot.

limiting devices
To qualify as a means for restricting the work envelope, these devices must stop all motion of the robot independent of control logic.

limiting operation
The operation with the least capacity in a total system with no alternative routings. The total system can be effectively scheduled by simply scheduling the limiting operation. Synonymous with bottleneck.

limit switch
An electrical switch that is actuated when the limit of a certain motion is reached and the actuator causing the motion is deactivated.

line
On a terminal, one or more characters entered before a return to the first printing or display position. A circuit connecting two or more devices. A thin connected set of points contrasting with neighbors on both sides. Line representations are extracted from edges.

linear
The quality of an input/output relationship in which there is direct proportionality.

linear array
A solid-state video detector, consisting of a single row of light-sensitive semiconductor devices. This is used in linear array cameras.

linear array camera
A solid-state television camera that has only one row of photosensitive elements.

linear coefficient of thermal expansion
Extent to which a material elongates when heated. Unit increase in length per unit rise in temperature over specified temperature range. Slope of the temperature to dilation curve at a specified temperature. Mean coefficient is mean slope between two specified temperatures.

linear dimension entity
An annotation entity used to represent a distance between two locations.

linear discontinuities
Ragged lines of variable width that may appear as a single jagged line or exist in groups. They may or may not have a definite line of continuity, often originate at a casting surface, and usually become smaller as a function of depth.

linear interpolation
A function automatically performed in the control that defines the continuum of points in a straight line based on only two taught coordinate positions. All calculated points are automatically inserted between the taught coordinate positions upon playback. The ability to control the motion of one or more axes linearly. The type of interpolation in which straight lines are employed to approximate a required surface, by controlling the motions of one or more machine axes. A method of contouring curved shapes by a series of straight line segments.

linearity
The degree to which an input/output is a directly proportional relationship.

linear language
A language that is automatically expressed as a linear representation. For example, FORTRAN is a linear language; a flowchart is not.

linear programming
The concept of expressing the interrelationship of activities of a system in terms of a set of linear constraints in nonnegative variables. A program, that is, values of the variables, is selected that satisfies the constraints and optimizes a linear objective function in these variables.

line balancing
An assembly-line process can be divided into elemental tasks, each with a specified time requirement per unit of product and a sequence relationship with the other tasks. Line balancing is the assignment of these tasks to workstations so as to minimize the number of workstations and to minimize the total amount of unassigned time at all stations. Line balancing can also mean a technique for determining the product mix that can be run down an assembly line to provide a fairly consistent flow of work through that assembly line at the planned line rate. For example, if an automotive assembly line happened to be scheduled one day with nothing but convertibles, some workers would be standing idle, while others would not be able to keep pace with the line.

line control block
A storage area containing control information required for scheduling and managing line operations. One line control block is maintained for each line in the data communication system.

line detectors
Oriented operators for finding lines in an image.

line entity
A geometric entity consisting of a straight segment connecting two points in space.

line followers
Technique for extending lines currently being tracked on a CAD system.

line font definition entity
A structure entity that defines a line font.

line fonts
Repetitive pattern used to give meaning to a line, for instance, solid, dashed, dotted, and so forth. Repetitive pattern used in CAD to give a displayed line appearance characteristics that make it more easily distinguishable, for example, a solid, dashed, or dotted line.

line hit
A disturbance causing a detectable error on a communications line.

line number
The identification number of an instruction or statement in a sequential program.

line printer
The computer output peripheral that prints an entire line of characters as a unit. This principle is largely responsible for the high printing speed.

line production
A method of plant layout in which the machines and other equipment required, regardless of the operations they perform, are arranged in the order in which they are used in the process.

line scan
A one-dimensional image sensor.

line smoothing
An automated mapping capability for the interpolation and insertion of additional points along a linear entity, yielding a series of shorter linear segments to generate a smoothed curve appearance to the original linear component. The additional points or segments are created only for display purposes and are interpolated from a relatively small set of stored representative points. Thus, data storage space is minimized.

line speed
The maximum data rate that can be transmitted over a communications line. The rate at

which signals can be transmitted over a channel, usually measured in bauds or bits per second.

line synchronization
The ability to synchronize the operation of an industrial robot with a moving production line so that the robot automatically compensates for variations in line speed.

linetype
An attribute that indicates the style of the image of certain visible lines (e.g., solid or dashed).

line (video screen)
A horizontal row of characters on a display or terminal screen. Most screens are either 40- or 80-width column displays.

line weight
An entity attribute that is used to determine the line display thickness for that entity.

linewidth
Also known as bandwidth, this is the wavelength range over which most of the laser's energy is distributed.

linewidth scale
An attribute that indicates the relative width of the image of a visible line. The linewidth scale factor is applied to a workstation-dependent nominal value.

link
In computer programming, the part of a computer program, in some cases a single instruction or an address, that passes control and parameters between separate portions of the computer program; also in computer programming, to provide such a link. A communications path between two nodes in a network.

linkage
A means of communicating information from one routine to another.

linkage editor
An operating system program that prepares the output of language translators for execution. It combines separately produced modules; resolves cross references among them; replaces, deletes, and adds control sections, and produces an executable load module.

link redundancy level
The ratio of actual number of links to the minimum number of links required to connect all nodes of a network.

liquid crystal display
A reflective visual readout of alphanumeric characters. Since its segments are displayed only by reflected light, it has extremely low power consumption—as contrasted with a LED display that emits light.

liquid honing
Polishing metal by bombardment with an air-ejected liquid containing fine solid particles in suspension. If an impeller wheel is used to propel the suspension, the process is called wet blasting.

liquid laser
A laser in which the active element is either an organic dye or an inorganic liquid. Not used for metalworking applications.

liquid temperature
Temperature at which an alloy finishes melting during heating or starts freezing during cooling, under equilibrium conditions. For pure metals, same as solidus temperature and known simply as melting point. Effective liquids temperature is raised by fast heating and lowered by fast cooling.

liquidus
In a constitution or equilibrium diagram, the locus of points representing the temperatures at which the various compositions in the system begin to freeze on cooling or to finish melting on heating.

LISP
Popular programming language for use in artificial intelligence. LISP is the acronym for LISt Processing language. LISP was the first language to concentrate on working with symbols instead of numbers. Although it was introduced by John McCarthy in the early 1960s, continuous development has enabled LISP to remain dominant in artificial intelligence. Lately, LISP has proved to be an outstanding language for systems programming as well.

LISP machine (or "AI workstation")
A single-user computer designed primarily to expedite the development of AI programs.

list
An ordered set of items of data. A data structure in which each item of data can contain pointers to other items. Any data structure can be represented in this way, which allows the structure to be independent of the storage of the items. The term also means to print data.

listing
A printout, usually prepared by a language translator, that lists the source-language statements of a program.

list processing
A method of processing data in the form of lists. Usually, chained lists are used so that the logical order of items can be changed without altering their physical limitations.

live load
The load being moved, not including the weight of equipment being used to effect the movement.

living hinge
A hinge formed by injection-molding as an integral part of the key and bezel.

load
The power delivered to a machine or apparatus. The weight (force) applied to the end of the robot arm. A device intentionally placed in a circuit or connected to a machine or apparatus to absorb power and convert it into the desired useful form. To insert data into memory storage. In computer operations, the amount of scheduled work, usually expressed in terms of hours of work. In programming, to feed data or programs into the computer.

load-and-go
An operating technique in which there are no stops between the loading and execution phases of a program and that may include assembling or compiling.

load capacity
The maximum total weight that can be applied to the end of the robot arm without sacrifice of any of the applicable published specifications of the robot.

load center
A group of workstations that can all be considered together for purposes of loading and scheduling. The horizontal longitudinal distance from the intersection of the horizontal load-carrying surface and vertical load-engaging face of the forks (or equivalent load-positioning structure) to the center of gravity of the load.

load cycle time
The length of time in the manufacturing process required for unloading the last workpiece and loading the next one.

load deflection
The difference in position of some point in a body between a nonloaded and an externally loaded condition. The difference in position of a manipulator hand or tool, usually with the arm extended, between a nonloaded condition and an externally loaded condition. Either or both static and dynamic loads may be considered.

loader
A program that operates on input devices to transfer information from offline memory to online memory.

load factor
The ratio of the average load over a designated period of time to the peak load occurring in that period.

loading
The process of applying expected requirements against capacity.

load leveling
Spreading orders out in time or rescheduling operations so that the amount of work to be done in the time period tends to be distributed evenly.

load profile
The projected load pattern over several periods of time. A display of future capacity requirements based on planned and released orders over a given span of time.

load sharing
The distribution of a given workload among a number of computers on a network.

local
An image operator that is variable where the variation is a function of the values of the pixel nearby to wherever the operator is being applied.

local area network
A system for linking terminals, programs, storage, and graphic devices at multiple workstations over relatively small geographic areas for rapid communication.

local batch
Offline batch processing. Users send raw data by courier to service bureaus to be processed and returned.

local modeling transformation
The modeling transformation of the current structure, composed during structure transversal with the global modeling transformation inherited from the parent structure, to form the current composite modeling transformation.

local operation
Transformers of the gray-scale value of the picture elements, according to the gray-scale values of the element itself and its neighbors

Local Area Network
CAD/CAM Local Area Network. Reprinted with permission of McGraw-Hill Book Company, from *CAD/CAM Handbook* by Eric Teicholtz.

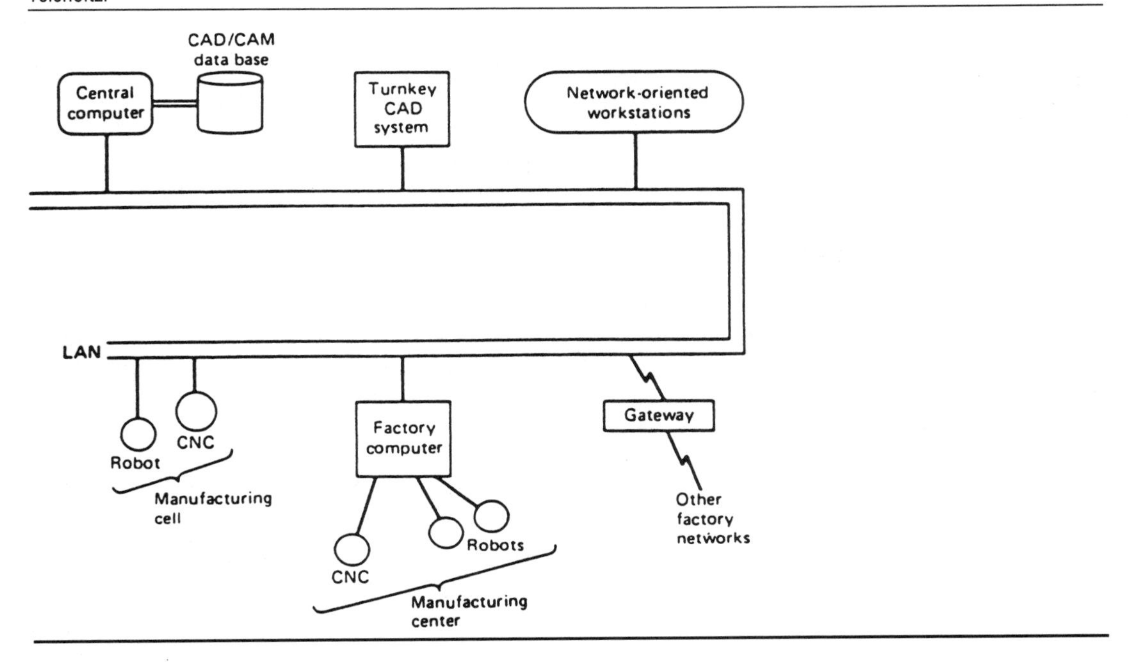

in a given neighborhood. Examples are gradient, sharpening, smoothing (noise filtering), edge extraction, and so on.

locating surfaces
Machined surfaces on a part that are used as reference surfaces for precise locating and clamping of the part in a fixture.

location
Part of the main memory in which an instruction or item can be stored, usually identified by an address. The coordinates of an image or object relative to an observer. An image location is defined by the *X-Y* coordinates of its centroid. A three-dimensional object's location is defined by the *X*, *Y*, and *Z* coordinates of some point, such as its center of gravity.

location code
The alphanumeric code that conveys the location of inventory in a stockroom or warehouse.

location ring
A ring that serves to align the nozzle of an injection cylinder with the entrance of the sprue bushing and the mold to the platen of a molding machine.

locator
An input class providing positioning information.

locking shift
An invocation of a code set into the in-use table that remains in effect until another code set is invoked in its place.

locomotion
Some means of moving around in a specified environment.

log
A record of values and/or action for a given function.

logger
A functional unit that records events and physical conditions, usually with respect to time.

logic
A means of solving complex problems through the repeated use of simple functions that define basic concepts. Three basic logic functions are AND, OR, and NOT. A technique

for drawing inferences that relies on formal rules for manipulating symbols.

logical comparison
A logic operation to determine whether two strings are identical.

logical data independence
The capacity to change the overall logical structure of the data without changing the application programs.

logical device coordinates
A device-independent Cartesian coordinate system in which the composition of images is specified at the workstation level. Thus, workstation window and workstation viewport are both located in logical device coordinates space.

logical expression
In assembler programming, a conditional assembly expression that is a combination of logical terms, logical operations, and paired parentheses.

logical graphic functional/logical operations
A CAD capability that applies Boolean operations (AND, OR AND/NOT, XOR) to areas of graphic entities. This provides the designer with interactive tools to create new figures from existing ones or to expand a design-rules checking program, for example, in IC design.

logical inferences per second
A means of measuring the speed of computers used for AI applications.

logical operation
Execution of a single computer instruction.

logical operations
Nonarithmetical operations such as selecting, sorting, matching, comparing, and so forth.

logical picture element
A geometric construct associated with the drawing point whose size determines the stroke width of graphics primitives. Although the terms pixel and pel are synonymous in common usage, the standard uses pixel for physical picture elements and pel for logical picture elements.

logical record
A record whose scope, direction, or length is governed by the specific nature of the information or data that it contains, rather than by some feature or limitation of the storage device that holds it. Such records differ in size from the physical records in which they are contained.

logical representation
Knowledge representation by a collection of logical formulas (usually in first-order predicate logic) that provide a partial description of the world.

logical shift
A shift that equally affects all of the characters of a computer word.

logical terminal
A device addressable by its logical function, rather than its physical address. Translation from logical to physical addresses is achieved by a common routine using a table. The table can be updated during online operation by the network manager to alter the physical terminal assigned to a logical function. The routine can be incorporated into that used to achieve the concept of a standard terminal.

logic circuit
An electronic circuit that performs information processing. Usually encoded on a chip, it is composed of logic gates that are the Boolean logic building blocks. See logic gates and Boolean logic.

logic design
A design specifying the logical functions and interrelationships of the various parts of an electrical or electronic system using logic symbols. Also called a logic diagram.

logic diagram
A drawing that represents the logic functions AND, OR, NOT, and so forth.

logic element
A symbol that has logical meaning. May also be called a logic symbol, for example, a gate, a flip-flop, and so forth.

logic elements
Various forms of electrical switches within the computer. The term is used interchangeably with switching elements.

logic gate
A combination of transistors that detects the presence or absence of electrical pulses, which in turn represent binary digits (1s and 0s).

logic level
The voltage magnitude associated with signal pulses representing ones and zeros (1s and 0s) in binary computation.

logic symbol
A symbol used in electrical or electronic design to represent a logic element graphically (i.e., a gate or a flip-flop).

log normal distribution
A continuous probability distribution where the logarithms of the variable are normally distributed.

log-off
The procedure by which a user ends a computer terminal session.

log-on
The procedure by which a user begins a computer terminal session.

longitudinal redundancy check
A data communications error-checking technique. A longitudinal redundancy check character is accumulated at both the sending and receiving stations during the transmission and is compared for an equal condition that indicates a good transmission of the previous block.

long-range resource planning
A planning activity for long-term capacity decisions, based on the production plan and perhaps on even more gross data (e.g., sales per year) beyond the time horizon for the production plan. This activity is to plan long-term capacity needs out to the time period necessary to acquire gross capacity additions, such as a major factory expansion.

long-term repeatability
Closeness of agreement of repeated position movements, under the same conditions, to the same location. The degree to which an industrial robot or other programmable mechanism can repeatedly locate either of the end points, the program path, or a cycle over a long period of time, under the same conditions.

look-ahead analysis
A CAD/CAM capability that enables the user to detect and prevent part gouging once the part under design is actually manufactured. Part gouging can happen if the NC tool is too large to navigate the part geometry.

lookup table
Memory that sets the input and output values for gray-scale thresholding, windowing, inversion, and other display or analysis functions. Input values are for the storage of video input pixels into image memory, and output values are for the display of stored pixels on the monitor.

loop
Sequence of instructions that are executed repeatedly, with or without modifications during each iteration, until a terminating condition is satisfied.

loss angle
Measure of electrical power losses in insulating material subjected to alternating current. The arctangent of dissipating factor and, thus, the angle between a material's parallel resistance and its total parallel impedance in a vector diagram when the material is used as dielectric in a capacitor. Complement of phase angle. Sometimes called phase defect angle.

loss factor
Measure of electrical power loss in insulating material subjected to alternating current. Product of dissipation factor and dielectric constant. Loss factor is expressed in the same units as dissipation factor. Low loss factor is generally desirable. Loss factor generally increases with humidity, weathering, deterioration, and exponentially with temperature.

lost-wax process
An investment casting process in which a wax pattern is used.

lot
Batch or group of workpieces processed at one time.

lot number
A unique identification assigned to a homogeneous quantity of material.

lot size
The quantity of an item ordered from the shop or a vendor; may be more than the actual requirement quantity depending on lot-sizing rules.

lot size code
A code that indicates the lot sizing technique selected for a given item.

lot size inventory
Inventories that are maintained whenever quantity price discounts, shipping costs, or setup costs, and so forth make it more economical to purchase or produce in larger lots than are needed for immediate purposes.

lot sizing
The process of, or techniques used in, determining lot size.

low-level circuit
Low-energy circuit, usually in the 5- to 12-V, 10- to 100-mA range.

low-level code
Identifies the lowest level in any bill of material at which a particular component may ap-

pear. Net requirements for a given component are not calculated until all the gross requirements have been calculated down to that level. Low-level codes are normally calculated and maintained automatically by the computer software.

low-level features
Pixel-based features such as texture, regions, edges, lines, corners, and so on.

low-level language
A programming language in which statements translate on a one-for-one basis. See also assembly language and machine language.

low-order position
The rightmost position of a number or word.

LTTP
A simplified parts programming language for two-axis turning applications and part of the manufacturing software and services TOOLPATH processor. LTTP is a subset language and may be interchanged with ADAPT and other features of the TOOLPATH processor in the same part program.

lubricant
Any substance used to reduce friction between two surfaces in contact.

lumen
The unit of luminous flux. It is equal to the flux through a unit solid angle (steradian) from a point source of 1 candela or to the flux on a unit surface of which all points are at a unit distance from a point source of 1 candela.

lumen/FT2
A unit of incident light. It is the illuminance on a surface 1 square foot in area on which a flux of 1 lumen is uniformly distributed, or the illuminance at a surface, all points of which are at a distance of 1 foot from a uniform source of 1 candela.

luminance
Luminous intensity (photometric brightness) of any surface in a given direction per unit of projected area of the surface as viewed from that direction, measured in footlamberts.

luminous directional reflectance
Ratio of brightness to brightness that an ideally diffusing, completely reflecting light surface would have when illuminated and viewed in the same manner. Luminous directional reflectance is commonly determined for an opaque material illuminated as 45° and viewed normal to the surface.

luminous reflectance
Measure of brightness. Ratio of light reflected to light incident. Luminous reflectance of a material of such thickness that any increase in thickness would fail to change the value is called luminous reflectivity. Calculated from spectral reflectance (including spectral component) and spectral luminosity, it is a function of spectral and angular distribution of incident light energy. Also called total luminous reflectance to distinguish it more readily from luminous directional reflectance.

luminous transmittance
Ratio of light transmitted by material to light incident on it. Calculated from spectral transmittance and spectral luminosity, it is a function of spectral distribution of incident light energy.

lux
International System (SI) unit of illumination in which the meter is the unit of length; 1 lux equals 1 lumen per square meter.

machine
Module at which a part can be altered. A computer, CPU, or other processor.

machine attention time
That portion of a machining operation during which the worker performs no physical work, yet must watch the progress of the work and be available to make necessary adjustments, initiate subsequent steps or stages of the operation at the proper time, and so on.

machine center
A specific area or department of one or more machines that is considered as an entity for production planning and reporting purposes. A group of similar machines that can all be considered together for purposes of loading.

machine center template
A machine template to which additional floor area has been added to provide space necessary for operating services and maintaining the machine, as well as for material.

machine code
An operation code that a machine is designed to recognize.

machine-controlled time
That part of a work cycle that is entirely controlled by a machine and therefore is not influenced by the skill or effort of the worker.

machine cycle
The basic operating cycle of a CPU during which a single instruction is fetched, decoded, and executed. A set period of time in which the computer can perform a specific machine operation.

machine downtime
Any time during a regular working period that a machine cannot be operated.

machined surface
A group of faces on a part that can be machined by a single process.

machine element
A work cycle subdivision that is distinct, describable, and measurable, the time for which is entirely controlled by a machine and therefore is not influenced by the skill or effort of a worker.

machine-hour
A unit for measuring the availability or utilization of machines.

machine idle time
That portion of a regular working period during which a machine that is capable of operation is not being used.

machine-independent
Pertaining to procedures or programs created without regard for the actual devices that will be used to process them. Often refers to high-level programming in COBOL, FORTRAN, and so on.

machine instruction
An instruction that a machine (computer) can recognize and execute.

machine interference
The time that a machine is idle because the operator is servicing another machine in the group.

machine language
A computer term referring to exact digital instructions to a computer that it can execute directly, without modification or translation. Most programmers prefer not to program in machine language, which requires keeping track of all memory locations as well as being difficult to identify, since only numbers are generally used. The "first-level" computer language, as compared to a "second-level" assembly language, or a "third-level" compiler language.

machine learning
An area of AI research that investigates techniques for creating computer programs that can learn from their own experience.

machine loading
The accumulation by work station(s), machine, or machine group of the hours generated from the scheduling of operations for released orders by time period. Machine loading differs from capacity planning in that it does not use the planned orders from MRP but operates solely from scheduled receipts. As such, it has very limited usefulness.

machine logic
All of the complex work that a data-processing machine performs is based on four fundamental operations—AND, OR, NOT, and MEMORY. In electronic computers, these functions are executed by circuits called gates and triggers. The organization of these building

blocks into circuits that can carry out programs is called machine logic or logical design. AND gates are devised so that they emit a 1 bit as output only when all the input bits are 1s. OR gates produce a 1 output whenever at least one of the inputs is a 1. A 0 output occurs only if both inputs are 0. The NOT operation is performed by a device called the inverter. This component reverses the signal that is applied at its input. If a 0 bit enters, a 1 output occurs. If a 1 bit enters, there is a 0 output. AND gates and OR gates may have more than two outputs. However, the same logical principles apply regardless of the number of inputs. An AND gate with three inputs performs a function equivalent to three relays in series. OR gates behave in the same way as relays in parallel.

machine-oriented language
A programming language that is more like a machine language than a human language, for instance, assembly language, as opposed to COBOL.

machine sensible
The ability of a medium and its content to be processed by computer equipment.

machine template
A two-dimensional scale representation of the maximum floor area occupied by a machine.

machine tool
A power-driven machine (not hand-held) used to cut, form, or shape metal. A powered machine used to form a part, typically by the action of a tool moving in relation to the workpiece.

machine tool program
A processor for in-house minicomputer system that utilizes a flexible disk drive media for storing workpiece programs and data files.

machine translation
An area of AI research that attempts to use computers to translate text from one language to another. These programs often use a combination of natural-language understanding and natural-language generation techniques.

machine utilization
The amount of time (percentage) a machine is doing work compared to its total available time. Work time plus idle time equals available time.

machine utilization factor
The number of parts produced during a day times the number of hours per part divided by 24.

Machine Vision
Block Diagram of Basic Machine Vision Conversions

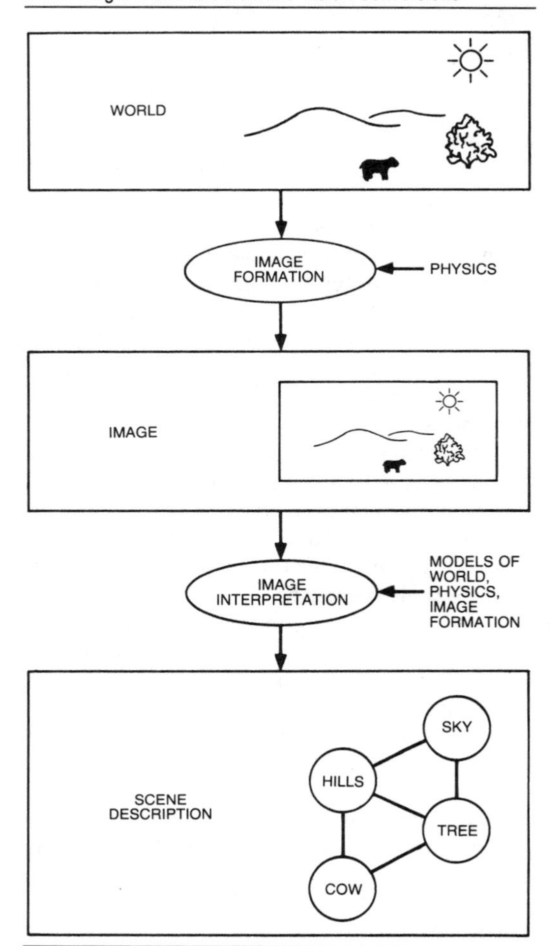

machine vision
Computer perception, based on visual sensory input, to develop a concise description of a scene depicted in an image. The term is used synonymously with image understanding and scene analysis. The automatic acquisition and analysis of images to obtain desired data for controlling an activity. The ability of an automated system to perform certain visual tasks normally associated with human vision, including sensing, image formation, image analysis, and image interpretation or decision making.

machining
Removing material in the form of chips from work, usually through the use of a machine.

machining center
A module that is capable of performing a variety of metal removal operations on a part including drilling, milling, boring, and so forth, usually under automatic control. A numerically controlled machine tool, such as a milling machine, capable of performing a variety of operations, such as milling, drilling, tapping, reaming, and boring. Usually also included are arrangements for storing 10 to 100 tools and mechanisms for automatic tool change.

machining parameters
Parameters that control machining operations, including feed rate, cutting speed, and depth of cut.

machining process
Any particular machining operation viewed as an indivisible activity for planning purposes.

macro
A programming symbolic-language instruction that will generate more than one absolute language instruction. In assembler programming, an assembler language statement that causes the assembler to process a predefined set of statements called a macro definition. The statements normally produced from the macro definition replace the macro instruction in the program. Synonymous with macro call. A user-defined, linked series of commands executed with a single keystroke. Permits a menu to be customized to perform a complex but frequently used function efficiently. A locally stored string of presentation code represented by a single-character name. When the macro-name character is received, the locally stored string is processed in its place.

macro body
The portion of a macro definition containing statements that define the action of the macro.

macro definition entity
The structure entity, containing the macro body within its parameter data section, used to define a specific macro.

macro-etch
Etching of a metal surface for accentuation of gross structural details and defects for observation by the unaided eye or at magnifications not exceeding 10 diameters.

macrogenerator
A computer program that replaces macro instructions in the source language with the defined sequence of instructions in the source language.

macrograph
A graphic reproduction of the surface of a prepared specimen at a magnification not exceeding 10 diameters. When photographed, the reproduction is known as a photomacrograph.

macro instance entity
A structure entity that will invoke a macro defined by a macro definition entity.

macro instruction
A symbolic instruction in a source language that produces a number of machine-language instructions. It is made available for use by the programmer through an automatic programming system. A macro instruction is a method of describing in a one-line statement a function, or functions, to be performed by the object program. Macro instructions enable the programmer to write one instruction such as "read a tape" and the processor will then automatically insert the corresponding detailed series of machine instructions. In this manner, the programmer avoids the task of writing one instruction for every machine step.

macroscopic
Visible at magnifications from 1 to 10 diameters.

macrostructure
The structure of metals as revealed by examination of the etched surface of a polished specimen at a magnification not exceeding 10 diameters.

magenta
The hue attribute evoked by a wavelength combination that is the approximate complement of 515 nanometers. It is the complement of green.

magnesite wheel
A grinding wheel bonded with magnesium oxychloride.

magnetic-bubble memory
A main storage technology characterized by densities of 10 million bits per square inch and with realized densities of up to 1 billion bits per square inch conceivable. Since bubble memory is magnetized, it does not lose its charge when electricity is disabled. Therefore, it is nonvolatile main memory.

magnetic card
A card with a magnetic surface on which data can be stored by selective magnetization of portions of the flat surface.

magnetic core
A small doughnut-shaped piece of ferromagnetic material, about the size of a pin head, capable of storing one binary digit represented by the polarity of its magnetic field. Thousands of these cores strung on wire grids form an internal memory device. Cores can be individually charged to hold data and sensed to issue data.

magnetic delay line
A delay line whose operation is based on the time of propagation of magnetic waves.

magnetic disk
A flat circular plate with a magnetic surface on which data can be stored in the form of magnetized spots. These data are arranged in circular tracks around the disks and are accessible to reading and writing heads on an arm that can be moved to the desired track as the disk rotates.

magnetic disk storage
Data stored as magnetized spots arranged in binary form in concentric tracks on each face of the disk. The magnetic disk is a thin metal disk resembling a phonograph record; it is coated on both sides with a magnetic recording material. A characteristic of disk storage is that data is recorded serially bit-by-bit, eight bits per byte, along a track rather than by columns of characters. Disks are normally mounted on a stack on a rotating vertical shaft. Enough space is left between each disk to allow access arms to move in and read or record data. A single magnetic disk unit is capable of storing several million characters. Usually, more than one disk unit can be attached to a computer.

magnetic-domain memory
An experimental memory method that uses magnetic bubbles to store data. The bubbles can also be read optically.

magnetic drum
A device that stores information in the form of magnetized spots on a continuously rotating cylinder. A magnetic reading and writing head is associated with each track, so that the desired track can be selected by electrical switching. Provides faster access to information than disks.

magnetic energy product curve
Curve obtained by plotting product of induction and demagnetizing force against corresponding values of induction; that is, product of coordinates of demagnetizing curve as abscissa versus induction as ordinate.

magnetic energy product, maximum
Maximum external energy that contributes to magnetization of material. Corresponds to maximum value of the abscissa for magnetic energy product curve.

magnetic field strength
The measured intensity of a magnetic field at any given point. It is usually expressed in oersteds.

magnetic film storage
Functions similarly to core storage; however, instead of individual cores strung on wires, magnetic film is made of much smaller elements in a form. One type of magnetic film, known as planar film (thin film), consists of very thin, flat wafers made of nickel-iron alloy. These metallic spots are connected by ultrathin wires and are mounted on an insulating base, such as glass or plastic. Magnetic film may also be in the form of plated wire. This is a type of cylindrical film, essentially the same as planar film, except that the film is wrapped around a wire. Another type of cylindrical film is the plated rod. This technique, known as thin-film rod memory, consists of an array of tiny metal rods only 0.1 inch (2.54 mm) long. The operation of a magnetic film memory unit is similar in principle to that of a magnetic card unit, as the storage elements in both are formed into planes that may be stacked.

magnetic head
An electromagnet that can perform one or more functions of reading, writing, and erasing data on a magnetic data medium.

magnetic hysteresis
In a magnetic material, such as iron, a lagging in the values of the resulting magnetization due to a changing magnetic force.

magnetic ink
An ink that contains particles of a magnetic substance whose presence can be detected by magnetic sensors.

magnetic ink character recognition
The character recognition of characters printed with ink that contains particles of a magnetic material. Referred to as MICR.

magnetic-inspection flaw indications
Accumulation of ferromagnetic particles along areas of flaws or discontinuities due to dis-

tortion of magnetic lines of force in those areas.

magnetic-particle inspection
A nondestructive method for detecting cracks or other discontinuities at or near the surface of ferromagnetic material.

magnetic storage
Any device that stores data, using magnetic properties such as magnetic cores, tapes, and films.

magnetic tape
Flexible plastic tape, often 0.5 inch wide with seven or nine channels or horizontal rows that extend the length on the tape. One side is uniformly coated with magnetic material on which data is stored. It is used for registering images, sound, or computer data. Magnetic tape is a "sequential" medium with a very low cost per bit, used for archival storage, sorts, and so forth.

magnetic tape storage
A storage system that uses magnetic spots, representing bits, on coated plastic tape.

magnetic thin film
A layer of magnetic material, usually less than 1 micron thick, often used for logic or storage elements.

magneto optical
Also called thermomagneto optic, is the most promising method for creating erasable optical disk drives. This method, currently being used by Verbatim Corporation, uses the laser to heat a given spot on the media, which lowers coercivity and increases its susceptibility to magnetic change. A surrounding bias field then causes the magnetic domain in this region to change orientation (i.e., from down to up), thus denoting a written bit. Erasing is accomplished by reversing the orientation.

magnification
The relationship of the length of a line in the object plane to the length of the same line in the image plane. It may be expressed as image magnification (image size/object size) or its inverse, object magnification.

main control unit
In a computer with more than one instruction control unit, that instruction control unit to which, for a given interval of time, the other instruction control units are subordinated. An instruction control unit may be designated as the main control unit by hardware or by hardware and software. A main control unit at one time may be a subordinate unit at another time.

mainframe
A term referring to the CPU (central processing unit) part of the hardware. The mainframe does not encompass any of the input/output devices. The largest type of computer, usually capable of serving many users simultaneously, with a processing speed about 100 times faster than that of a microcomputer.

main memory
The primary storage facilities forming an integral physical part of the computer and directly controlled by the computer. In such internal facilities, all data are automatically accessible to the computer. Contrast with external storage.

main program
A program unit not containing a function, subroutine, or block data statement and containing at least one executable statement. A main program is required for program execution.

main storage
Program-addressable storage from which instruction and other data can be loaded directly into registers for subsequent execution or processing. The general-purpose program-addressable storage of a computer from which instructions may be executed and from which data can be loaded directly into registers.

maintained contract switch
An alternate-action switch.

maintenance
Any activity intended to eliminate faults or to keep hardware or programs in satisfactory working condition including tests, measurements, replacements, adjustments, and repairs.

maintenance breakdown
Emergency maintenance, including diagnosis of the problem and repair of a machine or facility, after a malfunction has occurred.

maintenance prevention
Maintenance work to be repeated at intervals. It includes both minor operations like lubrication and inspection and major jobs like the overhaul of a press.

maintenance programmer
Makes and tests necessary changes to existing applications programs.

maintenance programming
Makes and tests necessary changes to existing applications programs.

maintenance time
Preventive and corrective time required for hardware maintenance. Contrast with available time.

major axis
The axis of a maximum elongation of a blob in a plane. Derived from second moment calculations.

majority
A logic operator having the property that if P is a statement, Q is a statement, R is a statement, . . . , then the majority of $P, Q, R, \ldots$, is true if more than one-half of the statements are true, false if one-half or less are true.

major motion axes
May be described as the number of independent directions the robot arm can move the attached wrist and end-effector relative to a point of origin of the manipulator, such as the base. The number of robot arm axes required to reach world coordinate points is dependent on the design of the robot arm configuration.

make-or-buy decision
A logical decision resulting from a careful evaluation of relevant factors concerning inhouse capabilities and vendor capabilities. Factors considered often include cost, delivery time, capability, quality, the proprietary nature of the product, and management philosophy.

make-to-order
Products manufactured after receipt of a customer order specifying the final configuration and/or delivery time of the product.

make-to-order products
The end item is finished after receipt of a customer order. Frequently, long lead-time components are planned prior to the order arriving in order to reduce the delivery time to the customer. Where options or other subassemblies are stocked prior to customer orders arriving, the term *assemble to order* is frequently used.

make-to-stock product
The end item is shipped from finished goods, *off the shelf* and therefore is finished prior to a customer order arriving.

makeup time
Available time used for reruns due to malfunctions or mistakes.

malfunction
Any incorrect functioning within electronic, electrical, or mechanical hardware.

malleability
The characteristic of metals that permits plastic deformation in compression without rupture.

management
The process of utilizing material and human resources to accomplish designated objectives. It involves the activities of planning, organizing, directing, coordinating, and controlling. That group of people who perform the functions described above.

management audit
A systematic appraisal or evaluation of the worth or quality of the various management functions or activities or procedures in an organization, made by comparing them to established norm for good management or theoretical measures of management perfection.

management engineer
One who has the necessary education, training, and experience to perform the functions of management engineering.

management engineering
The application of engineering principles to all phases of planning, organizing, and controlling a project or enterprise.

management information system
A data-processing system designed to furnish management and supervisory personnel with current information to aid in the performance of management functions. Data is recorded and processed for operational purposes, problems are isolated and referred to upper management for decision making, and information is fed back to reflect progress in achieving major objectives.

mandrel
The core around which paper, fabric, or resin-impregnated fibrous glass is wound to form pipes or tubes. In extrusion, the central finger of a pipe or tubing die.

man-hour
A unit for measuring work. It is equivalent to one person working at normal pace for 60 minutes, two people working at normal pace for 30 minutes, or some similar combination of people working at normal pace for a period of time.

manipulation
The process of controlling and monitoring data table bits or words by means of the user's program in order to vary application functions. Grasping, releasing, moving, transporting, or otherwise handling an object.

manipulator
A mechanism usually consisting of a series of segments, jointed or sliding relative to one another, for the purpose of grasping and moving objects usually in several degrees of freedom. It may be remotely controlled by a computer, PC, or by a human.

manipulator-oriented language
Programming language for describing exactly where a robot's arm and gripper should go and when. To be contrasted with task-oriented languages for describing what the effect of robot action should be.

man-minute
A unit used for measuring work. It is equivalent to one person working at normal pace for 1 minute, two people working at a normal pace for 30 seconds, or an equivalent combination of people working at normal pace for a period of time.

mantissa
The positive fractional part of the representation of a logarithm. In the expression, log 643 – 2.808, .808 is the mantissa and 2 is the characteristic.

manual control
A device containing controls that manipulate the robot arm and allow for the recording of locations and program motion instructions.

manual data input
A means of inserting complete format data manually into the control system. These data are identical to information that could be inserted by a means of a tape (including all auxiliary functions, feedrate number, etc.).

manual element
A distinct, describable and measurable subdivision of a work cycle or operation performed by one or more human motions that are not controlled by process or machine.

manual feedrate override
Enables the operator to reduce or increase the feedrate. This feature generally consists of a dial on the operator's console that will enable the operator to adjust or "override" the programmed feedrate. The percentage of override may vary from approximately 1 to 150 percent of the programmed feedrate.

manual input
Data entered manually by the operator or programmer to modify, continue, or resume processing of a computer program.

manual manipulator
A manipulator controlled by a human operator.

manual programming
A means of teaching a robot by physically presetting the cams on a rotating stepping drum, setting limit switches on the axes, arranging wires, or fitting air tubes.

manual rescheduling
The most common method of rescheduling open orders. Under this method, the MRP system provides information on the part numbers and order numbers that need to be rescheduled. Due dates and/or order quantity changes required are then analyzed and changed by material planners or other authorized persons.

manufacturing
A series of interrelated activities and operations involving the design, material selection, planning, production, quality assurance, management, and marketing of discrete consumer and durable goods.

manufacturing automation protocol (MAP)
Defines a network communications structure for multivendor factory automation systems. MAP is based on the OSI (Open Systems Interconnection) reference model developed by ISO and CCITT and uses current and emerging international communication standards. It is intended for reference in factory automation purchased specifications. It will facilitate user companies in selecting a common set of communication protocols that will result in off-the-shelf products supporting open multivendor communications. The MAP protocol selections corresponding with the OSI layers are as follows:

Layer Seven – ISO CASE kernel, ISO FTAM, directory service, network management, MMFS
Layer Six – Null
Layer Five – ISO session kernel
Layer Four – ISO transport class four
Layer Three – ISO connectionless internet

Layer Two – IEEE 802.2 class one
Layer One – IEEE 802.4 broadband

manufacturing calendar
A calendar, used in inventory and production planning functions, that consecutively numbers only the working days so that the component and work order scheduling can be based on the actual number of working days available.

manufacturing lead time
The complete time necessary to manufacture an item. Includes order paperwork time, setup time, runtime, queue time, move time, inspection time, stock pickout, and stocking time.

manufacturing lease time
The total time required to manufacture an item; included here are order, preparation time, queue time, setup time, runtime, move time, inspection, and put-away time.

manufacturing order
A document or group of documents conveying authority for the manufacture of specified parts or products in specified quantities.

manufacturing planning
The function of setting the limits or levels of manufacturing operations in the future, consideration being given to sales forecasts and the requirements and availability of personnel, machines, materials, and finances. The manufacturing plan is usually in fairly broad terms and does not specify in detail each of the individual products to be made but usually specifies the amount of capacity that will be required.

manufacturing process
The series of activities performed upon material to convert it from the raw or semifinished state to a state of further completion and a greater value.

manufacturing-related parts
Parts that undergo similar processing steps (e.g., stamped parts or milled parts).

manufacturing resource planning
A method for the effective planning of all the resources of a manufacturing company. Ideally, it addresses operational planning in units, financial planning in dollars, and has a simulation capability to answer "what-if" questions. It is made up of a variety of functions, each linked together: business planning, production planning, master production scheduling, material requirements planning, capacity requirements planning, and the execution systems for capacity and priority. Outputs from these systems would be integrated with financial reports such as the business plan, purchase commitment report, shipping budget, inventory projections in dollars, and so forth. Manufacturing resource planning is a direct outgrowth and extension of material requirements planning. Often referred to as MRP II.

manuscript
A form used by the part programmer for listing the detailed instructions that can be transcribed directly by the tape preparation device or fed to a computer for further calculation.

map
To establish a set of values having a defined correspondence with the quantities or values of another set.

MAP/EPA system
A system capable of both OSI communication and non-OSI PROWAY-type communication. The OSI communication is identical to MAP end system communication, making use of all seven layers of standardized protocols. The non-OSI communication is streamlined for time-critical applications and bypasses the standardized upper-layer protocols and provides a direct interface to the link layer. The MAP/EPA system is one of the allowable systems that make up the MAP cell architecture.

map generalization
An automatic mapping process for reducing the amount of graphic and nongraphic information displayed on a map. Often employed in the creation of composite maps from a series of large-scale maps. The process may employ line filtering, symbol revision, reclassification, and other generalizing techniques.

MAP network
A set of MAP network segments connected by means of bridges viewed logically as a single network. A single MAP subnetwork is the simplest occurance of a MAP network.

MAP network segment
A single physical-medium system supporting MAP.

marginal analysis
An economic concept concerned with those elements of costs and revenue associated directly with a specific course of action, nor-

MAP/EPA System
MAP/EPA System Block Diagram. *Source*: Boeing Company.

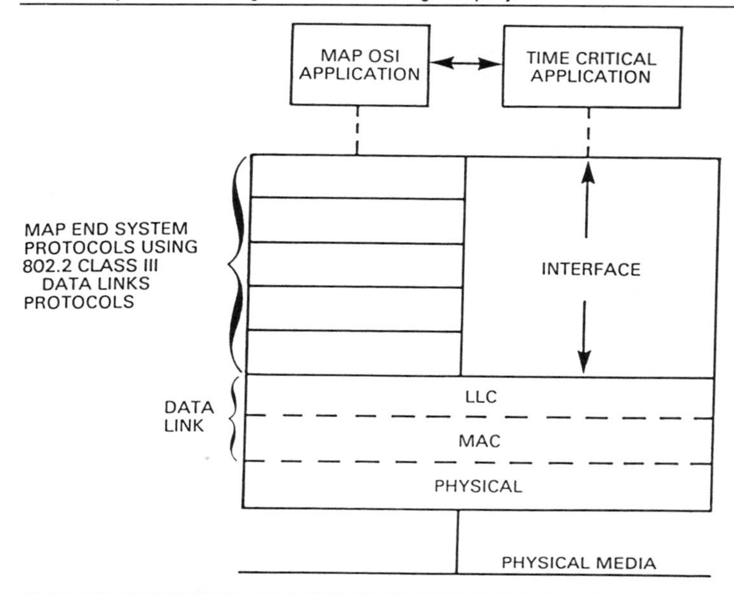

mally using available current costs and revenue as a base and usually independent of traditional accounting allocation procedures.

marginal cost
The additional out-of-pocket costs incurred when the level of output of some operation is increased by one unit.

mark
A symbol or symbols that indicate the beginning or the end of a field, a word, an item of data, or a set of data such as a file, a record, or a block.

marker
A glyph with a specified appearance, used to identify a particular location. Markers are a means by which two-dimensional and three-dimensional points manifest themselves on an output display surface.

marker-size scale factor
An attribute controlling the size of a marker. The size is specified as a multiple of the nominal marker size, a workstation-dependent constant.

marker type
An attribute that defines the glyph used to denote the position of a visible marker primitive.

Markov chain
A probabilistic mode of events in which the probability of an event is dependent only on the event that precedes it.

mark sensing
An operation by which data in machine-readable form can be written or marked on special oblong areas called bubbles. The markings can then be sensed by a machine and converted into punches in the same card.

mask
A pattern of characters used to control the retention or elimination of portions of another pattern of characters. An interchangeable sheet often made of plastic. It fits over a function board, menu pad, or graphics tablet. A patterned plate used to shield sections of silicon chip surface during the manufacture of integrated circuits.

mask design
The final phase of IC design by which the circuit design is realized through multiple masks corresponding to multiple layers on the IC. The mask layout must observe all process-related constraints and minimize the area the circuit will occupy.

masking
The process of creating an outline around a standard image and then comparing this outline with test images to determine how closely they match. An operation in which regions of an image are set to a constant value, usually white or black.

mass production
The large-scale production of parts or material in a continuous process uninterrupted by the production of other parts or material. A method of quantity production in which a high degree of planning, specialization of equipment and labor, and integrated utilization of all productive factors are the outstanding characteristics. Continuous processing of like parts.

mass properties
Calculation of physical engineering information about a part, for example, perimeter, area, volume, weight, and moments of inertia.

mass-properties calculation
This CAD/CAM capability automatically calculates physical/engineering information (such as the perimeter, area, center of gravity, weight, and moments of inertia) of any 3-D part under design.

mass storage
Massive amounts of online secondary storage, readily accessible to the CPU of a computer, providing lower cost than main memory for a given amount of information stored. Includes devices such as magnetic disk, drum, data cells, and so on.

master
A particular type of gage, usually of very simple geometry, used to set a dimension.

master control program
An operating system, especially on Burroughs machines, designed to reduce the amount of intervention required of the human operator by providing the following functions: schedules programs to be processed, initiates segments of programs, controls all input/output operations to ensure efficient utilization of each system component, allocates memory dynamically, issues instructions to the human operator and verifies that his or her actions were correct, and performs corrective action on errors in a program or system malfunction.

master control relay
A mandatory hardwired relay that can be de-energized by any hard-wired series-connected emergency stop switch. Whenever the master control relay is de-energized, its contacts must open to de-energize all applications I/O devices and power source.

master file
A main reference file of information used in a computer system. It provides information to be used by the program and can be updated and maintained to reflect the results of the processing operation.

master production schedule
For selected items, a statement of what the company expects to manufacture. It is the anticipated build schedule for those selected items assigned to the master scheduler. The master scheduler maintains this schedule and, in turn, it becomes a set of planning numbers that "drives" MRP. It represents what the company plans to produce expressed in specific configurations, quantities, and dates. The MPS should not be confused with a sales forecast that represents a statement of demand. The master production schedule must take forecast plus other important considerations (backlog, availability of material, availability of capacity, management policy and goals, etc.) into account prior to determining the best manufacturing strategy.

master schedule
A highly ambiguous term that differs very widely from company to company. In some companies, it is a broad schedule of the total number of units to be produced that does not specify the detail of accessories and specific products. In this sense, it is almost synonymous with the production plan; where a materials planning inventory control system is being used, the master schedule often specifies the individual end products and the time periods in which they are to be manufactured. It is then the basis for the scheduling of components. It can best be described as a high-level schedule from which detailed schedules are made.

master schedule item
A part number selected to be planned by the master scheduler. The item would be deemed critical in terms of its impact on lower-level components and/or resources such as skilled labor, key machines, dollars, and so forth. Therefore, the master scheduler, not the computer, would maintain the plan for these items.

A master schedule item may be an end item, a component, a pseudonumber, or a planning bill of material.

master scheduler
The job title of the person who manages the master production schedule. This person should be the best scheduler available, as the consequences of the planning done here has great impact on material and capacity planning. Ideally, the person would have substantial product and shop knowledge.

master-slave manipulator
A class of teleoperator having geometrically isomorphic master and slave arms. The master is held and positioned by a person; the slave duplicates the motions, sometimes with a change of scale in displacement or force.

master table of detail time studies
A master record of time-study data, arranged so that times for the same elements can be compared.

master terminal
A terminal designated as reserved for the system manager and thus privileged to initiate conversations for network management not available to user terminals. The identity of the master terminal may be changed when necessary by entering a special network management entry, after sign-on by a user recognized to be a system manager.

material flow
The progressive movement of material, parts, or products toward the completion of a production process between workstations, storage area, machines, departments, and the like.

material handling
The movement of materials, parts, subassemblies, or assemblies either manually or through the use of powered equipment.

material-handling robot
A robot designed to grasp, move, transport, or otherwise handle parts of materials in a manufacturing operation.

material-handling system
System or systems used to move and store parts and the materials used in processing the parts (e.g., tools, coolant, wastes).

material management
Control of the complete flow of materials from purchase and production planning through finished goods delivery to the customer.

material-processing robot
A robot designed and programmed so that it can machine, cut, form, or in some way change the shape, function, or properties of the materials it handles between the time the

Material-Handling System
Volvo Material-Handling System

materials are first grasped and the time they are released in a manufacturing process.

material requirements planning
A system that uses bills of material, inventory and open order data, and master production schedule information to calculate requirements for materials. It makes recommendations to release replenishment orders for material. Further, since it is time-phased, it makes recommendations to reschedule open orders when due dates and need dates are not in phase. Originally seen as merely a better way to order inventory, today it is thought of primarily as a scheduling technique, that is, a method for establishing and maintaining valid due dates on orders.

material requisition
An authorization that identifies the type and quantity of materials to be withdrawn from inventory.

materials control
The function of maintaining a constantly available supply of raw materials, purchased parts, and supplies required for the production of products.

materials management
The grouping of management functions related to the complete cycle of material flow, from the purchase and internal control of production materials to the planning and control of work-in-process to the warehousing, shipping, and distribution of the finished product. Differs from materials control in that the latter term traditionally is limited to the internal control of production materials.

mathematical induction
A method of proving a statement concerning terms based on natural numbers not less than N by showing that the statement is valid for the term based on N and that, if it is valid for an arbitrary value of N that is greater than N, it is also valid for the term based on $(N + 1)$.

mathematical model
A mathematical representation of the behavior of a process, device, or concept.

Material Requirements Planning
A Computerized Material Control System Using Material Requirements Planning (MRP). Reprinted with Permission of Reston Publishing Company, Prentice-Hall, from *Robotics and Automated Manufacturing* by Richard Dorf.

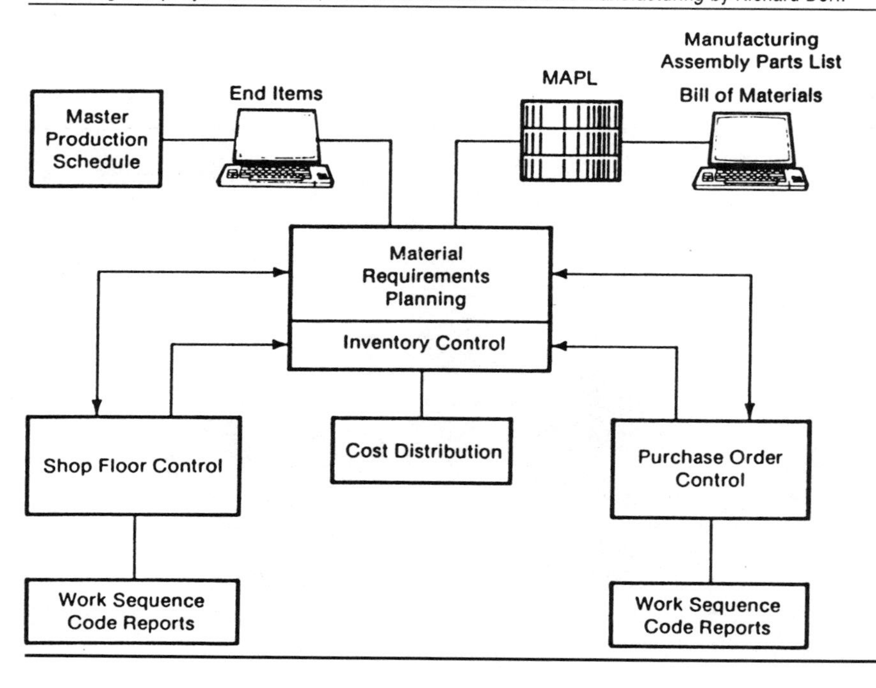

mathematical programming
In operations research, a procedure for locating the maximum or minimum of a function subject to constraints. Certain techniques that are useful for solving allocation problems, where limited resources must be allocated between competing demands.

matrix
A two-dimensional array of circuit elements such as wires, diodes, and so on that can transform a digital code from one type to another. A group of numbers organized on a rectangular grid and treated as a unit. The numbers can be referenced by their position in the grid. An arrangement of elements (numbers, characters, dots, diodes, wires, etc.) in perpendicular rows.

matrix array
An image sensor that produces a matrix image.

matrix-array camera
A solid-state television camera that has a rectangular array of photosensitive elements.

matrix bill of material
A chart made up from the bills of material for a number of products in the same or similar families. It is arranged in a matrix with parts in columns and assemblies in rows (or vice versa) so that requirements for common components can be summarized conveniently.

matrix printer
An output device that prints each character by means of a specially placed series of dots. Usually, the dots are made by small wires that press against a ribbon and paper.

maximum inventory
Sum of economic lot size plus twice the reserve stock. Especially of interest in allocating stockroom size.

maximum operate point
Sensitivity setting above which control will not respond to the dark signal produced by object to be detected.

maximum operating temperature
The maximum temperature at which any point within an application system may be safely maintained during continuous use.

maximum order quantity
An order quantity modifier, applied after the lot size has been calculated, that limits the order quantity to a preestablished maximum.

maximum speed
The greatest rate at which an operation can be accomplished according to some criterion of satisfaction. The greatest velocity of movement of a tool or end-effector that can be achieved in producing a satisfactory result.

maximum working area
On the horizontal plane, the area at the workplace bounded by the imaginary arch drawn by the worker's fingertips moving in the horizontal plane with the arm fully extended and moving about the shoulder as a pivot. The section where the maximum area of the right and left hands overlap constitutes the maximum working areas for the two hands. In the vertical plane, the space on the surface of the imaginary sphere generated by rotating, about the worker's body as an axis, the arc traced by the fingertips of the worker's right or left hand, when the arm is fully extended and is moved vertically about the shoulder as a pivot.

MDCAPT
An APT system that provides enhanced part programming capabilities for machined parts ranging from the most simple to the ultimate in complexity.

MDEDIT
A system that creates and manages master files of coordinate-point data and provides the capability to update, retrieve, merge, edit, print, and make inventories of point data files.

MDI
A method of operation where data is entered manually by keyboard or other means instead of by program tape.

mean
The arithmetic average of a group of numbers.

mean absolute deviation
The average of the absolute differences between the planned (or forecasted) and actual values in a group of numbers.

means-ends analysis
A problem-solving approach in which problem-solving operators are chosen in an iterative fashion to reduce the difference between the current problem-solving state and the goal state.

mean time between failure
The average time the system or a component of the system works without failure.

mean time to repair
The average time a device is expected to be out of service after failure.

measure
A value (associated with a logical input device) determined by one or more physical input devices and a mapping from the values delivered by the physical devices. The logical input value delivered by a logical input device is the current value of the measure. The current or last-requested value of a logical input device (LID).

measurement line
A line in the work zone of a machine along which measurements are taken.

measurement point
A point in the work zone of a machine from which measurements are made.

measure of effectiveness
The criterion used in evaluating alternative solutions to a problem to select the optimum one.

mechatronics
A term coined by the Japanese to describe the integration of mechanical and electronic engineering. The concept specifically refers to a multidisciplined, integrated approach to product and manufacturing system design and encompasses the next generation of machines, robots, and smart mechanisms for advanced manufacturing technology. The environments for mechatronics are primarily factory automation, office automation, and home automation.

media
The vehicles by which information is stored or transmitted; can be classified as source input and output.

media access
Methodology by which a node gains access to the network to transmit a message.

median
The middle value in a set of measured values when the items are arranged in order of magnitude. If there is no middle value, the median is the average of the two middle values. The central value of a group of numbers, that is, the average of the two extreme values.

medium
A drawing material that can normally be reproduced. Vellum and mylar (thin plastic film) are the most popular for conventional drafting. The material on which computer data is recorded, for example, magnetic tape, floppy diskette, Winchester disk, and so on.

medium access control
Techniques for allocating the use of a shared communications medium to a particular node wishing to transmit a message. Two basic control methods are used in local area networks, polling techniques and contention techniques. Token passing represents polling techniques (token ring or token bus); CSMA/CD is a contention access method.

medium-scale integration
Usually less than 100 circuits on a single chip of semiconductor. It was widely used in third-generation systems.

megabyte
Literally 1,000,000 bytes; usually 1024 × 1024 bytes.

megahertz
Abbreviated as MHz, 1,000,000 (10^6) cycles per second.

megassembly systems
Multistation, multiproduct assembly systems containing at least 10 robots.

melt index
Amount, in grams, of a material (usually a thermoplastic) that, when subjected to a 2160-gram force for 10 minutes at 190° C, can be forced through a 0.0825-inch orifice.

melting point
For a pure metal, the temperature at which liquefaction occurs on heating or solidification occurs on cooling under equilibrium conditions. Alloys and other materials have a melting range, and melting point is often desired as a temperature near the bottom of this range at which observable change caused by melting occurs. In general, melting point is determined by heating a small specimen to a temperature not too far below the melting point, then raising temperature slowly (a few degrees a minute) and watching closely for the first indication of liquefaction.

melting range
Range of temperature between solidus temperature and allowable liquidus temperature.

memory
Any device into which data can be input, retained, and later retrieved for use. That part of a computer that retains data or program information. Synonymous with storage and per-

tains to a computer device into which data can be entered, held, and/or retrieved. Usually refers to the internal capacity of the computer and is handled by a magnetic core or semiconductor arrangement in contrast to external storage that refers to such mediums as magnetic tape, disks, or drum.

memory address
Specifies memory word location in memory. Memory words are indicated by this address. The memory address takes the form of a code number. In a program, memory addresses are referred to rather than the memory word itself. Accessing a memory word means obtaining its contents from memory by telling the computer the address of its location, which could be in one of several of the computer storage areas. The contents of the memory word are then taken electronically to the computer processing unit to be operated on. There are several kinds of memory location whose contents are another address. An indirect address specifies a memory location whose contents are another address. This is used in cases where an instruction contains many bits. A direct address specifies a memory location that contains the data to be operated on. An immediate address means that what would be an address in other addressing modes are the data and/or instructions to be operated on.

memory capacity
The number of actions that a robot can perform in a program.

memory chip
A chip whose components form thousands of cells, each holding a single bit of information.

memory cycle time
The minimum time between two successive data accesses from a memory.

memory location
The smallest position in a computer memory to which the computer can refer.

memory management
The allocation of main memory space on a multiprogramming system.

memory map
The graphic representation of all of the memory locations a computer can address. A listing of addresses or symbolic representation of addresses that define the boundaries of the memory address space occupied by a program or a series of programs. Memory maps can be produced by a high-level language such as FORTRAN.

memory module
A processor module consisting of memory storage and capable of storing a finite number of words.

memory protection
In data processing, an arrangement for preventing access to storage for either reading or writing or both.

menu
A list of available responses or computer commands to be input as instructions in a computer program. A common CAD/CAM input device consisting of a checkerboard pattern of squares printed on a sheet of paper or plastic placed over a data tablet. These squares have been preprogrammed to represent a part of a command, a command, or a series of commands. Each square, when touched by an electronic pen, initiates the particular function or command indicated on that square.

menu-driven
Programs that use menus or a list of commands available to the user. The user simply selects the desired option. Compare to command-driven.

menu pad
Also known as a menu tablet. An input device having a flat surface on which functions are selected using a stylus or puck. Similar to a function board without buttons.

menu switching
Changing the available functions that can be executed on the CRT. A corresponding card is also changed. The interchangeable mask may be placed over a function keyboard, menu pad, or graphics tablet.

merge
To combine two or more sets of related data into one, usually in a specified sequence. This can be done automatically on a CAD/CAM system to generate lists and reports. An operation combining two or more files of data into one in a predetermined sequence.

message
A sequence of characters used to convey information or data. In data communication, messages are usually in an agreed format with a heading that controls the destiny of the message and text that consists of the data being carried.

message switching
A method of receiving a message over communications networks, transmitting it to an intermediate point, storing it until the proper outgoing line and station are available, and then transmitting it again toward its destination. The destination of each message is indicated by an address integral to the message.

meta
Prefix designating reflexive applications of the associated concept.

metacognition
The capability to think about one's own thought processes.

metadata base
An instance of a DIMS conceptual schema including the tables produced by the schema compilers, precompilers, and translators.

metafile
A mechanism for retaining and transporting graphical data and control information. The information contains a device-independent description of one or more pictures.

metafile element
A functional item that can be used to construct a picture or convey information.

metaknowledge
Knowledge about knowledge.

metalanguage
A language used to specify itself or another language.

metal-arc welding
Arc welding with metal electrodes. Commonly refers to shielded metal-arc welding using covered electrodes.

metal inert-gas welding
A method of joining two ferrous metal parts by passing a heavy electrical current from a metal rod to the grounded parts. The resulting electric discharge melts the metal rod and the part joints together to form a weld. This process normally is conducted with a shielding gas that prevents oxidation of the molten joint and thus increases weld integrity.

metallizing
Applying a thin coating of metal to a nonmetallic surface. May be done by chemical deposition or by exposing the surface to vaporized metal in a vacuum chamber or by metal powder spraying through a flame curtain.

metal-oxide semiconductor
A technology for constructing integrated circuits with layers of conducting metal, semiconductor material, and silicon dioxide as an insulator.

metarule
In artificial intelligence, refers to the manner in which rules should be employed. A higher-level rule used to reason about lower-level rules.

methodology for unmanned manufacturing
A Japanese program to develop an unmanned factory by the mid-1980s. This will depend heavily on robots. The Japanese government has contributed $116 million and many private companies have donated funds and technical assistance. Plans includes a 250,000-square-foot factory that will produce 50 mechanical components such as gear boxes, motors, and so forth. Automatic technology will include forging, heat treating, welding, press work, machining, inspection, assembly, and finishing, as well as automated packaging and storage.

methods engineer
The title given a member of that subclassification of industrial engineering comprised of individuals qualified by training, education, and experience to establish methods and the means by which they can be made most effective.

methods engineering
The technique that subjects each operation of a given piece of work to close analysis to eliminate every unnecessary element or operation and to approach the quickest and best method of performing each necessary element or operation.

methods study
The analysis of the sequence of motions used or proposed for use in performing an operation and of the tools, equipment, and workstation layout used or proposed for use.

methods time measurements
A system of time standards for predetermined motions. A processor analyzes any operation into certain classifications of human motions required to perform it and assigns to each motion controlled by the individual performing it a predetermined time standard determined by the nature of the motion and the conditions under which it is made.

methods training
Detailed instruction and guided practice given workers to ensure that they use the proper methods to perform their jobs. Courses or pro-

grams of instruction given in the techniques of scientific management as related to methods engineering.

metrology
The science of measurement.

micro
Short for microscopic; a detailed or close-range view.

MICROAPT
A flexible hardware/software system tailored to meet sophisticated numerical control and data-processing needs. MICROAPT features an implementation of the APT language for operation on a low-cost, in-house minicomputer.

micro-array computer
A special-purpose multiprocessor system designed for high-speed calculations with arrays of data.

microchip
A popular nickname for the integrated circuit chip.

microcode
A computer program at the basic machine level.

microcomputer
A complete tiny computing system, consisting of hardware and software, that usually sells for less than $5000 and whose main processing blocks are made of semiconductor integrated circuits. In function and structure, it is somewhat similar to a minicomputer, with the main difference being price, size, speed of execution, and computing power.

microfiche
A rectangular transparency approximately 4 by 6 inches, containing multiple rows of greatly reduced page images of reports, catalogs, rate books, and so forth. Data reductions range from 13 up to several hundred times smaller than the originals.

microfilm
A roll of photographic film, small in size, which when developed and projected onto a screen produces a legible copy of the item or form photographed.

micro-floppy diskette
Storage medium developed by Sony, enclosed in a plastic case, $3\frac{1}{2}$ inches in diameter, with capacity to store 278K characters when formatted (437.5K unformatted).

micrograph
A graphic reproduction of the surface of a prepared specimen, usually etched, at a magnification greater than 10 diameters. If produced by photographic means, it is called a photomicrograph.

micrometer
Also known as micron, this is a unit of length equal to one-millionth of a meter.

micron
One millionth of a meter. Also called micrometer.

microprocessor
The central unit of microcomputer that contains the logical elements for manipulating data and performing arithmetical or logical operations on it. A single chip may contain RAM, ROM, and PROM memories, clocks, and interfaces for memory and I/O devices.

microprogramming
A method of operation of the CPU in which each complete instruction starts the execution of a sequence of instructions, called microinstructions, at a more elementary level.

microsecond
One-millionth of a second.

microshrinkage
A casting defect, not detectable at magnifications lower than 10 diameters, consisting of interdendritic voids. This defect results from contraction during solidification, where there is not an adequate opportunity to supply filler material to compensate for shrinkage. Alloys with a wide range in solidification temperature are particularly susceptible.

micro-Winchester disk
A $3\frac{1}{2}$-inch Winchester disk.

middle infrared
The portion of infrared spectrum between 3 and 6 microns.

migration path
Methodology to allow a practical implementation of MAP-specified functionality as it is developed.

millesecond
One thousandth of a second.

milling
Essentially the same as sawing, except the cutting operation leaves a trough of some specific shape, depending on the form of the cutter used.

milling machine
Milling machines use tools such as drills and routers that rotate and have a cutting edge on their periphery. Originally, the term *milling* implied that the rotating tool was held sta-

tionary and the workpiece was moved against it to effect the cutting or routing. In many modern milling machines, the tool can also move in and out along one or more axes.

MIMS

Computer program that incorporates capabilities for planning and tracking materials flow, maintaining inventory and production controls, processing bills of material, as well as material requirements planning functions. It also is used for purchasing and receiving, sales order tracking, capacity planning, and cost-accounting activities.

minicomputer

A small, programmable, general-purpose computer typically used for dedicated applications. The term often refers only to the mainframe, which typically sells for less than $25,000. Usually, it is a parallel binary system with 8-, 12-, 16-, 18-, 24- or 36-bit word length incorporating semiconductor or magnetic core memory and offering from 4K words to 264K words of storage and a cycle time of 0.2 to 8 microseconds or less. Minicomputers are used nearly everywhere large computers were used in the past and cost much less. As minicomputer prices have dropped, performance has increased and, as faster memories and logic are achieved, is likely to continue increasing, thereby further broadening the applications base.

mini-floppy diskette

A flexible storage medium; each diskette is $5\frac{1}{4}$ inches in diameter, with a capacity of approximately 45 to 60 text pages (calculated at 4000 characters per single-spaced page).

minimal simulation

The minimal required action necessary when a workstation cannot draw an output primitive as intended.

mini-MAP system

One of the allowable systems that make up the MAP cell architecture. Mini-MAP systems may be reasonable for unintelligent devices, such as sensors that never need communication outside their local network. It is possible to build a device that supports only the time-critical protocols and eliminates standardized upper-layer protocol implementation. Such a system is non-OSI because it is not capable of open systems communication. However, this device is compatible with the MAP/EPA systems when communicating at the data link level.

Mini-MAP System

Mini-MAP System Block Diagram. *Source*: Boeing Company.

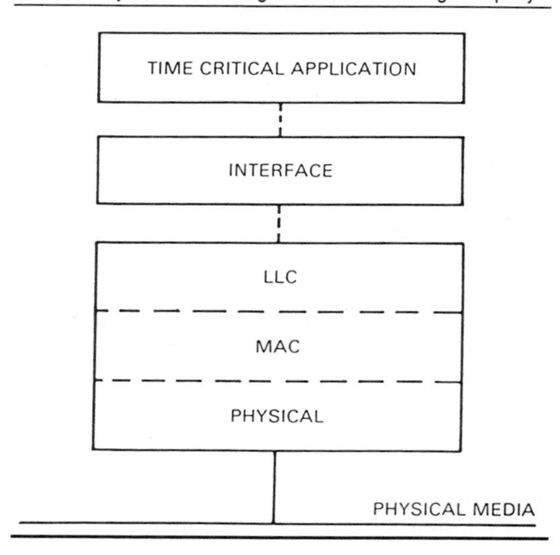

minimum inventory

The minimal allowable inventory for an independent demand item.

min-max system

An ordering algorithm that orders the quantity needed to bring the inventory back to its maximum when the inventory quantity reaches a minimum.

minor axes

With robots, the number of independent attitudes the wrist can orient the attached end-effector, relative to the mounting point of the wrist assembly on the arm. In machine vision, the axis of minimum elongation of a blob in a plane. Derived from second moment calculations.

mirror

To reflect through an axis.

mirror-image programming

A feature of some machine control units that provides for the reversing of all instructions programmed for a specific axis by means of a switch.

mirroring

Graphic construction aid. Ability to create a mirror image of a graphic entity. A CAD design aid that automatically creates a mirror image of a graphic entity on the CRT by flipping the entity or drawing on its X or Y axis.

misrun
A casting not fully formed, resulting from the metal solidifying before the mold is filled.

mnemonic
A technique for assisting human memory. A mnemonic term is often an abbreviation designed to help programmers remember instructions, for example, ART for arithmetic operation or MPY for multiply.

mnemonic code
Instructions for a computer written in a form easy for the programmer to remember, which must later be converted into machine language.

mobile robot
A robot mounted on a movable platform. The motions of the robot about the workplace are controlled by the robot's control system.

mobot
An acronym for modular robot.

modal
A type of change mechanism that extends its effect until another change specification supersedes it.

modal attribute settings
The specification of attributes during structure traversal using a current value that applies to subsequent output primitives until changed.

mode
A part of a computer that allows the user to perform a certain type of functions (e.g., point mode). A selected method of operation (e.g., RUN, TEST, or PROGRAM). The value that occurs most often in a group of numbers. A description of the cross-sectional shape of a laser.

model
A geometrically accurate and complete representation of a real object stored in a CAD/CAM data base. An approximate mathematical representation that simulates the behavior of a process device or concept so that an increased understanding of the system is attained.

model-based expert system
A type of expert system, usually intended for diagnostic purposes, that is based on a model of the structure and behavior of the device it is designed to understand.

model-based vision system
A system that utilizes a priori models to derive a desired description of the original scene from an image.

model driven
A top-down approach to problem-solving in which the inferences to be verified are based on the domain model used by the problem-solver.

modeling coordinates
A local coordinate system in which graphical objects are defined by the application program using output primitives.

modeling system
A high-level system for defining (constructing) and manipulating objects. A modeling system describes objects using modeling coordinates. Modeling coordinate objects are mapped into world coordinates using the modeling transformation.

modeling transformation
A transformation applied to all primitives to position them in world coordinate space.

modelocking
The modulation of the laser cavity at a frequency that is equal to the natural spacing of the modes of the laser resonator. The result is a train of short pulses at the modulation frequency, with a pulse width equal to the inverse of the laser's natural line width. This allows selective bursts of high peak power and short duration.

model space
A right-handed three-dimensional Cartesian coordinate space in which the product is represented.

modem
Stands for modulator-demodulator and is an electronic device that sends and receives digital data using telecommunication lines. To transmit data, the digital signals are used to vary (modulate) an electronic signal that is coupled into the telecommunication lines. To receive data, the electronic signals are converted (demodulated) to digital data.

modern control
A general term used to encompass both the description of systems in terms of state variables and canonical-state equations and the ideas of optimal control.

modes of operation
A major consideration because new modes of operation have a large impact on company op-

erations and EDP operations. Computers can be made to operate in many different ways to meet differing company needs. Plans to go to new modes may mean significant change and cost.

modifier
A quantity used to alter the address of an operand.

modular
Made up of subunits that can be combined in various ways. With robots, a robot constructed from a number of interchangeable subunits, each of which can be one of a range of sizes or have one of several possible motion styles (rectangular, cylindrical, etc.) and number of axes. *Modular design* permits assembly of products or software or hardware from standardized components.

modular bill of material
A type of planning bill arranged in product modules or options. Often used in companies where the product has many optional features, for example, automobiles.

modular system
A system design methodology that recognizes that different levels of experience exist in organizations and thereby develops the system in such a way so as to provide for segments or modules to be installed at a rate compatible with the users' ability to implement the system.

modulation
Impressment of information on a carrier signal by varying one or more of the signal's basic characteristics; frequency, amplitude, and phase. Differential modulation carries the information as the change from the immediately preceding state, rather than the absolute state.

modulation analysis
Instrumentation method used in electromagnetic testing that separates responses due to factors influencing the total magnetic field by separating and interpreting individually, frequencies or frequency bands in the modulation envelope of the (carrier frequency) signal.

modulation transfer function
A measure of the resolving capability of an imaging system. Shows the sine-wave spatial frequency amplitude response of a system.

module
An interchangeable "plug-in" item, containing electronic components, that may be combined with other interchangeable items to form a complete unit. A mechanical component having a single degree of freedom that can be combined with other components to form a multiaxis manipulator or robot. A program unit that is discrete and identifiable with respect to compiling, combining with other units, and loading.

module block time
Period of time during which a module is prevented from processing a part because a part cannot leave a workstation due to a full off-shuttle queue and/or a blocked or failed parts-handling system.

module downtime
Period of time during which a module cannot process parts due to a failure in the module (a module's downtime plus uptime equals total time).

module idle time
Period of time in which a module is not processing a part because a part is not available; does not include module block time or part-transfer times.

module operation
Aggregation of all activities taking place during a single visit of a part to a module.

module operation time
Period of time required by a module to complete a single operation on a part; does not include part-transfer times into and out of module. Total time that a part spends at a module, including queuing time.

module testing or unit testing
The verification of a single program module, usually in an isolated environment (i.e., isolated from all other modules). Module testing also occasionally includes mathematical proofs.

module tool set
Set of all tools that can be accessed by a given module in its automatic mode of operation.

module uptime
Period of time during which a module can process parts (i.e., it is not in the failed state). A module's uptime plus downtime equals total time.

module utilization
The quantity total time minus module idle time minus module downtime divided by the

quantity total time minus module downtime.

modulo-*n* counter
A counter in which the number represented reverses to 0 in the sequence of counting, after reading a maximum value of $n - 1$.

modulus of elasticity
A measure of the rigidity of metal. Ratio of stress, within proportional limit, to corresponding strain. Specifically, the modulus obtained in tension or compression in Young's modulus, stretch modulus, or modulus of extensibility; the modulus obtained in torsion or shear is modulus of rigidity, shear modulus, or modulus of torsion; the modulus covering the ratio of the mean normal stress to the change in volume per unit volume is the bulk modulus. The tangent modulus and secant modulus are not restricted within the proportional limit; the former is the slope of the stress-strain curve at a specified point; the latter is the slope of a line from the origin to a specified point on the stress-strain curve. Also called elastic modulus and coefficient of elasticity.

modulus of rupture
Nominal stress at fracture in a bend test or torsion test. In bending, modulus of rupture is the bending moment at fracture divided by the section modulus. In torsion, modulus of rupture is the torque at fracture divided by the polar section modulus.

modus ponens
A mathematical form of argument in deductive logic. It has the form: if *A* is true, then *B* is true.

moire
The pattern (image) resulting from the combination of two periodic signals or scenes.

moisture-vapor transmission
Rate at which water vapor penetrates film over a given time at specified temperature and relative humidity (e.g., gm-mil/24 hr/100 in.^{2}).

molding cycle
The period of time occupied by the complete sequence of operations on a molding press requisite for the production of one set of moldings. The operations necessary to produce a set of moldings, without reference to the time taken.

molding powder
Plastic material in varying stages of granulation and comprising resin, filler, pigments, plasticizers, and other ingredients ready for use in the molding operation.

molecular weight, average
Molecular weight of polymeric materials determined by viscosity of the polymer in solution at a specific temperature. Gives average molecular weight of molecular chains in a polymer, independent of specific chain length.

momentary-action switch
A switch that returns from the operating condition to its normal circuit condition when its actuating force is removed, usually by a human operator.

monadic operation
An operation with one and only one operand.

monitor
A dedicated device used to display computer-generated information. Monitors may be black and white, green, or color CRT displays. A closed-circuit television system that allows the vision system operator to view the image of the object being investigated. A true monitor does not incorporate channel selector or audio components. An operating programming system that provides a uniform method for handling the real-time aspects of program timing, such as scheduling and basic input/output functions.

monitoring controller
A controller used in an application where the process is continually checked to alert the operator to possible malfunctions.

Monitor
NEC Multisync High-Resolution Color Monitor

monochromatic
Of light, that consisting of one wavelength only. In practice, a laser beam consists of a very narrow band of wavelengths around a central wavelength.

monochromator
A device for isolating narrow portions of the spectrum by dispersing light into its component wavelengths.

monochrome
Literally, "one color," but in graphics, a display that shows a light image on dark (or vice versa) background.

monocular
Pertaining to an image taken from a single viewpoint.

monospaced font
A text font in which every character cell has the same width.

monotectic
An isothermal reversible reaction in a binary system, in which a liquid on cooling decomposes into a second liquid and a solid. It differs from a eutectic in that only one of the two products of the reaction is below its freezing range.

Monte Carlo technique
Any procedure involving random sampling techniques to obtain a probabilistic approximation to the solution of a mathematical or physical problem. A procedure by which can be obtained approximate evaluations of mathematical expressions that are built up of one or more probability functions.

mosaic
A rectangular matrix of predefined elements that can be used to construct block-style graphic images.

most-significant character
The extreme left-hand character in a group of characters. Contrast to least-significant character.

motherboard
A relatively large piece of insulating material on which components, modules, or other electronic subassemblies are mounted and interconnections made by welding, soldering, or other means, using point-to-point or matrix wire or circuitry fabricated integrally with the board.

motion axis
The line defining the axis of motion, either linear or rotary, of a machine element. Such an axis can be defined using either an external reference or Type M or Type F straightness measurements.

motion cycle
A complete series of motion elements involved in performing an operation, beginning with a motion connected with the production of the unit and ending when the same motion is about to be repeated with the next unit.

motion hold
A means of externally interrupting continuance of motion of the robot from any further sequence or action steps without dissipating stored energy.

motion sensing
The ability of a vision system to form an image of an object in motion and to determine the direction and speed of that motion.

motion study
The analysis of the hand and eye movements occurring in an operation or work cycle, for the purpose of eliminating wasted movements and establishing a better sequence and coordination of movements.

motion/time analysis
A system of predetermined motion/time standards, used for describing and recording an operation in terms of its motion.

motor controller
A device or group of devices that serves to govern, in a predetermined manner, the electrical power delivered to a motor.

mouse
A hand-held device, separate from a keyboard, used to control cursor position on a display screen. As the mouse rolls along a table top, its relative position approximates the position of the cursor. Introduced by Xerox in the Star 8010 professional workstation in 1981, it is now quite popular.

movement inventory
A type of in-process inventory that arises because of the time required to move goods from one place to another.

move time
The actual time that a job spends in transit from one operation to another in the shop.

moving average
The series of arithmetic averages obtained by averaging the last k successive terms in a time series, these terms being spaces at equal intervals. Thus, each average is usually ob-

tained by dropping from the computation the oldest term used in the preceding group of *k* terms and adding in its place the newest term to be included in the new *k* group. Weights may be assigned to each term of the group of *k* successive terms.

moving-beam bar code reader
A device that dynamically searches for a bar code pattern by sweeping a moving optical beam through a field of view.

multiaccess
The ability for several users to communicate with the computer at the same time, each working independently on his or her own job.

multibus
A bus developed by Intel Corporation, widely used in industrial computer systems.

multidrop
A communication system configuration using a single channel or line to serve multiple terminals, often at different geographical locations. Use of this type of line normally requires a polling mechanism and a unique address for each terminal. Also called a multipoint line.

multileaving
A technique for allowing simultaneous use of a communications line by two or more terminals.

multilevel bill of material
A multilevel bill shows all the components directly or indirectly used in an assembly, together with the quantity required of every component. If a component is a subassembly, all the components of the subassembly will also be exhibited in the multilevel bill.

multilevel where used
Multilevel where used for a component lists all the assemblies in which that component is directly used and the next higher-level assemblies into which the parent assembly is used.

multipass sort
A sort program that is designed to sort more items than can be in main storage at one time.

multiple-activity process chart
A synchronized graphic representation of operations performed simultaneously by two or more people, two or more machines, or a combination of people and machines.

multiple font
A scanner having the capability to recognize more than one type font but only one at a time.

multiple precision
Pertaining to the use of two or more computer words to represent a number in order to enhance precision.

multiplex
The transmission of multiple data bits through a single transmission line by means of a "sharing" technique. To interleave or simultaneously transmit two or more messages on a single channel.

multiplexing
The time-shared scanning of a number of data lines into a single channel. Only one data line is enabled at any instant. The division of a transmission facility into two or more channels either by horizontally splitting the frequency band transmitted by the channel into narrower bands, each of which is used to constitute a distinct channel (frequency division multiplexing), or by allotting this common channel to several different vertical information channels, one at a time (time division multiplexing).

multiprocessing system
A computing system employing two or more interconnected processing units, each having access to a common jointly addressable memory, to execute programs simultaneously. Also, loosely refers to parallel processing.

multiprocessor
A computer network consisting of two or more central processors under a common control.

multiprocessor control
A control scheme that employs more than one central processing unit in simultaneous parallel computation.

multiprogramming
A technique used to balance the CPU's speed with the slower peripherals by allowing several programs to run on the computer system at the same time. The goal is to make more efficient use of the system by keeping more parts of it busy more of the time. The difficulty is that this increases greatly the complexity and cost of the operating system and the overall computer system operation. This interleaving of the execution of two or more programs results in time-sharing of machine components.

multistroke line
On a vector display, a line created by drawing several overlapping or adjacent segments.

multitasking
An IBM term for multithreading. Used in the literature for IBM's real software package. Refers to the concurrent execution of one main task and one or more subtasks in the same partition or region.

multithread operation
A program construction technique that allows more than one logical path through the program to be executed simultaneously.

multiuser
A computer that can support several workstations working simultaneously.

N

NAND
Basic logical circuit used in designing digital hardware. The acronym for Not AND.

nanometer
Unit of length equal to 10-9 meter.

nanosecond
One-billionth of a second; a common unit of measure of computer operating speed.

NAPLPS
An abbreviation for North American Presentation Level Protocol Syntax, a standard for encoding and displaying graphics that was jointly developed in the United States and Canada. Although image creation tends to be slow, the standard allows devices of different resolution and capability to each show the image in the best form it can.

narrowband channels
Subvoice grade channels characterized by a speed range of 100 to 200 bits per second.

***n*-ary**
Pertaining to a selection, choice, or condition that has *n* possible different values or states.

National Electrical Code
A consensus standard for the construction and installation of electrical wiring and apparatus, established by the National Fire Protection Association.

natural aging
Spontaneous aging of a supersaturated solid solution at room temperature.

natural binary
A number system to the base (radix) two, in which the 1s and 0s have weighted value in accordance with their relative position in the binary word. Carries may affect many digits.

natural deduction
Informal reasoning.

natural language
The conventional method for exchanging information between people, such as English as a means of communication for human speakers and various formal written systems as a means of representing intentions in technical disciplines (chemical graphs, DNA sequences, engineering diagrams, and so on).

natural language generation
The part of natural language processing research that attempts to have computers present information to their users in a "natural" language, such as English.

natural language interface
A program that allows the user to communicate with a computer in a "natural" language, such as English. An NLI may incorporate both natural language understanding and natural language generation capabilities. An NLI is sometimes called a *natural language front end.*

natural language processing
An area of AI research that allows computers to use a "natural" language, such as English. Natural language processing is divided into natural language understanding and natural language generation. (*Illus. p. 242.*)

natural language understanding
The part of natural language processing research that investigates methods of allowing computers to understand a "natural" language, such as English.

natural number
One of the numbers, 0, 1, 2, . . .

***n*-bit byte**
A byte composed of *n* binary elements.

***n*-core-per-bit storage**
A storage device in which each storage cell uses *n* magnetic cores per binary character.

NDC space
A finite region within the normalized device coordinate system. It defines the maximum region usable by an application program.

NDC system
A device-independent two-dimensional or three-dimensional Cartesian coordinate system based on normalized device coordinate (NDC) values in the range zero to one.

near distance
The distance from the front clipping plane to the view reference point.

nearest neighbor
The technique used in the SRI algorithm to match features of an observed object with corresponding features of a learned object.

near infrared
Infrared radiation from 0.75 to 3 microns.

necessary condition
A condition that must be met if a particular problem is to be solved.

Natural-Language Processing
Types of Natural-Language Processing Problems

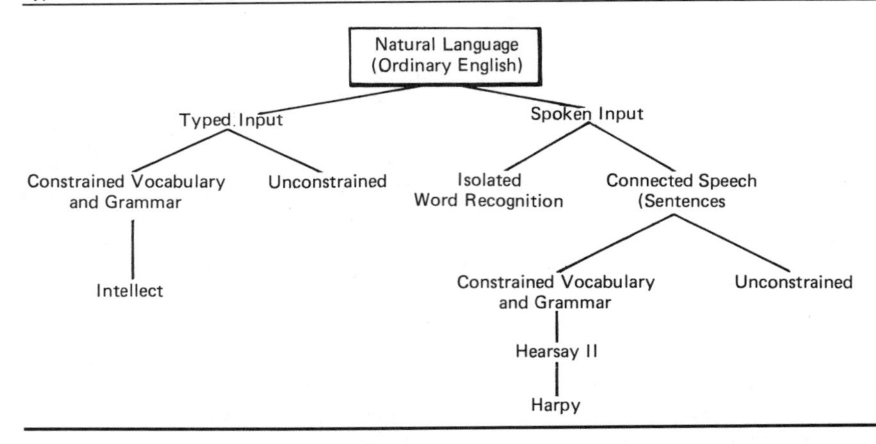

necessary-condition set
The set of all necessary conditions relating to a given development effort.

necking
The operation of reducing the opening of a drawn cup or shell.

needle blow
A specific blow molding technique where the blowing air is injected into the hollow article through a sharpened hollow needle that pierces the parison.

negate
To change a proposition into its opposite.

negation
The monadic Boolean operation, the result of which has the Boolean value opposite to that of the operand.

negative acknowledgment
In data communication, indicates that the previous transmission was in error and the receive is ready to accept a retransmission. NAK is also the "not ready" reply to a station selection (multipoint) or to an initialization sequence (line bid).

negative-bounded planar portion
A hole, as defined in IGES specification.

negative image
A picture signal having a polarity that is opposite to normal polarity and that results in an image in which the white and black areas are reversed.

NEMA standards
Consensus standards for electrical equipment approved by the majority of the members of NEMA (National Electrical Manufacturers Association).

NEMA type 12
A category of industrial enclosures intended for indoor use primarily to provide a degree of protection against dust, falling dirt, and dripping noncorrosive liquids. They are designed to meet drip, dust, and rust-resistance tests. They are not intended to provide protection against conditions, such as internal condensation.

Neper
A unit of measuring power. The number of Nepers is the base e logarithm of the ratio of the measured power levels.

nest
To incorporate a structure or structures of some kind into a structure of the same kind. For example, to nest one loop (the nested loop) within another loop (the nesting loop). To embed subroutines or data at different hierarchical levels so that routines can be executed or accessed iteratively.

nesting
Organizing design data into levels (i.e., a hierarchical structure) for greater efficiency in storing and processing repetitive design elements. In nesting, a design is first broken into

major components or building blocks that are then further subdivided into smaller components, and so on. With such a nested design, identical components need to be represented only once in the data base, saving memory storage and simplifying data retrieval and modification. In computer-aided mechanical design, a programming technique for arranging multiple parts on a larger sheet or plate for optimum use of material. The parts are cut or burned out of the larger sheet. The objective of nesting is to minimize material costs and scrap.

net
A set of logically related connection points (pins) comprising a particular signal in an electrical or electronic design. Also called a logical net.

net change MRP
An approach by means of which the material requirements plan is continually retained in the computer. Whenever there is a change in requirements, open order or inventory status, or engineering usage a partial explosion is made only for those parts affected by the change. Net changes systems may be continuous and totally transaction-oriented or done in a periodic (often daily) batch.

net list
A list, by name, of each signal and each symbol component and pin logically connected to the signal (or net). No physical order is implied. A net list can be generated automatically by a computer-aided design system.

net load capacity
The additional weight or mass of a material that can be handled by a machine or process without failure over and above that required for a container, pallet, or other device that necessarily accompanies the material.

net requirements
The actual production, purchasing requirements, and net requirements for a particular component. The net requirements are calculated by deducting the available inventory from the gross requirements.

net weight
The weight of an article exclusive of the weights of all packing materials and containers.

network
A computer communications system consisting of one or more terminals communicating with a single host computer system, which acts as the network control component through internal programming or perhaps through a front-end processor. The chief characteristic of a network is the single controlling host computer system that may include multiple processors. The general use of the word network to mean a collection of interconnected components is no longer precise, just as the word system no longer carries the connotation of close proximity of components. For example, a CAD/CAM system might be connected to a mainframe computer to off-load heavy analytic tasks. Also refers to a piping network in computer-aided plant design. *Local networking* is the communications network internal to a robot. *Global networking* is the ability to provide communications connections outside of the robot's internal system.

network analog
The expression and solution of mathematical relationships between variables, using a circuit or circuits to represent these variables.

network control program
The interface program that communicates with the network on one side and with user programs in the host computer on the other side.

network (data base)
A method of organizing a data base with linkages external to the physical organization.

network interface sublayer (NISL)
Provides a mapping function between the facilities of actual communications network (i.e., a LAN) and the services required of layer four.

network layer
The purpose of the MAP network layer is to provide message routing between end nodes, either on the same network or any other subnetwork, regardless of the physical distance separating the subnetworks.

network management
The facility by which the network communication is observed and controlled.

network management entity
A logically singular instance of a network manager. For example, there may be several people doing NM on a single MAP segment but the totality of their effort is the NM entity. In particular, they share common data about

Network Layer
MAP Specifications Summary by Layer. *Source*: Boeing Company.

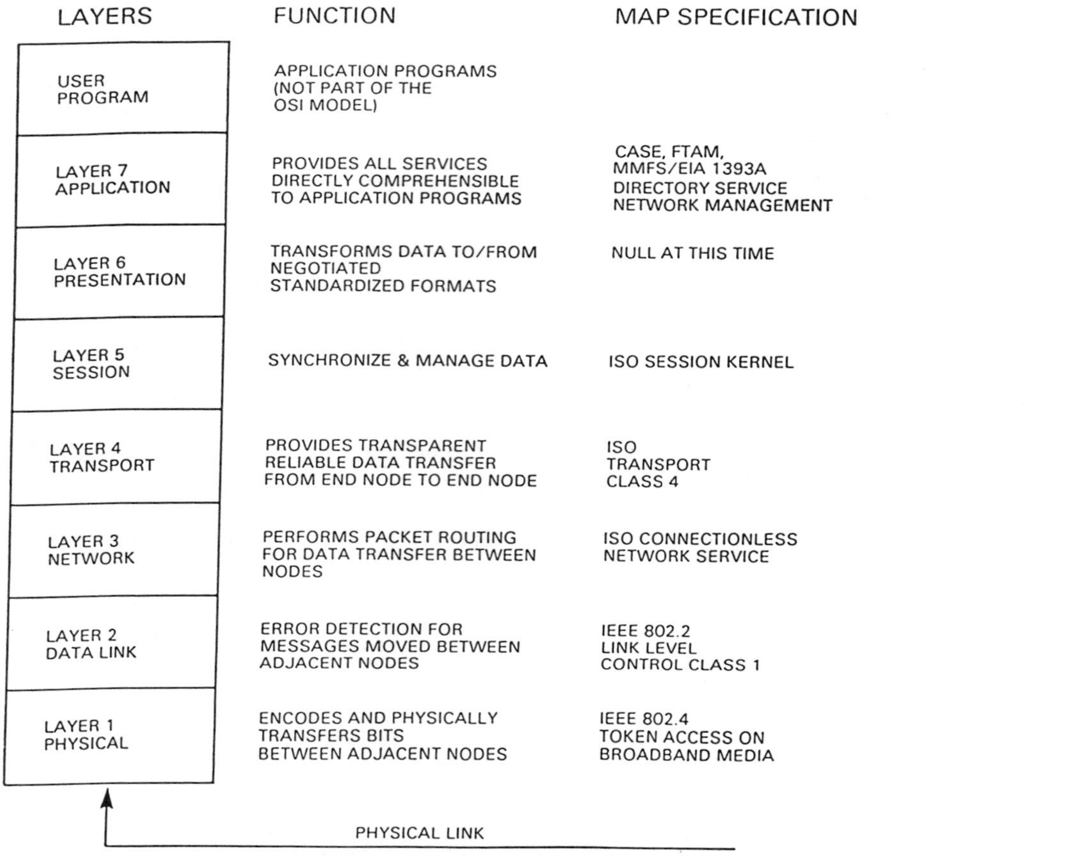

LAYERS	FUNCTION	MAP SPECIFICATION
USER PROGRAM	APPLICATION PROGRAMS (NOT PART OF THE OSI MODEL)	
LAYER 7 APPLICATION	PROVIDES ALL SERVICES DIRECTLY COMPREHENSIBLE TO APPLICATION PROGRAMS	CASE, FTAM, MMFS/EIA 1393A DIRECTORY SERVICE NETWORK MANAGEMENT
LAYER 6 PRESENTATION	TRANSFORMS DATA TO/FROM NEGOTIATED STANDARDIZED FORMATS	NULL AT THIS TIME
LAYER 5 SESSION	SYNCHRONIZE & MANAGE DATA	ISO SESSION KERNEL
LAYER 4 TRANSPORT	PROVIDES TRANSPARENT RELIABLE DATA TRANSFER FROM END NODE TO END NODE	ISO TRANSPORT CLASS 4
LAYER 3 NETWORK	PERFORMS PACKET ROUTING FOR DATA TRANSFER BETWEEN NODES	ISO CONNECTIONLESS NETWORK SERVICE
LAYER 2 DATA LINK	ERROR DETECTION FOR MESSAGES MOVED BETWEEN ADJACENT NODES	IEEE 802.2 LINK LEVEL CONTROL CLASS 1
LAYER 1 PHYSICAL	ENCODES AND PHYSICALLY TRANSFERS BITS BETWEEN ADJACENT NODES	IEEE 802.4 TOKEN ACCESS ON BROADBAND MEDIA

PHYSICAL LINK

the network and know of each other's actions. This term is useful when discussing the impacts of interconnected LANs and broken LAN networks.

network manager
The consciousness of the network management facility. The network manager is the initiator of network manager (NM) action. The network manager may be human or machine.

network operations center
A specialized center that assists in network operations, monitoring of network status, supervision and coordination of network maintenance, accumulation of accounting and usage data, and user support.

network planning
A broad generic term for techniques used to plan complex projects. Two of the most popular techniques are PERT and CPM, that seek to identify the controlling or limiting paths in a complex process.

network redundancy
Describes the property of a network that has additional links beyond the minimum number necessary to connect all nodes.

network representation
A data structure consisting of nodes and labeled connecting arcs.

network security
The measures taken to protect a network from an unauthorized access, accidental or willful

interference with normal operations, or destruction, including protection of physical facilities, software, and personnel security.

network service access point
The means of communication between the network service component and the end-to-end service component.

network topology
The geometric arrangement of links and nodes in a network.

neutral density filter
An optical device that reduces the intensity of light without changing the spectral distribution of light energy.

new-frame action
The elimination of all temporary information and the redrawing, if necessary, of all visible retained information. On a hard-copy device, the recording medium is advanced to a fresh drawing area. The action is implicit in several functions, such as making a retained segment invisible.

new-line character
A formal effector that causes the print or display positions to move to the first position of the next line.

Newtonian liquid
A liquid in which the rate of flow is directly proportional to the force applied. The viscosity is independent of the rate of shear; there is no yield value in Newtonian flow.

NEWVICON
Trade name for a image pickup tube. The photoconductive target is a heterojunction structure characterized by high sensitivity, nonblooming of high brightness details, relative freedom from image burn, and good resolution.

nibble
Popular name for four bits or half a byte. Also spelled nybble.

nibbling
The operation of removing flat material by means of a rapidly reciprocating punch to form a contour of a desired shape.

nil
A value returned for some parameters to indicate that an open structure, for example, is nonexistent.

nines complement
The diminished radix complement in the decimal numeration system.

***n*-level address**
An indirect address that specifies *n* levels.

node
A point (representing such aspects as the system state or an object) in a graph connected to other points in the graph by arcs (usually representing relations). Any station, terminal, terminal installation, communications computer, or communications computer installation in a computer network. A point in space used to define a finite element topology.

node address
A unique address for each machine on the local network(s) where local control of addressing is maintained.

noise
Any unwanted disturbance within a dynamic electrical or mechanical system. Any unwanted electrical disturbance or spurious signal that modifies the transmitting, indicating, or recording of desired data. In a computer, extra bits or words that have no meaning and must be ignored or removed from the data at the time it is used. Random variations of one or more characteristics of any entity such as voltage, current, or data. A random signal of known statistical properties of amplitude, distribution, and spectral density.

noise equivalent power
Also called noise equivalent exposure. The radiant flux necessary to give an output signal equal to the detector noise.

noise-free environment
A space theoretically containing neither electrostatic nor electromagnetic radiation. Since this is an industrial impossibility, users attempt to approximate it in direct relation to application requirements.

noise immunity
The ability of the computer or robot controller to reject unwanted noise signals.

noise spike
Voltage or current surge produced in the industrial operating environment.

nominal
Identifying value of a measurable property. This standard value is halfway between maximum and minimum limits of the tolerance range.

nominal black
The color black (all 0s) in color mode 0 or the color that is at color map address 0 in color modes 1 and 2.

nominal value
The natural workstation-dependent value for an attribute, such as linewidth, to which other possible values on that workstation are relative.

nominal white
The color white (all 1s) in color mode 0 or the color that is at color map address 011 . . . 1 in color modes 1 and 2.

nonbrowning
Descriptive of lens glass, faceplate glass, and glass envelopes for image pickup tubes, used in radiation-tolerant imaging systems. Nonbrowning glass will not discolor (turn brown) when irradiated with atomic particles and waves.

nonconjunction
The dyadic Boolean operation, the result of which has the Boolean value 0 if and only if each operand has the Boolean value 1.

nondedicated queue
Queue used both by pallets entering and pallets leaving a station.

nondestructive inspection
Inspection by methods that do not destroy the inspected part.

nondisjunction
The dyadic Boolean operation, the result of which has the Boolean value 1 if and only if each operand has the Boolean value 0.

nonequivalence operation
The dyadic Boolean operation, the result of which has the Boolean value 1 if and only if the operand has different Boolean values.

nonidentity operation
The Boolean operation, the result of which has the Boolean value 1 if and only if all the operands do not have the same Boolean value. A nonidentity operation on two operands is a nonequivalence operation.

noninterlaced
Refers to the process of scanning or displaying every line of an image in sequence. A camera that reads every electron line of pictures from a sensor device is noninterlaced.

nonlinear programming
In operations research, a procedure for locating the maximum or minimum of a function of variables that are subject to constraints, when the function or the constraints or both are nonlinear.

non-monotonic logic
A logic in which results are subject to revision as more information is gathered.

nonrepetitive
A descriptive term applied to a type of work, operation, part, or the like that does not recur frequently or in any reasonable regular sequence.

nonretained data
Graphical data that is not stored by the graphics system for use in subsequent pictures.

nonretained segment
A nameless segment, efficient in storage utilization, that cannot be modified; used primarily for plotting applications efficient in storage utilization.

nonretained structure
The single structure that contains nonretained data. The nonretained structure is deleted each time that an implicit regeneration occurs.

nonreturn-to-reference recording
The magnetic recording of binary digits in such a way that the patterns of magnetization used to represent 0s and 1s occupy the whole storage cell, with no part of the cell magnetized to the reference condition.

non-servo control
The control of a robot through the use of mechanical stops that permit motion between two end points.

non-servo robot
A robot that can be programmed by its motions only to the end points of each of its axes of motion.

nonspacing
A character that does not cause the cursor to be automatically advanced after the character has been received and displayed.

nonswitched line
A communications link that is permanently installed between two points.

nontransparent
Pertaining to the mode of binary synchronous transmission in which all control characters are treated as control characters (i.e., not treated as text).

nontransparent mode
In data comunications, transmission of characters in a defined format, for example, ASCII or EBCDIC, in which all defined control characters and control character sequences are recognized and treated as such.

nonvolatile memory
A memory that is designed to retain its information while its power supply is turned off.

nonvolatile random-access memory
A special kind of RAM that will not lose its contents due to loss of power. No battery backup is needed with this kind of memory.

no-operation instruction
An instruction whose execution causes a computer to do nothing other than to proceed to the next instruction to be executed.

NOR
A logic operator having the property that if *P* is a statement, *Q* is a statement, *R* is a statement, . . . , then the NOR of *P, Q, R*, . . . , is true if all statements are false, false if at least one statement is true. *P* NOR *Q* is often represented by a combination of OR and NOT symbols, such as PVQ.

normal capacity
The maximum rate at which a task can be performed with standard workforce and a standard work configuration.

normal distribution
A statistical distribution with a mean of zero and a standard deviation of one, typified by a bell-shaped curve.

normal elemental time
The selected or average elemental time, adjusted by leveling or other methods to calculate the time required by a qualified worker to perform a single element or operation.

normalization transformation
A transformation that maps the boundary and interior of a window in world coordinates to the boundary and interior of a viewport in NDC coordinates. Also known as a viewing transformation or window-to-viewport transformation.

normalized device coordinate system
A device-independent two-dimensional or three-dimensional Cartesian coordinate system whose coordinates are in the range zero to one. Normalized device coordinates are used in defining views of objects. In particular, they are used for specifying viewports, image transformations, and input obtained from stroke and locator devices.

normalizing
Heating a ferrous alloy to a suitable temperature above the transformation range and then cooling in air to a temperature substantially below the transformation range.

normally closed contact
A contact pair that is closed when the coil of relay is not energized.

normally open contact
A contact pair that is open when the coil of relay is not energized.

NOT
A logic operator having the property that if *P* is a statement, then the NOT of *P* is true if *P* is false and false if *P* is true.

notation
A set of symbols, and the rules for their use, for the representation of data.

not element
A logic element that performs the Boolean operation of negation.

nuclear teleoperator
A device used for manipulation or inspection operations in a radioactive environment, sometimes incorporating some mobility capability by means of a wheeled or tracked vehicle, and controlled continuously by a remote human operator.

null
Empty. Having no memory. Not usable.

null character
A control character that is used to accomplish media-fill, or time-fill, and that may be inserted into or removed from a sequence of characters without affecting the meaning of the sequence; however, the control of equipment of the format may be affected by this character.

null layer
A layer that provides no additional services. It exists only to provide a logical path for the flow of network data and control.

number crunching
The rapid processing of large quantities of numbers.

numerical analysis
A method of obtaining useful quantitative solutions to problems that have been expressed mathematically and involve complex processes or relationships (e.g., integration) by means of trial and error.

numerical control machine
A technique for the control of machine tools or factory processes in which information on

the desired actions of the system is prerecorded in numerical form and used to control the operation automatically. The NC system consists of all elements of the control system and of the machine being controlled. As most numerically controlled devices have very limited logical or arithmetic capability (to keep costs low), they rely on their input tapes for detailed and explicit guidance. It is common for a computer to prepare the control tapes, using information presented in a manageable and concise form.

numerical control system
A system in which actions are controlled by the direct insertion of numerical data at some point. The system must automatically interpret at least some portion of these data.

numerical data
Data in which information is expressed by a set of numbers or symbols that can only assume discrete values or configurations.

numerical geometry system
A graphics-based, partly interactive, three-dimensional system for computer-aided design and manufacture. The system permits the design process to be carried out either by direct entry of numerical data or from existing design systems.

numeric character
A character that belongs to one of the set of digits 0 through 9.

numeric character set
A character set that contains digits and may contain control characters, special characters, and the space character, but not letters.

numeric code
A code according to which data are represented by a numeric character set.

numeric data base
This type of source data base typically contains numeric values from original sources and/or data that has been summarized or otherwise statistically manipulated. Most often presented in the form of time series, numeric data bases range from simple balance-sheet data to complex econometric models. Economic, demographic, and financial data are often depicted in numeric data bases.

Numerical Control Machine
Block Diagram of a Numerically Controlled Machine Tool. Reprinted with Permission of Reston Publishing Company, Prentice-Hall, from *Robotics and Automated Manufacturing* by Richard Dorf.

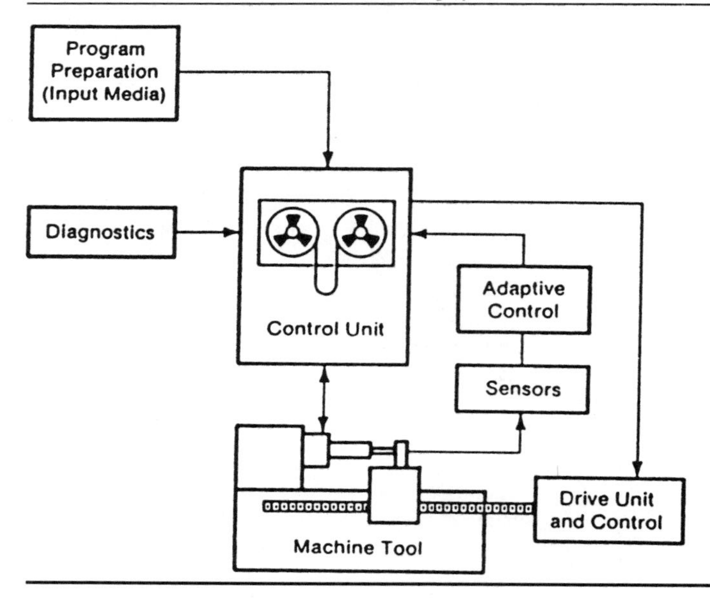

numeric processing
The traditional use of computers to manipulate numbers.

numeric word
A word consisting of digits and possible space characters and special characters.

numeripower
A modular approach to NC tape preparation and two-axis positioning and profiling. Postprocessing function is user-definable, eliminating the need for discrete postprocessors. Plotting verification capability at both the geometric level and at the completed tape level.

nybble
Half a byte, of four bits.

O

object
Any element of a drawing. Synonym for primitive and entity. In CAD, a pictorial element that can be treated as a unit, allowing it to be moved or scaled on screen, associated with attributes and data in the data base, and replicated; in displays, on-screen images that are defined in sections of memory and then moved across the screen by supplying positioning values to the graphics-display system rather than redrawing the image at a new location. The most common form of graphic object is the sprite, a screen object under the control of the graphic system that reports any collisions as it is moved over the screen. A three-dimensional figure or shape that is the subject of investigation for a vision system. An object forms the basis for creating an image. A conceptual graphics entity in an application program. In PHIGS, objects are described in terms of structure elements defined in modeling coordinates and attributes, all of which can be organized into hierarchical structures. In Core and GKS, objects are described in world coordinates in terms of output primitives and primitive attributes.

object-centered viewing
A viewing situation in which the object is considered fixed and the most likely dynamic behavior is for the viewer to move around the object.

object code
Absolute language output from a compiler or assembler that is itself executable machine code or is fully compiled and is ready to be loaded into the computer. Relocatable machine-language code.

object deck
A deck of tabulating cards containing the condensed computer instructions for handling a specific general processor. The programmer usually writes only one instruction per card. The instructions are then processed by the computer and condensed to an object deck, where a multitude of instructions are included on a single card.

objective
A desired end result, condition, or goal that forms a basis for managerial decision making.

objective function
An equation defining a scalar quantity, to be minimized under given constraints by an optional controller in terms of such performance variables as error, energy, and time. The objective function defines a trade-off relationship between these cost variables. The goal or function that is to be optimized in a model. Most often, it is a cost function that one is attempting to minimize, subject to some restrictions, or a profit function that one is trying to maximize, subject to some restriction.

objective lens
The optical element that receives light from an object or scene and forms the first or primary image.

object language
The output of a translation process. Contrast with source language. Synonymous with target language.

object module
The primary output of an assembler or compiler, which can be linked with other modules and loaded into memory as a runnable program. The object module is composed of the relocatable machine-language code, relocation information, and the corresponding global symbol table defining the use of symbols within the module.

object-oriented language
In robotics, a synonym for task-oriented language. In general use, a programming language in which procedures for doing things are accessed through descriptions of the things to be worked on.

object-oriented programming
A programming approach focused on objects that communicate by message passing. An object is considered to be a package of information and descriptions of procedures that can manipulate that information.

object plane
An imaginary plane that is focused onto a lens' specified image plane.

object program
An absolute language program of instructions for the computer made from a source program written in symbolic language. The program-

mer writes the source program and a processor program translates it into an object program.

oblique view
A parallel projection whose projectors are not perpendicular to the view plane.

obsolescence
A decrease in the value of an asset brought about by the development of new and more economical methods, processes, or machinery.

occlusion
The result in an image of an object in the optical path that is not part of the scene but blocks off some portion of the image. The inability to see a portion of an object because an obstruction has been placed between the object and the observer.

octal
Pertaining to the number base of eight. In octal notation, octal 214 is two times 64, plus one times 8, plus four times 1, and equals decimal 140.

octal numbers
As with the binary system that operates to a base two, this system operates to a base eight. Its advantage is that fewer digits are required than for the binary code. It is also more readily convertible to the decimal system than is the binary system and is used as an intermediate step, within the computer, to convert from decimal to binary and vice versa. This is less cumbersome than going directly from decimal to binary and vice versa.

odd parity
Condition existing when the sum of the number of 1s in a binary word is always odd.

offline
Pertaining to equipment or devices not under direct control of the central processing unit. For example, the computer might generate a magnetic tape that would then be used to generate a report offline while the computer was doing another job. May also be used to describe terminal equipment that is not connected to a transmission line.

offline operation
Usually applies to peripheral equipment that operates independently of the central computer. This equipment is generally utilized where it is not necessary to operate the full capabilities of the major portion of the computer, thus saving time and expense.

offline programming
Defining the sequences and conditions of actions on a computer system that is independent of the robot's "on-board" control. The prepackaged program is loaded into the robot's controller for subsequent automatic action of the manipulator.

offline programming language teaching
The method of "teaching" the robot involving control and positioning through a computer program. It is only relevant for programmable robots. Coordinates for positions and attitude of limbs are present by the program so that the robot is not required to be led through motions.

offline storage
A storage device not under control of the central processing unit.

offset
A displacement in the axial direction of the tool that is the difference between the actual tool length and that established by the programmer. The count value output from an A/D converter resulting from a zero input analog voltage. Used to correct subsequent nonzero measurements.

off-shuttle queue
Pallet queue at exit from a module.

off the shelf
Refers to production items that are available from current stock and need not be either newly purchased or immediately manufactured. Also relates to computer software or equipment that can be used by customers with little or no adaptation, thereby saving the time and expense of developing their own.

one-dimensional
Refers to a language whose expressions are customarily language represented as strings of characters, for example, FORTRAN.

one-plus-one address instruction
An instruction that contains two address parts, the plus-one address being that of the instruction that is to be executed next unless otherwise specified.

ones complement
The diminished radix complement in the pure binary numeration system.

one-shot molding
In the urethane foam field, indicates a system whereby the isocyanate, polyol, catalyst, and

other additives are mixed together directly and a foam is produced immediately (as distinguished from prepolymer).

one-way-only operation
A mode of data communication operation in which data are transmitted in a preassigned direction over a channel. Synonym, simplex circuit.

on hand
The balance shown in perpetual inventory records as being physically present at the stocking locations.

online
Pertaining to devices under direct control of the central processing unit. Operation where input data are fed directly from the measuring devices into the CPU, or where data from the CPU are transmitted directly to where it is used. Such operation is in real time. Refers to direct interactions with a computer where there is immediate response to an inputted command. Opposite of batch. The operation of peripherals or terminals in direct interactive communication and under control of the central processing unit by means of a communication channel. May also be used to describe terminal equipment connected to a transmission line. Also pertaining to a user's ability to interact with a computer either by means of the console or a terminal.

online debugging
The act of finding and fixing program bugs while the program is loaded in the computer.

online operation
Applies to computer operations and calculations that are performed by the computer itself or the major portion of the computer.

online processing
A general data-processing term concerning access to computers, in which the input data enters the computer directly from the point of origin or in which output data is transmitted directly to where it is used. The process usually requires random-access storage. In online processing, a user has direct and immediate access to the computer system by means of terminal devices. Information and instructions are entered by way of a terminal, processing by the computer is begun virtually immediately, and a response is received as soon as possible, often within seconds.

online programming
A means of programming a robot on a computer that directly controls the robot. The programming is performed in real time.

online services
A term describing computer functions available to end-users not in possession of a host computer. Such services include the use by many users of one computer's processing power (time-sharing); access to stored files of information (reference and source data bases); the use of storage media for private data (archival storage); the use of prepared programs (online software packages); and peripheral output (printed or recorded representations of data). Two or more of these services are often combined for the convenience of the end-user. See also online services company, time-sharing, online processing, and online.

online services company
Typically provides computer services to end-users connected by means of a terminal with the host computer. These services include access to processing power (time-sharing); prepared programs (online software packages); research information (source and reference data bases); storage media (archival storage); and peripheral output (printed or recorded representations of data, in either a raw or a processed state). See service bureau.

online software package
A service offered by online services companies. Online services companies offer the use of these up-and-running programs to end-users, usually for a fee. The end-user inputs his or her own data (usually via telecommunications) and utilizes the online services company's software to manipulate said data. Examples include payroll and data base management systems. In the latter case, the online services company provides both the software and the storage necessary to create private data bases.

online storage
A storage device under direct control of the central processing unit.

online system
A system in which the input data enter the computer directly from the point of origin or in which output data are transmitted directly to where they are used.

on-machine
Error compensation (that could be either active or precalibrated) in which the workpiece is not removed from the machining station for gaging.

on-module gauging
Gauging performed at a module as an integral step in the processing of parts by the module.

on order
Of stock, the quantity represented by the total of all outstanding replenishment orders. The on-order balance increases when a new order is released, and it decreases when material is received to fill an order or when an order is canceled.

on-shuttle queue
Pallet queue at entrance to a module.

opacity
Degree to which material or coating obstructs transmittance of visible light. Term is used primarily with nearly opaque materials, for which opacity is generally reported by contrast ratio. Opacity of light-transmitting materials is usually expressed by its opposite or complementary property; that is, total transmittance of diffuse light transmission factor.

opaque
Of a medium, one that does not transmit light.

opcode
A one-byte character that initiates the execution of a locally stored geometric primitive or control operation. An opcode may be followed by zero or more operands.

open code
In assembler program, that portion of a source module that lies outside of and after any source macro definitions that may be specified.

open loop
A control system whereby data flows only from the control to the mechanism but does not flow from the mechanism back to the control. A method of control in which there is no self-correcting action for the error of the desired operational condition. Without feedback.

open-loop control
A system of robot control that does not rely on a feedback loop for measuring performance. In open-loop control, communication is in one direction only.

open-loop robot
A robot that incorporates no feedback, that is, no means of comparing actual output to commanded input of position or rate.

open-loop system
A control system that has no means for comparing the output with the input for control purposes (i.e., there is no feedback).

open order
A released and unfilled order, either to manufacturing or purchasing.

open shop
A computer facility in which programming is performed by the user, rather than by a group of computer programmers. The computer operation itself may be described as open shop if the user/programmer also serves as the operator, rather than a full-time trained operator.

open subroutine
A subroutine of which a replica must be inserted at each place in a computer program at which the subroutine is used.

open system
A computer processor or connected set of processors with software support that allows an application entity running in the open system to communicate with other application entities running in the same or other open systems.

open systems interconnect
A seven-layer model for an open system proposed by ISO.

open systems interconnection (OSI) reference model
The network architecture chosen for the technical and office protocols (TOP) is the open systems interconnection (OSI) reference model. The OSI reference model is a generalized layered description of the functions needed to perform reliable data communications and provide network functions to the user. The purpose of the international standard reference model for open systems interconnection is to provide a common basis for the coordination of standards development for the purpose of systems interconnection, allowing existing standards to be placed into perspective within the overall reference model. The ability to use the OSI mode to encompass existing and emerging standards for data communications and migration purposes is a major reason for choosing it for the TOP network architecture. (*Illus. p. 254*.)

Open Systems Interconnection Reference Model

LAYERS	FUNCTION	LAYERS
User Program	Application Programs (not part of the OSI model)	User Program
Layer 7 Application	Provides all services directly comprehensible to application programs	Layer 7 Application
Layer 6 Presentation	Transforms data to and from negotiated standardized formats	Layer 6 Presentation
Layer 5 Session	Synchronizes and manages dialogues	Layer 5 Session
Layer 4 Transport	Provides transparent, reliable data transfer from end-node to end-node	Layer 4 Transport
Layer 3 Network	Performs message routing for data transfer between nodes	Layer 3 Network
Layer 2 Data Link	Detects errors for messages moved between nodes	Layer 2 Data Link
Layer 1 Physical	Electrically encodes and physically transfers messages between nodes	Layer 1 Physical

Physical Link

operand
The part of a computer instruction that tells the computer where the data to be processed is stored. A single- or multiple-byte string from the numeric data field of the PDI code set that is used to specify control, attribute, or coordinate parameters required by the op-code.

operating range
The reach capability of a robot. Also called work envelope.

operating system
A more-or-less general-purpose computer program that controls the execution of the specific "applications" programs run on a computer. An operating system provides the critical interface between the computer hardware and the humans that use it, handling things like program scheduling, the "input" and "output" of data, detecting and diagnosing errors in the performance of application programs, and other housekeeping chores. A structured set of software programs that control the operation of the computer and associated peripheral devices in a CAD/CAM system, as well as the execution of computer programs and data flow to and from peripheral devices. May provide support for activities and programs such as scheduling, debugging, input/output control, accounting, editing, assembly, compilation, storage assignment, data management, and diagnostics. An operating system may assign task priority levels, support a file system, provide drives for I/O (input/output) devices, support standard system commands or utilities for online programming, process commands, and support both networking and diagnostics. Software that controls the operation of a data-processing system and that may provide the following services: (1) determine what jobs are running and what parts of the computer system are working on each job at any givven time, (2) impose standards and procedures on machine operation, (3) take care of the numerous little details lumped together as "housekeeping," and (4) invoke standard troubleshooting actions in cases of malfunction. They are usually very complex and use large quantities of core and disk storage. Sometimes called supervisor, executive, monitor, or master control program, depending on the computer manufacturer.

operating time
That part of available time during which the hardware is operating. It includes development time, production time, and makeup time. Contrast with idle time.

operation
A single defined action.

operational amplifier
A high-gain amplifier used as the basic element in analog computation.

operational module
A collection of executable modules that contribute to one or more jobs.

operational tooling
A set of the tools used in operation and maintenance of the system.

operation code
That part of a computer instruction that tells it what function (such as addition) to perform.

operation number
Reference to a distinct operation on a distinct part setup.

operation precedence diagram
Diagram that shows, for a given part number, which operations must be performed prior to other operations and which operations can be performed in any order.

operation process chart
A graphic, symbolic representation of the act of producing a product or providing a service, showing operations and inspections performed or to be performed with their sequential relationships and materials used. Operation and inspection time required and location may be included.

operation sequence
Ordered list of operations encountered by a part setup in a flexible manufacturing system.

operations manual
The manual that contains instructions and specifications for a given application. Typically includes components for operators as well as programmers. Sometimes also includes a log section.

operations research
Quantitative analysis of industrial and administrative operations to derive an understanding of the factors controlling operational systems, with the view of supplying management an objective basis to make decisions concerning the actions of people, machines, and other resources in a system involving repeatable operations. Frequently involves representing the operation or the system with a mathematical model. See also linear programming, numerical analysis.

operations sequence
The sequential steps that manufacturing engineering recommends that a given assembly or part follow in its flow through the plant. For instance, operation one may be to cut bar stock; operation two may be to grind bar stock; operation three, to shape; operation four, to polish; operation five, to inspect and return to stock. This information is normally defined on a document referred to as a route sheet.

operation start date
The date when an operation should be started, based upon the work remaining and the time remaining to complete the job.

operation table
A table that defines an operation by listing all permissible combinations of values of the operands and that indicates the result for each of these combinations.

operator
A general term for CAD users. Not all users are designers and drafters. A designer or drafter, however, may be an operator. The user of an interactive application program. The operator typically interacts with the application program through a display console consisting of a physical display screen and a collection of input devices. A symbol that represents the action to be performed in a mathematical operation.

operator console
The device that enables the operator to communicate with the computer, that is, it is used to enter data or information, to request and display stored data, to actuate various preprogrammed command routines, and so on.

operator inputs
In a man/machine system, information received (sensed) by the operator from the instructions or displays or directly from the environment.

operator outputs
In a man/machine system, the action taken by the operator; the result of a decision based on the input, for example, manipulation of controls and verbal communication.

operator process chart
A motion-study aid used to chart the time relationship of the movements made by the body members of a worker performing an operation.

opportunity cost
The return on capital that could have resulted had the capital been used for some purpose other than its present use. Sometimes refers to the best alternative use of the capital; at other times, to the average return from feasible alternatives.

optical axis
The straight line passing through the centers of the curved surfaces of a lens or lens system.

optical character recognition
The ability of a vision system to identify an alphanumeric character.

optical comparator
A gaging device that accurately measures dimensions of objects using backlighting. This

device does not analyze the image or make decisions.

optical coupler
A device that couples signals from one electronic circuit to another by means of electromagnetic radiation, usually infrared or visible. A typical optical coupler uses a light-emitting diode (LED) to convert the electrical signal of the primary circuit into light and a phototransistor in the secondary circuit to reconvert the light signal back into an electrical signal.

optical data recognition
The general term used to describe any form of optical recognition of data. Included in the ODR are optical mark readers, document readers, and page readers.

optical density
Term indicating absorption of light. Numerically equal to the logarithm to base 10 of reciprocal of transmittance of material.

optical distance
Product of actual length of light path and refractive index of material.

optical distortion
Any apparent alternation of geometric pattern of an object when seen through a plastic or as reflection from a plastic surface.

optical document reader
An optical reading device that scans or reads only a small portion of an input document and that is limited to a few type styles or fonts.

optical encoder
A device that measures linear or rotary motion by detecting the movement of markings past a fixed beam of light.

optical flow
The distribution of velocities of apparent movement in an image caused by smoothing changing brightness patterns.

optical mark reading
The process of identification of marks by an optical scanner.

optical mark recognition
Machine recognition of marks on paper in the form of dots, check marks, or other recognizable symbols.

optical page reader
An optical reading device that can scan an entire page or document and that may have the capability of reading many type styles.

optical pumping
The use of a flashlamp, arc lamp, or other light source rather than an electrical discharge to excite the lasing medium.

optical pyrometer
An instrument for measuring the temperature of heated material by comparing the intensity of light emitted with a known intensity of an incandescent lamp filament.

optical ROM
A type of optical disk technology in which the information (digital data and images) is laser-etched onto the surface of a master disk that is used to press copies. A laser optical system within the drive reads the data. This method of optical storage is currently the most pervasive and is used in the form of compact disks (CDs).

optical scanner
A device that optically scans printed or written data and generates its digital representations.

optic sensor
A device or system that converts light into an electrical signal.

optimal control
A control scheme whereby the system response to a commanded input is optimal according to a specified objective function or criterion of performance, given the dynamics of the process to be controlled and the constraints on measuring.

optimization
Achieving the best solution to a problem from an overall point of view, as opposed to optimizing the component parts of the overall problem (suboptimizing).

optimize
The rearrangement of instructions or data in NC or computer applications to obtain the best balance between operating efficiency and best use of hardware capacity.

optimum plant size
The plant capacity that represents the best balance between the economics of size and the cost of carrying excess capacity during the initial years of sales.

option
A choice or feature offered to customers for customizing the end product. In many companies, the term *option* means a mandatory choice—the customer must select from one of

the available choices. For example, in ordering a new car, the customer must specify an engine but need not necessarily select an air conditioner.

optional features
Additional capabilities available at a cost above a basic price.

optional pause instruction
An instruction that allows manual suspension of the execution of a computer program.

optional stop
A miscellaneous function command similar to a program stop, except that in this instance the control ignores the command unless the operator has previously pushed a button to recognize the command (generally denoted as an MO1 code in accordance with standards).

OR
A logic operator having the property that if *P* is a statement, *Q* is a statement, *R* is a statement, . . . , then the OR of *P, Q, R*, . . . , is true if at least one statement is true and false if all statements are false. *P* or *Q* is often represented by $P + Q$, PVQ.

order
To place in sequence according to some rules or standards. A request for an item.

order backlog
The quantity of outstanding orders received but not shipped.

order control
Control of the progress of each customer order or stock order through the successive operations in its production cycle.

ordered dither
The adding of dots to an image in computer graphics to soften jagged edges created by the limitations of showing objects as lines or dots. In ordered dithering, the dots are added by a deterministic rule. In random dithering, they are added by a statistical rule.

order entry
Information and authorization to purchase, sell, or supply goods or to perform work.

ordering cost
In calculating economic order quantities, refers to the costs that increase as the number of orders placed increases. Includes costs related to the clerical work of preparing, issuing, following, and receiving orders, the physical handling of goods, inspections, and machine setup costs, if the order is being manufactured.

order multiples
An order quantity modifier applied after the lot size has been calculated, that increments the order quantity to a predetermined multiple.

order number
A number assigned by production control to a given order in the shop.

order point
The inventory level such that if the total stock on hand plus on order falls to or below the order point, action is taken to replenish the stock. The order point is normally calculated as forecasted usage during the replenishment lead time plus safety stock.

order policy
That area of a company's management policy that concerns itself with the inventory and production control parameters that affect customer service level.

order preparation lead time
The time required to analyze requirements and open order status and to create the paperwork necessary to release a purchase requisition or a work order.

order priority
The value or importance an order has in relation to other orders.

order promising
The process of making a delivery commitment, that is, answering the question, When can you ship? For make-to-order products, this usually involves a check of uncommitted material and availability of capacity.

order quantity
The lot size or quantity ordered to replenish inventory. Usually determined by lot-sizing rules and subject to modification by a minimum, maximum, and so on.

order quantity modifiers
Order quantities are calculated based upon one of the lot-sizing rules. However, it may be necessary to adjust the calculated lot size due to some special considerations. These adjustments are called order quantity modifiers.

order release
The conversion of a planned order into a released (active) order that would go to manufacturing or purchasing. This action changes a planned order into a scheduled receipt. The connection between the planning and implementation phases. Its function is to change the

status of an order from "planned" to "released."

ordinate dimension entity
An annotation entity used to indicate dimensions from a common reference line in the direction of the *XT* or *YT* axis.

OR gate
A circuit designed to compare binary TRUE-FALSE (or on-off or one-zero) inputs and pass a resultant TRUE signal if any input is TRUE.

orientation
The movement or manipulation of an object consistently into a controlled position in space. The alignment of bars and spaces to the scanner. Often referred to as vertical (picket fence) or horizontal (ladder). The angle formed by the major axis of an image relative to a reference axis. For an object, the direction of the major axis must be defined relative to a three-dimensional coordinate system. The angular position of a blob or object.

orifice bushing
The outer part of the die in an extruder head.

origin
A reference point whose coordinates are all zero.

original equipment manufacturer
A purveyor of a product made for assembly into a final system or larger subassembly by another manufacturer. Often, OEMs make computer peripherals that are integrated into a complete system by a mainframe vendor.

orthogonal
Generalized squareness or perpendicularity. For example, on a rotary table the rotational and radial motions are orthogonal although they could not be defined as either square or perpendicular, except locally.

orthographic
A type of layout, drawing, or map in which the projecting lines are perpendicular to the plane of the drawing or map. In CAD, it is the commonly accepted way of showing mechanical objects. In computer-aided piping layouts, orthographic drawings are automatically generated on the system from projections of the 3-D, dimensioned piping data base. In computer-aided mapping, an orthographic map is an azimuthal projection in which the projecting lines are perpendicular to a tangent plane. Current CAD systems allow the fast conversion of surface models into maps of differing projections to serve specific needs.

orthonormal
A term describing two vectors that are orthogonal and of unit length.

oscillator
Another word to describe a laser cavity, which is an electromagnetic oscillator.

outlier
A datum that falls significantly away from other data for a similar phenomenon. For example, if the average sales for some product were 10 units per month, and 1 month had sales of 500 units, this might be considered an outlier.

out-of-line errors
The sum of straightness errors and errors of direction (nonorthogonality or parallelism).

out-of-pocket costs
Costs that involve cash payments, such as direct labor, as opposed to depreciation that does not.

output
The data returned by a computer, either directly to the user or to some form of storage. Data emitted from a storage device, transferred from primary to secondary storage, or that is the product of an information processing operation; reports produced by a computer peripheral device. Information transferred from the robot controller through output modules to control output devices. The end result of a particular CAD/CAM process or series of processes. The output of a CAD cycle can be artwork and hard-copy lists and reports. The output of a total design-to-manufacturing CAD/CAM system can also include numerical control tapes for manufacturing.

output contacts
Switch contacts operated by the robot. Output contacts provide a robot with some control over the application, such as turning on or off motors, heaters, grippers, welding equipment, and so on. Controlling output contacts becomes part of the robot routine. The robot is taught to close (or open) a contact at a particular point in the routine. Whenever that routine is replayed, that contact is closed (opened) at the same point at which it was taught. In addition to turning on or off individual output contacts, computer-controlled robots allow simultaneous operation of several contacts, pulsing of contacts, and "handshaking." The latter term means that after an

output contact is turned on or off, the robot waits for a specific input signal to acknowledge that action before performing the rest of the routine. This is especially useful when controlling a tool.

output coupler
The window through which a portion of the laser light passes after striking the partially transmissive mirror at one end of the laser activity.

output device
A computer peripheral such as a card punch that converts electrical signals into the form used by the output device, such as holes punched into cards, and so forth. Devices such as solenoids, motor starters, and others that receive data from the computer or robot controller.

output media
Reports, documents, and punched cards or tape generated as output from a computer.

output power
The measurement of the energy per unit time produced by the laser, measured in watts.

output primitive
A basic graphical element (e.g., a line or a text string) used to construct an object. The values of attributes determine certain aspects of the appearance of an output primitive. A picture element that the IPAD graphics system can create; used to create objects or images that result in real beam or pen movements.

overaging
Aging under conditions of time and temperature greater than those required to obtain maximum change in a certain property, so that the property is altered in the direction of the initial value.

overconstrain
To clamp by more than the minimum number of mechanical constraints necessary to prevent motion.

overflow
In an arithmetic operation, the generation of a quantity beyond the capacity of the register or location that is to receive the result.

overhead
Costs or expenses that are not directly identifiable with or chargeable to the manufacture of a particular part or product. General indirect costs that are aggregated and distributed to all units produced by some allocation method, such as direct or absorption costing; also referred to as burden. Nonproductive effort taking place when the operating system and the programs are performing administrative tasks but no production-work is getting done. In worst cases, overhead may eat up more machine time than data processing does.

overhead percentage
The percentage applied to a labor cost to calculate the overhead cost of performing work in that work center. It is used to distribute those costs which cannot be directly related to specific products or services.

overlapped schedule
The *overlapping* of successive operations, whereby the completed portion of a job lot at one work center is processed at one or more succeeding work centers before the pieces left behind are finished at the preceding work center.

overlapping
A procedure where parts of a lot that are through operation one will be started on operation two before the entire lot is through operation one.

overlay
A technique for bringing routines into high-speed memory from some other form of storage during processing, so that several routines will occupy the same storage locations at different times. It is used when the total memory requirements for instructions exceed the available high-speed memory. Generally, the sets of information are not related, except that they are needed in the same program at different times. The overlay concept thus permits the breaking of a large program into segments that can be used as required to implement problem solution. A multiregion arrangement that gives virtually unlimited memory space for a program. The number of passes through a computer required to process a complete task or program. A segment of code or data to be brought into the memory of a computer to replace existing code or data.

overlay region
Overlay segments that share the same physical memory location and address space form an overlay region.

overload
A load greater than that which a device is designated to handle.

overscan

In video systems, the practice of making the image slightly larger than the picture tube, thus preventing the showing of blank spaces at the edges of the screen. Computer displays are usually adjusted for underscan so even the edges of the active display area can be seen.

overshoot

The degree to which a system response to a step change in reference input goes beyond the desired value. The amount of overtravel beyond the command position. The amount of overshoot is related to factors such as system gain, servoresponse, mechanical clearances, and inertia factors relating to mass, feedrate, and strain.

overtravel

The distance that the plunger of a switch is driven past the operating point when the switch is actuated.

P

packed decimal
Representation of a decimal value by two adjacent digits in a byte. For example, in packed decimal, the decimal value 23 is represented by 0010 0011.

packed density
The number of storage cells per unit length, unit area, or unit volume; for example, the number of bits per inch stored on a magnetic tape track or magnetic drum track.

packet
A group of bits including data and control elements that is switched and transmitted as a unit. The data is arranged in a specified format.

packet-switched network
A data communications network that has a fixed internal topology and uses software to dynamically route messages from source to destination. Packet-switched networks are usually composed of a large number of geographically dispersed nodes connected by dedicated, high-speed data links. Users of the network access other resources attached to the network (computer, terminals, etc.) by connecting to one of these nodes. A characteristic of packet-switched networks is that much greater line utilization is possible (reducing costs of connection) by sharing high-speed links between multiple users.

packing density
The number of units of useful information contained within a given linear dimension, usually expressed in units per inch, that is, the number of characters stored on tape or drum per linear inch on a single track.

pad
An area of plated copper on a printed circuit board that provides (1) a contact for soldering component leads, (2) a means of copper-path transition from one side of the printed circuit board to the other, and (3) a contact for test probes.

page
In virtual storage systems, a fixed-length block of instruction, data, or both that can be transferred between real storage and external page storage; typically about 4K bytes. A program will be divided into pages in order to minimize the total amount of main memory storage allocated to the program at any one time. The pages will normally be stored on a fast direct-access store and can be moved into main memory by an operating system or hardware whenever the instructions of that subdivision need to be performed.

page fault
In virtual storage systems, a program interruption that occurs when a page that is marked "not in real storage" is referred to by an active page.

page fixing
In virtual storage systems, marking a page as nonpageable so that it remains in real storage.

paging
In virtual storage systems, the process of transferring pages between real storage and external page storage. If a page is not transferred from auxiliary storage until it is actually needed, then paging is said to be done by demand. Look-ahead schemes have been implemented with some success.

paging rate
In virtual storage systems, the average number of page-ins and page-outs per unit of time.

paint
To fill in a bounded graphic figure on a raster display, using a combination of repetitive patterns or line fonts to add meaning or clarity. To draw directly on a video screen, as opposed to writing programs that create images. Paint programs are usually easy to use but require some skill for precision work. They also are limited generally to the resolution of the screen, so they cannot be used for highly detailed work.

palette
The overall selection of colors or shades available in a graphics-display system. Due to the difficulty of making many colors available at a time, most graphic systems limit the numbers of colors shown on screen to a small fraction of the available palette. Device that serves as a standardized conveyance for the part in a flexible manufacturing system; may or may be an integral part of a cart.

pallet serial number
A unique number assigned to a specific graphics pallet.

pallet type
A distinct pallet design.

pan
Orientation of a view, as with video camera, in azimuth. Motion in the azimuth direction. In computer graphics, to move side to side or up and down, but in cinematography, to rotate the camera. Moving sideways is to dolly or to cross. The process of translating all elements of a picture with a workstation transformation to give the appearance of movement to a point or object of interest.

paper tape
A continuous strip of paper in which holes can be punched to represent data.

paper-tape punch/reader
A peripheral device that can read as well as punch a perforated paper tape generated by a CAD/CAM system. There tapes are the principal means of supplying data to an NC machine.

parabolic interpolation
A high order of interpolation; producing contoured shapes by having the cutting tool travel through parabolas or portions of parabolas. The simultaneous and coordinated control of two axes of motion such that the resulting cutter path describes a segment of a parabola. A convenience feature, similar to circular interpolation.

paragraph
A set of one or more COBOL sentences, making up a logical processing entity, and preceded by a paragraph name or a paragraph header.

parallax
The change in perspective of an object when viewed from two slightly different positions. The object appears to shift position relative to its background, and it also appears to rotate slightly.

parallel
Several bits traveling over separate pathways (grouped together) simultaneously.

parallel adder
A digital adder in which addition is performed concurrently on digits in all the digit places of the operand.

parallel beam
A beam of light that is optically controlled so the light travels in a parallel path. Generally used when the object is larger than the lens diameter.

parallel circuit
A circuit in which two or more of the connected components or symbolic elements are connected to the same pair of terminals so that current flow divides between the parts. Contrast with a series connection, where the parts are connected end-to-end so that the same current flows through all.

parallel code
A code configuration that is optically scanned in its entirety at one time. Since all code marks are read simultaneously, the code can pass by the reader in either direction.

parallel communication
A digital communication method that transmits the bits of a message several at a time (usually 8 to 17 bits at a time); usually only used over distances of a few feet, with electrical cables as the transmission medium.

parallel computer
A computer having multiple arithmetic or logic units that are used to accomplish parallel operations or parallel processing.

parallel conversion
A method of system implementation that overlaps with the operation of the system being replaced. It minimizes the risk consequences of a poor system.

parallel dimensioning
Showing the size of objects or regions with labeled lines with arrowheads extending between short segments set even with the edges of the object being dimensioned.

parallel interface
A type of digital interface using multiple data lines, each line transmitting one bit of data at a time.

parallel operation
Operates on all bits of a word simultaneously.

parallel output
Simultaneous availability of two or more bits, channels, or digits.

parallel processing
Processing more than one program at a time on a parallel basis, where more than one processor is active at one time (distinguished from multiprocessing, where only one processor is active on one program at a time). Simultaneous processing, as opposed to the sequential processing in a conventional (von Neumann) type of computer architecture. The processing of pixel data in such a way that a group of

pixels are analyzed at one time, rather than one pixel at a time.

parallel projection
The image of a three-dimensional object in which lines that are parallel in the object appear parallel in the image, without regard to relative distance or depth.

parameter
Used in its customary mathematical meaning, an unknown quantity that may vary over a certain set of values. In statistics, it most usually occurs in expressions defining frequency distributions (population parameters) or in models describing a stochastic situation (e.g., regression parameters). The domain of permissible variation of the parameter defines the class of population or model under consideration.

parameter data section
A section of a file consisting of specific geometric or annotative information about the entities or pointers to related entities.

parametric calibrations
The methodology of measuring all of the motions of each carriage of a machine and using the range of the measured deviations from ideal to tolerance the machine.

parametric element processor
Computervision's high-level applications language used to build variable 3-D geometric constructions and solve problems.

parametric spline curve entity
A geometric entity consisting of polynomial segments subject to certain continuity conditions.

parametric spline surface entity
A geometric entity that is a smooth surface made from a grid of patches. The patches are regions between the component parametric curves.

parent curve
The full curve on which a segment curve lies.

parent structure
A structure that invokes another structure, the child structure.

parity
A method of testing the accuracy of binary numbers used in recorded, transmitted, or received data.

parity bit
An additional bit added to a binary word to make the sum of the number of 1s in a word always even or odd. Used to check that data has been transmitted accurately; a receiving device counts the "on" bits of every arriving byte; if odd parity is specified, an error condition will be flagged any time an even number of "on" bits are detected.

parity check
Considering the numerical control standard RS-244-B, a hole punched in one of the track columns (channels) whenever the total number of holes representing a character is even. The net result is that every character is represented by an odd number of holes. This can be used as a check when reading the tape, since an even number of holes would indicate a source of error in the punching. The control system recognizes only an odd number of holes in the character and will automatically stop if an even number of holes in a character is read. The opposite applies with the RS-358-B standard code in that an even parity is sought, and the control system will automatically stop when an odd number of holes is read across the tape.

parse tree
The treelike data structure of a sentence, resulting from syntactic analysis, that shows the grammatical relationships of the words in the sentence.

parsing
Processing an input sentence to produce a more useful representation.

part
Normally refers to a material item that is used as a component and is not an assembly. Workpiece processed by a flexible manufacturing system. The graphic and nongraphic representation of a physical part designed on a CAD system. A product ready for sale; an assembly, subassembly, or a component.

part classification
A coding scheme, typically involving four or more digits, that specifies a discrete product as belonging to a part family.

part family
Generic family of parts based either on form (e.g., parts of rotation, prismatic parts) or manufacturing technology (e.g., stamped parts, milled parts). A set of discrete products that can be produced by the same sequence of machining operations. This term is primarily associated with group technology.

part-handling system
Network including carts, pathways, conveyors, chains, and storage areas for parts, fixtures, and pallets.

partial order
Any shipment received or shipped that is less than the amount ordered.

particle size
With powdered metal, the controlling lineal dimension of an individual particle, as determined by analysis with screens or other suitable instruments.

parting
In the recovery of precious metals, the separation of silver from gold. The zone of separation between cope and drag portions of mold or flask in sand casting. A composition sometimes used in sand molding to facilitate the removal of the pattern. Cutting simultaneously along two parallel lines or along two lines that balance each other in the matter of side thrust. A shearing operation used to produce two or more parts from a stamping.

parting agent
A lubricant, often wax, used to coat a mold cavity to prevent the molded piece from sticking to it and thus to facilitate its removal from the mold. Also called release agent.

parting line
A plane on a pattern or a line on a casting corresponding to the separation between the cope and drag portions of a mold.

partition
A portion of a computer's main memory set aside to hold a single program on a fixed-partition memory management system.

part number
An alphanumeric code that specifically and exactly identifies a part as unique.

part orientation
The angular displacement of a product being manufactured relative to a coordinate system referenced to a production machine, for example, a drilling or milling axis. Reorientation is often required as the product proceeds from one processing step to another.

part period balancing
A dynamic lot-sizing technique that uses the same logic as the least total cost method. The difference is that PPB employs a routine called "look ahead/look back." When the look ahead/look back feature is used, a lot quantity is calculated and before it is firmed up, the next or the previous periods' demands are evaluated to verify whether it would be economical to include them in the current lot.

part program
Set of instructions in the module control language and format necessary to perform the operations required to produce a given part. A complete set of NC instructions to produce a part on an NC machine.

part-program block
Logical subdivision of a part program, for example, the set of instructions for an elementary operation.

part programmer
A person who prepares the planned sequence of events for the operation of a numerically controlled machine tool. The programmer's principal tool is the manuscript on which the instructions are recorded. There are two types of part programming: (1) manual and (2) computer assisted. With manual part programming, the part programmer writes the instructions (blocks) that would go directly to the control system. With computer-assisted part programming, the part programmer writes instructions that are fed to the computer and then converted to block instructions required for the control system. A person who prepares computer programs that are to be used internally in the computer in order to solve the specific problems presented by the part programmer. A numerical control computer programmer generally requires a formal advanced mathematical background, in addition to a thorough knowledge of computer operations.

part serial number
Unique number assigned to a specific part.

part set
Set of all part numbers for a particular end item.

part set tool set
Set of all tools required to produce a given part set on a given flexible manufacturing system.

part setup
Entity consisting of a part mounted in a specific orientation on a fixture.

part setup precedence diagram
Diagram that shows, for a given part number, which part setups must be processed prior to other setups, and which setups can be performed in any order.

part setup route
Same as operation sequence.

parts explosion
A list or drawing of all parts used in all the subassemblies of an assembly or product.

part side
A portion of a part that is processed during one fixturing (setup).

part-transfer time
Period of time in which a part moves between a wait station (e.g., a queue position) and a workstation or between two workstations.

part type
A part within a bill of material may be defined, among others, as regular, phantom, or reference.

pass
One cycle of processing a body of data.

passivation
The changing of the chemically active surface of a metal to a much less reactive state.

passive accommodation
Compliant behavior of a robot's end point in response to forces exerted on it. No sensors, controls, or actuators are involved. The remote center compliance (RCC) provides this in a coordinate system acting at the tip of a gripped part.

passive device
Any graphics device that does not support graphical input, such as an output-only device.

passive mode
In computer graphics, a mode of operation of a display device that does not allow an online user to alter or interact with a display image.

passivity
A condition in which a piece of metal, because of an impervious covering of oxide or other compound, has a potential much more positive than when the metal is in the active state.

password
A unique word or string of characters that a program, computer operator, or user must supply to meet security requirements, before gaining access to the system.

password protection
A security feature of certain computer and CAD/CAM systems that prevents access to the system or to files within the system without first entering a password, that is, a special sequence of characters.

past due
Refers to late, delayed work or orders; work not done (or missed) in the period it was assigned to be done.

patch
A surface represented by parametric functions of two parameters that blend four given boundary curves.

path
In printed circuit board design, the copper interconnections between pins or the route that an interconnection takes between nodes. Also a copper-foil line. A particular track through a state graph.

path control
A programmed motion of a robot that is either point-to-point or continuous path motion.

pattern
One possible representation of the interior of polygon or fill-area primitives. The interior of the image is filled in with a two-dimensional cell pattern selected from a pattern table. In Core, the pattern is set by a pixel-array function.

pattern directed invocation
The activation of procedures by matching their antecedent parts to patterns present in the global data base (the system status).

pattern generation
Transforming CAD integrated circuit design information into a simpler format (rectangles only or trapezoids only) suitable for use by a photo- or electron-beam machine in producing a reticle. Using a pattern generator to physically produce an IC reticle.

pattern matching
An AI technique that recognizes similarities between patterns and objects or events. Matching patterns in a statement or image against patterns in a global data base, templates, or models.

pattern recognition
Description, classification of pictures or other data structures into a set of categories. It usually involves an algorithm for analyzing camera scenes. Identification of objects is accomplished by determining specific characteristics of the given object and comparing them to the corresponding characteristics of a prototype. A technique that classifies images or objects within an image into predetermined categories, usually using statistical methods. The abil-

ity of a robot or computerized system to determine certain characteristics or structures within a picture. The ability of a machine to recognize patterns of shapes.

pattern reference
The starting position for a polygon or fill-area pattern. In Core, the equivalent point is called the pixel-pattern origin.

pattern style
A method of filling closed figures with patterns. A pattern style consists of an array of variously colored or shaded points. The points may be the size of a pixel or larger.

pattern table
A workstation-dependent table of possible pattern values.

pause instruction
An instruction that specifies the suspension of the execution of a computer program; a pause instruction is usually not an exit.

pause-interrupt
A stopping of all involvement of the robot during its sequence in which all power is maintained on the robot arm.

payload
The mass that can be moved by a robot, given its specified performance in terms of accuracy, speed, repeatability, and the like. Larger payloads may sometimes be accommodated at reduced performance by the robot (slower speed, for example). Maximum weight carried at normal speed—also called workload.

peak
The maximum or minimum instantaneous value of a changing quantity.

peak to peak
The amplitude difference of a signal between the most positive and the most negative excursions (peaks) of the signal.

pearlite
A lamellar aggregate of ferrite and cementite, often occurring in steel and cast iron.

peel strength
A measure of the strength of an adhesive bond. The average load per unit width of bond line required to part bonding materials where the angle of separation is 180° and the separation rate is 6 inches per minute.

peer network device
Any device directly attached to the MAP local area network that contains a system manager and has transport layer functionality.

pegging
Refers to the process of showing the source of the original demand for a part or an order. Pegging differs from "where used" in that "where used" shows all possible sources of part demand, whereas pegging shows the specific source of the part order demand.

pegging requirements
Keeping track of the relationship between the requirements for specific end item orders and the requirements that those orders generate for lower-level components. Depending on the industry, pegging may have to be made in different ways: (1) for an individual end item in production on customer order or (2) for a batch of end items in repetitive production. Identifying directly the end item for which each component is due (full pegging) or identifying directly the next assembly level only, with the possibility to trace the requirement origin level by level through all levels (single-level pegging). Pegging connects all materials into networks that allow a kind of critical path scheduling—earliest dates, latest dates, and so on. It allows a simulation of component allocation in order to plan the production of assemblies and end items.

pendant
Any portable control device, including teach pendants, that permits an operator to control the robot from within the work envelope of the robot.

pendant control
A control panel mounted on a pendant cable that enables the human operator to stand in the most favorable position to observe, control, and record the desired movements in the robot's memory.

penetrant
A liquid with unique properties that render it highly capable of entering small openings. This characteristic makes the liquid especially suitable for detecting discontinuities in material.

penetrant inspection
A method of nondestructive testing for determining the existence and extent of discontinuities that are open to the surface in the part being inspected. The indications are made visible through the use of a dye or fluorescent chemical in the liquid employed as the inspection medium.

penetration
In founding, a defect on a casting surface caused by metal running into voids between sand grains. In welding, the distance from the original surface of the base metal to that point at which fusion ceased.

pen plotter
An electromechanical CAD output device that generates hard copy of displayed graphic data by means of a ballpoint pen or liquid ink. Used when a very accurate final drawing is required. Provides exceptional uniformity and density of lines, precise positional accuracy, as well as various user-selected colors.

percent of fill
A measure of the effectiveness with which the inventory management system responds to actual demand. The percent of customer orders filled off the shelf can be measured in either units or dollars.

perception
A robot's ability to sense its environment by sight, touch, or some other means and to understand it in terms of a task—for example, the ability to recognize an obstruction or find a designated object in an arbitrary location. An active process in which hypotheses are formed about the nature of the environment or sensory information is sought to confirm or refute hypotheses.

percussion welding
Resistance welding simultaneously over the entire area of abutting surfaces with arc heat, the pressure being applied by a hammerlike blow during or immediately following the electrical discharge.

perforating
The operation of piercing, in rows, large numbers of small (usually round) holes in a part that may have been previously formed.

performance
The degree to which a machine or mechanism is able to achieve a desired result. The quality of behavior. The degree to which a specified result is achieved. A quantitative index of such behavior or achievement such as speed, power, or accuracy. One of the major factors on which the total productivity of a computer system depends, largely determined by a combination of three other factors: (1) throughput, (2) response time, and (3) availability.

period costs
All costs related to a period of time rather than a unit of product, for example, marketing costs or property taxes.

peripheral equipment
Units that may communicate with the programmable controller but are not part of the programmable controller. Usually called simply *peripherals*. These are external (to the CPU) devices performing a wide variety of input, output, and other tasks. Online peripherals are connected electronically to the CPU. Others are offline (not connected). Examples are printers, monitors, and storage devices.

peripheral interchange program
A short utility that allows a file(s) to transfer to another device.

peritectoid
An isothermal reversible reaction in which a solid phase reacts with a second solid phase to produce yet a third solid phase on cooling.

permanent mold
A metal mold (other than an ingot mold) of two or more parts that is used repeatedly for the production of many castings of the same form. Liquid metal is poured in by gravity.

permanent set
Plastic deformation that remains upon releasing the stress that produces the deformation.

permanent storage
A method or device for storing the results of a completed program outside the CPU, usually in the form of magnetic tape or punched cards.

permanent tool
A tool with a relatively long life, such as a jig or fixture. It is usually expensive, is designed to make one specific item or several items of a similar nature. Replacement inventories of permanent tools are not usually maintained, and there is a long lead time for replacement. Requirements for such a tool can be planned by regarding it as a facility during capacity requirements planning.

permeability
Ratio of maximum value of induction to maximum value of magnetizing force for material in a symmetrically, cyclically magnetized condition. In founding, the characteristics of molding materials that permit gases to pass through them. Permeability number is determined by a standard test. With powdered

metal, the property measured as the rate of passage under specified conditions of a liquid or gas through a compact. In magnetism, a general term used to express various relationships between magnetic induction and magnetizing force. These relationships are either *absolute permeability,* which is the quotient of a change in magnetic induction divided by the corresponding change in magnetizing force or *specific* (relative) *permeability,* the ratio of the absolute permeability to the permeability of free space. Passage or diffusion of gas, vapor, liquid, or solid through a barrier without affecting it. Rate of such passage.

permutation
An ordered arrangement of a given number of different elements selected from a set.

perpetual inventory
A record that tracks the on-hand balance of a part and all incoming and outgoing transactions for the part that affect its in-stock balance.

persistance
The period of time a phosphor continues to glow after excitation is removed.

personal computer
A small, relatively inexpensive computer well-suited for use in the home, office, school, or some smaller-scale industrial applications.

perspective projection
The image of a three-dimensional object in which parallel lines appear to converge in the image as a function of relative distance or depth of the object from the position of the viewer.

phase angle
Angle by which voltage leads current in material subjected to AC current. Angle between the parallel reactance of the material and its total parallel impedance in a vector diagram, when the material is used as dielectric in capacitor. Complement of loss angle.

phase-change recording
A method whereby the optical sensitive layer of the disk has two states, amorphous and crystalline. The laser changes a spot to the amorphous state to write a bit. The drawback to this optical recording method is temperature instability; typically, at 100° C the media switches to the crystalline state, thus destroying written data.

phase modulation
Angle modulation in which the phase angle of a sinusoidal carrier is caused to vary from a reference carrier phase angle by an amount proportional to the instantaneous amplitude of the modulating signal.

phase reviews
In information system planning and design, a method for timely evaluation of a project's progress.

phase transition
Abrupt change in physical properties as temperature is changed continuously. Freezing of water at 0° C is an example.

phonemes
The fundamental speech sounds of a language.

phosphor
A substance capable of luminescence.

photocathode
An electrode used for obtaining photoelectric emission.

photoconductivity
A property of some materials that causes the electrical conductivity of the material to change as the result of absorbing photons.

photoelectric proximity sensors
A version of the photoelectric tube and light source. These sensors appear to be well-adapted for controlling the motion of a manipulator. They consist of a solid-state light-emitting diode (LED) that acts as a transmitter of infrared light and a solid-state photodiode that acts as a receiver. Both are mounted in a small package.

photoisolator
A solid-state device that allows complete electrical isolation between the field wiring and the controller.

photometric stereo
An approach in which the light source illuminating the scene is moved to different known locations and the orientation of the surfaces deduced from the resulting intensity variations.

photon
An elemental unit or particle of light that has energy but no mass or charge. A photon has properties of both a particle and a wave.

photopattern
Production of an IC mask by exposing a pattern of overlapping or adjacent rectangular areas.

Personal Computer
3M Machine Vision System Using IBM Personal Computer

photopic vision
Vision that occurs at moderate and high levels of luminance and permits distinction of colors. This is light-adapted vision. It is attributed to the retinal cones in the eye. The contrasting capability is twilight or scotopic vision.

photoplotter
A CAD output device that generates high-precision artwork masters photographically for printed circuit board design and IC masks.

photoresistor
An electronic component which changes an electrical characteristic called resistance by an amount proportional to the amount of light striking the component.

phrase-structure grammar
Also referred to as context-free grammar. Type two of a series of grammars defined by Chomsky. A relatively natural grammar, it has been one of the most useful in artificial intelligence natural-language processing.

physical
Layer one of the OSI model—the physical layer defines the electrical and mechanical aspect of interfacing to a physical medium for transmitting data, as well as setting up, maintaining, and disconnecting physical links. Physical hardware in this layer include interface devices, modems, and communication lines. Software to control the communication devices (device drivers) is also included.

physical data independence
Means that the physical layout and organization of the data may be changed without changing either the overall logical structure of the data or the application programs.

physical decentralization
The geographical dispersing of facilities and activities. The management control can either be centralized or decentralized.

physical distribution
The combination of activities associated with the movement of material, usually finished products, from the manufacturer to the customer. In many cases, this movement is made through one or more levels of field warehouses.

physical inventory
The determination of inventory quantity by actual account. Physical inventories can be taken on a continuous, periodic, or annual basis.

physical picture element
The smallest displayable unit on a given display device.

physical record
A basic unit of data that is read or written by a single input/output command to the computer. It consists of data that are recorded between gaps on tape or address markers on disk. One physical record often contains multiple logical records or segments.

physical storage organization
Concerned with the physical representation and layout and organization of the data on the storage units. It is concerned with the indices, pointers, chains, and other means of physically locating records and with the overflow areas and techniques used for inserting new records and deleting records.

pick
A class of logical input devices that provides identifying information for selected output primitives.

pick-and-place robot
A simple robot, often with only two or three degrees of freedom. A pick-and-place robot transfers items from place to place by means of point-to-point moves. Little or no trajectory control is available. Often referred to as a bang-bang robot.

pick aperture
An implementation-dependent and workstation-dependent rectangle used for testing objects during a pick input operation. To be picked, at least some portion of the object must be within the pick aperture. The coordinate system used to define the pick aperture is implementation-dependent and workstation-dependent.

pick class
An element of a pick set. A primitive will be picked if one of the pick classes that are members of the pick set associated with the primitive is also contained in the pick filter for a pick device.

pick date
The start date of the picking activity for a work order. On this date, the system produces a list of orders due to be picked and a pick, list, tags, and turnaround cards.

pick device
A logical input device that returns the pick identifier attached to a selected output primitive.

pick filter
A collection of pick classes (inclusion pick set and exclusion pick set) associated with a pick interaction. Is used to identify primitives that are eligible or ineligible for picking.

pick identifier
An attribute specifying the name of one or more primitives returned by a pick logical input device.

picking
Taking materials out of their bins to satisfy requisitions.

picking list
A list of items to be picked in stores. For the sake of efficiency, it groups together a number of requisitions or orders on the basis of factors supplied by management, such as order priority, maximum number of picks per list, multiple orders for the same item, geographical sequence, and so forth.

pick path
The traversal path to a picked primitive from its root structure, represented as a series of structure elements.

pick path depth
The number of levels of the pick path returned on a pick, as specified when the pick device is initialized.

pick set
An attribute of an output primitive that specifies as a set of pick classes, the primitive's eligibility for picking.

pickup tube
A television camera image pickup tube.

picosecond
One-trillionth (10^{-12}) of a second. One-thousandth of a nanosecond.

pictorial information
The display information resulting from the application of geometric primitives and mosaics.

picture
The collection of images that is visible at any moment on a display surface.

picture description instruction
A command composed of an opcode followed by zero or more operands that constitutes an

executable picture drawing or control command.

picture descriptor prefix elements
Elements used to set the interpretation modes of attribute elements for the entire picture.

picture element
An element of the matrix of gray-scale values $f(i,j)$.

piece part
A part that cannot be easily subdivided; an elementary part.

piecework programming
The method of using an outside service organization to prepare a program for which payment is arranged by accomplishment, rather than on a time basis.

pie chart
A chart that shows the relative values of various quantities as arc-shaped sections of a circle. It is one of the more popular displays in many business graphics packages.

piercing
Essentially the same as blanking, except that the operation may be done on a previously formed piece part and that portion that is separated is usually scrap.

piezoelectric
The property of certain crystalline salts to produce an electrical charge on the surface of the crystal, as a function of mechanical pressure.

pilot lot
An experimental or preliminary order for a product.

pilot plant
A plant devoted to the production of pilot lots or to the continuous production of small quantities of a product for the purpose of experimenting with its design and production methods.

pilot-type device
A device used in a circuit for control apparatus that carries electrical signals for directing the performance but does not carry the main power current.

pin
The conductive post, contact, or fitting for each wire within a connector. A connection point on the edge of a printed circuit board. A mechanical device for blocking motion.

pinboard
A perforated board in which pins are manually inserted to control the operation of equipment.

pinch point
Any point where it is possible for a part of the body to be injured between the moving or stationary parts of a robot and the moving or stationary parts of associated equipment or between the material and moving parts of the robot or associated equipment.

pincushion distortion
An effect that makes the sides of an image appear to bulge inward on all sides like a pincushion. Caused by an increase in effective magnification, as points in the image move away from the image center.

piping and instrumentation diagram
Schematic 2-D drawing that shows the major equipment, pipelines, and components to be used in process plant design.

pitch
The angular rotation of a moving body about an axis perpendicular to its direction of motion and in the same plane as its top side. Rotation of the end-effector in a vertical plane around the end of the robot manipulator arm.

pixel
A small element of a scene, or *picture element*, in which an average brightness value is determined and used to represent that portion of the scene. Pixels are arranged in a rectangular array to form a complete image of the scene. The individual elements in a digitized image array. Each pixel is like a tile in a mosaic. A 512 × 512 resolution screen contains 262,144 pixels.

pixel aspect ratio
Generally used in conveyor vision where the pixel length is matched to the belt travel (width). Typically 1:1.

placement
The assignment of printed circuit components to permanent fixed positions on the PC board layout. This can be done automatically on a CAD system.

placement angle
The orientation of an object moved on screen from a library compared to its orientation in that library.

plain
A polygon fill style in which all of the pixels enclosed by the polygon are assigned the same color or intensity value.

plan
A view of an object or image as directly down from top or up from the bottom. A sequence

of actions to transform an initial situation into a situation satisfying the goal conditions.

plane entity
A geometric entity that is a surface, with the property that the straight line passing through any two distinct points on the surface lies entirely on the surface.

planetary drive
A gear reduction arrangement consisting of a sum spur gear, two or more planetary spur gears, and an internally toothed ring gear.

planing
Essentially the same as shaping, except the operation is primarily intended to produce flat surfaces.

planned order
A suggested order quantity and due date created by MRP processing, when it encounters net requirements. Planned orders are created by the computer; exist only within the computer; and may be changed or deleted by the computer during subsequent MRP processing if conditions change. Planned orders at one level will be exploded into gross requirements for components at the next lower level. Planned orders also serve as input to capacity requirements planning, along with released orders, to show the total capacity requirements in future time periods.

planning
The process whereby an individual or an organization identifies opportunities, needs, strategies, objectives, and policies that are used to guide and manage the organization through future periods. All planning consists of (1) accumulation of information, (2) sorting and relating bits of information and beliefs, (3) establishing premises, (4) forecasting future conditions, (5) establishing needs, (6) identifying opportunities, (7) establishing objectives and policies, (8) structuring alternative courses of action, (9) ranking or selecting total systems of action that will achieve the best balance of ultimate (future) and immediate objectives, (10) establishing criteria and means for measuring adherence to the selected program of action, and (11) managing the organization to achieve the objectives. Some kinds of planning include short- and long-range planning, product planning, financial planning, and so on.

planning and decision support
An area of AI research that applies AI techniques to planning and decision-making processes, primarily to assist managers who have decision-making responsibilities.

planning bill (of material)
An artificial grouping of items, in bill-of-material format, used to facilitate master scheduling and/or material planning.

planning horizon
In a MRP system, the planning horizon is the span of time from the current to some future date for which material plans are generated. This must cover at least the cumulative purchasing and manufacturing lead time and usually is quite a bit longer.

plant layout
The physical arrangement, either existing or in plans, of industrial facilities.

plasma
A metal vapor that forms above the point at which a laser beam interacts with the surface of a metal. This plasma consists of free electrons that can interfere with the processing of the metal by the laser.

plasma panel
A type of CRT utilizing an array of neon bulbs, each individually addressable. The image is created by turning on points in a matrix (energized grid of wires) comprising the display surface. The image is steady, long-lasting, bright, and flicker-free; selective erasing is possible.

plaster molding
Molding wherein a gypsum-bonded aggregate flour in the form of a water slurry is poured over a pattern and permitted to harden; after removal of the pattern it is thoroughly dried. The technique is used to make smooth nonferrous castings of accurate size.

plastic deformation
A change in dimensions of an object under load that is not recovered when the load is removed; opposed to elastic deformation.

plasticity
The quality of being able to be shaped by plastic flow.

plasticize
To soften a material and make it plastic or moldable, either by means of a plasticizer or the application of heat.

plated wire memory
A memory consisting of wires that are coated with a magnetic material. The magnetic material may be magnetized in either of two directions to represent ones and zeros.

platen
The sliding member or slide of a hydraulic press. A part of a resistance welding, mechanical testing, or other machine with a flat surface, to which dies, fixtures, backups, or electrode holders are attached, and which transmits pressure.

plating
Forming an adherent layer of metal upon an object.

playback accuracy
Difference between a position command recorded in an automatic control system and that actually produced at a later time, when the recorded position is used to execute control. Difference between actual position response of an automatic control system during a programming or teaching run and that corresponding response in a subsequent run.

playback robot
A manipulator that can produce, from memory, operations originally executed under human control. A human operator initially operates the robot in order to input instructions. All the information relevant to the operations (sequence, conditions, and positions) is put in memory. When needed, this information is recalled (or played back, hence, its name) and the operations are repetitively executed automatically from memory.

plot
To show data in graphic form using a set of scaled axes. The most common form is the ordinary line graph on orthogonal (right-angle) axes, but data also can be plotted on pie charts, polar charts, and others.

plotters
Devices that convert computer output into drawings on paper or on display-type terminals instead of a printed listing. They can produce line graphs, bar charts, maps, engineering drawings, and so forth.

plug
A rod or mandrel over which a pierced tube is forced. A rod or mandrel that fills a tube as it is drawn through a die. A punch or mandrel over which a cup is drawn. A protruding portion of a die impression for forming a corresponding recess in the forging. A false bottom in a die. Also called a peg.

Plotters
Plotter Used for CAD Output Presentation. *Courtesy:* Calcomp Computer Systems.

plugboard
A removable panel containing an array of terminals that can be interconnected by short electrical leads in prescribed patterns to control various machine operations.

plug forming
A thermoforming process in which a plug or male mold is used to partially preform the part before forming is completed using vacuum or pressure.

plug tap
A tap with chamfer extending from three to five threads.

pocketing
Mass removal of material within a predetermined boundary by means of NC machining. The NC tool path is automatically generated on the system. Machining begins at an inner point of the pocket and continues to the outer boundary in ever-widening machining passes.

point
The pick-out objects already on the screen for an operation such as movement, revision, and deletion.

point constraint
A setting or mode that forces a newly entered point to the position of an existing point. It is

used to extend lines from the end of existing lines or to add features to existing objects.

point-dimension entity
An annotation entity consisting of a leader, text, and an optional circle or hexagon enclosing the text.

point entity
A geometric entity that has no size but possesses a location in space.

pointer
A number that indicates the location of an entity within an IGES file.

point operation
Transformation of the gray-scale value of the picture elements according to a given function. Examples are thresholding, contrast enhancement, and so on.

point system
A method of job evaluation in which a range of point values is assigned to each of several job factors. The wage rates are then determined by comparing the total points each receives with the point values and wage rates and key jobs.

point-to-point
The same as positioning. Machining done at a number of locations with no concern of how the cutting tool gets to the location, since it is not in contact with the workpiece when travel takes place. A servo or non-servodriven robot with a control system for programming a series of points without regard for coordination of axes. A limited communications network configuration with communication between two terminal points only, as opposed to multipoint and multidrop.

point-to-point control
A control scheme whereby the inputs or commands specify only a limited number of points along a desired path of motion. The control system determines the intervening path segments. A system in which controlled motion is required only to reach a given end point, with no path control during the transition from one end point to the next. Since contouring control systems can also perform point-to-point operations, and since the cost of contouring systems has been greatly reduced due to CNC, practically all systems now being offered are of the contouring type.

point-to-point motion
A type of robot motion in which a limited number of points along a path of motion is specified by the controller, and the robot moves from point to point rather than in a continuous smooth path.

point transformation
Replacement of one pixel value by another where the new value is a function of only the previous value.

Poisson distribution
A statistical distribution similar to normal distribution, except that the standard deviation is always assumed to be equivalent to the square root of the mean.

Poisson's ratio
The absolute value of the ratio of the transverse strain to the corresponding axial strain in a body subjected to uniaxial stress; usually applied to elastic conditions.

polar coordinate system
A coordinate system, two of whose dimensions are angles, the third being a linear distance from the point of origin. These three coordinates specify a point on a sphere.

polarization
A means of restricting the vibrations of the electromagnetic field to a single plane. In a laser, this prevents optical loss and produces a higher-quality beam. The restrictions of the vibrations of electric or magnetic field vector to one plane. In electrolysis, the formation of a film on an electrode such that the potential necessary to get a desired reaction is increased beyond the reversible electrode potential.

poles
The number of completely independent circuits that are built into a switch; that is, a single-pole switch controls one circuit, and a double-pole switch controls two circuits, and so on.

polishing
The operation of smoothing the surface of a piece part by action of abrasive particles, adhered to the surface of moving belts or wheels, coming in contact with the piece part being polished.

polling
A technique by which each of the terminals sharing a communications line is periodically interrogated to determine whether it requires servicing. The multiplexer or control station sends a poll that, in effect, asks the terminal selected, "Do you have anything to transmit?"

polygon
One way of representing regions of a drawing or map in computer graphics. A polygon is an enclosed shape defined by arcs (boundaries), nodes (intersections with other objects), and a centroid (the central reference point for the polygon). An output primitive consisting of an area enclosed by a sequence of straight lines. In PHIGS, an output primitive consisting of a collection of polygonal areas that are considered as one area.

polygonal area
A closed planar figure defined by the vertices of its piecewise linear boundary (or edge). The boundary may cross itself.

polygonal edge
The boundary of a polygon that may be displayed independently or as part of the interior.

polyhedron
A three-dimensional object made up of surfaces connected along edges.

polyline
An output primitive consisting of a set of connected lines.

polyline bundle table
A table associating specific values for all workstation-dependent aspects of a polyline primitive with a polyline bundle index. The table contains entries for line type, line width scale factor, and color index.

polymarker
An output primitive consisting of a set of locations, each indicated by a marker.

polymarker bundle table
A table associating specific values for all workstation-dependent aspects of a polymarker primitive. The table contains entries for marker type, marker-size scale factor, and color index.

polyphase sort
An unbalanced merge sort in which the distribution of sorted subsets is based on a Fibonacci series.

population
The entire set of items from which a sample is drawn.

population inversion
A state in which more atoms or molecules of a lasing medium are at a high energy level than at some lower energy level, so that photons can be released for the lasing process to occur.

pop-up window
Temporary windows shown on top (in the foreground) of existing material on screen to show the user menu choices, help information, messages, or the like.

porosity
Relative extent of volume of open pores in a material. Ratio of pore volume to overall volume of the material in percent.

port
A functional unit of a mode through which data can enter or leave a data network. The place where another device is connected to the computer. Ports can be serial or parallel. See serial and parallel. A connecting unit between a data link (I/O channel, data bus, interface module) and a device (computer, data terminal, CRT).

portability
The ease with which a computer program developed in one programming environment can be transferred to another.

position
The definition of an object's location in space, its orientation, and its velocity. To specify the location of an element in a drawing, either by keying-in coordinates or by using a mouse, light pen, or other pointing device. In a string, each location that may be occupied by a character or binary element and that may be identified by a serial number.

position control
Control by a system in which the input command is the desired position of a body.

position error
In a servomechanism that operates a manipulator joint, the difference between the actual position of that joint and the commanded position.

positioning
The movement or manipulation of an object consistently into a controlled position in space.

positioning accuracy
A term used to denote the difference between the true displacement of a machine and that recorded by the machine measurement system (scale). The term is ambiguous in that the line of measurement must be specified because of Abbe offsets on any real machine. Accuracy is a measure of the robot's ability to move to a programmed position. Repeatability is its

ability to do this time after time. With the pick-and-place robot, accuracy and repeatability are interchangeable. With a programmable robot, repeatability can be improved by fine-tuning the controls.

position transducer
A device usually geared to a precision ball-nut lead screw for measuring position and converting this measurement to an electrical signal for transmission back to the control cabinet, where a comparison is made with the input instruction. This cylindrical device measures approximately $1\frac{1}{2}$ inches long by 1 inch in diameter.

positive image
Screen image made up of dark lines on a lighter background.

post
The act of identifying a structure for display on a particular workstation.

postamble
A sequence of binary characters recorded at the end of each block on phase-encoded magnetic tape for the purpose of synchronization when reading backward.

postdeduct inventory transaction processing
A method of doing inventory bookkeeping in which the book (computer) inventory of components is reduced only after completion of activity on their upper-level parent or assembly. This approach has the disadvantage of a built-in differential between the book record and what is physically in stock.

post field
Usually a 16-byte field attached to a data record on a WORM drive. Because data cannot be erased or overwritten on a WORM drive, this field contains pointers to the updated information.

post priority
The priority used in displaying structures.

postprocess gaging with feedback
A specific example of error compensation by precalibration, where the workpiece is used as a master.

postprocessor
A software program or procedure that formats graphic or other data processed on the system for some other purpose. For example, a postprocessor might format cutter centerline into a form that a machine controller can interpret. A set of computer instructions that transforms tool centerline data (CL) into machine commands using the proper tape codes and format required by the specific machine/control system. The computer instructions that convert cutter location (CL) data, based on a part's geometry, to the proper specifications necessary to run a particular piece of NC machinery. Each type of NC machine has its own unique postprocessor. A program that translates product definition data from the form represented by an IGES file into the data base form of a specific CAD/CAM system.

potentiometer
An encoder that provides a resistance proportional to position. This is achieved by running a brush over a resistant material and mechanically increasing or decreasing the amount of resistant material between the brush and the electrical resistive element.

power cylinder
A linear mechanical actuator consisting of a piston in a cylindrical volume and driven by high-pressure hydraulic or pneumatic fluid.

power density
Laser output power per unit area, such as watts per square centimeter.

power factor
Ratio of power expended in circuit to the product of electromotive force (EMF) acting in circuit and current in it; that is, ratio of watts to volt-amperes. Because energy loss is directly proportional to frequency, low power factor is essential in materials used at high frequencies.

power supply
The function of the power supply is to provide energy to the manipulator's actuators. In the case of electrically driven robots, the power supply functions basically to regulate the incoming electrical energy. Power for pneumatically actuated robots is usually supplied by a remote compressor that may also serve other equipment.

power tool
Any powerful programming device that dramatically increases programmer productivity.

pragmatics
The relationship of characters or groups of characters to their interpretation and use. The study of the use of language context.

preamble
A sequence of binary characters recorded at the beginning of each block on a phase-coded

magnetic tape for the purpose of synchronization.

precalibrated error compensation
A means of error compensation in which the error is measured once and compensated for during subsequent operations.

precious metal
One of the relatively scarce and valuable metals: gold, silver, and the platinum-group metals.

precipitation hardening
Hardening caused by the precipitation of a constituent from a supersaturated solid solution.

precipitation heat treatment
Artificial aging in which a constituent precipitates from a supersaturated solid solution.

precision
Degree of mutual agreement among individual measurements. Relative to a method of test, precision is the degree of mutual agreement among individual measurements made under prescribed like conditions. The degree of accuracy. Generally refers to the number of significant digits of information to the right of the decimal point for data represented within a computer system. Thus, the term denotes the degree of discrimination with which a design or design element can be described in the data base. See also accuracy. The standard deviation, or root-mean-squared deviation of values around their mean.

precompiler program
A program that detects and provides source-program correction before the preparation of the object program.

predefined associatives
Associatives that are defined within graphics standard.

predetermined motion time
There are several systems of motion times that list all motions that a human can perform in accomplishing factory or office tasks and a standard performance time for each of the motions. The most widely used system is methods time measurement.

predicate
That part of a proposition that makes an assertion (e.g., states a relation or attribute).

predicate logic
A modification of propositional logic to allow the use of variables and functions of variables.

prefix notation
A list representation (used in LISP programming) in which the connective, function, or predicate is given before the arguments. A method of forming mathematical expressions in which each operator precedes its operands and indicates the operation to be performed on the operands or the immediate results that follow it.

premise
A first proposition on which subsequent reasoning rests.

preparatory function
A command on the tape changing the mode of operation of the control that is generally noted at the beginning of a block and consists of the letter character ''G'' plus a two-digit number.

preplaced line
A run (or line) between a set of points on a PC board layout that has been predefined by the designer and must be avoided by a CAD automatic routing program.

preposition
The basic element employed when the transporting device or the object transported is prepared for the next basic element, which is usually position.

preprinted symbol
A symbol that is printed in advance of application, either on a label or on the article to be identified.

preprocessing
Operations used in machine vision to enhance an image or to obtain information about an image.

preprocessor
In emulation, a program that converts data from the format of an emulated system to the format accepted by a target system. In general programming, a preprocessor is a program that examines the source program for preprocessor statements that are then executed, resulting in the alteration of the source program. A program that translates product definition data from the data base form of a specific CAD/CAM system into the form represented by an IGES file.

presence-sensing device
A device designed, constructed, and installed to create a sensing field or area around a robot that will detect an intrusion into such field or area by a person, robot, and so forth.

presence-sensing safeguarding device
A device designed, constructed, and installed to create a sensing field or area to detect an intrusion into such field or area by personnel, robots, or other objects.

presentation graphics
Charts, slide shows (both on-screen and transferred to photographic film), and similar graphics intended to present information such as financial data, organizational charts, and strategies.

presentation layer
The sixth of seven layers defined by ISO's reference model for open systems interconnection.

present value
The value today of future cash flows. For example, the promise of $10.00 a year in hand today.

preset
The limit established for a counter or timer line; the preset is entered into the C element of each line. The current count or time available from the register referred to in the D element cannot exceed this limit. At the present value, the logic line's coil is energized.

pressure forming
A thermoforming process wherein pressure is used to push the sheet to be formed against the mold surface, as opposed to using a vacuum to suck the sheet flat against the mold.

presumptive instruction
An instruction that is not an effective instruction until it has been modified in a prescribed manner.

pretravel
The distance through which the plunger of a switch moves from its free position to the operating point.

preventive maintenance
Precautionary measures taken on a system to forestall failures, rather than to eliminate them after they have occurred, by providing for systematic inspection, detection, and correction of incipient problems before they develop into major defects.

primal sketch
A primitive description of the intensity changes in an image. It can be represented by a set of short line segments separating regions of different brightnesses.

primary colors
Three colors wherein no mixture of any two colors can produce the third. The most common groups of primary colors are red, green, and blue or cyan, magenta, and yellow.

primary station
In a network, the station that has the priority to select and transmit information to a secondary station and has the responsibility to ensure information transfer. The assignment of primary status to a station is temporary and governed by standardized procedures.

primary storage
The "memory" of a computer, where instructions and data being worked upon are contained. Most primary storage today is made up of small iron rings or cores that can be electrically charged; therefore, primary storage is often called core storage.

prime costs
Direct costs of material and labor; does not include general sales and administrative costs.

prime operations
Critical or most significant operations whose production rates must be planned. Sometimes referred to as *pinch points* in the shop, gateway operations, key work centers, bottlenecks.

primer
A coating applied to a surface prior to the application of a plastic coating, adhesive, lacquer, enamel, or other material, to improve the performance of the bond.

primitive
A design element at the lowest stage of complexity. A fundamental graphic entity. It can be a vector, a point, or a text string. The smallest definable object in a display processor's instruction set. Applied to computer displays and graphics, an operation that the display chip or subsystem can be called upon to do. It is analogous to an operating system call for a program. Primitives range from simple display of a single character through rotation and displacement of screen objects. Also called entities. The basic design elements, such as a line or arc, that are joined to make symbols or groups.

primitive attribute
A general characteristic or aspect of an output primitive.

primitive solid
The basic elements that can be individually sized and located. The solids are combined to make another solid.

principal axis
The axis passing through the centroid of the degenerated envelope and perpendicular to the plane of the circular or regular polygonal cross section of the degenerated envelope.

principal plane
An imaginary plane in or near a lens where the light rays appear to have bent.

principles of motion economy
The rules and their corollaries applying to human motions, that guide toward development of the optimum way of accomplishing a given job.

printed circuit board
A baseboard made of insulating materials and an etched copper-foil circuit pattern on which are mounted ICs and other components required to implement one or more electronic functions. PC boards plug into a rack or subassembly of electronic equipment to provide the brains or logic to control the operation of a computer or a communications system, instrumentation, or other electronic systems. The name derives from the fact that the circuitry is connected not by wires but by copper-foil lines, paths, or traces actually etched onto the board surface. CAD/CAM is used extensively in PC board design, testing, and manufacture.

printer
Any output device that produces *printouts*. Very slow compared to the CPU's electronic speed.

printout
A printed sheet giving all the data of a program that has either been manually or computer processed. The printout is used for reference and visual checking.

priority
In a general sense, refers to the relative importance of jobs, that is, which jobs should be worked on and when. It is a separate concept from capacity.

priority indicator
In data communications, a group of characters that indicate the relative urgency of a message and thus its order of transmission.

priority modes
The organization of the flow of work through a computer, varying from a normal noninterrupt mode to a system in which there are several depths of interrupt.

priority rules
Rules that allow someone to know what job to do next, that is, which job is most important in relation to other existing ones.

priority scheduling rule
A technique for assigning an objective priority sequence number that can then be utilized for the scheduling or production jobs.

prismatic part
A rectangular or box-shaped part, usually processed on a machining center.

private line
A communication channel for private use; a leased, owned, or otherwise dedicated channel.

problem description
A statement that describes a problem domain.

problem domain
The set of concerns and symptoms to be jointly addressed by the system under development.

problem-oriented language
A programming language that reflects the type of problem being solved, rather than the computer on which the program is to be run. A source language, suited to describing procedural steps, that is designed for convenience of program specification in a general problem area rather than for easy conversion to machine instruction code. The components of such a language may bear little resemblance to machine instructions. POLs are generally machine independent (FORTRAN, BASIC, COBOL, PL/1, and others).

problem program
Any program that is executed when the central processing unit is in the problem state; that is, any program that does not contain privileged instructions. This includes language translators and service programs, as well as programs written by a user.

problem reduction
A problem-solving approach in which operators are used to change a single problem into several subproblems that are usually easier to solve.

problem-solving
A procedure using a control strategy to apply operators to a situation to try to achieve a goal.

problem state
The condition of the problem at a particular instant.

procedural knowledge representation
A representation of knowledge about the world by a set of procedures—small programs that know how do specific things (how to proceed in well-specified situations).

procedure
The course of action taken for the solution of a problem.

procedure block
A collection of statements, headed by a procedure statement and ended by an end statement, that is a part of a program, especially PL/1.

procedure division
One of the four main component parts of a COBOL program. The procedure division contains instructions for solving a problem. The procedure division may contain imperative statements, conditional statements, paragraphs, procedures, and sections.

procedure manual
A formal organization and indexing of a firm's policies and practices. They are usually printed and distributed to the appropriate functional areas.

procedure-oriented language
A problem-oriented language that facilitates the expression of a procedure as an explicit algorithm, for example, FORTRAN, PL/1, or ALGOL.

process
Series of continuous actions, with an hierarchical system of levels divided into activities that are accomplished by executing one or more jobs. Continuous and regular production executed in definite uninterrupted manner. A computer application that primarily requires data comparison and manipulation. The CPU monitors the input parameters in order to vary the output values. (As generally contrasted with a machine, a process of a CPU does not cause mechanical motion.)

process chart
A graphic presentation of events occurring during a series of actions or operations and of information pertaining to those events.

process chart symbols
Graphic symbols or signs used in process charts to depict the types of events that occur during a process.

process control
Pertaining to systems whose purpose is to provide automation of continuous operations, and characterized by in-line adjustments to regulate an operation. This is contrasted with numerical control, which provides automation of discrete operations.

process cost system
A costing system in which the costs are collected by time period and averages over all the units produced during the period. This system can be used with either actual or standard costs in the manufacture of large number of identical units (cf. job order costing).

process design
In computer-aided plant design, the activity of stringing together a number of processing units, such as reactors and distillation columns, into a process flow sheet with the objective of schematically representing a means of converting raw material into desired product(s).

process detailing
Planning for the details of machining. Process detailing includes cutting parameters selection, cutter path generation, and so forth.

processing
The act of prescribing the production process to produce a product as designed. This may include specifying the equipment, tools, fixtures, machines, and the like required; the methods to be used; the workers necessary; and the estimated or allowed time. The carrying out of the production process.

processing element
A computing system consisting of hardware and software elements attached to the local network to perform certain specified data-processing operations.

processing program
A general term for any program that is not part of the operating system. This includes language processors, application programs, service programs, and user-written programs.

processing speed
The time required for a vision system to analyze and interpret an image. Typical vision systems can inspect from 2 to 15 parts per second.

process instruction
The part of a process plan that specifies the operations on a part and the sequences of these operations.

process layout
A method of plant layout in which the machines, equipment, and areas for performing the same or similar operations are grouped together.

process model
A decision model for the selection of processes.

processor
A machine language program that accepts a source program written in a symbolic form and translates it into an object program acceptable to the machine for which the source program was written. A processor frequently called a compiler or translator is a program usually supplied by the equipment manufacturer for creating machine-language programs. A processor previously stored in the computer receives the source program written in symbolic language by the programmer and translates the instructions into machine-language instructions that are acceptable to the computer. This machine-language program is called an object program. A unit in the computer or robot controller that scans all the inputs and outputs in a predetermined order. The processor monitors the status of the inputs and outputs in response to the user-programmed instructions in memory, and it energizes or de-energizes outputs as result of the logical comparisons made through these instructions. A device capable of receiving data, manipulating it, and supplying results, usually of an internally stored program. A program that assembles, compiles, interprets or translates. Also used synonymously with CPU.

processor program
A software program used to convert computer instructions written in symbolic language into absolute language. Assemblers and compilers are both processors; assemblers usually convert one symbolic instruction into one machine instruction, while a compiler can convert one symbolic instruction into a number of machine instructions.

process piping
A category of design and drafting falling within the giant field of industrial plant design.

process plan
A detailed plan for the production of a piece part or assembly. It includes a sequence of steps to be executed according to the instructions in each step and consistent with the controls indicated in the instructions. A detailed plan for the production of a piece part or assembly. It includes a sequence of steps to be executed according to the instructions in each step and consistent with the controls indicated in the instructions.

process planning
Specifying the sequence of production steps from start to finish and describing the state of the workpiece at each workstation. Recently, CAM capabilities have been applied to the task of preparing process plans for the fabrication or assembly of parts.

process sequencing
Determining a suitable ordering of processes for machining a part.

process sheet
A sketch, diagram, or listing of the operations in the sequential order necessary to accomplish the desired result, such as transforming material from one state to another. The detailed part routing that shows each step in the manufacture of a product.

process simulation
A program utilizing a mathematical model created on the system to try out numerous process design iterations with real-time visual and numerical feedback. Designers can see on the CRT what is taking place at every stage in the manufacturing process. They can therefore optimize a process and correct problems that could affect the actual manufacturing process downstream.

process time
The time required to complete a specified series of progressive actions or operations on one unit of production. That portion of a work cycle during which the material or part is being machined or treated according to a specification or recipe designed to produce the desired action or result.

procurement
The act of obtaining items from outside vendors.

product
End item resulting from manufacturing.

product cycle
The total of all steps leading from the design concept of a part to its final manufacture. How many of these steps can be aided or automated by a CAD/CAM system depends on the particular features or capabilities provided by the system.

product definition
Data required to describe and communicate the characteristics of physical objects as manufactured products.

product group
A group of products having common classification criteria. Design engineering may classify items by function, size, shape, or material in order to retrieve all items having common characteristics, when required for a specific design purpose. This avoids duplication of design, routings, items, stock accounting, and so on. The sales department may classify items by product groups according to potential users, function, size, and the like.

production
The manufacturing of goods. The act of changing the shape, composition, or combination of materials, parts, or subassemblies to increase their value. The quantity of goods produced. The fabrication and assembly of durable goods.

production capacity
The highest sustainable output rate that can be achieved with the current product specifications, product mix, worker effort, plant, and equipment.

production capacity planning
The function of setting the limits or levels of manufacturing operations in the future, consideration being given to sales forecasts and the requirements and the availability of workers, machines, materials, and money.

production center
A group of productive facilities that, for administrative and accounting purposes, are considered a unit. The area containing a machine or machines operated by a worker or workers, as well as the space required for the storage of materials at the machine and for loading and unloading it; auxiliary tools, benches, jigs, and the like; and the free and safe movement of the worker while working.

production characteristic
An attribute of the production process affecting the design of a component.

production control
The function of directing or regulating the movement of goods through the entire manufacturing cycle, from the requisitioning of raw materials to the delivery of the finished product. The function of directing or regulating the orderly movement of goods through the entire production cycle, from the requisition of raw materials to the delivery of the finished product.

production cycle
The elapsed time required to produce a product.

production department
The part of a manufacturing organization responsible for the actual processing of materials or parts. That subdivision of management responsible for planning how, where, and what cost to manufacture or assemble a product.

production engineer
An individual qualified by education, training, and experience to perform production engineering functions and who specialized in this work.

production engineering
The function of planning where and when to perform work necessary to produce a product and of coordinating internal and external orders, delivery dates, workers, machines, and the like, thereby promoting efficient operation. A term used as a synonym for industrial engineering, methods engineering, and manufacturing engineering. Designing products to be manufactured utilizing materials, equipment, methods, processes, and skills that are available.

production limitation
A feature of the production process constraining the design of a component.

production load
The demand for output established by scheduling, based on consumer orders or sales forecasts.

production materials requisition
An authorization either in the form of a slip of paper, punched card, or CRT transaction that identifies the type and quantity of materials to be purchased or withdrawn from stores. Synonymous with material order, materials requisition.

production monitoring
Checking the status and progress of production activities.

production order
A document conveying authority for the production of specified parts or products. The production order sometimes shows the date when the job must be completed and sometimes shows due dates for each individual operation that has been assigned by the schedule.

production order control
Control of the progress of each customer order or stock order through the successive operations in its production cycle.

production order packet
Manufacturing order that travels with the job and includes a group of documents like the routines, blueprints, materials requisition, move tickets, time tickets, and the like. Usually, many of the documents are in the form of punched cards.

production part set
A subset of the system part set planned for production within the present planning horizon.

production plan
Information required to produce a desired system output; information includes part numbers, quantities, tools, modules, schedules, and so on.

production planning
The systematic scheduling of workers, materials, and machines by using lead times, time standards, delivery dates, workloads, and similar data for the purpose of producing products efficiently and economically and meeting desired delivery dates. Routing and scheduling. Planning of the time and resource for the production of a manufacturing system.

production rate mix
Production part set and associated quantities, expressed as rates.

production rates
The quantity of production usually expressed in units, hours, or some other broad measure, expressed by a period of time, that is, per hour, per shift, day, week, and so forth.

production report
A formal written statement giving information on the output of an organization or one or more of its subdivisions for a specified period.

production rule
A rule in the form of an if-then or condition-action statement used in the knowledge base of an expert system. A production rule typically represents a single heuristic.

production schedule
A plan that authorizes the factory to manufacture a certain quantity of a specific item. Usually initiated by the production planning department.

production study
A detailed record, often in the form of a time study or work sampling study, of an activity, operation, or group of activities or operations over a period of time; used to obtain reliable data concerning working time, idle time, downtime, personal time, machine breakdowns, amount produced, and so on.

production system
A knowledge base that relies on knowledge represented in the form of production rules.

production unit
The workers, equipment, and areas involved in performing a given task. A measure of a product expressed in terms of weight, volume, quantity, dollar value, and the like.

productive time
Elapsed time during which useful work is performed in a manufacturing process. That portion of an operation cycle during which the worker's time is utilized effectively.

productivity
The measure of the amount of output, in either goods or services, per unit of input. The higher the productivity, the higher the output versus input.

productivity ratio
A widely accepted means of measuring CAD/CAM productivity (throughput per hour) by comparing the productivity of the design/engineering group before and after installation of

Productivity
Comparison of Manual Labor vs. Robot Productivity

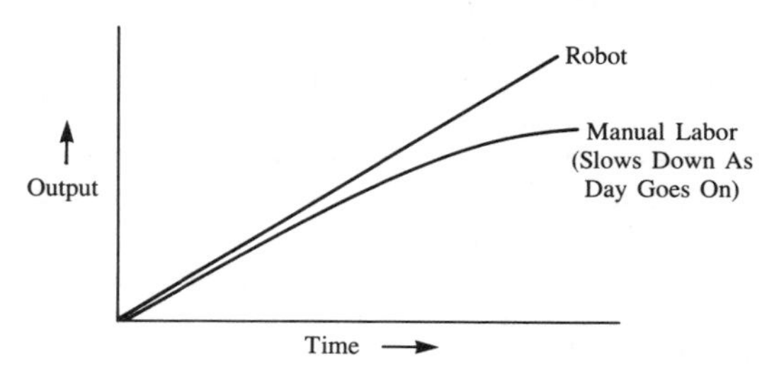

Productivity Representative for Compatible Cycles

the system or relative to some standard norm or potential maximum. The most common way of recording productivity is by actual man-hours/actual CAD hours, expressed as 4:1, 6:1, and so on.

product load profile
A statement of the key resources required to manufacture one unit of a selected item. Often used to predict the impact of the item scheduled in the master production schedule on these resources. Synonymous with bill of labor, bill of resources, resource profile.

product mix
The distribution of the various items in a production plan. The combination of individual product types and the volume product that make up the total production volume.

product structure
A representation of the way components go into a product during its manufacture. A typical product structure would show, for example, raw material being converted into fabricated components, components being put together to make subassemblies, subassemblies going into assemblies, and so on.

product structure record
A computer record defining the relationship of one component to its immediate parent and containing fields for quantity required, engineering effectivity, scrap factor, application selection switches, and so forth.

profiling
Removing material around a predetermined boundary by means of numerically controlled (NC) machining. The NC tool path is automatically generated on the system.

program
A plan for the solution of a problem. A complete program includes plans for the transcription of data, coding for the computer, and plans for the absorption of the results into the system. The list of coded instructions is called a routine. To plan a computation or process from the asking of a question to the delivery of the results, including the integration of the operation into an existing system. Thus, programming consists of planning and coding, including numerical analysis, systems analysis, specification of printing formats, and any other functions necessary to the integration of a computer in a system.

program button
Located on a function board. Pressing a button (e.g., LINE) connects the software program enabling that function (e.g., draw a line) to be performed.

program counter
A register that indicates the memory address of the next instruction in the program to be executed by the central processing unit.

program editing
Through the use of a computer, a keyboard, and CRT, it is possible to implement easily a program editing feature. The CRT enables the operator to display the functional data stored at previously programmed points. The operator may delete or modify these data and restore them. He or she may also insert points between previously stored points by using the teach pendant.

program evaluation and review technique
A time-event network analysis system in which the various events in a program or project are identified, with the planned time for each, and are placed in a network showing the relationships of each event to other events.

program generator
Generally, a program that permits a computer to write other programs automatically.

program interrupt
The term applied to the automatic interrupt of regular program operation, whenever an I/O unit becomes available. Processing is said to be asynchronous.

program item number
A number that relates an item of work to the work breakdown structure. It is used as a primary index to work items and for cost collection.

program library
A collection of programs and routines.

program listing
A printed list of computer instructions, usually prepared by an assembler or compiler.

program loop
A series of instructions that may be executed repeatedly in accordance with the logic of the program.

programmable
Capable of being instructed to operate in a specified manner of accepting setpoints or other commands from a remote source. A feature of a robot that allows it to be instructed to perform a sequence of steps and then to perform this sequence in a repetitive manner. It can then be reprogrammed to perform a different sequence of steps, if desired.

programmable controller
A solid-state control system that has a user-programmable memory for storage of instructions to implement specific functions such as I/O control logic, timing, counting, arithmetic, and data manipulation. A personal computer consists of a central processor, input/output interface, memory, and programming device that typically uses relay-equivalent symbols. The personal computer is purposely designed as an industrial control system that can perform functions equivalent to a relay panel or a wired solid-state logic control system.

programmable device
In the MAP document, this term is used in reference to plant-floor devices such as programmable logic controllers, robots, CNC machines, and so on.

programmable logic controller
A stored-program device intended to replace relay logic used in sequencing, timing, and counting of discrete events. Instead of physical wiring relay, push buttons, limit switches, and so on, a programmable logic controller is programmed to test the state of input lines, to set output lines in accordance with input state, or to branch to another set of tests. The instruction sets of these machines generally exclude all arithmetic and Boolean operators but do include vital decision instructions such as skip, transfer unconditional, transfer conditions, and even transfer and link.

programmable manipulator
A device capable of manipulating objects by executing a stored program resident in its memory.

programmable read-only memory
A read-only memory that can be modified by special electronic procedures. A semiconductor device that may be written to once and continually read from; normally not volatile.

programmed learning
Refers to an instructional methodology based upon alternating material with questions in book form (*programmed text*) or on an interactive terminal (computer assisted-instruction).

programmed logic arays
A logic device alternative to ROM that uses a standard logic network programmed to perform a specific function. Programmed logic arrays are implemented in either MOS or bipolar circuits and are particularly suited to decoding and/or logic with advantage over ROMs. As such, they are likely to be used in code converters, computer instruction decoding, and I/O device command decoding.

programmer
A person involved in designing, writing, and testing computer programs. Depending upon the philosophy of the particular institution, programming can include substantial amounts of analysis, and so forth. One who "teaches" a robot a complete set of instructions to accomplish a specific task. In NC, there are two important types of programmers: (1) a workpiece programmer who writes the instructions for the computer to act upon to develop a specific program tape and (2) a computer programmer who develops the routines that give the computer the basic intelligence to act upon instructions when they are prepared by the workpiece programmer.

programmer's apprentice
System that helps programmers program by keeping track of decisions, recalling program skeletons, automatically testing revised programs, supporting natural-language interaction, and translating to and from various program representations.

programming
Preparing a list of instructions for the computer to use in the solution of a problem. In industrial robots, it is necessary to program the robot to perform the required tasks. This can be accomplished in the following ways: (1) *lead through*—the robot is placed in the "teach" mode of operation and points in space are recorded as the robot is led through the desired sequence, (2) *walk through*—the robot is placed in the "teach" mode of operation, manually walked through the desired sequence of movements and operations, and the program is recorded and played back in the "operate" mode during actual operation; (3) *plug in*—a prerecorded program is physically transferred into the robot control, usually through the use of magnetic tape; and (4) *computer programming*—these robots are programmed through the use of computer programs written especially for the computers being used to control the robot actions.

programming environment
The total programming setup that includes the interface, the languages, the editors, and other programming tools. A collection of tools used

to support construction and debugging of a module.

programming flowchart
A chart showing the sequence of operations in a program.

programming in logic
A logic-oriented AI language developed in France and popular in Europe and Japan.

programming languages
A language, other than machine language, used for expressing computer programs. The major kinds of programming languages are as follows: (1) assembly or symbolic machine languages one-to-one equivalence with computer instructions, but with symbols and mnemonics as an aid to programming; (2) macroassembly languages that are the same as assembly or symbolic machine languages but permit macroinstructions to be used for coding convenience; (3) procedure-oriented languages for expressing methods in the same way as expressed by algorithmic languages; and (4) problem-oriented languages for expressing problems. Procedure-oriented languages may be further divided into (1) algebraic languages (numerical computation), (2) string-manipulating languages (text manipulation), (3) simulation languages (such as GPSS or DYNAMO), and (4) multipurpose languages (such as PL/1).

programming modules
A limited set of instructions handled as a unit for development, testing, and implementation.

programming RPQ (PRPQ)
A custom alteration or addition to the operating system programming or program products. The PRPQ may be used to solve unique data-processing problems.

programming support representative
An individual responsible for field maintenance of vendor-supplied software.

program panel
A device for inserting, monitoring, and editing a program in a computer controller.

program product
A licensed, charged-for program provided by a mainframe vendor that performs a function for the user. A program product contains logic usable or adaptable to meet the user's specific requirements.

program scan
The time required for the processor to execute all instructions in the program once. The program scan repeats continuously.

program sharing
In networks, the ability for several users or computers to utilize a program at another node.

program stop
A miscellaneous function to stop the coolant, spindle, and feed after completion of other commands in the block. The operator may restart the cutting cycle without loss of accuracy.

program temporary fix
A temporary solution or bypass of a problem diagnosed by vendor field engineering, as the result of a defect in a current unaltered release of the program.

program unit
A simple and flexible device that permits programs for automatic card-punching operations to be easily prepared and inserted. The program unit controls automatic skipping over columns not to be punched, automatic duplicating of repetitive information, and the shifting from numerical to alphabetic punching positions and vice versa.

progressive fixture
A fixture that simultaneously holds two or more sides of the same part type, for example, side one and side two.

prohibited area
Any location within the total machine area to which access is prohibited during normal working cycles of the system.

project
Sequence of tasks and subtasks to be performed during an associated design and/or analysis effort.

projected available balance
In MRP, the inventory balance projected out into the future. It is the running sum of on-hand inventory minus requirements plus scheduled receipts.

projected finish date
The date at which a stop order will be completed, calculated by using the scheduling rules and today's date. By subtracting the required completion date from the projected finish date, one can define how early or late a job is running and use that to determine relative priority between jobs.

projection
A transformation function used to map three-dimensional coordinates to a constrained three-dimensional coordinate space. In particular, the intersections of all of the projectors with the view plane is called the projection of that object on the view plane. External feature that can be seen in silhouette.

project management
An organizational form that is generally superimposed upon a traditional functional organizational structure. Integrated teams of specialists work under the coordination and direction of a project manager to accomplish a project of limited duration. The project manager coordinates and manages across functional and organizational lines to complete a specific project or program.

project model
A time-phased project planning and control system that itemizes major milestones and points of user approval.

projectors
Lines that pass through enough points of an object and intersect the view plane to define the image of the object on the view plane.

project planning
The preliminary activity before the commencement of the actual study, in which areas to be studied and objectives desired are defined.

projection reference point
A world coordinate point specified relative to the view reference point that determines the direction of all projectors in a parallel projection or from which all projectors in a perspective projection emanate.

projection type
The type of projection to be used in view mapping, such as parallel projection or perspective projection.

PROLOG
Programming language based on formal logic. PROLOG is the language the Japanese have adopted as the main language for their fifth-generation computer project.

PROM
A programmable read-only memory. This device can be used to store binary information in a form accessible to digital circuitry. The information is stored permanently and not affected when power is removed. It is programmable only in the sense that it is blank when purchased and must be programmed by the end-user. Once programmed, it can never be changed.

promo
A grouping of processor languages that includes: (1) IFAPT, which is an APT subset devoted to turning application; (2) SURF-APT, which is specifically designed for the handling of sculptured surfaces; (3) PROTOUR, also devoted to turning applications but requiring only a minicomputer; and (4) PROPOINT for drilling, boring, tapping, and similar positioning operations and requiring only a minicomputer capability.

prompt
Output to the operator indicating that a specific logical input device is available. Also used to request input. Provides two-axis contouring with three-, four-, and five-axis positioning for machining centers, two- and four-axis lathes, flame and plasma arc burners, punch presses, wire EDMs. Interactive processor allows for creation, manipulation, and storage of contours and point and punch patterns, with full offset capability including access to CALPROMPT, a proprietary calculating program. A message or symbol generated automatically by the system and appearing on the CRT to inform the user of (1) a procedural error or incorrect input to the program being executed or (2) the next expected action, option(s), or input. See also tutorial.

prompting
In systems with time-sharing, a function that helps a terminal user by reminding the user to supply operands necessary to continue processing.

pronation
Orientation or motion toward a position with the back, or protected side, facing up or exposed.

proof
An attempt to find errors in a program without regard to the program's environment. Most proof techniques involve stating assertions about the program's behavior and then deriving and proving mathematical theorems about the program's correctness.

proof stress
The stress that will cause a specified small permanent set in a material. A specified stress to

be applied to a member or structure to indicate its ability to withstand service loads.

propagation delay
The time between when a signal is impressed on a circuit and when it is recognized at the other end; of great importance in satellite channels.

properties
Nongraphic data that may be associated, that is, linked in a CAD/CAM data base, with their related entities such as parts or components. Properties in electrical design may include component name and identification, color, wire size, pin numbers, lug type, and signal values. User-established attributes that dictate the significance of an entity or subfigure within the model.

properties data base
This type of source data base contains dictionary or handbook-type chemical or physical data.

property entity
A structure entity that allows numeric or text information to be related to other entities.

property list
A knowledge representation technique by which the state of the world is described by objects in the world by means of lists of their pertinent properties. A construct in LISP that associates with an object called an atom a set of one or more pairs, each composed of a "property" and the "value" of that property for that object.

proportional control
Control scheme whereby the signal that drives the actuator is monotonically related to the difference between the input command (desired output) and the measured actual output.

proportional-integral derivative control
Control scheme whereby the signal that drives the actuator equals a weighted sum of the difference, time integral of the difference, and time derivative of the difference between the input and the measured actual output.

proportional limit
The maximum stress at which strain remains directly proportional to stress.

proportionally spaced font
A text font in which character cells may vary in width as a function of the characters' visible representation.

proposition
A statement (in logic) that can be true or false.

propositional logic
An elementary logic that uses argument forms to deduce the truth or falsehood of a new proposition from known propositions.

proprietary program
A program controlled by an owner through the legal right of possession and title. Commonly, the title remains with the owner and its use is allowed with the stipulation that no disclosure of the program can be made to any other party without prior agreement between the owner and user. This applies to privately sold programs, program products offered for sale by mainframe vendors, and no-charge software provided by mainframers.

prosthetic robot
A controlled mechanical device connected to the human body, providing a substitute for human arms or legs when their function is lost.

protectable memory
A memory that will not lose information upon the loss of power.

protected memory
Storage (memory) locations reserved for special purposes, in which data cannot be entered directly by the user.

protection time
A number of days used as a safety buffer between the date demands are due and supply orders are to be completed.

protocol
A formal definition that describes how data is to be formatted for communication (transmitting or receiving) between two devices. A formal set of conventions governing the format and relative timing of message exchange in a communications network. A set of formats, rules, and procedures governing the exchange of information between peer processes.

protocol migration
The OSI model architecture used for TOP that allows for an orderly migration from vendor proprietary protocols to OSI protocols. The vendor's computer requiring the use of proprietary protocols would have two implementations of the upper-layer protocols (i.e., layers five to seven). Data communications between the vendor's computers could use the proprietary protocol upper layers when needed. When multivendor open communication is needed, the OSI path is taken using

the OSI protocol upper layers. Note that the lower layers (layers one to four) for both the vendor proprietary and OSI protocols are the OSI protocols only. This is required because of the many differences that can be encountered in addressing at the network layer (layer three) and the transport layer (layer four).

prototype
An initial model or system that is used as a base for constructing future models or systems.

proximal
Close to the base, away from the robot end-effector of the arm.

proximity
A method of object detection in which the source and detector are located on the same side; the detector senses energy from the source that is bounced back by the object being detected.

proximity detector
A device that may be used to sense that an object is close to a robot or to measure how far an object is from a robot.

proximity sensor
A device that senses that an object is only a short distance (e.g., a few inches or feet) away and/or measures how far away it is. Proximity sensors work on the principles of triangulation of reflected light, elapsed time for reflected sound, intensity-induced eddy currents, magnetic fields, back pressure from air jets, and others. A noncontact sensor that determines when one object is close to another. Devices that sense and indicate the presence or absence of an object without requiring physical contact. Five of six major types of proximity sensors available commercially are radio frequency, magnetic bridge, ultrasonic, permanent-magnet hybrid, and photoelectric. Noncontact sensors have widespread use, such as for high-speed counting, indication of motion, sensing presence of ferrous materials, level control, reading of coding marks, and noncontact limit switches.

proximity switch
A generic term. That means that the proximity switch is a device capable of acting as an electronic switch when in the presence or close proximity of an object. The important distinction that differentiates it from a mechanical switch is that it does not require physical contact with anything else to operate. The methods of achieving this operation have been far-ranging, making use of the effect of objects on RF fields, magnetic fields, capacitive fields, acoustic fields, and light rays. RF fields are usually altered by the presence of ferrous materials that absorb energy by any currents produced in them by the field. Magnetic fields have been utilized to simply close reed switches by bringing the magnet material between the magnet and the switch. Magnets have also been used to alter electric fields in devices making use of the Hall effect. Capacitive devices make use of the change in capacity that occurs when the object to be sensed acts as a plate of a capacitor for which

Protocol Migration
Migration of Proprietary Protocols. *Source*: Boeing Company.

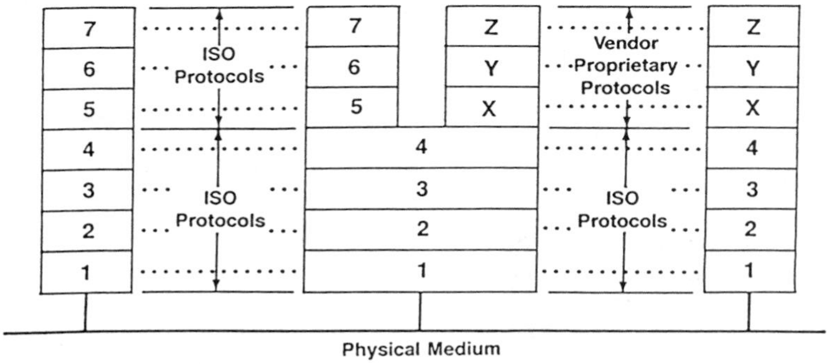

the sensor acts as the other plate when detecting metallic objects or as an alteration in the dielectric between plates when detecting nonmetallic objects. Sonic devices utilize sound fields that are either interrupted by the object to be detected or detect the reflection of sound from objects. Photoelectric devices work in a similar manner, except that light rays are detected rather than sound waves. All of these methods of detection have inherent strengths and weaknesses in actual application.

pseudo (OP) instruction
A symbolic representation of information to a compiler or interpreter. A group of characters having the same general form as a computer instruction but never executed by the computer as an actual instruction.

pseudorandom number sequence
An ordered set of numbers that has been determined by some defined arithmetic process but is effectively a random number sequence for the purpose for which it is required.

pseudoreduction
An approach to solving the difficult problem case where multiple goals must be satisfied simultaneously. Plans are found to achieve each goal independently and then integrated using knowledge of how plan segments can be intertwined without destroying their important effects.

puck
A hand-held, manually controlled input device that allows coordinate data to be digitized into the system from a drawing placed on the data tablet or digitizer surface. A puck has a transparent window containing cross hairs. On a digitizing tablet, a round palm-sized disk with a cross hair in the middle that is used to trace over a drawing to be entered into the computer.

pull distribution system
A system for replenishing field warehouse inventories wherein replenishment decisions are made at the field warehouse itself, not at the central supply warehouse or plant.

pulse
A signal that can be wholly described by a constant amplitude and the duration time. This signal form is typically used internally in computers, terminals, and other business machines but is also used by some communications facilities. A single burst of energy from a laser.

pulsed laser
A laser that emits light in a series of pulses rather than continuously.

pulse energy
The total energy contained in a single laser pulse.

pulse modulation
Transmission of information by varying the basic characteristics of a sequence of pulses; width (duration), amplitude, phase, and number.

pulse width
Also known as pulse length. This is the time or duration of the pulse emitted by a pulsed laser, in seconds.

pump
An energy source (light or electricity) used to excite the lasing medium and create a population inversion.

punched card
A heavy stiff paper of uniform size and shape suitable for being punched with a pattern of holes to represent data for being handled mechanically. Standard cards measure $7\frac{3}{8}$ inches (18.7 cm) by $3\frac{1}{4}$ inches (8.3 cm) and are divided into 80 vertical areas called card columns. These are numbered from 1 to 80, from the left-hand side of the card to the right. Each column is then divided into 12 punching positions called rows, which are designated from the top to the bottom of the card. Each column of the card is used to accommodate a digit, a letter, or a special character. Digits are recorded by holes punched in the appropriate positions of the card from 0 to 9. The top 3 punching positions of the card (12, 11, and 10) are called the zone positions. In order to accommodate any of the 26 letters of the alphabet in one column, a combination of a zone and a digit punch is used. In accordance with the designated rows on a punched card, the top edge is known as the *12-edge* and the bottom edge is known as the *9-edge*.

punched tape
A tape, usually paper, on which a pattern of holes or cuts is used to represent data. Since paper tape is a continuous recording medium, it can be used to record data in records of any

length. Tapes using five, six, seven, and eight channels are available. The most popular types used are the five-channel and eight-channel codes. In the five-channel code, data are recorded (punched) in five parallel channels along the length of the tape. Each row of punches across the width of the tape represents one letter, digit, or symbol. Since the five punching positions allow only 32 possible combinations of punches, a shift system is used to double the number of available codes. In eight-channel code, data are recorded (punched) in eight parallel channels along the length of the tape.

punching
The operation of separating or forming a flat piece of material in some particular shape.

purchased parts
Those parts that are purchased from an outside vendor and not fabricated within the user plant.

purchase order
A document completely describing what a buyer is ordering from a vendor. It includes information on part descriptions, price, quantity, need date, shipping terms, credit terms, and the like.

purchase requisition
An internal order to the purchasing department authorizing the purchase of a certain amount of any item by a certain date.

purchasing lead time
The total lead time required to obtain a purchased item. Included here are procurement lead time, vendor lead time, transportation lead time, receiving, inspection, and put-away time.

pure binary numeration system
The fixed-radix numeration system that uses the binary digits and the radix 2; for example, in this numeration system, the numeral 110.01 represents the number "six and quarter," that is, $1 \times 2^2 + 1 \times 2^1 + 1 \times 2^{-2}$.

push distribution system
A system for replenishing field warehouse inventories, wherein replenishment decision making is centralized, usually at the manufacturing site or central supply facility.

pushdown list
A list control procedure used in memory, where the next item to be retrieved and removed is the most recently stored item still in the list, that is, last-in, first-out. Synonymous with pushdown stack.

pushup list
A list control procedure used in memory where the next item to be retrieved and removed is the oldest item still in the list, that is, first-in, first-out (FIFO).

pyramid
A hierarchical data structure that represents an image at several levels of resolution simultaneously.

pyrometallurgy
Metallurgy involved in winning and refining metals where heat is used, as in roasting and smelting.

pyrometer
A device for measuring temperatures above the range of liquid thermometers.

Q

Q-bus
A set of electrical conductors that carry specific signals to several other electrical circuits.

Q-switch
A device that acts as a shutter to move in and out of the beam path at regular intervals so that a large amount of energy is stored and then released in a burst of energy. Q-switching is often provided as an option on continuous wave lasers.

quadrant
Defines the fourth part of a circle or a fourth part of a graph in reference to the *X* and *Y* axes.

quadtree
A representation obtained by recursively splitting an image into quadrants, until all pixels in a quadrant are uniform with respect to some feature (such as gray level).

quality assurance
The actions necessary to provide suitable confidence that an item will perform satisfactorily in actual operation. The function that has the responsibility to see that adequate product quality is ensured in the design process. Reflects the principle that quality is designed into a part or product.

quality control
The procedure of establishing acceptable limits of variation in size, weight, finish, and so forth for products or services and of maintaining the resulting goods or services within these limits. The function that has the responsibility to see that parts and products are made exactly to the specifications shown on the engineering drawings.

quality engineering
An engineering discipline for the establishment of quality tests and quality acceptance criteria and for the interpretation of quality data. The establishment and execution of tests to measure product quality and adherence to acceptance criteria.

quality factor
Reciprocal of dissipation factor or loss tangent. Also called storage factor. When materials have approximately the same dielectric constant, a higher quality factor indicates less power loss.

quantity per
The quantity of a component to be used in the production of its parent. Quantity per is used when calculating the gross requirements for components.

quantization
The subdivision of the range of values of a variable into a finite number of nonoverlapping and not necessarily equal subranges or intervals, each of which is represented by an assigned value within the subrange.

quantum efficiency
For a radiating source, the ratio of the number of photons emitted per second to the number of electrons flowing per second. For detectors, the ratio of the number of electron/hole pairs generated to the number of incident photons.

quartet
A byte composed of four binary elements.

quartz crystal
A thin slice of quartz that vibrates at a very steady frequency in response to an electrical current.

quasistatic error
An error produced by a process that proceeds so slowly that the system(s) may be said to remain arbitrarily close to equilibrium. The term is loosely used to categorize machine errors that vary in time at rates much less than 1 hertz.

quench aging
Aging induced by rapid cooling after solution heat treatment.

quench annealing
Annealing an austenitic ferrous alloy by solution heat treatment.

quench bath
The cooling medium used to quench molten materials.

quench hardening
Hardening a ferrous alloy by austenitizing and then cooling rapidly enough so that some or all of the austenite transforms to martensite.

quenching
Rapid cooling. When applicable, the following more specific terms should be used: direct quenching, fog quenching, hot quenching, interrupted quenching, selective quenching, spray quenching, and time quenching.

quench time
In resistance welding, the time from the finish of the weld to the beginning of temper.

query
A request for data entered while the computer system is processing. In data communication, the process by which a master station asks a slave station to identify itself and to give its status.

query language
A self-contained language allowing users access to a data base, using relational external schemas to represent the user's view of the data base.

query mode
Mode in which a user requests data base retrieval from a system by means of specific query language.

queue
Waiting lines resulting from temporary delays in providing service. A series of elements, one waiting behind the other; a waiting line.

queue time
The amount of time a job waits at a work center before setup or work is performed on the job. Queue time is one element of total manufacturing lead time. Increases in queue time result in direct increases to manufacturing lead time.

queuing
The mathematical technique that is applied to problems involving waiting lines or queues.

queuing theory
A collection of mathematical techniques and models designed to estimate the length and duration of waiting lines, given a probabilistic description of arrivals and servicing times.

quick kill
A sorting technique that eliminates additional handling of documents by early identification of those items that can be dropped from the sorting sequence.

QUICKPATH
Provides a two- or three-axis contouring programming capability on a minicomputer for inhouse use. QUICKPATH has full contouring geometry capability plus a simplified path-generation feature.

quiescing
The process of bringing a multiprogrammed system to a halt by rejection of new jobs.

quintet
A byte composed of five binary elements.

quotient
The number or quantity that is the value of the dividend divided by the value of the divisor and that is one of the results of a division operation.

QWERTY keyboard
A standard typewriter alphanumeric keyset. As carried over from the printing industry, named for the first six keys of the third row from the bottom.

R

radian
An arc of a circle equal in length to the radius; 2 pi radians equals 360 degrees, and one radian equals 57.3 degrees.

radiant energy
The energy contained in electromagnetic radiation as it travels.

radiant flux
The time rate of flow of radiant energy per unit area.

radio frequency
A frequency or wave within the range of radio transmission.

radio-frequency preheating
A method of preheating used for molding materials to facilitate the molding operation or reduce the molding cycle. The frequencies most commonly used are between 10 and 100 MHz.

radio-frequency welding
A method of welding thermoplastics using a radio-frequency field to apply the necessary heat. Also known as high-frequency welding.

radiograph
A permanent visible image on a recording medium produced by penetrating radiation passing through the material being tested.

radiographic
The reciprocal of the thickness of a given equivalence factor material taken as a standard. It not only depends on the standard but also on the radiation quality.

radiographic contrast
The difference in density between an image, or part of an image, and its immediate surroundings on a radiograph. Radiographic contrast depends upon both subject contrast and film contrast.

radiographic inspection
The use of X-rays or nuclear radiation, or both, to detect discontinuities in a material, and to present their images on a recording medium.

radiographic sensitivity
A measure of quality of radiographs, whereby the minimum discontinuity that may be detected on the film is expressed as a percentage of the base thickness. It depends on subject and film contrast and on geometrical and film graininess factors.

radiography
A nondestructive method of internal examination in which metal or other objects are exposed to a beam of X-ray or gamma radiation. Differences in thickness, density, or absorption caused by internal discontinuities are apparent in the shadow image, either on a fluorescent screen or on photographic film placed behind the object.

radius-dimension entity
An annotation entity that is a measurement of the radius of a circular arc.

radix
The fundamental number in a number system, for example, 10 in the decimal system, 8 in the octal system, and 2 in the binary system. Synonymous with base. In a radix-numeration system, the positive integer by which the weight of the digit place is multiplied to obtain the weight of the next higher digit place—for example, in the decimal-numeration system, the radix of each digit place is 10; in a binary code, the radix of each position is 2.

radix complement
A complement obtained by subtracting each digit of the given number from the number that is one less than the radix of that digit place, then adding one to the least-significant digit of the result and executing any carries required; for example, 830 is the tens complement, that is, the radix complement of 170 in the decimal-numeration system using three digits.

radix-numeration system
A positional representation system in which the ratio of the weight of any one digit place to the weight of the next lower digit place is a positive integer. The permissible values of the character in any digit place range from zero to one less than the radix of that digit place.

radix point
In a representation of a number expressed in a radix-numeration system, the location of the separation character associated with the fractional part.

random
Occurrences that have no regular or forecastable pattern, where the chance of any one occurrence is no greater than that of any other.

random access
Random-access devices do not have to be read from the beginning to find a specific address, as is necessary with paper tape and magnetic tape. See direct-access storage.

random access memory (RAM)
Memory from which data can be retrieved regardless of input sequence. A memory whose information media are organized into discrete location, sectors, and so on. Each uniquely identified by an address. Data may be obtained from such memory by specifying the data address(es) to the memory, for example, core, drum, disk, cards.

random access storage
A storage device such as a magnetic core, magnetic disk, and magnetic drum in which each record has a specific predetermined address that may be reached directly; access time in this type of storage is effectively independent of the location of the data.

random dispatching rule
The sequencing of jobs to be run on some basis that is unrelated to a measure of effectiveness for that operation. For example, an operation may choose to run those jobs that have the loosest time standards.

random dither
A way of adding dots to an image to soften jagged lines created by the way raster-scan displays show diagonal lines. With random dithering, the dots that are added are positioned by statistical rather than deterministic rules.

random file
Data records stored in a file, without regard to the sequences of the key or control field.

random interlace
A technique for scanning that is often used in machine vision systems where there is no fixed relationship between adjacent lines in successive fields.

random number
A number selected from a known set of numbers in such a way that the probability of occurrence of each number in the set is predetermined.

random sample
A selection of observations taken from all of the observations of a phenomenon in such a way that each chosen observation has the same possibility of selection.

random variation
A fluctuation in data that is due to uncertain or random occurrences.

range
A characterization of a variable or function. All the values that a function may possess. Distance from the imaging device to elements or objects in a scene.

range-imaging sensors
Measures the distance from itself to a raster of points in the scene. Different range-imaging sensors have been applied to scene analysis in various research laboratories. These sensors may be classified into two types, one based on the trigonometry of triangulation and the other based on the time of flight of light (or sound).

range sensing
The measurement of the distance to an object.

ranking method
A job evaluation system, wherein each job as a whole is ranked with respect to all the other jobs and no attempt is made to establish a measure of value.

raster
Parallel horizontal lines drawn by an electron gun on the face of a video monitor to display a video image. There are approximately 240 noninterlaced raster lines or 420 interlaced raster lines visible on a CRT.

raster display
A display in which the entire display surface is scanned at a constant refresh rate.

raster graphics
A method of creating images by controlling the writing of a beam or pen as it sweeps out a

Raster
Raster Display System

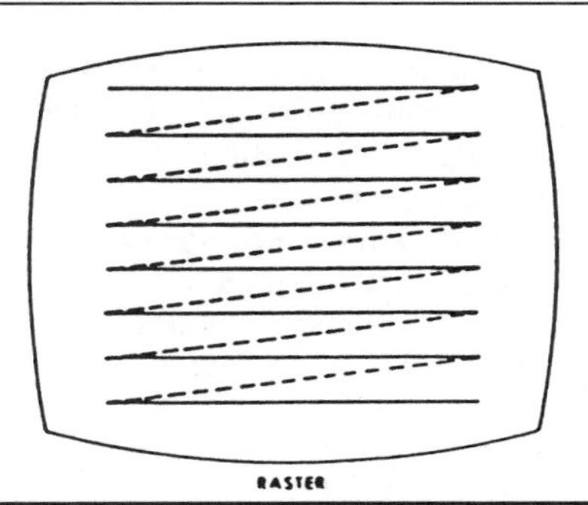

pattern of adjacent parallel lines over the screen or paper. By choosing when to write and when not to write, an operator uses the written parts of the raster to form an approximation of the image being drawn. The finer the raster (the more lines per image and the faster the writing can be turned on or off), the more raster operations done on pixel data to allow several images to be combined on a display. For example, to allow the correct box to end up on top in an overlapped window display. Common raster operations include logical operations (AND, OR, Exclusive OR) plus inversion and blink.

raster scan (video)
Currently, the dominant video scanning technology in CAD graphic displays. Similar to conventional television, it involves a line-by-line sweep across the entire CRT surface to generate the image. Raster-scan features include good brightness, accuracy, selective erase, dynamic motion capabilities, and the opportunity for unlimited color. The device can display a large amount of information without flicker, although resolution is not as good as with storage tube displays. The process used by television cameras to create a video image.

rate control
Control system in which the input is the desired velocity of the controlled object.

rated load
The amount of weight or mass for which a system was designed and with which a system or manipulator can operate with a specified margin of safety.

rated load capacity
A specified weight or mass of a material that can be handled by a machine or process that allows for some margin of safety relative to the point of expected failure.

rate voltage
That maximum voltage at which an electrical component can operate for extended periods without undue degradation.

rational B-spline curve
A parametric curve expressed as the ratio of two linear combinations of B-spline basis functions. Each basis function in the numerator is multiplied by a scalar weight and a vector B-spline coefficient. Each corresponding basis function in the denominator is just multiplied by the weight.

rational B-spline surface
A parametric surface expressed as the ratio of two linear combinations of products of pairs of B-spline basis functions. Each product of basis functions in the numerator is multiplied by a scalar weight and a vector B-spline coefficient. Each corresponding product of basis functions in the denominator is just multiplied by the corresponding weight.

rational number
A real number that is the quotient of an integer divided by an integer other than zero.

rat's nest
A feature on PC design systems that allows users to view all the computer-determined interconnections between components. This makes it easier to determine whether further component placement improvement is necessary to optimize signal routing.

raw material
Material in or nearly in its natural or original state—unprocessed material. The basic stock (steel, aluminum, wood, and others) purchased in order to make parts.

ray tracing
Computing the visible objects and intensities in a scene by following light rays from each screen pixel back through their reflections in the scene to the light sources.

reach
Defines the robot's arm movement or work envelope. The work envelope usually has one of three shapes—cylindrical, spherical, or spheroidal—depending on the basic configuration of the arm and on the major axes of motion. For practical purposes, the description of the work envelope can be simplified by citing only its three major parameters: (1) degrees of rotation about the center axis (horizontal arm sweep); (2) vertical motion at both minimum and maximum arm extension; and (3) radial arm extension, measured from the center axis.

reactive power specific incremental
Specific reactive power in a magnetic material when subjected simultaneously to a unidirectional (*biasing*) and an alternating magnetizing force. This can be determined by the test used for incremental arc permeability.

read
To copy, usually from one form of storage to another, particularly from external or second-

ary storage to internal storage. To sense the meaning of arrangements of hardware. To sense the presence of information in a recording medium.

read around ratio
The number of times a specific spot, digit, or location in electrostatic storage may be consulted before a spillover of electrons causes a loss of data stored in surrounding spots. The surrounding data must be restored before the deterioration results in any loss of data.

reader
A device capable of sensing information stored in an offline memory medium (cards, paper tape, or magnetic tape) and generating equivalent information in an online memory device (register or memory locations).

read in
To sense information contained in some source and to transfer this information by means of an input to internal storage.

read-only memory (ROM)
A memory that cannot be altered during normal computer use. Information is permanently stored and can be read from any location at high speed but can never be altered. Users have to be sure that their program is right before they enable it into a ROM. On most computers, once enabled the program gives no flexibilities. Because programming ROMs is expensive, relatively short programs are usually enabled in a ROM.

read-only storage
Special circuitry in a computer that allows it to process commands written for some other type of computer. Some of the newer computers also use ROS as an integral part of their own command circuitry.

readout
To transfer data from internal storage to an external storage device or to display processed data by means of a printer, automatic typewriter, and the like. A visual display of the cutting tool coordinate axes locations in relation to the established zero.

readout, command
Generally applies to contouring systems and to the display of absolute position as derived from the position commands within the control system. In effect, the display is an accumulative readout of the incremental movements that have been fed into the control system. Assuming that the machine tool is operating in accordance with the control system, it can be reasonably assumed that the readout represents the actual position of the table.

readout, position
Display of absolute position as derived from a position feedback device such as a transducer. The transducer is normally attached to a lead screw. Although theoretically more reliable than the command readout, it is generally more expensive since a separate feedback circuit is required.

readout, slide location
Display of absolute position as derived from a sliding-scale measurement device, such as an inductosyn, which is attached directly to the sliding moving part, that is, the table. Theoretically, it is the most reliable of the three readout devices (command, position, slide location). It is also the most expensive of the three.

read/write head
The mechanism that writes data to or reads data from a magnetic recording medium.

read/write memory
A data storage device that can be both read from and written into (i.e., modified) during the normal execution of a program.

real constant
A string of decimal digits that must have either a decimal point or a decimal exponent and may have both.

realization
With regard to time standards, refers to the ratio of the allowed standard time to the actual time taken to perform a task.

realized value
An actual value used by a workstation.

real number
A number that may be represented by a finite or infinite numeral in a fixed-radix numeration system.

real storage
In virtual storage systems, the storage of a computing system from which the central processing unit can directly obtain instructions and data and to which it can directly return results. Same as processor storage.

real time
An operation that takes place in a short time so that its use does not affect the time required to perform any other operations. Refers to

tasks or functions executed so rapidly by a CAD/CAM system that the feedback at various stages in the process can be used to guide the user completing the task. In solving a problem, a speed sufficient to give an answer within the actual time the problem must be solved. Pertaining to the actual time during which a physical process transpires. Pertaining to the performance of a computation during the actual time that the related physical process transpires in order that results of the computation can be used in guiding the physical process. The processing of transactions as they occur, rather than batching them. Pertaining to an application in which response to input is fast enough to affect subsequent inputs and/or guide the process and in which records are updated immediately. Online processing is used for real-time systems; however, not all online processing is real time. An online system may be shared by many users so that response time is not always immediate.

real-time clock
A program-accessible clock that indicates the passage of actual time. The clock may be updated by hardware or software.

real-time control
A control system in which the calculations or control functions necessary for operation are conducted at the time that control is occurring, instead of preprocessing or predetermining the control responses.

real-time processing
The ability of a vision system to interpret an image in a short enough time to keep pace with most manufacturing operations. A computer that operates at a rate compatible with the operation of the physical equipment or processes. The computations are performed while the operation is taking place and may be used to guide the operation.

reaming
The operation of enlarging and finishing a hole to an accurate dimension.

recast
A layer of metal that melts and then resolidifies along the walls of a hole or cut edge of a laser-processed metal.

receiving
Often a responsibility of the purchasing or materials function due to the direct relationship of their duties. The receiving function is generally responsible for the following activities in a manufacturing company: (1) physical receipt of incoming materials, (2) inspection of shipment for conformance with purchase order (quantity and obvious damage), (3) identification and delivery to destination, and (4) reporting and notification of necessary personnel.

receiving device
Equipment that can receive coded bit combinations by means of, for example, telecommunication or physical interchange of storage media.

recognition
A match between a description derived from an image and a description obtained from a stored model.

record
A collection of files. The information relating to one area of activity in a data-processing activity, that is, all information on one inventory item. Sometimes called item.

recording density
The quantity of bits recorded per unit area on magnetic media; usually measured in bits per inch (bpi). The closeness with which data are stored on magnetic tape. The most common densities are 200, 556, 800, and 1600 characters per inch.

record-playback robot
A manipulator for which the critical points along desired trajectories are stored in sequence by recording the actual values of the joint position encoders of the robot as it is moved under operator control. To perform the task, these points are played back to the robot servosystem.

recovery
Restoration of normal processing after hardware or software malfunction through detailed procedures for file backup, file restoration, and transaction logging.

recrystallization
The change from one crystal structure to another, as occurs on heating or cooling through a critical temperature. The formation of a new, strain-free grain structure from that existing in cold-worked metal, usually accomplished by heating.

recrystallization annealing
Annealing cold-worked metal to produce a new grain structure, without phase change.

recrystallization temperature
The approximate minimum temperature at which complete recrystallization of a cold-worked metal occurs within a specified time.

rectangular array
Insertion of the same entity at multiple locations on a CRT, using the system's ability to copy design elements and place them at user-specified intervals to create a rectangular arrangement or matrix. A feature of PC and IC design systems.

rectangular coordinate robot
A robot whose manipulator arm moves in linear motions along a set of Cartesian, or rectangular axes. The work envelope forms the outline of a three-dimensional rectangular figure.

rectangular coordinate system
Same as Cartesian coordinate system but applied to points in a plane.

rectification
A general-purpose, computer-aided mapping tool for adjusting the geometric information in a data base, based on updated values of a number of points after their position has been more accurately determined, typically, after a map has been digitized.

recursive
Pertaining to a process in which each step makes use of the results of earlier steps.

recursive function
A function whose values are natural numbers that are derived from natural numbers by substitution formulas in which the function is an operand.

recursive operations
Operations defined in terms of themselves.

reduction of area
Commonly, the difference, expressed as a percentage of original area, between the original cross-sectional area of a tensile test specimen and the minimum cross-sectional area measured after complete separation. The difference, expressed as a percentage of original area, between original cross-sectional area and that after straining the specimen.

redundancy
The policy of building duplicate components into a system to minimize the possibility of a failure disabling the entire system. For example, redundancy of hardware performing steps critical to the performance of a CAD/CAM facility guarantees minimum interruption of service. An equally important function of redundancy is to maintain continuous throughput. Duplication of information or devices in order to improve reliability.

redundant modules
Identical modules in the same flexible manufacturing system.

redundant processor
A processor that duplicates or partially duplicates the operation of another processor to substantially reduce the possibility of total system failures.

redundant system
FMS configuration in which any single element (e.g., tool, tool cluster, station, or track intersection) can fail and the system part set can still be produced.

redundant tooling
Two or more identical tools in the same flexible manufacturing system.

reed switch
A sealed device used as a switch, actuated by an external magnetic field.

reentrant
A characteristic of software programs that allows interruption at any time before completion and subsequent restart. Thus, the designer can make modifications in the middle of a CAD/CAM automatic program without having to start it again later from the beginning.

reference data base
This type of data base contains information that directs users to a primary source (printed or nonprinted) for additional details or for the complete text. There are two types of reference data bases, bibliographic and referral. Reference data base contrasts with source data base. See data base, bibliographic data base, and referral data base.

reference designator
Text (i.e., a component name) used in CAD to identify a component and associate it with a particular schematic element.

references
Four-digit numbers used in the construction of the user's logic. Every element of each logic line uses a single reference number. References can be either discrete (logic line's coils, inputs, or latches) or register (input or holding).

referral data base
This type of reference data base contains citations to nonprint material. Nonprint materials include individual, organization, audiovisual data, radio or television programs, unpublished proceedings data, and others. Abstracts are often included with the citations.

reflectance
The ratio of energy that is reflected from a surface to the energy that is incident upon the surface.

reflex
A method of object detection in which the source and detector are located on the same side; a retroflector on the far side returns the energy from the source to the detector.

refraction
The bending of light rays as they pass from one medium into another.

refractory
Material of very high melting point with properties that make it suitable for such uses as furnace linings and kiln construction. The quality of resisting heat.

refractory alloy
A heat-resistant alloy. An alloy having an extremely high melting point. An alloy difficult to work at elevated temperatures.

refractory metal
A metal having an extremely high melting point. In the broad sense, it refers to metals having melting points above the ranges of iron, cobalt, and nickel.

refresh
A CAD display technology that involves frequent redrawing of an image displayed on the CRT to keep it bright, crisp, and clear. Refresh permits a high degree of movement in the displayed image, as well as high resolution. Selection erase or editing is possible at any time without erasing and repainting the entire image. Although substantial amounts of high-speed memory are required, large complex images may flicker.

refreshable
The attribute of a load module that prevents it from being modified by itself or by any other module during execution. A refreshable load module can be replaced by a new copy during execution by a recovery management routine, without changing either the sequence or results of processing.

refresh display
A form of CRT display where only the line elements that make up a drawing are projected and connected on the face of the screen. This type of display must be "refreshed" 30 to 40 times per second to avoid flicker. It can show shades of color but not full color and also has good dynamic capability.

refresh rate
The rate at which the graphic image on a CRT is redrawn in a refresh display, that is, the time needed for one refresh of the displayed image.

regeneration
The redrawing of the display surface to update its contents and remove the nonretained structure, deleted structures, and unposted structures. May be explicit or implicit. On some devices, it may also be automatic. This case is different from implicit due to the requirement to support nonretained data.

regeneration MRP
An approach where the master production schedule is totally reexploded down through all bills of material, at least once per week to maintain valid priorities. New requirements and planned orders are completely "regenerated" at that time.

regenerative track
Part of a track on a magnetic drum or magnetic disk used in conjunction with a read head and a write head that are connected to function as a circulating storage.

region
The bounded area enclosed by a closed curve or a combination of curves. An area of an image over which the light intensity is constant. A region ends when it meets another region with a different intensity level. Regions are formed by such object characteristics as flat surfaces, areas of constant colors, or shadows.

regional network
A computer network whose nodes cover a defined geographical area.

region growing
Process of initially partitioning an image into elementary regions with a common property, such as gray level, and then successively merging adjacent regions having sufficiently small differences in the selected property, until only regions with large differences between them remain.

region of interest
The area inside defined boundaries that the user wants to analyze.

register
A memory device capable of containing one or more computer bits or words. A register has zero memory latency time and negligible memory access time.

registration
The degree of accuracy in the positioning of one layer or overlay in a CAD display or artwork, relative to another layer, as reflected by the clarity and sharpness of the resulting image. Processing images to correct geometrical and intensity distortions, relative translational and rotational shifts, and magnification differences between one image and another or between an image and a reference map. When registered, there is a one-to-one correspondence between a set of points in the image and in the reference.

regression analysis
The use of models to determine the mathematical expression that best describes the functional relationship between two or more variables. Regression models are often used in forecasting.

regular element
An element of an operation or process that either occurs at every cycle of that operation or process or occurs frequently and in a fixed pattern with the cycles of that operation or process, for instance, once every third cycle.

rejected inventory
Inventory that does not meet quality requirements but has not yet been sent to be reworked, scrapped, or returned to a vendor.

rejection
The act of rejecting an item by the buyer's receiving inspection department because the item does not meet the purchase quality specification.

rejection (discrimination) accuracy
The degree to which voice recognition equipment will not accept unwanted or invalid inputs, including extraneous noise; high rejection accuracy allows operation in a high-noise environment.

relation
An aspect or quality that connects two or more things or parts as being or belonging or working together or as being of the same kind.

relation address
An address expressed as a difference, with respect to a base address.

relational data base
A method of organizing a data base to permit association of information contained in separate records by placing data associated with each key in separate tables.

relational data model
A data model representing data as a collection of relations, where relationships between relations are represented in terms of comparable fields.

relation character
In COBOL, a character that expresses a relationship between two operands.

relative coordinates
An ordered pair or three-tuplet of signed numbers between −1 (inclusive) and 1 (noninclusive) that specifies (in twos complement arithmetic) either the new location of the drawing point with respect to the old location of the drawing point, when used within a geometric primitive, or the dimensions of a given field when used with one of the control commands.

relative coordinate system
A coordinate system whose origin moves relative to world or fixed coordinates.

relative data
In computer graphics, values in a computer program that specify displacements from the actual coordinates in a display space or image space.

relative error
The ratio of an absolute error to the true, specified, or theoretically correct value of the quantity that is in error.

relative order
In computer graphics, a display command in a computer program that causes the display device to interpret the data bytes following the order as relative data, rather than as absolute data.

relaxation approach
An iterative problem-solving approach in which initial conditions are propagated utilizing constraints, until all goal conditions are adequately satisfied.

relay
An electrically operated device that mechanically switches electrical circuits, by opening and/or closing, by variations in the conditions

of one electric circuit. A relay can thereby affect the operation of other devices in the same or other electric circuits.

relay logic
A representation of the program or other logic in a form normally used for relays.

relay symbology
The use of the symbols developed for the hard-wired relay.

release
To convert a planned order into an open order that is then known as a scheduled receipt.

relevant backtracking
Backtracking (during a search) not to the most recent choice point but to the most relevant choice point.

reliability
The probability that a device will function without failure over a specified time period or amount of usage. The percentage of time during which the robot can be expected to be in normal operation (i.e., not out of service for repairs or maintenance). This is also known as the uptime of the robot.

relocatability
A capability whereby programs or data may be located to different places in main memory at different times without requiring modification to the program.

relocatable addresses
The addresses used in a program that can be positioned at almost any place in primary storage. Usually, however, once the program is link edited, the addresses used are absolute for the remainder of that processing run. Some programs are self-relocating; they can be located at any storage position at any particular time. Addresses are assigned by the use of base address and displacement or paging.

relocation factor
The algebraic difference between the assembled origin and the loaded origin of a computer program.

reluctivity
Reciprocal of magnetic normal permeability of a material.

remanence
Magnetization remaining when ferromagnetic material is removed from a magnetic field. Conversely, a material that exhibits remanence is ferromagnetic.

remote axis admittance
A device that performs the fine motions of parts mating. It can be mounted on the end of a robot arm and will perform the final phases of assembly when the parts are in close proximity and when they are in contact.

remote batch
A method of entering jobs into the computer from a remote terminal in a conversational mode for processing later in a batch-processing mode. In this mode, a plant or office geographically distant from the central computer can load in a batch of transactions, transmit them to the computer and get back the results by mail or by means of direct transmission to a printer or other output device at the remote site.

remote batch processor
A terminal device located some distance from the central computer, used to process data by batches of records.

remote center compliance
A compliant device used to interface a robot or other mechanical workhead to its tool or working medium. The remote center compliance allows a gripped part to rotate about its tip or to translate without rotating when pushed laterally at its tip. The remote center compliance thus provides general lateral and rotational "float" and greatly eases robot or other mechanical assembly in the presence of errors in parts, jigs, pallets, and robots. It is especially useful in performing very close clearance or interference insertions.

remote data concentration
The use of communications processors for the multiplexing of data from many low-speed lines or terminals (or low-activity lines or terminals) onto one or more higher-speed lines.

remote I/O
The capability of locating the input/output device at a considerable distance from the processor to conserve field wiring.

remote job entry
Input of a batch job from a remote site and receipt of the output by way of a line printer or card punch at a remote site. The technique allows various systems to share the resources of a batch-oriented computer by giving the user access to centrally located data files and access to the power necessary to process those files.

Repeatability
Repeatability of Robot Gripper Mechanism

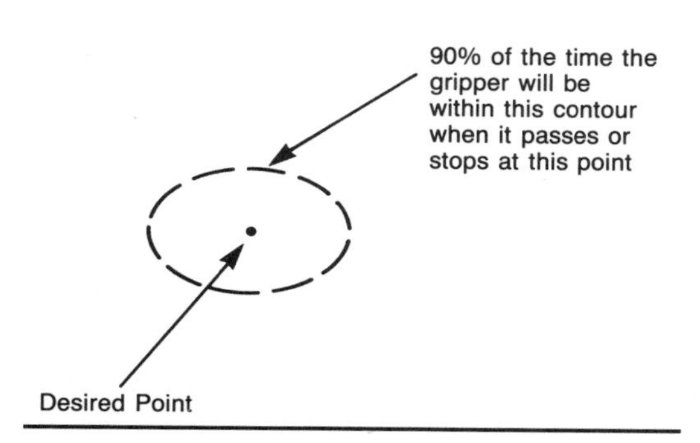
Repeatability
90% of the time the gripper will be within this contour when it passes or stops at this point
Desired Point

repaint
A CAD feature that automatically redraws a design displayed on the CRT. Redraw a display image on a CRT to reflect its updated status.

repair
To restore the system to operating condition after damage or malfunction.

repeatability
Repetitive accuracy; the closeness of agreement of repeated position movement under the same conditions to the same location. (As opposed to accuracy, a measurement of the deviation between a taught location point and the played-back location.) Under identical conditions of load and velocity, this deviation will be finer than accuracy tolerance. The closeness of agreement among the number of consecutive movements made by the robot arm to a specific point. Closeness of agreement of repeated position movements to the same indicated location and under the same conditions. The ability of the manipulator arm to position the end-effector at a particular location within a specified distance from its position during the previous cycle.

repeater
In TOP, a repeater is a transparent device used to connect segments of an extended network at the physical layer. This implies that only networks of one type may be interconnected using this method (i.e., physical layers must be identical). The primary use of repeaters is to extend networks in which the media access control protocol imposes limitations on the size of a network.

repeating pattern
An ordered sequence of items (elements) that, after a certain point, repeats itself.

repetition rate
The number of pulses per second produced by a pulsed laser.

repetitive addressing
A method of implied addressing, applicable to zero-address instructions, in which the operation part of an instruction implicitly addresses the operands of the last instruction executed.

replacement theory
Methods of analysis that deal with two types of situations involving the life pattern of the equipment involved. One pertains to equipment that gradually deteriorates in value or efficiency; the other involves equipment that breaks down permanently.

replenishment lead time
The total period of time that elapses from the moment it is determined that a product is to be reordered until the product is back on the shelf, available for use.

replica master
A teleoperator control mechanism that is kinematically equivalent to the slave manipu-

Repeater
Repeater Architecture. *Source*: Boeing Company.

USER
7
6
5
4
3
2
1
1
USER
7
6
5
4
3
2
1
NETWORK (IEEE 802.3)
NETWORK (IEEE 802.3)

lator or other device that is being controlled (i.e., the master has the same kind of joints in the same relative positions as does the slave). As a human operator moves the master by hand, the control system forces the slave to follow the master's motions. A replica master may be larger than, smaller than, or the same size as the slave device it controls. The control system may reflect back to the joints of the master any forces and torques that are applied to corresponding joints of the slave (bilateral force feedback) to allow the operator to "feel" objects remotely through the slave. The replica master (and the slave) may or may not have a geometry similar to that of a human arm.

replicate
To generate an exact copy of a design element and locate it on the CRT at any point(s) and in any size or scale desired.

reported symptom
An observed operational failure in the existing system.

report generator
A programming system for producing a complete report, given only a description of the desired content and format of the output reports and certain information about the input file and hardware available.

report program generator
A processing program that can be used to generate object programs that produce reports. RPG has powerful and relatively simple input/output file manipulation (including table lookup) but is relatively limited in algorithmic capabilities.

representation
Formalization and structuring of knowledge in a computer so that it can be manipulated by the knowledge base management system. A symbolic description or model of objects in the image or scene domain.

reproducibility
The ability to be precise or repeatable over a relatively long period of time.

reprogramming
Changing a program written for one computer so that it will run on another.

request input
One of three input modes. In request mode, the application program requests graphical input and then waits for an input value to be returned.

requirements alteration
Processing a revised master production schedule through MRP in order to review the impact of the changes. Not to be confused with net change, which, in addition to changes to the MPS, processes changes to inventory balances, bills of material, and so forth through MRP.

requirements planning
The planning of requirements for components based upon requirements for higher-level assemblies. The production schedule is exploded or extended through the use of the bills of material, and the results are netted against inventory. Materials planning is usually done level-by-level and frequently in time series.

requirements planning, capacity
A determination and projection of the workload (capacity requirements) by time period for each work center, department, plant, and so on. It enables management to determine future bottlenecks, critical areas, and the like. It is one of the main functions of manufacturing activity planning.

requirements planning, material
One of two basic approaches to planning and controlling inventories. It has two main logical steps: (1) a simulation of future inventory position computed from present inventory and planned or known future issues and receipts and (2) explosion of requirements into lower-level requirements through bills of material. These two steps are repeated at each assembly level, starting with end items. In this way, requirements and shop and purchase orders for lower-level components are computed from the end item requirements. The other basic approach, order-point technique, looks at the past and uses historical data, while requirements planning looks to the present or the future by using actual or forecast orders to generate requirements.

requirements regeneration
Regenerating requirements and taking into account all changes to the master production schedule since the last plan was generated. Planned orders are recomputed; however, existing release forms are in progress.

requisition
An authorization for specific materials or parts to be obtained either from a parts vendor or the stockroom in a specified quantity at a certain time.

rerun
A repeat of a machine run, usually because of a correction, an interrupt, or a false start.

rerun point
In a computer program, one of a set of carefully selected points designed into a computer program such that if an error is detected between two such points, the problem may be rerun by returning to the last such point instead of returning to the start of the program.

rescheduling
The process of changing the wanted data for an existing replenishment order.

rescheduling assumption
A fundamental piece of MRP logic that assumes that existing open orders can be rescheduled in to nearer time periods far more easily than new orders can be released and received. As a result, planned order receipts are not created until all scheduled receipts have been applied to cover gross requirements.

reservation
The process of designating stock for a specific customer order.

reserved word
A word of a source language whose meaning is fixed by the particular rules of that language and cannot be altered for the convenience of any one computer program expressed in the source language; computer programs expressed in the source language may also be prohibited from using such words in other contexts in the computer program.

reserve for depreciation
That part of the first or base cost of an asset that has been periodically written off as an expense and accumulated in the accounting records because of wear, tear, obsolescence, or inadequacy of the asset.

reset
To return a register or storage location to zero or to a specified initial condition.

reset key
Erases all programs and restores the computer to its original mode.

resident
Existing permanently in computer memory.

residual method
A method of magnetic-particle testing in which the indicating medium is applied after the magnetizing force has been discontinued.

residue check
A validation check in which an operand is divided by a number to generate a remainder that is then used in checking.

resist
A material used to shield parts of a chip during etching.

resistance brazing
Brazing by resistance heating, the joint being part of the electrical circuit.

resistance welding
Welding with resistance heating and pressure, the work being part of the electrical circuit. Examples are resistance spot welding, resistance seam welding, projection welding, and flash butt welding.

resistivity
Measure of electrical resistance properties of a conducting material, given by the resistance of a unit length of unit cross-section area. Reciprocal of conductivity. ASTM tests include D-193 for electrical conductor materials; D-63 for materials used in resistors, heating elements, and electrical contacts; and A-344 for magnetic materials. The terms resistivity and volume resistivity are used interchangeably for conductors.

resistor
An electronic component that impedes the flow of current in an electronic circuit.

resolution
The least interval between two adjacent discrete details that can be distinguished from one another. The smallest increment of distance that can be read and acted upon by an automatic control system. A measure of the smallest possible increment of change in the variable output of a device. The number of visible distinguishable units in the device coordinate space, as contrasted to the number of addressable points. The smallest feature of an image that can be sensed by a vision system. Resolution is generally a function of the number of pixels in the image, with a greater number of pixels allowing better resolution.

For the visual clarity of a display screen, the result of the number of dots per square inch in a particular screen matrix. The smallest increment of distance that will be developed by the numerical control system in order to actuate a machine tool. The resolution of a digital image is given by the value M and N and the number of bits by which the gray-scale values of $f(i, j)$ are represented.

resolved motion rate control
A control scheme whereby the velocity vector of the end point of a manipulator arm is commanded and the computer determines the joint angular velocities to achieve the desired result. Coordination of a robot's axes so that the velocity vector of the end point is under direct control. Motion in the coordinate system of the end point along specified directions or trajectories (line, circle, etc.) is possible. Used in manual control of manipulators and as a computational method for achieving programmed coordinate axis control in robots.

resolver
A transducer that converts rotary or linear mechanical position into an analog electrical signal by means of the interaction of electromagnetic fields between the movable and the stationary parts of the transducer.

resonator
The laser cavity, lasing medium and rod, and the two reflecting mirrors. The mirrors reflect photons of light back and forth (i.e., cause them to resonate), and the laser beam is amplified.

resource
Any means available to network users, such as computational power, programs, data files, storage capacity, or a combination of these.

resource availability
An assessment of the resources available to development of a module.

resource requirements planning
The process of converting the production plan and/or the master production schedule into the impact on key resources, such as man-hours, machine hours, storage, standard cost dollars, shipping dollars, inventory levels, and so forth. Product load profiles or bills of resources could be used to accomplish this. The purpose of this is to evaluate the plan prior to attempting to implement it. Sometimes referred to as a rough-cut check on capacity. Capacity requirements planning is a detailed review of capacity requirements. Synonymous with rough-cut capacity planning, rough-cut resource planning (cf. closed-loop MRP). Rough-cut planning based on product family planning bills of labor, materials, or other resources performed to ensure that the production plan is achievable before it is disaggregated to form the detailed master production schedule. This resource planning does not take into account existing open shop orders.

response time
The amount of time elapsed between generation of an inquiry at a data communications terminal and receipt of a response at that same terminal. Response time, thus defined, includes transmission time to the computer, processing time at the computer (including access time to obtain any file records needed to answer the inquiry), and transmission time back to the terminal. Elapsed time from the last user keystroke until the first meaningful system character is displayed at the user's terminal.

responsivity
The relative sensitivity of a photodetector to different wavelengths of light.

restart
To resume a computer program interrupted by operator intervention.

restore
To bring back to its original state a design currently being worked on in a CAD/CAM system after editing or modification that the designer now wants to cancel or rescind.

restricted envelope
The set of points representing the maximum extent or reach of the robot hand or working tool in all directions. The work envelope can be reduced or restricted by limiting devices that establish limits that will not be exceeded in the event of any foreseeable failure of the robot or its controls. The maximum distance that the robot can travel after the limit device is actuated will be considered the basis for defining the restricted (or reduced) work envelope.

restricted work envelope
That portion of the work envelope to which a robot is restricted by limiting devices that es-

tablish limits that will not be exceeded in the event of any reasonably foreseeable failure of the robot or its controls. The maximum distance that the robot can travel after the limiting device is actuated shall be considered the basis for defining the restricted work envelope of the robot.

resume
A feature of some application programs that allows the designer to suspend the data-processing operation at some logical break point and restart it later from that point. See also reentrant.

retained data
Graphical data that is stored (retained) in the graphics data storage for use in subsequent editing or display.

retained image
Also called image burn. A change produced in or on the image sensor that remains for a larger number of frames after the removal of a previously stationary light image and that yields a spurious electrical signal corresponding to that light image.

retained segment
A named segment that has associated retained-segment dynamic attributes that may be modified to modify the segment's image. In order to change the primitives in a retained segment, the segment itself must be deleted and recreated.

retained structure
A structure containing retained data.

retentivity
The maximum value of residual induction in magnetic material (B_{rs}). A measure of the permanence of magnetization. Low values are desirable for electromagnetic devices; high values for permanent magnets, relays, and magnetos.

reticle
A pattern mounted in the focal plane of a system to measure or locate a point in the image.

retrofit
Stands for RETROactively FITting and applies to the updating or significant modification of a piece of equipment that is in the field.

retroflective
Characteristic of material causing it to reflect light back to its source regardless of angle of incidence. Characteristic of material causing it to reflect light back to the source.

retroflector
A reflector, specially constructed, that reflects energy back to the source from which it came (independent of angle for small angles). It is also known as a *corner reflector*.

return key
The carriage return key on a terminal keyboard, that, when struck, places the cursor at the left margin one line below its previous horizontal position.

return-to-reference recording
The magnetic recording of binary characters in which the patterns of magnification used to represent 0s and 1s occupy only part of the storage cell, the remainder of the cell being magnetized to a reference condition.

reverse channel
Provision of simultaneous data path in the reverse direction over a half-duplex facility; normally it has a much lower bandwidth (transmission speed) than the main data path. Most commonly, it is used for positive/negative acknowledgments of previously received data blocks.

reverse image
A symbol in which the normal dark areas are represented in the light areas.

reverse video
A mode of displaying selected characters on a CRT screen in a manner that is exactly the opposite of that screen's normal display color. For example, on a screen with light characters against a dark background, reverse video would show dark characters against a light background.

revision
A change that does not greatly affect the main function of the basic system.

rework
The process of remedying an incorrectly manufactured part by reprocessing it either in the normal manufacturing system or offline.

rework lead time
The time required to rework a lot of material in-house or at a supplier.

rework routing
A routing that shows additional work that must be done in order to rework items that failed to meet their quality specifications. Such routings are often developed only when the problem occurs. In some cases, however,

because of a high rejected rate during a particular process, the rework routing is held permanently in the routing data.

RGB monitor

Commonly used to refer to the color space, mixing system, or monitor in color computer graphics. In RGB, a color is defined as percentages of red, green, and blue, with 0, 0, 0 equivalent to black and 100, 100, 100 equivalent to white. Shades of gray have equal percents of red, green, and blue between 0 and 100, that is, 25, 25, 25 is a dark gray.

right-handed Cartesian coordinate system

A coordinate system in which the axes are mutually perpendicular and are positioned in such a way that, when viewed along the positive Z axis toward the origin, the positive X axis can be made to coincide with the positive Y axis by rotating the X axis 90 degrees in the counterclockwise direction.

right-handed system

A three-dimensional coordinate system in which, if the positive X axis (thumb) points to the right, and the positive Y axis (index finger) points up, then the positive Z axis (middle finger coming out of the palm of the right hand) points toward the viewer.

right justified

A field of numbers (decimal, binary, etc.) that exists in a memory cell, location, or register possessing no significant zeros to its right is considered to be right justified; 0001200000 is considered to be seven-digit field, right justified; 000001200 is a two-digit field, not right justified.

ring network

A computer network where each computer is connected to adjacent computers.

robomation

A contraction of the words "robot" and "automation," meaning the use of robots to control and operate equipment or machines automatically.

robot

A reprogrammable multifunctional manipulator designed to move material, parts, tools, or specialized devices through variable programmed motions for the performance of a variety of tasks.

Robotic Industries Association (RIA)

The only trade association in North America organized specifically to promote the use and development of robotic technology. Founded in 1974 as the Robot Institute of America, RIA now represents more than 330 leading U.S. robot manufacturers, distributors, corporate users, accessory equipment and systems suppliers, consultants, service companies, and research institutes. RIA member companies benefit from a wide variety of services designed to further the exchange of technical and trade-related information. Trade-related meetings, workshops, seminars, and expositions sponsored by RIA showcase the latest developments in robotics to worldwide audiences. RIA also plays an active role in establishing industry standards, reporting statistics, publishing and distributing current robotics information and representing the robotics industry to government, industry, academia, and the press. In 1984, RIA established two new trade groups: the National Personal Robot Association and the Automated Vision Association. Both groups reflect RIA's goal of serving all industries that are active in developing robotic technology.

robotics

The study of robots or the practice of using robots. The science of designing, building, and applying robots.

Robotics International of SME

The largest individual member association that leads in the generation and exchange of information in robotics to advance the technology for improved productivity and quality of life. It was founded in 1980 to provide comprehensive and integrated coverage of the field of robotic technology. RI/SME is the organizational "home" for scientists, engineers, and managers concerned with robotics. It has more than 10,000 individual members and chapters in many cities worldwide. Robotics International of SME is both applications- and research-oriented and covers all phases of robot research, design, installation, operation, human factors, and maintenance within the plant facility. Members benefit through participation in numerous educational activities, seminars, and conferences such as ROBOTS 11. RI/SME publications (including the bimonthly *Robotics Today* magazine), a certification program, and chapter activities keep members informed of state-of-the-art developments. RI/SME is an association affiliated

Robot System
Principle Components of Robot System. *Courtesy:* Cincinnati Milacron.

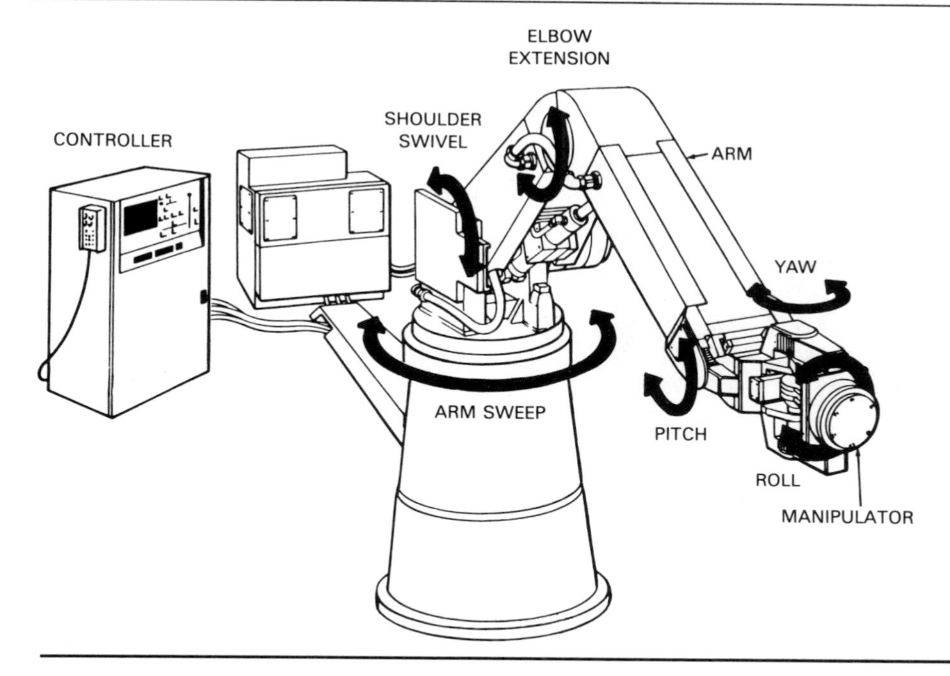

with the Society of Manufacturing Engineers and is, in a very real sense, industry's partner and a vital constructive force in the professional advancement in robotics. The parent SME has more than 80,000 members in 70 countries.

robot motions
Four types of work motions: (1) anthropometric motion—motions of a robot as in a shoulder, elbow and a wrist, developing a modified spherical work envelope; (2) cylindrical motion—motion of a robot's arm when mounted on cylindrical axis; (3) polar motion—motions of a robot by two axes or rotation that create a modified spherical work envelope; and (4) rectilinear motion—motions of a robot in three dimensions along straight lines (slides or channels).

robot programming language
A computer language especially designed for writing programs for controlling robots.

robot systems
Includes the robot hardware and software, consisting of the manipulator, power supply, and controller; the end-effector(s); any equipment, devices, and sensors the robot is directly interfacing with; any equipment, devices, and sensors required for the robot to perform its task; and any communications interface that is operating and monitoring the robot, equipment, and sensors.

robot vision
The use of a vision system to provide visual feedback to an industrial robot. Based upon the vision system's interpretation of a scene, the robot may be commanded to move in a certain way. (*Illus. p. 310.*)

Rockwell hardness test
A test for determining the hardness of a material, based upon the depth of penetration of a specified penetrator into the specimen under certain arbitrarily fixed conditions of test.

rod memory
A thin-film memory in which the metallic film is deposited on short metallic rods. The rods are then strung into planes and stacked according to the coding structure of the system.

Robot Vision
PRAB Robot Equipped with Vision Sensor to Control Gripper

roll
Circular motion at an axis. Rotation of the robot end-effector in a plane perpendicular to the end of the manipulator arm. The angular displacement of a moving body around the principle axis of its motion.

rollback
A programmed return to a prior checkpoint or rerun point.

roll forming
The operation of shaping flat-strip material into a desired form by passing it through a series of shaped rolls. Each roll brings the shape closer to its final form.

ROM
A data storage device generally used for a control program, whose content is not alterable by normal operating procedures.

root node
The initial (apex) node in a tree representation.

root segment
That segment of an overlay program that remains in main storage at all times during the execution of the overlay program; the first segment in an overlay program.

root structure
A structure that has been posted to a workstation (identified for display).

rotate
To turn an object around its central axis or reference point. Along with scaling and translation, it is one of the basic graphic transformations. To turn a displayed 2-D or 3-D construction about an axis through a predefined angle relative to the original position.

rotation
Turning an object or part or all of a display image about an arbitrary point in a two-dimensional or axis in three-dimensional coordinate space. Movement of a body around an axis, that is, such that (at least) one point remains fixed.

rotational motion
A degree of freedom that defines motion of rotation about an axis.

rotational part
A part with at least one axis of rotation, usually machined on a lathe.

rotational symmetry
A part possesses rotational symmetry about a specified axis if its orientation is repeated by rotating it through a certain angle about the axis.

rough machining
Machining without regard to finish, usually to be followed by a subsequent operation.

roughness
Relatively finely spaced surface irregularities, of which the height, width, and direction establish the predominant surface pattern.

route
The path followed by a man, material, part, or the like in a particular production process. To prescribe the above path.

router
Program that automatically determines the routing path for the component connections on a printed circuit board.

routers
Similar devices to bridges, except allowing access to multiple LANs. Routers require going up through the first three layers of the OSI model. Map routers are used to connect multiple networks together at a common point. A router provides path selection and alternate routing based on destination network layer addresses and status of connected networks.

routine
A sequence of a computer or robot controller instructions that monitors and controls a specific application function. Set of instructions arranged in proper sequence to cause a computer to perform a desired operation.

routing

A form listing for the manufacture of a particular item the sequence of operations, transportations, storages, and inspections to be used, and usually also the standard times applicable and the machines, equipment, tools, workstations, number of workers, materials, parts, and the like that are to be used. Establishing the sequence of processes, operations, transportations, and storages to be followed and the machines, tools, workstations, and miscellaneous equipment that will be used in producing a particular part, product, or job lot. In production, the sequence of operations to be performed in order to produce a part or an assembly. In telecommunications, the assignment of the communications path by which a message can reach its destination. A function within a layer that translates the name of an entity or the service access point address to which it is attached into a path by which the entity can be reached. The list of instructions of sequential operations to manufacture a part. Usually showing each workstation, operation, setup time, runtime, and machines, tools, jigs, and fixtures necessary for each production step. The operation of cutting away material overhanging a template by means of a rotating cylindrical cutter engaging the edge of the material to be profiled or cut. (The term routing is also applied to the operation of making shaped cavities and is more commonly known as end milling.) Placement of interconnections on a printed circuit board. In production, the sequence of steps to be performed in the production of a part or assembly.

Routers

Router Architecture. *Source*: Boeing Company.

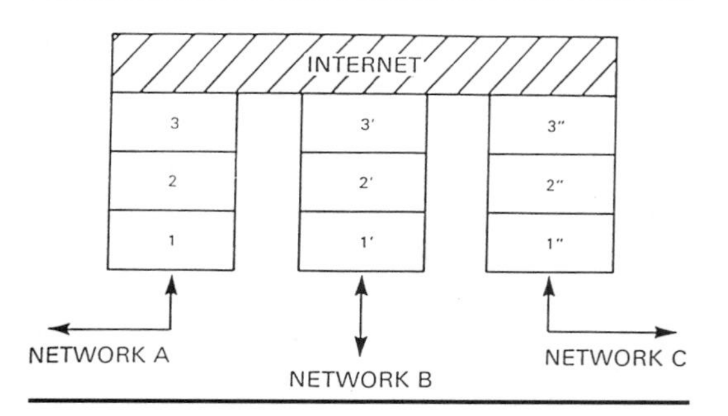

routing sheet

A list of the sequence of moves between operations or work centers.

row

A path perpendicular to the edge of the tape along which information may be stored by presence or absence of holes or magnetized areas. A character would be represented by a combination of holes and no holes across the tape that would fall along a row.

row binary

A method of encoding binary numbers on a card, where each bit in sequence is represented by the presence or absence of punches in successive positions in a row as opposed to a series of columns. Row binary is especially convenient for 40 or fewer bit words, since it can be used to store 12 binary words on each half of an 80-column card.

RPG

A computer language that can be used on several computers. The language stresses complete output reports based upon information that describes the input files, operations, and output format. RPG is an acronym for Report Program Generator.

RS-232-C, RS-422, RS-423, RS-449

Standard electrical interfaces for connecting peripheral devices to computers. EIA standard RS-449, together with EIA standards RS-422 and RS-423, are intended to gradually replace the widely used EIA standard RS-232-C as the specification for the interface between data terminal equipment (DTE) and data circuit-terminating equipment (DCE) employing serial binary data interchange. Designed to be compatible with equipment using RS-232-C, RS-449 takes advantage of recent advances in integrated circuit design, reduces cross talk between interchange circuits, permits greater distance between equipment, and permits higher data signaling rates (up to 2 million bits per second). RS-449 specifies functional and mechanical aspects of the interface, such as the use of two connectors having 37 pins and 9 pins instead of a single 25-pin connector. RS-422 specifies the electrical aspects for wideband communication over balanced lines at data rates up to 10 million bits per second. RS-423 does the same for unbalanced lines at data rates up to 100,000 bits per second.

Rule-Based Expert System
Comparison of Traditional Computer Program vs. Rule-Based Program

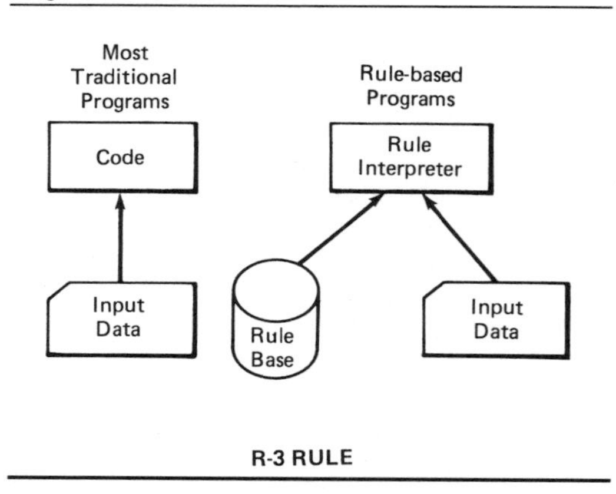

R-3 RULE

rubber banding
A CAD capability that allows a component to be tracked (dragged) across the CRT screen, by means of an electronic pen, to a desired location, while simultaneously stretching all related interconnections to maintain signal continuity. During tracking, the interconnections associated with the component stretch and bend, providing an excellent visual guide for optimizing the location of a component to best fit into the flow of the PC board or other entity, minimizing total interconnect length and avoiding areas of congestion. A technique for displaying a straight line that has one end fixed and the other end following a stylus or some input device.

ruby laser
A solid-state laser that uses a synthetic ruby crystal doped with a chromium impurity as the lasing medium.

rule
An on-screen line of a specified width, especially those used to divide areas in forms or page layouts. In artificial intelligence, a pair composed of an antecedent condition and a consequent proposition that can support deductive processes such as back-chaining and forward-chaining.

rule-based expert system
A type of expert system where the knowledge is represented in the form of production rules.

rule-based program
A computer program that explicitly incorporates rules or ruleset components.

rule-based system
System in which knowledge is stored in the form of simple if-then or condition-action rules.

ruled surface entity
A surface generated by connecting corresponding points on two space curves by a set of lines.

rule-interpreter
The control structure for a production rule system.

ruleset
A collection of rules that constitutes a module of heuristic knowledge.

run
The single and continuous execution of a program by a computer or a given set of data.

run-length encoding
A data compression technique in which an image is raster-scanned and only the lengths of runs of consecutive pixels with the same property are stored. A method of reducing data storage requirements for cell arrays. A run of identical cells is coded by an ordered pair: length of run and color-index value. A method of denoting the length of a string of elements, each of which has the same property. Usually used in denoting the length of strings of picture elements with the same brightness. A data compression technique for reducing the amount of information in a digitized binary image. It removes the redundancy that arises from the fact that such images contain large regions of adjacent pixels that are either all white or all black (i.e., black-white transitions are relatively infrequent). The brightness information is replaced by a sequence of small integers that tell how many consecutive black and white pixels are encountered while traversing each scan line. For gray-white imagery, some compression can be achieved by considering the first n high-order bits of the brightness information to represent n different binary images and then transforming each into run-length format (the low-order bits will vary so much that there will be little redundancy to remove).

running fit
Any clearance fit in the range used for parts that rotate relative to each other. Actual values of clearance resulting from stated shaft and hole tolerances are given for nine classes of running and sliding fits for 21 nominal shaft sizes in ASA B4.1-1955.

Rule Interpreter
Block Diagram Showing Relationship Between Rule Interpreter, Knowledge Base, and Global Data Base

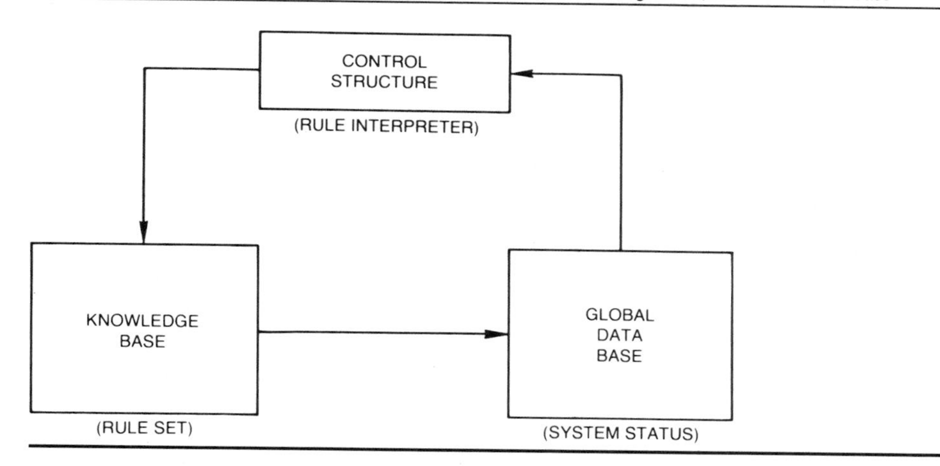

running in parallel
Where two central processors are used jointly for the same operation. If one central processor fails, the second central processor is used for backup.

running time
The time during which a machine is actually producing. For example, the running time would include execution but would not include setup, maintenance, and waiting for the operator.

runout
The unintentional escape of molten metal from a mold, crucible, or furnace. The defect in a casting caused by the escape of metal from the mold.

runout time
Time required by machine tools after cutting time is completed before the tool and material are completely free of interference so that the next sequence of operations can proceed.

run sheet
A log-type document used in continuous processes to record raw materials used, quantity produced, in-process testing results, and so on. May serve as an input document for inventory records.

runtime
The standard time allowed for the processing of one piece of a specific item for a specified machine and operation. The time required to make a complete execution of a single computer program.

run-unit
The CODASYL word for a single application program execution or task.

rupture strength
The nominal stress developed in a material at rupture. This is not necessarily equal to ultimate strength. Since necking is not taken into account in determining rupture strength, it seldom indicates true stress at rupture.

S

safeguard
A barrier guard, device, or procedure designed for the protection of personnel.

safety capacity
The planning or reserving for excess manpower and equipment above known requirements for unexpected demand. This reserve capacity is in lieu of safety stock.

safety stock
The average amount of stock on hand when a replenishment quantity is received. Its purpose is to protect against the uncertainty in demand and in the length of the replenishment lead time. The size of the safety stock is dependent upon the expected deviation of the actual lead-time usage from the forecasted usage. Safety stock and cycle stock are the two main components of any inventory.

safety time
The difference between the requirement date and the planned in-stock date. In an MRP system, material can be ordered to arrive ahead of the requirement date.

sample
A portion of a universe of data chosen to estimate some characteristic(s) about the whole universe. The universe of data could consist of historical delivery cycles, unit costs, sizes of customer orders, number of units in inventory, and the like.

sample input
One of three input modes. In sample mode, the application program obtains the measure of a logical input device without waiting for any external operator action.

sampling
The practice of selecting a small portion of the total group under consideration for the purpose of inferring the value of one or several characteristics of the group.

sampling distribution
The distribution of values of a statistic calculated from samples of a given size.

sand blasting
Same as blasting, except that the abrasive material is fine sand of some definite grit.

satellite
A remote system connected to another, usually larger, host system. A satellite differs from a remote intelligent workstation in that it contains a full set of processors, memory, and mass-storage resources to operate independently of the host.

satellite computer
A processor connected locally or remotely to a larger central processor and performing certain processing tasks—sometimes independent of the central processor, sometimes subordinate to the central processor.

Saticon
Trade name for a vision system image pickup tube of the direct readout type. Has low lag, near-unity gamma, and well-balanced spectral sensitivity.

satisfice
To achieve a solution, in artificial intelligence, that satisfies all imposing constraints (opposed to "optimize"). Developing a satisfactory, but not necessarily optimum, solution.

satisficing level
A level of performance for which one is willing to settle even though it is not the maximum or best performance level.

saturation
The extremes of operating range wherein the output is constant regardless of changes in input. Attribute of a color that determines the extent to which it differs from a gray of the same lightness. The degree to which a color is monochromatic.

saturation magnetization
The maximum value of magnetization. If a ferromagnetic material is magnetized by an external field, magnetization cannot increase indefinitely with increasing external field.

Saturn Project
A development program for a new family of subcompacts to be produced in this country, using start-to-finish innovation to become cost-competitive with small Japanese cars. Saturn Corporation is an independent, wholly owned subsidiary of General Motors Corporation, created as a highly autonomous subsidiary of GM to produce a line of small cars. As GM's first new domestic nameplate since Chevrolet in 1918, Saturn will use advanced technology, electronic business systems, and improved human relations to reduce the cost

advantage foreign manufacturers now have. In doing so, Saturn will also serve as a learning laboratory for improved competitiveness for all of GM. The subsidiary will eventually operate its own new, highly integrated manufacturing and assembly complex, employing unique technologies in both product and processing, advanced management methods, and a just-in-time inventory philosophy. Roger B. Smith, chairman of GM, has stated, "These cars and the innovation they represent in integrated design, engineering, manufacturing, assembly, materials management, and human relations will be an historic step toward overcoming the Japanese cost advantage in small cars. As Saturn's concepts then spread throughout our entire product line, they will help ensure that the American auto industry remains competitive and is able to provide secure, good-paying American jobs."

sawing
The operation of separating material by teeth performing successive cuts in material. Operation may follow a straight line or contour.

scale
To enlarge or reduce the size of an image or element. Scaling is usually done proportionately along all dimensions at once but it can be done on fewer. To enlarge or diminish the size of a displayed entity without changing its shape, that is, to bring it into a user-specified ratio to its original dimension. Scaling can be done automatically by a CAD system. Used as a noun, scale denotes the coordinate system by representing an object.

scaled drawing
A drawing in which lengths on the drawing are proportional to the size of objects in the real world. Architectural plans, for example, are almost always scaled, while electronic circuit diagrams are not.

scale factor
The number used as a multiplier, so chosen that it will cause a set of quantities to fall within a given range of values. To scale the values 856, 432, -95 and -182 between -1 and $+1$, a scale factor of 1/1000 would be suitable.

scale ratio
The number that indicates how much larger or smaller the shape of an object is shown. For example, a 2 means that a drawing shows an object twice as large as actual.

scaling
Enlarging or reducing all or part of a display image by multiplying the coordinates of display elements by a constant value. For different scaling in two orthogonal directions, two constant values are required. In assembler programming, indicating the number of digit positions in object code to be occupied by the fractional portion of a fixed-point or floating-point constant.

scan
The search for a symbol that is to be optically recognized. A search for marks to be recognized by the recognition unit of the optical scanner. To move a sensing point around an image. To examine signals or data point-by-point in logical sequence.

scan line
One scanned line of an image.

scanner
A device that examines a spatial pattern one part after another and generates analog or digital signals corresponding to the pattern. Scanners are often used in mark sensing, pattern recognition, or character recognition. A device that turns an image into digital graphics form for input to a computer. Some are combination cameras and video digitizers, while others use a light beam to trace the image. A laser-operated device that can recognize bar code or alphanumeric font and convert information from these items to digital data.

scanner, analog input
A device that will, upon command, connect a specified sensor to measuring equipment and cause the generation of a digit count value that can be read by the computer.

scanning raster
The pattern of adjacent parallel lines that makes up a raster-scanned video image. In most cases, the raster is a series of almost horizontal lines, starting at the top of the screen and working across and down.

scan time
The time necessary to completely execute the entire PC program once, including I/O update.

SCARA robot (selective compliance assembly robot arm)
A new low-cost, high-speed assembly robot moving almost entirely on a horizontal plane. The mechanism originated with the Japanese. (*Illus., p. 316.*)

SCARA Robot
Adept Technology's Adept one™ SCARA Robot System

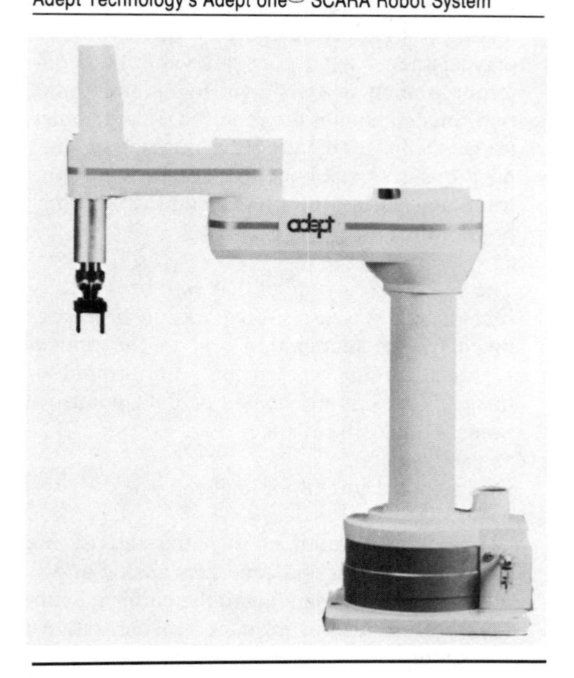

scatter chart
A graph showing the actual observed relationships between two variables by the use of plotted points.

scattering
The process by which light passing through a medium or reflecting off a surface is redirected throughout a range of angles.

scene
The three-dimensional environment from which the image is generated. The portion of space being investigated by a vision system. A scene may be very complex, or it may consist simply of a single item.

scene analysis
The process of seeking information about a 3-D scene from information derived from a 2-D image. It usually involves the transformation of simple features into abstract descriptions. The ability of a camera-equipped robot to recognize an arbitrary object placed in an arbitrary scene.

schedule
A listing of jobs to be processed through a work center, department, or plant and their respective start dates as well as other related information.

scheduled receipts
Within MRP, open production orders and open purchase orders are considered as *scheduled receipts* on their due date and will be treated as part of available inventory during the netting process for the time period in question. Scheduled receipt dates and/or quantities are not normally altered automatically by the MRP system. Further, scheduled receipts are not exploded into requirements for components, as MRP logic assumes that all components required for the manufacture of the item in question have been either allocated or issued to the shop floor.

scheduling
Determining the order of activities for execution, usually based on control heuristics.

scheduling rules
Basic rules that are spelled out ahead of time so that they can be used consistently in a scheduling system. Scheduling rules usually specify the amount of calendar time to allow for a move and for queue, how load will be calculated, and so on.

schema
Definition of data elements and their relationships to form the global logical structure of a data base.

schematic
A drawing showing relationship or movement. It is meant to show the organization of a topic, not necessarily the direct appearance. However, dimensioned schematics do show parts in proportion to their actual size.

scissor
To apportion a drawing into segments that can be viewed on a CRT screen.

scleroscope test
A hardness test when the loss in kinetic energy of a falling metal "tup," absorbed by indentation upon impact of the tup on the metal being tested, is indicated by the height of rebound.

scotopic vision
Vision that occurs in faint light or in dark adaptation. It is attributed to the retinal rods in the eye. The contrasting capability is daylight or photopic vision.

scrap
Fragments of stock removed from a part during manufacture. A rejected or discarded part.

scrap allowance
The factor that expresses the quantity of a particular component that is expected to be scrapped while that component is being built into a given assembly. Also, a factor that expresses the amount of raw material needed in excess of the exact calculated requirement to reduce a given quantity of a part. The factor, dependent upon the type of assembly or part, is carried in the product structure segment and is used to increase the requirements as the component requirements are exploded.

scrap rate
The percentage difference between the amount or number of units of product started in a manufacturing process and that amount or number of units completed at an acceptable quality level.

scrap usage
The expected average quantity of an item to be scrapped each period.

scratch-pad memory
A high-speed memory used by the CPU to temporarily store a small amount of data so that the data can be retrieved quickly when needed. Interim calculations are stored in a scratch-pad memory.

scratch tape
A tape that is available for writing on. The previous contents are obsolete and can be "scratched."

screw-plasticating injection molding
A technique in which the plastic is converted from pellets to a viscous melt by means of an extruder screw that is an integral part of the molding machine. Machines are either single stage (in which plastication and injection are done by the same cylinder) or double stage (in which the material is plasticated in one cylinder and then fed to a second for injection into a mold).

script
A technique for representing knowledge that stores the events that take place in familiar situations in a series of "slots." A script is composed of a series of scenes that are, in turn, composed of a series of events. Frame-like structures for representing sequences of events.

scroll
A graphic display technique, whereby the generation of a new line of alphanumeric text at the bottom of a display screen automatically regenerates all other lines of text one line higher than before and deletes the top line. In computer graphics, the continuous vertical or horizontal movement of the display elements within a window, in a manner such that new data appear at one edge of the window as old data disappear at the opposite edge. The window may include the entire display surface.

scrolling
Moving through information on a computer display, either vertically or horizontally, to view information otherwise excluded.

seam tracking
A mechanical probe with feedback mechanism to sense and allow for changes in a given taught path or a vision system to look at a given path (or setpoint) and determine if it has changed its location with respect to the robot, or a voltage within the welding arc that is read on each side of the welding arc when that arc is oscillated. The difference (or changes) in voltage send a signal to the robot to change its path accordingly.

seam welding
Arc or resistance welding in which a series of overlapping spot welds is produced with rotating electrodes, rotating work, or both. A longitudinal weld in sheet metal or tubing.

search
The process of attempting to proceed from an *initial state* to a *goal state* by systematically evaluating possible alternative solution paths.

search function
Feature that enables a robot system to adjust the position of data points within an existing cycle, based on changes in external equipment and workpieces. One use of the search function is in stacking operations, especially when the stacked items are fragile or have irregular thicknesses. The time delay inherent in deceleration from the input signal activation will permit some movement beyond the robot's receipt of the signal; so, if the signal originates through a limit switch that is closed upon contact with the stack, some compliance must be built into the robot gripper. A fragile workpiece would also require a slow velocity during the search segment.

search models
Operations research models that attempt to find optimal solutions with adaptive searching approaches.

search routine
A robot function by which an axis or axes move in one direction until terminated by an external signal. Used in stacking and unstacking of parts or to locate workpieces.

search space
All of the possibilities that might be evaluated during a search. The search space often is represented as a tree structure, called a *search tree*. The implicit graph representing all the possible states of the system that may have to be searched to find a solution. In many cases, the search space is infinite. The term search space is also used for nonstate-space representations.

secondary station
In a network, a station that has been selected to receive a transmission from the primary station.

second-generation computer
A computer utilizing solid-state components. In COBOL, a logically placed sequence of one or more paragraphs. A section must always be named. In computer graphics, to construct and to position a bounded or unbounded intersecting plane with respect to one or more displayed objects and then generate and display the intersection on a display. In the NC field, the period of technology associated with transistors (solid state).

section display symbol
An arrangement of fonted straight lines in a repetitive planar pattern at a specified spacing and angle.

section entity
A pattern used to distinguish a closed region in a diagram. It is represented as a form of the copious data entity.

sectioning and paneling
Computer-aided mapping functions that allow the cartographer to create new maps by joining a number of smaller-scale maps. The two functions trim extraneous data, allow maps with common boundaries to be merged, and correct small inconsistencies.

sector
That part of a track or band on a magnetic drum, a magnetic disk, or a disk pack that can be accessed by the magnetic heads in the course of a predetermined rotational displacement of the particular drive.

security
The general subject of making sure the computerized data and program files of the company cannot be accessed, obtained, or modified by unauthorized personnel, and cannot be fouled up by the computer or its programs. Security is implemented by special software, special hardware, and the computer's operating procedure.

seek time
The time that is needed to position the access mechanism of a direct-access storage device at a specified position.

segment
A set of data that can be placed anywhere in a memory and can be addressed relative to a common origin. A segment contains one or more data items (usually more) and is the basic quantum of data that passes to and from the application programs under control of the data base management software. A collection of elementary operations performed contiguously at a single workstation. A collection of display elements that can be manipulated as a unit.

segmentation
The process of breaking up an image into regions (each with uniform attributes) usually corresponding to surfaces of objects or entities in the scene. The process of dividing a scene into a number of individually defined regions or segments. Each segment can then be analyzed separately.

segment attribute
A general characteristic of a segment that may be static (fixed throughout the lifetime of the segment) or dynamic (alterable during the lifetime of the segment). Segment attributes are visibility, highlighting, detectability, segment priority, and segment transformation.

segmented program
A program that has been divided into parts (segments) in such a manner that each segment is self-contained, interchange of information between segments is by means of data tables in known memory locations, and each segment contains instructions to cause (or request a monitor to cause) the transfer to the next segment.

segment (or image)
A transformation that causes the display elements defined by a segment to appear with varying position (translation), size (scale), and/or orientation (rotation) on the display surface.

segment priority
A segment attribute used to determine which of several overlapping segments takes precedence for graphic output and input.

selected elemental time
The raw or unadjusted time chosen as being representative of the actual times taken to perform a single element of an operation.

selected time
The time value chosen by the time study observer from those obtained for an element of a time-studied operation as being representative of the time used by the worker when he or she performed the element correctly.

selecting
A communication network technique of inviting another station or node to receive data.

selective erase
A CAD feature for deleting portions of a display without affecting the remainder or having to repaint the entire CRT display.

self-checking
A bar code or symbol using a checking algorithm that can be applied to each character to guard against undetected errors. Nonself-checked codes may employ a check digit or other redundancy in addition to the data message.

self-diagnostic
The hardware and firmware within a controller that allows it to continuously monitor its own status and indicate any fault that might occur within it.

self-relocating program
A program that can be loaded into any area of main storage and that contains an initialization routine to adjust its address constants so that it can be executed at that location. See also relocatable program.

semantic
Pertaining to the meaning, intention, or significance of a symbolic expression, as opposed to its form.

semantic grammar
A grammar for a limited domain that, instead of using conventional syntactic constituents such as noun phrases, uses meaningful components appropriate to the domain.

semantic interpretation
Producing an application-dependent scene description from a feature set, or representation, derived from the image.

semantic network
A technique for representing associations between objects and events. A representation of objects and relationships between objects as a graph structure of nodes and labeled arcs.

semantic primitives
Basic conceptual units in which concepts, ideas, or events can be represented.

semantics
The discipline dealing with the relationship between signs and symbols (including words), their meaning, and human behavior.

semiconductor
A solid crystalline substance whose electrical conductivity falls between that of a metal and an insulator.

semiconductor memory
A method of storing data (binary digits) using semiconductors.

semikinematic
A mount that approximates a kinematic mount.

sensitive direction
The direction in which error motions of a rotary axis produce one-to-one form errors (in machining) or measuring errors (in metrology) on the part being machined or measured.

Semantic Interpretation
Block Diagram of Semantic Interpretation Elements from Image to Description

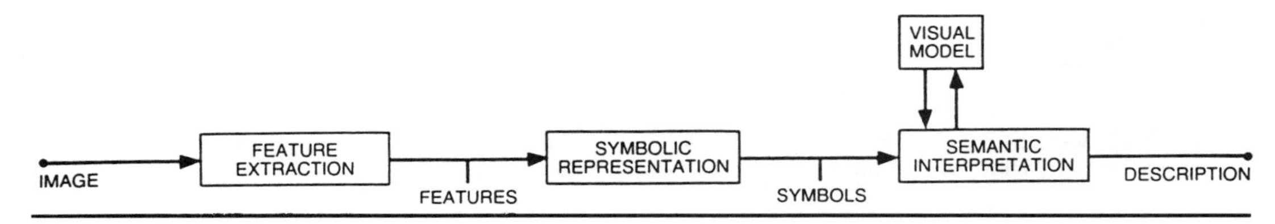

sensitivity
A factor expressing the incident illumination upon the active region of an image sensor required to produce a specified signal at the output.

sensor
A transducer whose input is a physical phenomenon and whose output is a quantitative measure of that physical phenomenon. A "transducer" that takes in information about the physical state of things and converts it to an electrical signal that can be processed by a control system. Sensors can be simple, such as a temperature monitor on a machine tool, or highly complex, such as a machine vision system. Other sensors monitor things like torque on a robot wrist, the pressure exerted by robot grippers, or the vibrations in a workpiece being machined.

sensor array
Typically, sensor elements that are mounted in such a way that they form rows and columns.

sensor-based system
An organization of components, including a computer, whose primary source of input is data from sensors and whose output can be used to control the related physical process.

sensory control
Control of robot based on sensor readings. Several types can be employed: (1) sensors used in threshold tests to terminate robot activity or to branch to another activity, (2) sensors used in a continuous way to guide or direct changes in robot motions (see also active accommodation), (3) sensors used to monitor robot progress and to check for task completion or unsafe conditions, and (4) sensors used to retrospectively update robot motion plans prior to the next cycle.

sensory-controlled robot
A robot whose program sequence can be modified as a function of information sensed from its environment. Robot can be servoed or nonservoed.

sensory hierarchy
A relationship of sensory processing elements, whereby the results of lower-level elements are utilized as inputs by higher-level elements.

sepia
A translucent medium. If a translucent reproduction is made from an original, the reproduction also becomes reproducible.

sequence
A single ladder diagram configuration programmed to perform a single operation. A sequence has one output (coil) and is identified by a unique number (address).

sequence checking
Checking items in a file to ensure that they are all in ascending, or descending, order.

sequence codes
Numbers assigned to a list of items in a straight sequence, starting with one, without regard to classification or order of the subjects being coded. It is useful for any short list of names, products, or accounts where the only object is the application of simple code numbers and where the arrangement of data is not important.

sequence number
A number identifying the relative location of blocks or groups of blocks on a tape. The sequence number is identified by the letter "n" as the address and usually falls at the beginning of a block. A control system incorporating a sequence number display would "read out" the number of the block being read by the tape reader that would correspond to the number following the address letter. This feature enables the operator to identify the tape location with respect to the position of the machine. An additional feature called block search allows the operator to manually enter a block, or sequence number, and the control system will automatically search for the number. Sometimes referred to as tape search.

sequencer
A controller that operates an application through a fixed sequence of events.

sequence robot
A robot whose physical motion and trajectory follows a preprogrammed sequence.

sequencing
Specifying the order of performance of tasks so that available production facilities are utilized in an optimal manner. Ordering in a series or according to rank or time.

sequential
Pertaining to the occurrence of events in time sequence, with no overlap of events. In numeric sequence, normally in ascending order.

sequential access
A method of storage on external media in which information is stored one item after another. Tapes are sequentially accessed.

sequential-access memory
An auxiliary memory device that lacks any addressable data areas. A specific piece of data can only be found by means of a sequential search through the file.

sequential batch processing
A mode of operating a computer in which a run must be completed before another run can be started.

sequential control
A mode of computer operation in which instructions are executed in an implicitly defined sequence until a different sequence is explicitly initiated by a jump instruction.

sequential data set
A data set whose records are organized on the basis of their successive physical positions, such as on magnetic tape. Contrast with direct data set.

sequential processing
The traditional computer processing technique of performing actions one at a time, in sequence.

sequential storage
Secondary storage where data are arranged in ascending or descending order, usually by item number. This type of storage is usually associated with magnetic tape. Sequential storage usually is processed in a batch.

serial
Pertaining to data or instructions that are processed in sequence, one bit at a time, rather than in parallel (several bits at a time). The handling of data in a sequential fashion such as to transfer or store data in a digit-by-digit time sequence or to process a sequence of instructions one at a time.

serial access
Descriptive of a storage device or medium where there is a sequential relationship between access time and data location in storage—that is, the access time is dependent upon the location of the data.

serial arithmetic operations
A method by which the computer handles arithmetic fields one digit at a time, usually from right to left.

serial code
A bar code symbol typically used with a fixed-beam scanner where the scanning action is caused by the motion of the symbol past the scanning head. The bits of the symbol are evaluated one at a time (serially) as the symbol passes.

serial communications
A digital communication method that transmits the bits of a message one at a time. The most common long-distance transmission method, suitable for use with cable, radio, or modulated light as the transmission medium.

serial interface
An interface that transmits data bit by bit rather than in whole bytes. Serial interfaces are much slower than parallel interfaces but much cheaper. A method of data transmission that permits transmitting a single bit at a time through a single line. Used where high-speed input is not necessary.

serial operation
A type of information transfer performed by a digital computer in which all bits of a word are handled sequentially.

series circuit
A circuit in which components or symbolic elements are connected end-to-end so that the same current flows throughout the circuit.

serrating
The operation of producing teeth on the inside or outside edge of the workpiece by broaching. The workpiece is fixed and the tool moves.

server process
An internal IPEX process associated with devices, files, and server processes by means of the message router.

service
To make fit for use; adjust; repair; maintain.

serviceability
The ease with which hardware or software failures can be detected, diagnosed, and repaired.

service access point
The access means by which a pair of communicating entities in adjacent layers use or provide services.

service bureau
A company that supplies users with batch or interactive processing either online or offline.

service function
A mathematical relationship of the safety factor to service level, that is, to the fraction of demand that is routinely met from stock.

service level
A measure of the effectiveness of customer service, usually expressed as a ratio or percentage such as percentage of line items shipped versus ordered.

service parts
Parts used for the repair and/or maintenance of an assembled product. Typically, they are ordered and shipped at a date later than the shipment of the product itself.

service parts demand
The need for a component to be sold by itself, as opposed to being used in production to make a higher-level product.

service reference model (SRM)
A specification of the minimum set of features that must be implemented by a receiving device in order to meet the requirements for a particular service and the maximum set of features that should be assumed by an information source when encoding text and pictorial information.

services
The functions offered by an open system to communicating entities in adjacent layers.

service time
The time required to serve the customer (i.e., fill his demand) after he places demand on an inventory.

service versus investment chart
A curve showing the amount of inventory that will be required to give various levels of customer service.

servocontrolled robot
The control of a robot through the use of a closed-loop servosystem, in which the position of a robot axis is measured by feedback devices and compared with a predetermined point stored in the controller's memory.

servomechanism
A mechanical or electromechanical device whose driving signal is determined by the difference between the commanded position and the actual position at any point in time. An automatic feedback control system for mechanical motion. A control system for the robot in which the computer issues commands, the air motor drives the arm, and a sensor measures the motion and signals the amount of the motion back to the computer. This process is continued until the arm is repositioned to the point requested. A powerful amplifying device that takes an input signal from some low-energy source and directs an output requiring large quantities of energy. A type of closed-loop control system in which mechanical position is the controlled variable.

servomotor
Controlled by feedback systems so that when the ultimate position is reached, the motor stops operating.

servosystem
A control mechanism linking a system's input and output, designed to feed back data on system output to regulate the operation of the system.

servovalve
A transducer whose input is a low-energy signal and whose output is a higher-energy fluid flow that is proportional to the low-energy signal.

session
The period of time during which a user engages in a dialog with a conversational time-sharing system; the time elapsed from when a terminal user logs-on the system until he logs-off the system.

set
A unit or units and the necessary assemblies, subassemblies, and basic parts connected or associated together to perform an operational function.

setpoint
The required or ideal value of a controlled variable, usually preset in the computer or system controller by an operator.

settling time
The time for a damped oscillatory response to decay to within some given limit.

setup
A task taking place at the beginning of an operation, for example, adjust machines, fasten fixtures on machine table, and so forth.

setup cost
The cost of preparing a production machine to run many production parts.

setup lead time
The time in hours or days needed to prepare before a manufacturing process can start. Setup time may include run and inspection time for the first piece.

setup operations
Aggregation of operations performed on a part during a single setup.

setup time
The time it takes to prepare a production machine to run many production parts. Usually includes time necessary to get tools and fixtures, set up the first part, and perhaps even run and inspect the first part produced.

set value
A value as provided by the application program to a workstation. The workstation may not be able to render the set value exactly.

seven-segment display
A display format consisting of seven bars so arranged that each digit from 0 to 9 can be displayed by energizing two or more bars. LED, LCD, and gas-discharge displays all use seven-segment display formats.

sexadecimal
Pertaining to a selection, choice, or condition that has 16 possible different values or states.

shade
Term used to characterize difference in lightness between two surface colors, the other attributes of color being essentially constant. A lighter shade of a color has higher lightness but about the same hue and saturation.

shaded
A polygon fill style based on an array of vertex color indices. Not supported by PHIGS.

shading
A method using seven levels of gray scale to differentiate a specific blob from neighboring black objects and the white background. A large area brightness gradient in an image, not present in the original scene. The variation of relative illumination over the surface of an object caused, for example, by curvature of the surface. Arranging the intensity and purity of color or darkness of patterning on objects in an image, usually to provide visual clues to three-dimensional form.

shaft encoder
A rotary encoder used to encode or determine the position of the rotary shaft.

shake
Vibration of a robot's "arm" during or at the end of a movement.

shallow drawing
Essentially the same as drawing, except that the operation is restricted to shallow pieces and operation is usually accomplished by a stamping single-action process.

Shannon
In informational theory, a unit of logarithmic measure of information equal to the decision content of a set of two mutually exclusive events expressed by the logarithm to base two, for example, the decision content of a character set of eight characters equals three Shannons.

shape fill
The solid painting-in within the boundaries of a shape, using graphics. The automatic painting-in of an area on an IC or PC board layout, defined by user-specified boundaries; for example, the area to be filled by copper when the PC board is manufactured. Can be done online by CAD.

shaping
The operation of removing material from the surface of a material mass by a series of horizontally adjacent cuts. The tool is moved in forward strokes while the material moves in present increments in a plane at right angles to the tool.

shared control
A facility that allows multiple concurrent interactions with a particular unit of data in a data base.

shared fixture
Fixture designed to accommodate two or more part numbers simultaneously.

shaving
Finalizing a dimensional requirement on the shape of a part by cutting away a very small portion around its configuration.

shear
That type of force that causes, or tends to cause, two contiguous parts of the same body to slide relative to one another in a direction parallel to their plane of contact.

shear fracture
A fracture in which a crystal (or a polycrystalline mass) has separated by sliding or tearing under the action of shear stresses.

shearing
The operation of cutting the material by means of straight blades, one of which is fixed and the other moving progressively toward it until the material between is finally cleaved. Shearing may be in a straight line, circular, or other shape.

shear plane
A confined zone along which shear takes place in metalcutting. It extends from the cutting edge to the work surface.

shear strength
The stress required to produce fracture in the plane of cross section, the conditions of loading being such that the directions of force and of resistance are parallel and opposite although their paths are offset a specified minimum amount.

shield
Any barrier to the passage of interference-causing electrostatic or electromagnetic fields. An electrostatic shield is formed by a conductive foil layer surrounding a cable core. An electromagnetic shield is a ferrous metal cabinet or wireway.

shielded-arc welding
Arc welding in which the arc and the weld metal are protected by a gaseous atmosphere, the products of decomposition of the electrode covering, or a blanket of fusible flux.

shielded metal-arc welding
Arc welding in which the arc and the weld metal are protected by the decomposition products of the covering on a consumable metal electrode.

shielding
The practice of confining the electrical field around a conductor to the primary insulation of the cable by putting a conducting layer over and/or under the cable insulation. External shielding is a conducting layer on the outside of the cable insulation. Strand or internal shielding is a conducting layer over the wire insulation.

shift
To move information serially right or left in the register(s) of a computer. Information shifted out of a register may be lost, or it may be reentered at the other end of the register.

shift register
A program, entered by the user into the memory of a programmable controller, in which the information data (usually single bits) are shifted one or more positions on a continual basis. There are two types of shift register: asynchronous and synchronous.

ship set
Set of part numbers in the proper quantity required to assemble a particular complete end item.

ship set, tool set
Set of all tools required to produce a given ship set on a given FMS.

shop calendar
The establishment of an agreed-upon company calendar denoting how the days of the week relate to production days when the factory is operating. Usually done on a daily basis, starting at production day one of a year and ending on the last production day of the year.

shop-floor control
A system for utilizing data from the shop floors as well as data-processing files to maintain and communicate status information on shop orders and work centers. The major subfunctions of shop-floor control are (1) assigning priority to each shop order, (2) maintaining WIP quantity information for MRP, (3) conveying shop order status information to the office, and (4) providing actual output data for capacity control purposes.

shop order
An order calling for the manufacture of a quantity of items by a certain date, according to its authorized process.

shop planning
The coordination of material handling, material availability, the setup and tooling availability so that a job can be done on a particular machine. Shop planning is often part of the dispatching function and the term shop planning is sometimes used interchangeably with dispatching, although dispatching does not have to necessarily include shop planning. For example, the selection of jobs might be handled by the centralized dispatching function while the actual shop planning might be done by the manager or one of his or her representatives.

shortage
An item that cannot satisfy the total demand placed upon it, within the wanted dates of each requirement.

shortest process time rule
A dispatching rule that directs the sequencing of jobs in ascending order by processing time. Following this rule, the most jobs per time period will be processed. As a result, the average lateness of jobs is minimized, but some jobs will be very late.

short-term repeatability
Closeness of agreement of position movements, repeated under the same conditions during a short time interval, to the same location.

shot peening
The operation of compressing a selective surface of a piece part by a rain of metal shot impelled from rotating blades of a wheel or air blast.

shoulder
The joint, or pair of joints, on a robot that connect the arm to the base.

shrinkage factor
A percentage factor in the item master that compensates for expected loss during the manufacturing cycle, either by increasing the gross requirements or by reducing the expected completion quantity of planned and open orders. The shrinkage factor differs from the scrap factor in that the former affects all uses of the part and its components. The scrap relates to only one usage.

shrink fit
A fit that allows the outside member, when heated to a practical temperature, to assemble easily with the inside member.

shrink wrapping
A technique of packaging in which the strains in a plastics film are released by raising the temperature of the film, thus causing it to shrink over the package. These shrink characteristics are built into the film during its manufacture by stretching it under controlled temperatures to produce orientation of the molecules. Upon cooling, the film retains its stretched condition, but reverts toward its original dimensions when it is heated. Shrink film gives good protection to the products packaged and has excellent clarity.

shutdown
The action of making a real-time system unavailable at the end of the real-time day. It includes disabling all terminals so that no more entries can be made, monitoring the completion of exchanges in progress, upon completion closing all files in an orderly fashion and terminating the real-time job.

sign
The symbol or bit that distinguishes positive from negative numbers.

signal
The event, phenomenon, or electrical quantity that conveys information from one point to another. The name associated with a net.

signal highlighting
A CAD editing aid to visually identify—by means of noticeably brighter lines—the pins (connection points) of a specific signal (net) on a PC design. Identifying the connection points of a net in a printed circuit board.

signal processing
Complex analysis of waveforms to extract information.

signal-to-noise ratio
Relative power of the signal to the noise. As the ratio decreases on a line, it becomes more difficult to distinguish between information and noninformation (noise).

sign bit
A bit or binary element that occupies a sign position and indicates the algebraic sign of the number represented by the numeral with which it is associated.

sign character
A character that occupies a sign position and indicates the algebraic sign of the number represented by the numeral with which it is associated.

significant digit
A digit that contributes to the precision of a numeral. The number of significant digits is counted beginning with the digit contributing the most value, called the "most-significant digit," and ending with the one contributing the least value, called the "least-significant digit."

significant digit arithmetic
A method of making calculations using a modified form of a floating-point representation system in which the number of significant digits in the result is determined with reference to the number of significant digits in the operands, the operation performed, and the degree of precision available.

significant digit codes
Coding type reduces the work of decoding by providing a code number that can be read directly. In this type of coding, all or part of the numbers are related to some characteristics of the data such as weight, dimension, distance, capacity, or other significant factors. This method is suitable for coding long lists of items

where complete decoding would be laborious or impractical.

significant part numbers
Part numbers whose alphanumeric arrangement convey information about such things as product class, size, and material.

silicon
An abundant semiconducting element from which computer chips are made.

silicon-controlled rectifier
An electronic device that is generally used in control systems for high-power loads, such as electronic heating elements.

silk screen
A type of artwork that can be generated by a CAD/CAM system for printing component placement and/or identification information on a PC board.

silver brazing
Brazing with silver-base alloys as the filler metal.

silver-brazing alloy
Filler metal used in silver brazing.

simple lens
A lens system with only one element.

simplex circuit
A communication line used in one direction only or in either direction but not at the same time. It is perhaps best not to use the term. The first case is rare and the word *channel* is available; in the second case, half-duplex is commonly used.

simplex fixture
A fixture designed to accommodate a single part number.

simulated
To represent certain features of the behavior of a physical or abstract system by the behavior of another system.

simulation
The use of programming techniques alone to duplicate the operation of one computing system on another computing system. In computer programming, the technique of setting up a routine for one computer to make it operate as nearly as possible like some other computer. Contrast with emulation. A CAD/CAM computer program that simulates the effect of structural, thermal, or kinematic conditions on a part under design. Simulation programs can also be used to exercise the electrical properties of a circuit. Typically, the system model is exercised and refined through a series of simulation steps until a detailed optimum configuration is reached. The model is displayed on a CRT and continually updated to simulate dynamic motion or distortion under load or stress conditions. A great variety of materials, design configurations, and alternatives can be tried out without committing any physical resources. An AI technique that attempts to program a computer to exhibit intelligent behavior by using a ''model'' of intelligent human behavior.

simulator
A device or computer program that represents certain features of the behavior of a physical or abstract system.

simultaneity
The facility of a computer to allow input/output on its peripherals to continue in parallel with operations in the central processor.

single-level bill of materials
Shows only those components that are directly used in an upper-level item. It does not show any relationships more than one level down.

single-level pegging
A routine that traces requirements for a part or order up the product structure or order records one level at a time to reveal the source of any part or order requirement.

single-level where used
For a component, lists each assembly in which that component is directly used and in what quantity. This information is usually made available through the technique known as *implosion*.

single-point control motion
A safeguarding method for certain maintenance operations. A single-point control of robot system motion, when used for entry into the restricted work envelope of the robot, shall be such that it cannot be overriden at any location in a manner that would adversely affect the safety of the person performing the maintenance function. Before the robot system can be returned to its ''automatic'' operation, it requires a deliberate, separate action involving the single-point control or by the person in possession of the single-point control.

single precision
Refers to the number of memory locations used for a number in a computer. Single pre-

cision indicates one location for each number. In order to store a number with decimal places, higher precision is necessary.

single scan
A user-initiated operation that commands the PC processor to execute the application program from start to end and then stop. I/O is also updated. Used normally as a debugging tool during start-up.

single shift
An invocation of a code, set into the in-use table that affects only the interpretation of the next bit combination received. Interpretation then automatically reverts to the previous contents of the table.

single threading
A program that completes the processing of one message before starting another message. See also multithreading.

sink
The terminal node at which data is used in a network.

sixty-day ordering rule
A periodic ordering technique where inventory is reviewed and the amount on hand and on order is brought up to the equivalent of 60 days' normal sales.

sizing
The operation of finalizing a dimension on a part by direct pressure, impact, or a combination of both.

skeletal code
A set of instructions in which some parts, such as addresses, must be completed to the reference edge of a data medium.

skeleton representation
A representation of a 2-D region by the medial line and the perpendicular distance to the boundary at each point along it.

sketch
To draw freehand on a graphics system. Because of the way the image is made, this may be more difficult on a vector system than on a raster graphic system. On dot-graphics systems, this mode is often called paint mode.

sketch map
A rough line drawing of a scene.

slab laser
A newly developed laser concept in which the beam is amplified through a series of reflections along the walls of the laser cavity. High-output powers in a very compact laser can be developed in this way.

slack time
The difference in calendar time between the scheduled due date for a job and the estimated completion date. If a job is to be completed ahead of schedule, it is said to have slack time; if it is likely to be completed behind schedule, it is said to have negative slack time. Slack time can be used to calculate job priorities using methods such as the critical ration. In the critical path method, total slack is the amount of time a job may be delayed in starting without delaying the start of any other job in the project.

slack-time rule
A dispatching rule that directs the sequencing of jobs based on (days left × hr/day) − std hr left = priority, or (5 × 8) − 12 = 28.

slave
An I/O or printer-driven module driven by a master unit. In some low-level word-processing applications, it is common to have a master unit plus a number of slaves automatically grinding out repetitive letters.

slew rate
The maximum velocity at which a manipulator joint can move; a rate imposed by saturation somewhere in the servoloop controlling that joint (e.g., by a value's reaching its maximum open setting). The maximum speed at which the tool tip can move in an inertial Cartesian frame. The maximum rate at which a system can follow a commanded motion.

slice architecture
A method by which a section of the register file and ALU in a computer is placed in one package. In some systems, the registers are all four bits wide; others accommodate two bits. Two or more of these "slices" can be cascaded together to form larger word sizes.

sliding action
Where contacts slide over one another, resulting in a continuous self-cleaning of the surfaces, helping to ensure low contact resistance. Same as wiping action.

sliding fit
A series of nine classes of running and sliding fits of 21 nominal shaft sizes, defined in terms of clearance and tolerance of shaft and hole in ASA B4.1-1955.

slitting
Essentially the same as milling, except the trough produced is usually very narrow, as in the case of screw heads.

slot
A feature or component description of an object in a frame. Slots may correspond to intrinsic features such as name, definition, or creator; or may represent derived attributes such as value, significance, or analogous objects.

slotting
Essentially the same as shaping, except the tool is moved vertically. Sometimes referred to as vertical shaping.

small feature
Features that are too small to be employed for orienting purposes.

SMALLTALK
Programming language developed at Xerox's Palo Alto Research Laboratory. SMALLTALK has popularized a style of programming, according to which procedures communicate by sending each other messages. SMALLTALK is considered especially good for graphics-oriented programming.

smart sensor
A sensing device whose output signal is contingent upon mathematical or logical operations that are based upon internal data or additional sensing devices.

smart terminal
A terminal having computational capability, that is, editing commands and graphic abilities.

smelting
Thermal processing wherein chemical reactions take place to produce liquid metal from a beneficiated ore.

smoothing
Averaging by a mathematical process or by curve fitting, such as the method of least squares or exponential smoothing.

smooth scroll
To move a display a single pixel (screen dot) at a time, rather than by a whole character or line. Because users perceive many small movements as smoother motion than fewer large jumps, moving by pixel creates a smoother scrolling display.

snap
The action taken by a CAD/CAM graphics program when it interprets a user-specified location as the nearest of a set of reference locations. Thus, when a point is input by digitizing, the system may snap to the closest point on the grid.

snug fit
A loosely defined fit implying the closest clearances that can be assembled manually for firm connection between parts and comparable to one or more of the 11 classes of clearance locational fits given in ASA B4.1-1955.

Sobel operator
A popular convolution operator for detecting edges. Similar to other difference operators, such as the Prewitt operator.

Society of Manufacturing Engineers
Provides a common ground for manufacturing engineers and managers to share ideas, information, and accomplishments. SME opportunities include: (1) chapter programs through over 200 local chapters that provide SME members and a foundation for involvement and participation; (2) educational programs including seminars, clinics, programmed learning courses, and video tapes; (3) conferences and expositions that enable engineers to see, compare, and consider the newest manufacturing equipment and technology; (4) publications including *Manufacturing Engineering,* the *SME Newsletter, Technical Digest,* and a wide variety of books; and (5) SME's Manufacturing Engineering Certification Institute formally recognizes manufacturing engineers and technologists for their technical expertise and knowledge acquired through years of experience. In addition, the society works continuously with the American National Standards Institute, the International Standards Organization, and other organizations to establish the highest possible standards in the manufacturing field.

soft copy
An image on a video or other electronic screen, as opposed to hard copy on paper.

softening point
Indication of maximum temperature to which a nonmetallic material can be heated without loss of its normal "body." Minimum temperature at which a specified deformation occurs under a specified load. Softening point is often determined for materials that have no definite melting point, that is, materials that gradually change from brittle or very thick and slow-flowing materials to softer materials or less viscous liquids.

softening range
Range of temperature in which a plastic changes from rigid to soft state. Values de-

pend on the method of test. Sometimes referred to as softening point.

soft sectored

A term used to describe a particular diskette format and way of recording on the diskette. Soft-sectored disks are preformatted, having data fields that are changed and updated. The first track of a soft-sectored diskette identifies the disk, the next four tracks store basic format information, such as the track and sector location of stored material. Since soft-sectored diskettes require a format, less storage capacity is available (250K bytes out of a possible 400K) for text storage than is available with hard-sectored diskettes. Soft-sectored diskettes may be initialized and used with most popular microcomputers and operating systems.

software

A term coined to contrast computer programs with the "iron" or hardware of a computer system. Software programs are stored sets of instructions that govern the operation of a computer system and make the hardware run. Software is a key determining factor in getting more computer power per dollar. The processor programs, library routines, manuals, and other service programs supplied by a computer manufacturer to facilitate the use of a computer. In addition, it may refer to other programs specially developed to fit user needs. All the documents associated with a computer. Set of programs, procedures, rules and associated documentation that directs the operation of a computer. The programs and routines used to extend the capabilities of computers such as compilers, assemblers, routines, and subroutines. Also, all documents associated with a computer, for example, manuals and circuit diagrams. All the instructions, programs, computer languages, and operations that happen between the user and the hardware. The program that controls the operation of a computer or robot controller. APT program is an example of an automatic programming software package.

software compensation

A specific example of error compensation in which the map of the errors is stored in computer memory.

software development tool

A program designed to assist programmers in the development of software. Intelligent software development tools incorporate AI techniques.

software house

A company that offers software support service to users. This support can range from simply supplying manuals and other information to a complete counseling and computer part-time programming service (job shop or body shop).

software library

The software tools supplied by manufacturers, required for the development of user application programs.

soldering

Similar to brazing, with the filler metal having a melting temperature range below an arbitrary value, generally 800° F. Soft solders are usually lead-tin alloys.

solenoid

An electromagnet with a movable core or plunger that, when it is energized, can move a small mechanical part a short distance.

solid

One possible representation of the interior of a fill-area or polygon primitive. The interior of the image is filled with the solid color specified by the fill-area or polygon color-index attribute.

solid modeling

A type of 3-D modeling in which the solid characteristics of an object under design are built into the data base so that complex inter-

Solid Modeling
Presentation of 3-D Solid Modeling Image

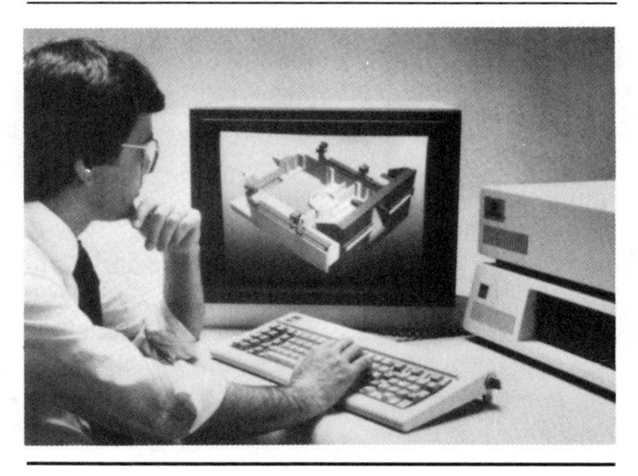

nal structures and external shapes can be realistically represented.

solid state
The logic of a computer or machine control unit made solely of solid-state electronic components. Pertaining to electronic devices, such as transistors, made from silicon and other solid substances as opposed to vacuum tubes or electromechanical relays.

solid-state camera
A camera that uses a solid-state integrated circuit to convert light into an electrical signal. A television camera that uses some type of solid-state integrated circuit instead of a vacuum tube to change a light image into a video signal. Solid-state cameras have the following advantages over vacuum-tube cameras: ruggedness; small size; no high voltages; insensitivity to image burn and lag (antibloom capability is possible with the proper readout technique); potentially very low cost, characteristic of solid-state technology; and a spatially stable precise geometry that effectively superimposes a fixed repeatable measurement grid over the object under observation, without the pincushion or barrel distortion introduced by the deflection systems of tube cameras.

solid-state device
Any element that can control current without moving parts, heated filaments, or vacuum gaps. All semiconductors are solid-state devices, although not all solid-state devices (e.g., transformers) are semiconductors.

solid-state laser
A laser that uses a glass or crystal as a host to an impurity, such as chromium or neodymium, which produces the lasing action.

solidus
In a constitution or equilibrium diagram, the locus of points representing the temperatures at which various compositions finish freezing or cooling or begin to melt on heating.

solidus temperature
Temperature at which an alloy begins to melt during heating or finishes freezing during cooling. For pure metals, same as liquids temperature and known simply as melting point.

solution path
A successful path through a search space.

solution structure
A framework consisting of the capabilities of a candidate solution and the relationships between those capabilities.

sonic sensor
A device or system that converts sound into an electrical signal.

sort
A processing run or operation to distribute data in numerical, alphabetic, or alphanumeric groups according to a given standard or rule. A key consisting of a prescribed uniform string of characters can be used as a means of making workable-size groups from a large volume of records.

sound-absorption coefficient, normal incidence
A measure of the effectiveness of material in absorbing sound energy. Fraction of normally incident sound energy absorbed by a material assumed to have an infinite surface.

sound-absorption coefficient, statistical
A measure of the effectiveness of a material in absorbing sound energy. Fraction of incident sound energy absorbed by a material under conditions where it is subject to equal sound from all directions over a hemisphere, that is, under reverberant sound conditions. Measurement is expensive and time-consuming, requiring a specially constructed reverberation chamber. Thus, property is sometimes estimated from specific normal sound-absorption coefficient or specific normal acoustic impedance. The normal coefficient is about half the statistical coefficient for very low values and approaches equality with it at very high values. Maximum numerical difference occurs at intermediate values and is about 0.25 to 0.35.

source
A text file written in a high-level language and containing a computer program. It is easily read and understood by people but must be compiled or assembled to generate machine-recognizable instructions. Also known as source code. The terminal node at which data enters a network.

source codes
Symbols and syntax easily understood by people; used to describe a procedure that a computer can execute.

source data automation (SDA)
The capturing of data for machine entry, as a key product of producing a required document

such as a purchase order, job order, or other source document. The byproduct then serves as input for a computer.

source data base
This type of data base contains a full representation of original information. This type of data base provides users with quantitative answers without referencing other sources. Contrast this with reference data base. There are four types of source data base: (1) numeric, (2) textual-numeric, (3) properties, and (4) full text. See data base, numeric data base, textual-numeric data base, properties data base, and full-text data base.

source document
The original paper on which are recorded the details of a transaction. An original record of some type that is to be converted into machine-readable form.

source language
The symbolic language comprised of statements and formulas used to specify computer processing. It is translated into object language by an assembler or compiler and is more powerful than an assembly language in that it translates one statement into many items. The original high-level language in which software was written. It is converted to object code by a compiler. Examples of source language are BASIC, COBOL, and PASCAL. A language used as input to a computer that is translated to the machine language that the computer requires.

source listing
A record of the computer instructions in program language.

source program
In a language, a program that is an input to a given translation process. A computer program written in symbolic language that will be converted into an absolute language object program using a processor program.

space character
A graphic character that is usually represented by a blank site in a series of graphics. The space character, though not a control character, has a function equivalent to that of a format-effector that causes the print or display position to move one position forward without producing the printing or display of any graphics. Similarly, the space character may have a function equivalent to that of an information reporter.

spacing
A type of character that causes the cursor to be automatically advanced after the character has been received and displayed.

spalling
The cracking and flaking of particles out of a surface.

span time
Actual time from part design to the completion of the finished product.

spares order
An order that has been placed by a customer for a specific part belonging to a finished product item. This part is normally used by the computer either for the immediate repair of the finished product item or for a backup spare part.

spares usage
That portion of an item's usage quantity that is the result of unplanned demand for spare and repair parts.

spatial frequency
The reciprocal of line spacing in an object or scene.

speaker-dependent
Class of voice-operated hardware using pattern recognition techniques; requires operator to first give equipment a sample of speech patterns before words can be recognized; equipment capabilities range from word recognizers to intelligent voice terminals.

speaker identification
An area of application wherein a previously spoken utterance is processed and analyzed according to its acoustic components for the purpose of isolating one from a group of speakers; not unlike fingerprint matching.

speaker verification
An area of voice equipment applications wherein the speaker's voice pattern is matched to previously entered patterns in storage to verify and authorize the speaker; typical usage is for facility access controls.

special
Refers to a part or group of parts that are unique to a particular order.

special character
A character that is neither numeric nor alphabetic. Special characters in COBOL include the space ( ), the period (.), as well as the following: + − */=$,;'')(.

specialized common carrier
A company that provides private-line communications services, for example, voice, teleprinter, data, and facsimile transmission.

special-purpose automation
A machining system or manufacturing line dedicated to very particular tooling characteristics and processing of parts or products in mass. See flexible automation, continuous production.

special-purpose computer
A computer that is designed to handle a restricted class of problems.

special-purpose logic
Those proprietary features of a controller that allow it to do things not normally found in relay ladder logic.

special robot
Usually, a custom-built robot. This is in contrast to a modular robot made up of standardized units.

special time allowance
A temporary time value applying to an operation in addition to or in place of a standard allowance to compensate for a specified, temporary, nonstandard production condition.

specification
A detailed characteristic used in describing a component. The set of specifications associated with a component is so complete that the component can be constructed or procured from it but has no specification in it that is not necessary to the description of the component.

specification proof
An association of a particular specification, with the test case that verifies system compliance.

specification set
The set of all specifications relating to a given development effort.

specific reactive power
For a specified normal induction in a material, the component of applied AC power that is "reactive"; that is, returned to the source when polarity is reversed. Product of induced voltage and reactive current per unit weight of material; that is, product of voltage, current, and sine of phase angle.

spectral
Of or relating to color.

spectral analysis
Interpreting image points in terms of their response to various light frequencies (colors).

spectral directional reflectance
Measure of brightness. The spectral reflectance that an ideal diffusing surface would need to appear the same as the specimen under test when illuminated and viewed the same way. It depends on the angular distribution of incident light and on the direction of viewing. The term spectral directional reflectivity is used to denote an inherent property of the material, as opposed to that of a particular object.

spectral reflectance
Measure of brightness. Ratio of light reflected by a specimen to homogeneous light energy incident on it. It depends on the angular distribution of the incident energy. The term spectral reflectivity is used to indicate the inherent property of a material of such thickness that any increase in thickness would fail to change the spectral reflectance. It may be expressed in the form of a curve of luminous directional reflectance versus wavelength.

spectrum
A range of frequencies or wavelengths.

specular
Reflection of light where the light is not dispersed when reflected.

specular reflection
Reflection of light from a surface at an angle equal but opposite to the angle of incidence.

speech recognition
Recognition by a computer (primarily by pattern matching) of spoken words or sentences. An area of AI research that attempts to allow computers to recognize words or phrases of human speech.

speech synthesis
The generation of speech by a computer. Developing spoken speech from text or other representations.

speech understanding
Speech perception by a computer.

speed
The maximum speed at which the tip of a robot is capable of moving at full extension. Term referring to the rotation of the spindle in revolutions per minute (rpm). Since the speed of an NC machine is programmed in rpm, it is often necessary to convert from the conventional form of describing the speed in feet per minute (fpm).

speed-payload tradeoff
The relationship between corresponding values of maximum speed and payloads with

which an operation can be accomplished to some criterion of satisfaction, with all other factors remaining the same.

speed rating
A method of performance rating that compares the speed or tempo with which a worker performs the options necessary to execute an operation against the observer's concept of standard or normal tempo.

speed-reliability tradeoff
The relationship between corresponding values of maximum speed and reliability with which an operation can be accomplished to some criterion of satisfaction, with all other factors remaining the same.

spherical
Mounted on a rotary base and resembling the turret of a tank, the arm not only extends and retracts but is pivoted so it can swing vertically, allowing rotary motion about a horizontal plane. The end-effector moves in a volume of space that is a portion of a sphere.

spherical aberration
A degration of an image due to the shape of the lens elements.

spherical coordinate robot
A robot whose construction consists of a horizontally rotating base, a vertically rotating shoulder, and a linear transversing arm connected in such a way that the envelope traced by the end of the robot arm at full extension defines a sphere in space.

spherical coordinate system
A coordinate system, two of whose dimensions are angles, the third being a linear distance from the point of origin. These three coordinates specify a point on a sphere.

spike
Voltage or current surge produced in the industrial operating environment.

spinning
The operation of forming sheet metal into circular shapes by means of a lathe, forms, and hand tools that press and shape the metal about the revolving form.

spiral mold cooling
A method of cooling injection molds or similar molds wherein the cooling medium flows through a spiral cavity in the body of the mold. In injection molds, the cooling medium is introduced at the center of the spiral, near the sprue section, as more heat is localized in this section.

spline
A common term used to describe the irregular curve concept in conventional drafting. A subset of a B-spline wherein a sequence of curves is restricted to a plane. An interpolation routine executed on a CAD/CAM system automatically adjusts a curve by design iteration until the curvature is continuous over the length of the curve. Piecewise continuous polynomial curves used to approximate a curve.

Spherical Coordinate Robot
Spherical Coordinate Robot Configuration

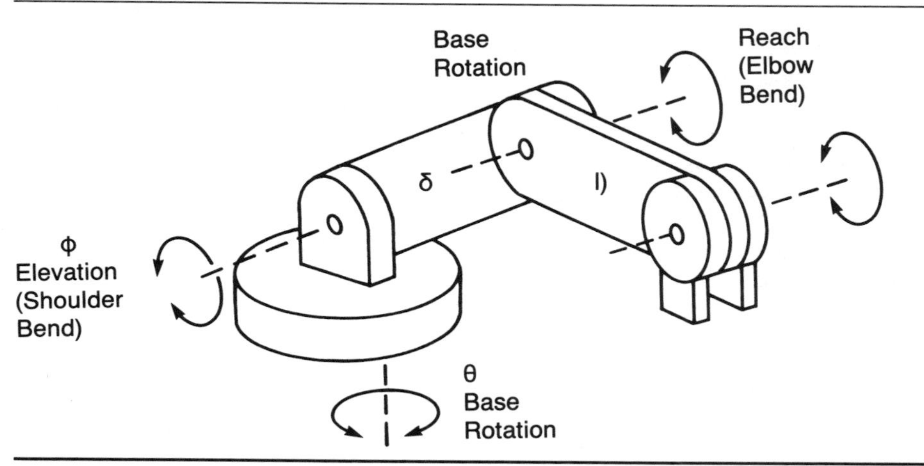

split
A general NC programming system for mills (including five axes), drills, lathes (including four axes), routers, and punches. The SWIFT lathe module aids in rapid programming with minimal input. A curve-fitting TABCYL feature is offered. The language uses Englishlike shop-oriented words and allows interactive processing and conversational part prove-out from the programmer's terminal or the machine tool.

split delivery
A method by which a larger quantity is ordered on a purchase order to secure a lower price but delivery is spread out over several dates to control inventory investment.

split-equipment cross-reference report
A hard-copy report that can be automatically generated by a CAD system, listing the exit and entry points of all common wiring appearing on two or more sheets of a wiring diagram.

split lot
A released shop order that has its original lot-size quantity divided into two or more parts, usually in order to expedite work on one portion of the lot.

split-ring mold
Mold in which a split cavity block is assembled in a chase to permit forming of undercuts in a molded piece. These parts are ejected from the mold and then separated from the piece.

sponge skrinkage
Found in heavier sections that are generally more than 2 inches thick. Sponge shrinkages appear on radiographs as dark areas, lacy in texture, usually with a diffuse outline.

spool drawing
A three-dimensional isometric schematic diagram of a portion of a piping project.

spooling
The reading and writing of input and output streams on auxiliary storage devices, concurrently with job execution, in a format convenient for later processing or output operations. Synonymous with concurrent peripheral operations.

spotfacing
The same as counterboring, except that the depth of the hole is restricted to just breaking through the surface of the material.

spot welding
A method of fastening sheet metal parts together in which a heavy electric current is passed through the plates at a spot. This current rapidly heats and melts the two sheet metal plates together, forming a small round spot weld. Welding of lapped parts in which fusion is confined to a relatively small circular area. Generally resistance welding but may also be gas-shielded tungsten-arc, gas-shielded metal-arc, or submerged-arc welding.

spray coating
Usually accomplished on continuous webs by a set of reciprocating spray nozzles traveling laterally across the web as it moves.

sprayed metal molds
Mold made by spraying molten metal onto a master until a shell of predetermined thickness is achieved. The shell is then removed and backed up with plaster, cement, casting resin, or other suitable material. Used primarily as a mold in sheet-forming processes.

spray-up
Covers a number of techniques in which a spray gun is used as the processing tool. In reinforced plastics, for example, fibrous glass and resin can be simultaneously deposited in a mold. In essence, roving is fed through a chopper and ejected into a resin stream that is directed at the mold by either of two spray systems. In foamed plastics, very fast-reacting urethane foams or epoxy foams are fed in liquid streams to the gun and sprayed on the surface. On contact, the liquid starts to foam.

springback
The deflection of a body when external load is removed. Usually refers to deflection of the end-effector of a manipulator arm.

squeezing
The operation of forming material by compressing it into a desired shape. The material may be confined in a shaped cavity or may be free-flowing.

stability
The ability of a laser system to maintain a beam with constant output characteristics, such as power levels.

stable state
In a trigger circuit, a state in which the circuit remains until the application of a suitable pulse.

stack
An area in memory for the temporary storage of data. Data in a stack is not retrieved by address but rather in chronological order, that is, last-in, first-out (LIFO).

stadimetry
The determination of distance, based upon the apparent size of an object in the camera's field of view.

staging
Putting on the material requirements for an order from inventory before the material is required. This action is taken as a protection from inaccurate inventory records but leads to increased problems in inventory records and availability.

stairstepping
The jagged lines on a low-resolution monitor trying to display a diagonal line or curve. Term jaggies is used as a synonym.

stand-alone machines
Typically a machining center or turning center with some method of automatic material handling, such as multiple pallets or chuck-changing arrangements. All of the features typical of the fully automated flexible system are available in the stand-alone machine including probing, inspection, tool monitoring, and adaptive control; however, these features are initiated and controlled at the machine control level. The major difference between the stand-alone machine and the standard machining or turning center is the pallet or chuck arrangement that allows for unattended operation for extended periods, with minimum operator assistance. The widest acceptance and use of any configuration of the flexible manufacturing concept has been in stand-alone machines. The stand-alone machine is well-suited for high variety, low-volume production. It will, in some cases, allow for long-range planning since it can be added as a component of a future FMC or FMS.

stand-alone program
A program that can be executed independently of an operating system.

standard
An acceptance criterion or an established measure for performance, practice, or design. A definitive rule, measure, or strict criterion used to direct a development activity.

standard allowance
The established or accepted amount by which the normal time for an operation is increased within an area, plant, or industry to compensate for the usual amount of fatigue or personal or unavoidable delays.

standard binary coded decimal interchange code
A computer code used by most second-generation computers; an expansion of the binary coded decimal system. The significant difference in the standard BCD code is the use of zone bits. The zone bits of an alphanumeric character perform a code function similar to

Stadimetry
Stadimetry (Direct Imaging) is a Technique for Measuring Distance with Machine Vision. Reprinted with Permission of McGraw-Hill for Tech Tran Corp., from *Machine Vision Systems: A Summary and a Forecast.*

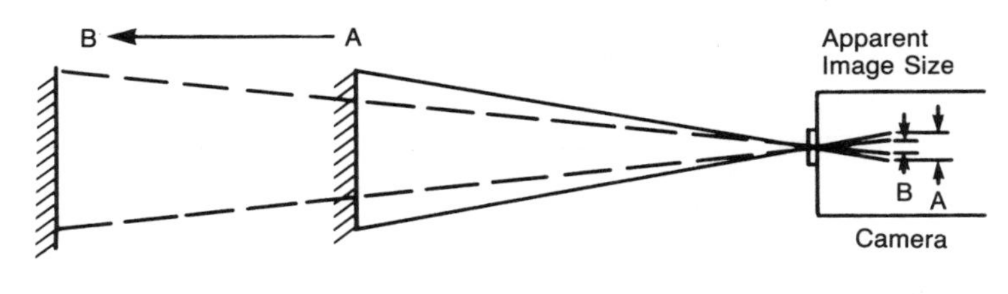

Stadimetry (Direct Imaging)

the zone positions of a punched card. They are used in combination with digits to represent the letters of the alphabet or special characters.

standard branching
Term for the way truly general-purpose industrial robot systems select or alter the programmed path and function, based on changes in the environment around them. The name given to such a facility may vary, but the purpose is the same. The robot reaches some point and interrogates an input signal to determine whether it is electrically active, or the robot is interrupted by activation of another input signal. In either case, the robot path "branches" to a section of the path/function program; if no signal is present at this decision point, or no interrupt occurs, the robot continues in a normal path sequence.

standard cost
The normal expected cost of an operation, process, or product—including labor, material, and overhead charges—computed on the basis of past performance costs, estimates, or work measurement.

standard elemental time
The normal elemental time plus allowances for fatigue and delays.

standard error
Applied to statistics, such as the mean, to provide a distribution within which samples of the statistics are expected to fall.

standard features
All capabilities included in the stated cost.

standard hour
An hour of time during which a specified amount of work of acceptable quality is or can be performed by a qualified worker following the prescribed methods, working at normal pace, and experiencing normal fatigue and delays.

standard hour plan
A wage incentive plan having standard times expressed as standard hours. The hourly base wage rate is paid for standard hours earned, rather than actual hours worked.

standardization
A management-sponsored program to establish criteria or policies that will promote uniform practices and conditions within the company and permit their control through comparisons.

standard performance
The performance that must be given by a worker to accomplish his or her work in the standard time allowed.

standard plan
A routine that lists a group of operations needed to produce a family of items. The items may have small differences in size, but they use the same sequence of operations.

standard practice
The established or accepted procedure used within an area, plant, or industry for carrying out a specified task or assignment.

standard ratio
A relationship based on a sample distribution by value for a particular company. When the standard ratio for a particular company is known, certain aggregate inventory predictions can be made, like the amount of inventory increase that would be required to give a particular increase in customer service.

standards
The industrial engineering time allowances measured and permitted to be utilized as the basic rate for an operation.

standard set
The set of all standards relating to a given development effort.

standard time data
A compilation of all the elements that are used for performing a given class of work, with normal elemental time values for each element.

standing-on-nines carry
In the parallel addition of numbers represented by decimal numerals, a procedure in which a carry to a given digit place is bypassed to the next digit place. If the current sum in the given digit place is nine, the nine is changed to zero.

star bit
In asynchronous transmission, the bit that indicates the beginning of a block of data.

star network
A computer network with peripheral nodes, all connected to one or more computers at a centrally located facility.

start-of-message code
A character or group of characters transmitted by the polled terminal and indicating to the stations on the line that what follows are addresses of stations to receive the answering message.

start section
The section of an IGES file containing a human-readable file prolog.

startup
The time between equipment installation and the full specific operation of the system.

state
The logic 0 or 1 condition in computer or robot controller memory or at a circuit's input or output.

state graph
A graph in which the nodes represent the system state and the connecting arcs represent the operators that can be used to transform the state from which the arcs emanate to the state at which they arrive.

state list
A defined data structure containing current values of system variables. Application programs can obtain state list (SL) contents through inquiry.

statement
In a programming language, a meaningful expression that may describe or specify operations and is usually complete in the context of that programming language.

state table
A computer-programming method in which the possible inputs to the computer and the appropriate responses are entered in the form of a table of statements of the form "IF (a possible condition) is true THEN (do the following things . . .)." the last entry in the table is always IF (anything else) THEN (call for help). There are a number of advantages to this procedure—it's an easy format for humans to understand, which simplifies the task of finding and correcting errors in the program.

static
Refers to a state in which a quantity does not change appreciably within an arbitrarily long time interval.

static accuracy
Deviation from true value when relevant variables are not changing with time. Difference between actual position response and position desired or commanded of an automatic control system, as determined in the steady state, that is, when all transient responses have died out.

static deflection
Load deflection considering only static loads, that is, excluding inertial loads. Sometimes static deflection is meant to include the effects of gravity loads.

static error
Deviation from true value when relevant variables are not changing with time. Difference between actual position response and position desired or commanded of an automatic control

State Graph
State Graph for a Simple AI Problem

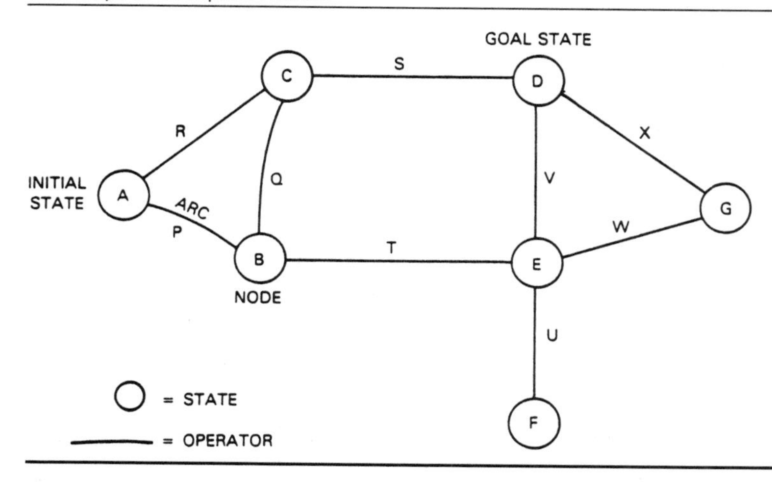

system, as determined in the steady state, that is, when all transient responses have died out.

static friction
The force required to initiate sliding or rolling motion between two contacting bodies; also called stiction.

static RAM
Random-access memory that requires continuous power but not continuous regeneration to retain its contents.

static shield
A device that protects internal circuits from static-discharge damage.

station
A physical location where a part normally stops either to have an operation performed on it or to wait for clearance to proceed to the next station. A data terminal device connected to a data link. It includes sources or sinks for the messages, as well as those elements that control the message flow on the link by means of data communication control procedures.

station control
A module in a control hierarchy that controls a workstation. The station control module is controlled by a cell control module.

statistical inventory control
Inventory control using a statistical or mathematical analysis of past history to determine future inventory requirements, order points, and so on.

statistical quality control
A means of controlling the quality of a product or process by the application of the laws of probability and statistical techniques to the observed characteristics of such product or process.

statistics
A field of scientific endeavor that is central to the scientific method in the proper design of experiments and the drawing of valid interferences about events in sampling from a given population or so-called universe.

status information
Information about the logical state of a piece of equipment. Examples are a peripheral device reporting its status to the computer or a network terminating unit reporting its status to a network switch.

steadiness
Relative absence of high-frequency vibration or jerk.

steady state
General term referring to a value that is not changing in time. Response of a dynamic system due to its characteristic behavior, that is, after any transient response has stopped; the steady-state response is either a constant or a periodic signal.

step
One operation in a computer program.

stepping motor
A bidirectional, permanent magnet motor that turns through one angular increment for each pulse applied to it. An electric motor that is designed to rotate in distinct steps with typically 200 steps per revolution. A stepping motor is driven by a digital pulse through a special driver circuit. In the absence of a driving pulse, a stepping motor is maintained in detent and will remain locked in that particular angular position.

stereopsis
The determination of distance through the use of binocular vision.

stereoscopic approach
Use of triangulation between two or more views, obtained from different positions, to determine range or depth.

stereotyped situation
A generic recurrent situation, such as "eating at a restaurant" or "driving to work."

stereo vision
An approach to image analysis that uses a pair of images of the same scene taken from different locations where depth information can be derived from the difference in locations of the same feature in each of the images.

stiction
Static friction; the force that tends to keep two mated surfaces from moving relative to one another.

stiffness
The amount of applied force per unit of displacement of a compliant body.

stimulated emission
The emission of a photon from an atom in an excited state when another photon encounters it. The resulting photon will have the same wavelength, phase, and spatial coherence as the original photon.

stitch
To interactively route a PC interconnect between layers and around obstructions with au-

Stereo Vision
Stereo Vision Technique for Measuring Distance with Machine Vision. Reprinted with Permission of McGraw-Hill for Tech Tran Corp., from *Machine Vision Systems: A Summary and a Forecast.*

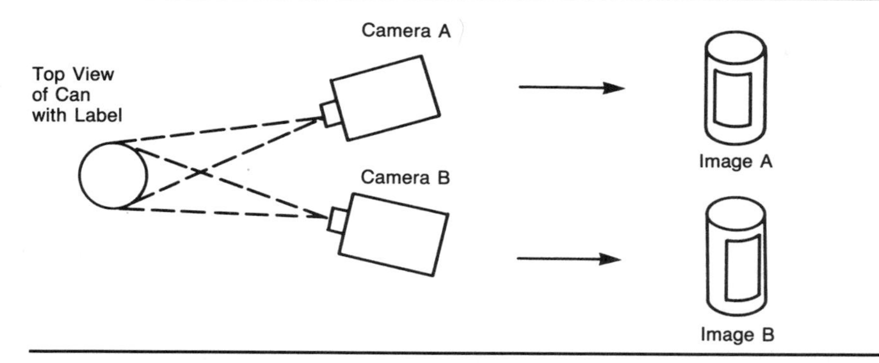

tomatic insertion of vias, when necessary, through the use of powerful editing commands available on a CAD system.

stock
Parts and material on hand.

stockkeeping unit
Represents an item at a particular location. For example, if product A is stocked at several locations, each combination of product A and a stocking location is a different SKU.

stockout
The lack of materials or components that are needed to be on hand in stock.

stockout percentage
A measure of the effectiveness with which the inventory management system responds to actual demand. The stockout percent can be a measurement of total stockouts to total line item orders or of line items incurring stockouts during a period to total line items in the system.

stockpoints
The points at which stockkeeping items can be stocked and subjected to management control.

stock status
A periodic report showing the inventory on hand and usually showing the inventory on order and some sales history for the products that are covered in the stock status reports.

stoke-writing
A CAD picture-generating technique. A beam is moved simultaneously in X and Y directions along a curve or straight path.

stop
A mechanical constraint or limit on some motion that can be set to stop the motion at a desired point.

stop bit
In asynchronous transmission, the bit that indicates the end of a block of data.

storage
A computer-oriented medium in which data is retained. Primary (main) storage is the internal storage area where the data and program instructions are retained for active use in the system (normally, core storage). Auxiliary or external storage is for less-active data, and the media may include magnetic tape, disk, or drum. The physical repository of all information relating to products designed on a CAD/CAM system. It is typically in the form of a magnetic tape or disk. Also called memory.

storage capacity
The amount of data that can be contained in a storage device or main memory, usually expressed in terms of "K" (1K = 1024 bytes). If the storage capacity of a computer is 16K,

the capacity of the computer is said to be 16,384 characters (bytes).

storage factor
Reciprocal of dissipation factor or loss tangent. Also called quality factor.

storage fragmentation
A phenomenon observed in systems using dynamic allocation of core storage that permit variability in the amount allocated.

storage life
Period of time during which resin or adhesive can be stored under specified temperature conditions and remain suitable for use. Storage life is sometimes called shelf life.

storage media
Materials on which data may be recorded.

storage protection
Methods of preventing access to storage; preventing the loss of stored data during power interrupts. Synonymous with memory protection.

storage tube
A CRT that retains an image for a considerable period of time without redrawing.

store and forward
The handling of messages or packets in a network by accepting the messages or packets completely into storage and then sending them forward to the next center. Also, the capture of transaction data on magnetic media for subsequent batch input to a computer.

stored-field read-only alterable memory
Memories generally read only in nature that may be reprogrammed in a limited fashion.

stored-program computer
A computer controlled by internally stored instructions that are treated as though they were data and that can subsequently be executed. A computer that has the ability to store, to refer to, and to modify instructions to direct its step-by-step operations.

stored-program numerical control
Usually the same as CNC. The distinguishing factor is an internal memory that can be altered by receiving new instructions.

stores control
Physical aspect of inventory management. Stores control considers how to enter materials into stores economically, keep control of their location, and pick material for issue.

stores ledger cards
Cards on which are maintained records of the material on hand and on order.

straight cut control system
A system that has feedrate control only along a single axis and, therefore, can control cutting action only along a path parallel to the linear (or circular) machine ways. Cannot coordinate two or more axes to produce true contouring. Also designated as *picture-frame* milling.

straight-line coding
A set of instructions in which there are no loops.

straightness error
Carriage straightness error—the motion of a carriage, designed for linear translation, perpendicular to the intended motion axis. For the purposes of this report, two types of straightness errors are defined: (1) Type M straightness, which is the movement one would measure with a stationary indicator reading against a perfect straightedge supported on the moving carriage and aligned with its motion axis and (2) Type F straightness, which is that movement measured by an indicator attached to the carriage reading against a fixed straightedge similarly aligned.

strain
A measure of the change in the size or shape of a body, referred to its original size or shape. Linear strain is the change per unit length of a linear dimension. True strain (or natural strain) is the natural logarithm of the ratio of the length at the moment of observation to the original gage length. Conventional strain is the linear strain referred to the original gage length. Shearing strain (or shear strain) is the change in angle (expressed in radians) between two lines originally at right angles. When the term strain is used alone, it usually refers to the linear strain in the direction of the applied stress.

strain aging
Aging induced by cold-working.

strain gage
A gage normally made up of fine wires that when cemented to a component can measure very small amounts of motion caused by flexing. Strain gages are utilized to measure strains and stresses in structural components. When cemented to an elastic material, strain gages can act as force sensors.

strain-gage rosette
Multiple strain gages cemented in a two- or three-dimensional geometric pattern, such

that independent measurements of the strain on each can be combined to yield a vector measurement of strain or force.

strain hardening
An increase in hardness and strength caused by plastic deformation at temperature below the recrystallization range.

stratified language
A language that cannot be used as its own metalanguage, for example, FORTRAN.

strength
The maximum total weight that can be applied to the end of the robot arm, without sacrifice of any of the applicable published specifications of the robot.

stress
Force per unit area, often thought of as force acting through a small area within a plane. It can be divided into components, normal and parallel to the plane, called normal stress and shear stress, respectively. True stress denotes the stress where force and area are measured at the same time. Conventional stress, as applied to tension and compression tests, is force divided by the original area. Nominal stress is the stress computed by simple elasticity formula, ignoring stress raisers, and disregarding plastic flow. In a notch-bend test, for example, it is bending moment divided by minimum section modulus.

stress-corrosion cracking
Failure of a material due to the combined effects of corrosion and stress. Generally, stress-corrosion cracking refers to the phenomenon by which stress increases the corrosion rate.

stress relieving
Heating to a suitable temperature, holding long enough to reduce residual stresses and then cooling slowly enough to minimize the development of new residual stresses.

stretch
A CAD design/editing aid that enables the designer to automatically expand a displayed entity beyond its original. To elongate or shorten an object or region of the drawing.

stretch forming
The operation of forming sheet or strip by gripping two sides or ends (depending on the shape of unformed blank or required contour) until the material takes a set.

string
A connected sequence of characters of bits that is treated as a single data item. A linear sequence of entities, such as characters or physical elements, in CAD. An input class that returns a character string.

string device
A logical input device providing a character string as its input measure.

strobe light
An electronic flashtube that produces hundreds or thousands of flashes of light per second. It is used to freeze images of moving objects for vision systems.

stroke
An input class that returns a sequence of coordinate positions.

stroke device
A logical input device providing a sequence of points as its input measure.

stroke edge
In character recognition, the line of discontinuity between a side of a stroke and the background, obtained by averaging, over the length of the stroke, the irregularities resulting from the printing and detecting process.

structural retrieval
Reading back structures from a structure archive.

structure
A grouping of elements that provides the necessary information to describe all or part of an object, excluding workstation state and control information.

structure archiving
The process of storing structure definitions in a file for later retrieval and manipulation.

structured light
Illumination that is projected in a particular geometrical pattern. Sheets of light and other projective light configurations used to directly determine shape and/or range from the observed configuration that the projected line, circle, grid, and so forth makes as it intersects the object. The determination of three-dimensional characteristics of an object from the observed deflections that result when a plane or grid pattern of light is projected onto the object. Usually used in robot vision systems. Vision systems can use a structured light system: it projects two parallel planes of light that look like two lines across the face of the object being illuminated. The apparent shape and length of these lines, as seen by a digital television camera, provide cues that the computer uses to deduce the shape and position of the object under observation.

structured programming
Planning and implementing a program as a series of linked logical modules, paying special attention to documentation, testability, and program clarity so as to simplify program debugging and maintenance.

structured walkthrough
A formalized technique whereby one programmer describes the step-by-step functioning of his or her program to other programmers.

structure element
An item of the data that comprises a structure. Structure data includes output primitives, attribute selections, labels, application data, pick set specifications, transformation selections, and structure invocations.

structure identifier
The numeric identifier by which a structure can be referenced.

structure invocation
The process of executing another structure.

structure light source
A light source that is luminated through fiber optics to a fixed head. In turn, the head is mounted to a position table, x, y, and z to move the light beam.

structure network
The hierarchical relationships among a family of related structures.

structure query
Communication of information on the current structure element to the application program, at the request of the application program.

structure reference
A reference to a structure created by executing a structure or by posting it as a root.

structure-state list
A data structure containing information for all structures (lists of workstations to which they are posted) defined during the execution of a PHIGS program.

structure traversal
The process of stepping through a structure to produce an image.

stub
A dummy program module used to test a higher-level module.

stud welding
Welding a metal stud or similar part to another piece of metal, the heat being furnished by an arc between the two pieces just before pressure is applied.

stylus
Pencillike input device to move the cursor on the screen. Pressing the point of the stylus to the tablet or drawing stops and starts a function.

subassembly
Two or more parts joined together to form a unit that is only a part of a complete machine, structure, or other article.

subcontracting
The farming-out of manufacturing work to outside vendors. Usually utilized as a way of solving a capacity problem in a plant.

subfigure
A part or a design element that may be extracted from a CAD library and inserted intact into another part displayed on the CRT. User-defined symbols that are used repetitively in a drawing.

subfigure definition entity
A structure entity that permits a single definition of a detail to be utilized in multiple instances.

subfigure instance entity
A structure entity that specifies an occurrence of the subfigure definition.

subnet
A collection of hardware and software interconnecting two or more host computers. A MAP-compatible, bridge-accessible subnetwork. Logically, both the backbone and the subnetwork look like one network.

suboptimization
A term describing a problem solution that is best from a narrow point of view but not from a higher or overall company point of view. For example, a department manager who would not work his or her department overtime in order to minimize the department's costs may be doing so at the expense of overall company profitability.

subordinate entity switch
A portion of the status number field of the directory entry of an entity. An entity is subordinate if it is an element of a geometric or annotative entity structure or is a member of a logical relationship structure. The terms subordinate and dependent are equivalent within this document.

subpixel resolution
Any technique that results in a measurement with a resolution (detection of change) less than one pixel.

subplan
A plan to solve a portion of the problem.

subproblems
The set of secondary problems that must be solved to solve the original problem.

subroutine
A series of computer instructions to perform a specific task for many other routines. It is distinguishable from a main routine in that it requires as one of its parameters a location specifying where to return to the main program after its function has been accomplished.

substitution error
An error seen in a misencodation, misread or human operator error, where characters that were to be entered are substituted with erroneous information.

subsurface discontinuity
Any defect that does not open onto the surface of the part in which it exists.

subsystem
An assemblage of elements within a system. A subsystem is often identified at a level of definition within a system at which an independent development effort is initiated. However, in its strictest interpretation, any assemblage of elements within a system may be referred to as a subsystem.

subtask
A sequence of jobs representing a meaningful step in a project.

summarized bill of material
A form of multilevel bill of material that lists all the parts and their quantities required in a given product structure. Unlike the indented bill of material, it does not list the levels of assembly and lists a component only once for the total quantity used.

summary records
Prepared as a result of totals accumulated by a punched-card accounting machine. The totals are automatically punched into unit records by the summary punch machine. Summary records can subsequently be used to prepare other reports that do not require the detail that provided the basis for the totals.

superalloy
An alloy developed for very-high-temperature service, where relatively high stresses (tensile, thermal, vibrator, and shock) are encountered and where oxidation resistance is frequently required.

superbill
A type of planning bill, located at the top level in the structure, that ties together various modular bills (and possibly a common-parts bill) to define an entire product or product family. The "quantity per" relationship of superbill to modules represents the forecasted percentage popularity of each module. The master scheduled quantities of the superbill explode to create requirements for the modules that also are master scheduled (cf. planning bill, modular bill, common-parts bill).

supercooling
Cooling below the temperature at which an equilibrium-phase transformation can take place, without actually obtaining the transformation.

superfinishing
A form of honing in which the abrasive stones are spring supported. Essentially the same as lapping except that the operation results in a higher degree of surface finish.

superheating
Heating a phase above a temperature at which an equilibrium can exist between it and another phase having more internal energy, without obtaining the high-energy phase. Heating molten metal above the normal casting temperature so as to obtain more complete refining or greater fluidity.

supervisor
The part of an operating system that coordinates the use of resources and maintains the flow of CPU operations, rather than processing data to produce results.

supervisory control
A control scheme whereby a person or computer monitors and intermittently reprograms, sets subgoals, or adjusts control parameters of a lower-level automatic controller, while the lower-level controller performs the control task continuously in real time.

supervisory-controlled robot
A robot incorporating a hierarchical control scheme whereby a device having sensors, actuators, and a computer, and capable of autonomous decision making and control over short periods and restricted conditions is remotely monitored and intermittently operated directly or reprogrammed by a person.

supervisory program
A computer program, usually part of an operating system, that controls the execution of

other computer programs and regulates the flow of words in a data-processing system.

supination
Orientation or motion toward a position with the front, or unprotected side, facing up or exposed.

suppression device
A unit that attenuates the high-intensity electrical noise caused by inductive loads operated through hard contacts.

surface
The visible outside portion of a solid object. Like a two-dimensional plane. It is an area having a thickness considered for practical purposes to be zero.

surface code
Group technology code designed to classify the details of a machining surface.

surface machining
Automatic generation of NC tool paths to cut 3-D shapes. Both the tool paths and the shapes may be constructed using the mechanical design capabilities of a CAD/CAM system. The ability to output three-, four-, and 5-axis NC toolpaths using 3-D surface definition capabilities, for example, ruled surfaces, tabulated cylinders, and surfaces of revolution.

surface-mount technology
An old process that has returned after 20 years to become the new darling of printed circuit board packaging. SMT involves the attachment of components to one or both sides of a PCB, without the use of through-holes for contact leads. Developed for dense electronic packaging where space limitations dictate, components are smaller, and board preparation costs are lower, SMT offers a variety of advantages over conventional board design. Space savings translate to shorter interconnection paths and correspondingly better high-frequency operation. Productivity is improved as well with the use of automated equipment to populate the boards. Surface-mountable components, leaded or leadless, are attached directly to the surface of a substrate with or without adhesive and are reflow soldered by means of wave, infrared reflow, vapor-phase reflow, or other methods.

surface of revolution
Rotation of a curve around an axis through a specified angle.

surface of revolution entity
A geometric entity consisting of a surface that is generated by rotating a curve, called the generatrix, about an axis, called the axis of rotation.

surface resistivity
A measure of the ability of surface of a dielectric material to resist flow of electric current. Ratio of potential gradient parallel to current along surface of material to current per unit width of surface. Together with volume resistivity, surface resistivity is often used to check purity of an insulating material during development and its uniformity during processing.

surge
A transient variation in the current and/or potential at a point in the circuit.

surge method
Inspection by first using a high surge of magnetizing force, followed by a reduced magnetic field during application of a powdered ferromagnetic inspection medium.

swap
In systems with time-sharing, to write the main storage image of a job to auxiliary storage and read the image of another job into main storage.

swapping
In systems with time-sharing, a process that writes a job's main storage image into main storage.

swarf cut
The removal of metal in such a manner as to generate a cutter path, whose axis is constantly changing in relation to the surface of the part.

swedging
The operation of forming or shaping by a single blow or squeeze of a die.

swing
The rotation about the center line of the robot.

switch
A point in a programming routine at which two courses of action are possible, the correct one being determined by conditions specified by the programmer; a branch point.

switch control
Control of a machine by a person through movement of a switch to one or two or a small number of positions. The device used for such control.

switch core
A core in which the magnetic material generally has a high residual flux density and a high ratio of residual to saturated flux density, with a threshold value of magnetizing force below which switching does not occur.

switched line
A telephone line that is connected to the switched telephone network.

switched network
A multipoint network that has circuit-switching capabilities. The telephone network is a switched network, as are telex and TWX.

switching
The action of turning on and off a device.

symbol
Any recognizable sign, mark, shape, or pattern used as a building block for designing meaningful structures. A set of primitive graphic entities (line, point, arc, circle, text, and others) that form a construction that can be expressed as one unit and assigned a meaning. Symbols may be combined or nested to form larger symbols and/or drawings. They can be as complex as an entire PC board or as simple as a single element, such as a pad. Symbols are commonly used to represent physical things. For example, a particular graphic shape may be used to represent a complete device or a certain kind of electrical component in a schematic. To simplify the preparation of drawings of piping systems and flow diagrams, standard symbols are used to represent various types of fittings and components in common use. Symbols are also basic units in a language. The recognizable sequence of characters END may inform a compiler that the route it is compiling is completed. A combination of characters and check characters, as required, that form a complete scannable entity. Cluster of objects or primitives that some programs treat as an entity, rather than a collection of entities.

symbol density
The number of characters per linear inch.

symbolic
Relating to the substitution of abstract representations (symbols) for concrete objects.

symbolic address
An address expressed in symbols that are convenient to the programmer.

symbolic coding
Broadly, any coding system in which symbols other than actual machine operations and address are used.

symbolic control
Pertaining to control by communication of discrete alphanumeric or pictorial symbols that are not physically isomorphic, with the variables being controlled, usually by a human operator.

symbolic description
Noniconic scene descriptions, such as graph representations.

symbolic inference
The process by which lines of reasoning are formed, for example, syllogisms and other common ways of reasoning step-by-step from premises. In the real world, knowledge and data—premises—are often inexact. Thus, some inferences procedures can use degrees of uncertainty in their inference making. In an expert system, the inference subsystem works with the knowledge in the knowledge base. The inference subsystem in an expert system is one of three necessary to achieve expert performance; the other two subsystems are the knowledge base management subsystem and the human interface subsystem.

symbolic language
A language that is convenient for the programmer because it uses mnemonic terms that are easy to remember. Once the program has been written in symbolic language, it must be converted to absolute language using a *processor program*

symbolic name
A data field identifier synonymous with symbolic address. The programmer creates symbolic names; the computer changes these symbolic names into storage addresses. FORTRAN and BASIC call the symbolic name a variable.

symbolic processing
Using computers to manipulate symbols, in contrast to traditional numeric processing. Symbolic processing is the basis of AI programming.

symbol length
The length of the symbol measured from the beginning of the quiet area to the start character to the end of the quiet area adjacent to a stop character.

symbol table
A table of labels and their corresponding numeric values.

symptom binding
In artificial intelligence, the association of a symptom to a problem domain.

symptom set
In artificial intelligence, the set of all symptoms relating to a given development effort. System may be referred to as a subsystem.

sync character
A character of defined bit pattern that is used by the receiving terminal to adjust its clock and achieve synchronization.

synchro
A shaft encoder based upon differential inductive coupling between an energized rotor coil and field coils positioned at different shaft angles.

synchronization
Process of adjusting a receiving terminal's clock to match the transmitting terminal's clock.

synchronous
Happening at the same time. Having the same period between movements or occurrences. Fixed-rate transmission of bits of data, synchronized by the same clock signal for both the sender and the receiver.

synchronous communications binary
A set of operating procedures for synchronous transmission used in IBM teleprocessing networks (BISYNC). With BISYNC, some system batch terminals automatically perform error checking on all incoming data and request retransmission of a message whenever it is not received exactly as sent. As a transmitting terminal, the system automatically retransmits messages when they are not accurately received by the remote station. Because of the reliability of data transmissions using binary synchronous methods, it becomes economical to collect and store large amounts of data at the processor using either cassettes or a mass-memory subsystem and to later transmit the data to computers or terminals, including other systems.

synchronous computer
A computer in which each event or the performing of any basic operation is constrained to start on signals from a clock and usually to keep in step with them.

synchronous data link control
A bit-oriented protocol for managing the flow of information on a data communications link. Supports full- or half-duplex, point-to-point, multipoint, and loop communications using synchronous data transmission techniques. Employs a cyclic redundancy error-check algorithm.

synchronous shift
Shift register that uses a clock for timing of a system operation and where one state change per clock pulse occurs.

synchronous transmission
A mode of data communications by which the bit stream and character stream are slaved to accurately synchronized clocks at the receiving and transmitting stations. Start and stop pulses are not required with each character.

syntactic
Pertaining to the form or structure of a symbolic expression in artificial intelligence, as opposed to its meaning or significance.

syntactic analysis
Recognizing images by a *parsing* process as being built up of primitive elements.

syntax
A set of rules describing the structure of statements allowed in a computer language. To make grammatical sense, commands and routines must be written in conformity to these rules. The structure of a computer command language, that is, the English-sentence structure of a CAD/CAM command language, for example, verb, noun, modifiers.

syntax error
A system response to a mistake in instruction, such as a transposition of characters or an omission of a character or word.

system
A collection of units combined to work as a larger integrated unit having the capabilities of all the separate units. An organized collection of parts or procedures united by regulated interaction and interconnected to perform a function. A combination of two or more sets, generally physically separated when in operation, and other such units, assemblies, and basic parts necessary to perform an operational function or functions. An arrangement of CAD/CAM data processing, memory, display and plotting modules, coupled with appropriate software to achieve specific objec-

tives. The term CAD/CAM system implies both hardware and software.

system control programming
Vendor-supplied programming that is fundamental to the operation and maintenance of the system. It serves as an interface with program products and user programs and is available without additional charge.

system degradation
The slowdown in execution of tasks that can occur when the CPU is given a task requiring a large amount of computation (i.e., a computer-bound task) or when two or more tasks require the same physical device at the same time (i.e., an input/output-bound task).

system design
The specification of the working relations between all the parts of a system in terms of their characteristic actions.

system element
A constituent part of a design structure that comprises capabilities and the technology disciplines by which those capabilities will be achieved.

system element relationship
A data or control link between two system elements.

system generation
The process of using an operating system to assemble and link together all of the parts that constitute another operating system.

system library
A collection of data sets in which the various parts of an operating system are stored.

system log
A data set in which job-related information, operational data, descriptions of unusual occurrences, commands, and messages to or from the operator may be stored.

system manager
Resides in all MAP nodes and is responsible for interfacing with network management to provide or to execute NM functions. The level of functionality resident with an SM is dependent on the class of node and the speciality resident within an SM is dependent on the class of node and the specific vendor implementation.

system part set
Set of all part numbers that can be produced by a given flexible manufacturing system without significant changes to its stations, tooling, part programs, and so forth.

system part set tool set
Set of all tools required to produce a given system part set on a given flexible manufacturing system.

system programmer
A programmer who plans, generates, maintains, extends, and controls the use of an operating system with the aim of improving the overall productivity of an installation. Also a programmer who designs programming systems and other applications.

system reliability
The probability that a system will perform its specified task properly under stated conditions of environment.

system requirements definition phase
The portion of system development whose purpose is to investigate a company, or part of a company, in sufficient depth to develop a firm business proposition involving a changed method of operation. It results in a statement of the functional requirements of new systems.

system resource
Any facility of the computing system that may be allocated to a task.

systems analyst
An individual who defines applications problems, determines system specifications, recommends equipment changes, designs data-processing procedures, and devises data verification methods. Prepares block diagrams and record layouts from which the programmer prepares flowcharts. May assist with or supervise the preparation of flowcharts.

systems and support software
The variety of software including assemblers, compilers, subroutine libraries, operating systems, application programs, and so forth.

systems approach
Looking at the overall situation, rather than the narrow implications of the task at hand; particularly, looking for interrelationships between the task at hand and other related functions. A general term for reviewing all implications of a condition or group of conditions, rather than the narrow implications of the problem at hand.

system ship set
That subset of a ship set that can be produced in a given FMS.

systems test

The running of the whole system against test data. A complete simulation of the actual running system for purposes of testing the adequacy of the system.

system testing

The verification and/or validation of the system to its initial objectives. System testing is a verification process when it is done in a simulated environment; it is a validation process when it is performed in a live environment.

T

table
A collection of data, each item being uniquely identified either by some label or by its relative position.

table lookup
The process of using a table of data stored in main memory during the running of the program to obtain a function value.

tablet
An input device on which a designer can ditize coordinate data or enter commands into a CAD/CAM system by means of an electronic pen. See also data tablet.

tablet mode
In CAD, entering location information by means of a graphics tablet, rather than with a keyboard. This allows drawings to be traced.

tab sequential format
A tape format that allows a *tab* code to be substituted in a given program word when the character is the same as that in the preceding word. Means of identifying a word by the number of tab characters preceding the word in a block. The first character of each word is a tab character. Words must be presented in a specific order but all characters in a word, except the tab character, may be omitted when the command represented by that word is not desired.

tabulated cylinder
Translation of a curve along a direction line with upper and lower limits on the distance of translation.

tabulated cylinder entity
A geometric entity that is a surface generated by moving a line parallel to itself along a space curve called the generatrix.

tabulator
A machine that processes punched cards.

tachometer
A speed-measuring instrument generally used to determine revolutions per minute. Tachometers may be used in conjunction with contouring systems as a supplemental control for governing feedrates. A rotational velocity sensor. A device capable of sensing the speed at which a shaft is rotating.

tactile
Perceived by the touch or having the sense of touch.

tactile feedback
The over-center sensation of a snap-action switch, as felt by the operator.

tactile sensors
Sensors that respond to contact forces that arise between themselves and solid objects. The object must actually be touched, unlike proximity sensors. Tactile sensors can be classified into two groups: touch sensors and stress sensors. Touch sensors indicate contact; stress sensors indicate the magnitude of the contact forces. The most common touch sensor is the microswitch, while the most common stress sensor is the strain gage. A device, normally associated with the hand or gripper part of an industrial robot, which senses physical contact with an object, thus giving an industrial robot an artificial sense of touch. A transducer that is sensitive to touch.

tag
To automatically attach an identifying mark (number or name) to a design element displayed on the CRT. Tags label geometric entities for easy identification.

tap density
Apparent density of a powder obtained by measuring volume in a receptacle that is tapped or vibrated during loading in a specified manner.

tape drive
Transmission of power from an actuator to a remote mechanism by means of flexible tape and pulleys. A mechanism for controlling the movement of magnetic tape past a reading or writing head.

tape punch
A peripheral device that generates (punches) holes in a paper tape to produce a hard copy of robot controller memory contents.

tape reader
A peripheral device for converting information stored on punched paper tape to electrical signals for entry into computer or robot controller memory.

tape row
That portion of tape on a line perpendicular to the reference edge, on which all binary characters may be either recorded or sensed simultaneously.

taper tap
A tap with a chamber of seven to nine threads in length.

tapping
The operation of cutting a screw thread into a round hole.

target
In image pickup tubes, a structure employing a storage surface that is scanned by an electron beam to generate an output signal corresponding to a charge density pattern stored thereon. The charge density pattern is usually created by light from an illuminated scene.

target language
A language into which statements are translated.

target program
A computer program in a target language that has been translated from a source language.

tarnish
Surface discoloration of a metal caused by formation of a thin of corrosion product.

task
A specific project that can be executed by a CAD/CAM software program. A specific portion of memory assigned to the uses for executing that project.

task-oriented language
Programming language for describing what the effect of robot action should be. To be contrasted with manipulator-oriented languages for describing exactly where a robot's arm and gripper should go and when.

TC-APT
The programming language TC-APT for minicomputers. It offers full APT capabilities, 2- or 3-D symbol format, automatic synonym assignments, extended macro technique, and a system macro library.

teach
To program a manipulator arm by guiding it through a series of points or in a motion pattern that is recorded for subsequent automatic action by the manipulator. To move a robot to or through a series of points to be stored for the robot to perform its intended task.

teach control
A hand-held control unit usually connected by a cable to the control system with which a robot can be programmed or moved.

teaching
The generation and storage of a series of positional data points, effected by moving the robot arm through a path of intended motions.

Teach Pendant
CIMROC Teach Pendant

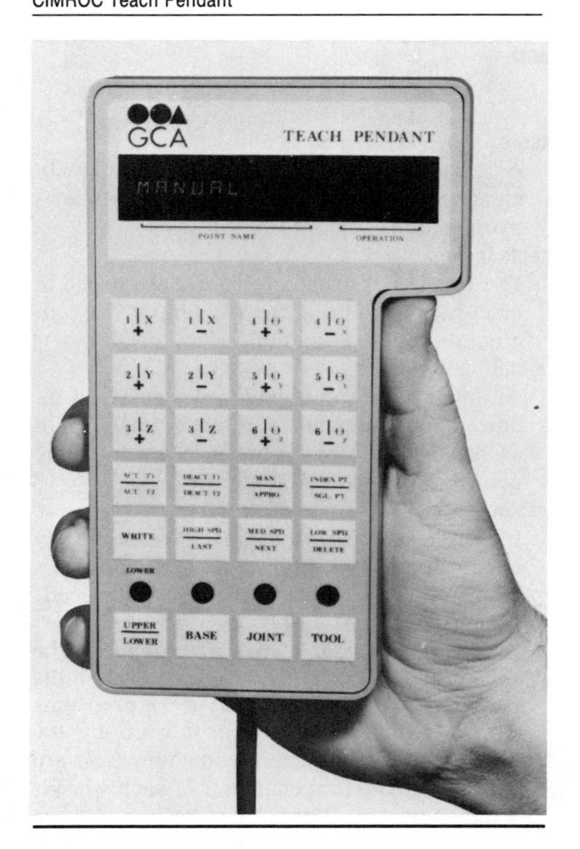

teaching interface
The physical configuration of the machine or the devices by which a human operator teaches a machine.

teach pendant
The control box that an operator uses to guide a robot through the motions of its tasks. The motions are recorded by the control memory for future playback.

teach restrict
A facility whereby the speed of movement of a robot during teaching, which during normal operation would be considered dangerous, is restricted to a safe speed.

tea laser
An acronym for transversely excited atmospheric laser. This gas laser uses a transverse flow of gas and operates at higher pressures than other gas lasers, generally near atmo-

spheric pressure. The result is a higher-energy beam.

teardown

The opposite of setup, taking place at the end of an operation: dismantling of assembly jigs, cleaning of vats or machines, and so on.

tear resistance

A measure of the ability of sheet or film materials to resist tearing.

technical and office protocol (TOP)

Details industry's approach to data communications within a technical and office environment, providing a standard framework for functional distributed processing across local and wide area networks. The principal goal of technical and office protocol (TOP) is to establish a uniform set of industry standards for technical and office data communications supporting computers and other programmable devices in a multivendor environment. TOP will accomplish several major architectural and operational objectives as follows: (1) *architectural*—allow for the interconnection of multiple office local area networks (LANs), and for connection to wide area networks (WANs), and digital private branch exchanges (PBXs) for long-distance communications; specify existing or emerging national, international, and/or industry standards; be based upon the ISO OSI reference model; and (2) *operational*—provide an environment for multiple vendors to participate on a standard user office communication network; facilitate free and easy data access and data interchange in a multivendor environment; lower office systems costs by reducing the need for multiple cables and customized networking software; improve flexibility and adaptability of production systems to meet changing demands.

technological implication

An implied constraint relating to use of a particular technology.

technology

A technical method of achieving a practical purpose.

technology discipline

A well-defined area of technology.

technology-sensitive industrial sector

An industry whose productivity, efficiency, and competitiveness—particularly in the world marketplace—is or is likely to be de-

TOP

Summary of TOP Model, Protocols, and Options. *Source*: Boeing Company.

	Layer and Function	Document Number (ISO)	Abbreviated Title (ISO)	Specific Options
7	Application	8571	File Transfer, Access, and Management (FTAM)	File Transfer
6	Presentation	-	-	Binary and ASCII*
5	Session	8327	Session	Full-Duplex, Basic Combined Subset and Session Kernel
4	Transport	8073	Transport	Class 4
3	Network	8473	Connectionless Network	Connectionless and Subnetwork Dependent Convergence Protocol for X.25 Wide Area Networks
2	Data Link	8802/2	Logical Link Control	Type 1, Class 1
1	Physical	8802/3	CSMA/CD**	CSMA/CD Medium Access Control, 10Base 5

* ASCII American Standard Code for Information Interchange
**CSMA/CD Carrier Sense Multiple Access With Collision Detection

pendent upon the utilization of more advanced manufacturing processes and methods.

telecommunication lines
Telephone and other communication lines that are used to transmit messages from one location to another.

telecommunications
Data transmission between a computing system and remotely located devices by means of a unit that performs the necessary format conversion and controls the rate of transmission.

teleoperator
A master-slave device that produces movements identical to or in direct proportion to actions or motions of a remotely located human operator. The device communicates certain feedback information such as position, forces, and the like to the human operator. A device having sensors and actuators for mobility and/or manipulation, remotely controlled by a human operator. A teleoperator can extend the human's sensory-motor function to remote or hazardous environments.

telephoto lens
A compound lens constructed so that its overall length, from rear focal point to front element, is less than its effective focal length.

teleprinter
A typewriterlike device capable of receiving or sending data in a communications system.

teleprocessing
The processing of data that is received from or sent to remote locations by way of telecommunication lines. Such systems are essential to hook up remote terminals or connect geographically separated computers.

Teletype
A registered trademark of a type of input/output device originally manufactured by Teletype Corporation; now often incorrectly used as a generic term to indicate any similar piece of equipment.

teletypewriter exchange service
A public teletypewriter exchange (switched) service in the United States and Canada, formerly belonging to AT&T but now owned by the Western Union Telegraph Company. Both Baudot and ASCII-coded machines are used.

telewriter
Typewriter-style keyboard device used to enter commands or to print out system messages.

telex
An automatic teleprinter exchange service provided worldwide by Western Union, similar to teletypewriter exchange service. Telex uses the public telegraph network. Only Baudot equipment is provided; business machines may also be used. Abbreviated TLX.

TEM
Acronym for Transverse Electromagnetic Mode, which is a means of describing the cross-sectional shape of a laser beam. Typical TEM modes for metalworking lasers include a Gaussian-shaped beam energy distribution that is the standard shape used in industry, a bimodal-shaped beam, and a ring-shaped distribution.

temper
In heat treatment, reheating hardened steel or hardened cast iron to some temperature below the eutectoid temperature for the purpose of decreasing the hardness and increasing the toughness. The process is also sometimes applied to normalized steel. In tool steels, "temper" is sometimes used, but inadvisedly, to denote the carbon content. In nonferrous alloys and in some ferrous alloys (steels that cannot be hardened by heat treatment), the hardness and strength produced by mechanical or thermal treatment, or both, and characterized by a certain structure, mechanical properties, or reduction in area during cold-working.

temperature resistance
Relationship between the temperature of a material and its electrical resistance, expressed by a multivalue table, by a graph, by a calculated temperature coefficient, or by calculated values for constants in a standard mathematical equation. Provides information needed for design and use of resistance heating elements and precision resistors in electrical and electronic circuits.

temper brittleness
Brittleness that results when certain steels are held within, or are cooled slowly through, a certain range of temperature below the transformation range. The brittleness is revealed by notched-bar impact tests at or below room temperature.

tempering
Reheating a quench-hardened or normalized ferrous alloy to a temperature below the trans-

formation range and then cooling at any rate desired.

template
Prototype model that can be used directly to match to image characteristics for object recognition or inspection. A prototype model or structure that can be used for sentence interpretation. The pattern of a standard commonly used component or part that serves as a design aid. Once created, it can be subsequently traced instead of redrawn whenever needed. The CAD equivalent of a designer's template might be a standard part in the data base library that can be retrieved and inserted intact into an emerging drawing on the CRT.

template matching
Correlating an object template with an observed image field—usually performed at the pixel level. The comparison of a picture or other data set against a stored pattern or template. The technique of comparing the image of a test object with that of a standard on a pixel-by-pixel basis for inspection or recognition purposes.

temporary segment
A nameless segment having no segment attributes. The image defined by a temporary segment remains visible only as long as information is being added to the displayed picture. A temporary segment's image disappears as soon as a new-frame action occurs.

temporary storage
Memory locations for storing immediate and partial results obtained during the execution of a program on the system.

tens complement
The radix complement in the decimal-numeration system.

tensile strength
In tensile testing, the ratio of maximum load to original cross-sectional area. Also called ultimate strength. The ultimate strength of a material subjected to tensile loading. The maximum stress developed in a material in a tension test.

tera (T)
Ten to the twelfth power (10^{12}), 1,000,000,000,000 in decimal notation. When referring to storage capacity, two to the fortieth power (2^{40}), 1,099,511,627,776 in decimal notation.

term
The smallest part of an expression that can be assigned a value.

terminal
A device equipped with a keyboard (or other input device) and an output device (e.g., display or printer) that is connected to a computer system for the input and/or output of data. A terminal may be as simple as a telephone or as complex as a small computer. Terminals are generally used for online systems. Any fitting attached to a circuit or device for convenience in making electrical connections. An interface device containing a cathode-ray tube and a keyboard for communication with a computer or robot controller.

terminal job
In systems with time-sharing, the processing done on behalf of one terminal user from log-on to log-off.

terminal user
In systems with time-sharing, anyone who is eligible to log-on.

terminate section
The final section of an IGES file, indicating the sizes of each of the preceding file sections.

termination
The load connected to the output end of a transmission line. The provisions for ending a transmission line and connecting to a bus bar or other terminating device.

ternary
Pertaining to a selection, choice, or condition that has three possible different values or states. Pertaining to a fixed-radix numeration system having a radix of 3.

ternary incremental representation
Incremental representation in which the value of an increment is rounded to one of three values, plus or minus one quantum or zero.

tesselation
The pixel pattern (as projected onto the object). Most tesselations are square (square pixels) and some are rectangular (rectangular pixels).

test case
A specific set of test data and associated detailed procedures developed for a particular test objective.

test data generator
Software for forming test data files holding desired or randomly generated values in nominated fields of nominated records.

test tool set
A collection of test tools used to perform testing.

testware design
A set of specifications used to construct testware for the system.

text
The information portion of a transmitted message, as contrasted with the header, check characters, and end-of-text characters. An output primitive consisting of a character string.

text adaptation
An attribute of text that specifies which text attributes may vary in determining how the text fits into the text-extent rectangle. The aspects involved are character height, character spacing, and character expansion factor.

text alignment
An attribute of text that specifies the mode of justification. This attribute has two components, one for horizontal justification and one for vertical justification.

text bundle table
A table associating specific values for all workstation-dependent aspects of a text primitive with a text bundle index. The table contains entries consisting of text font and precision, character expansion factor, character spacing, and color index.

text editing
Specific flexible editing facilities that have been designed into a computer program to permit the original keyboarding of textual copy, without regard for the eventual format or medium for publication. Once the copy has been placed in computer storage, it can be edited and justified easily and quickly into any required column width and for any specified type font, merely by specifying the format required.

text editor
An operating system program used to create and modify text files on the system.

text-extent rectangle
An attribute of text that defines a rectangle delimiting the extent of the text string.

text file
A file stored in the system in text format that can be printed and edited online as required.

text font
An attribute of text that indicates certain aspects of a visible character such as style, typeface, boldface, italic, and polygonal versus outline. Sets of typefaces of various styles and sizes. In CAD, fonts are used to create text for drawings, special characters such as Greek letters, and mathematical symbols.

text font and precision
A pair of attributes or a single two-component attribute that defines text font and precision. Together, the two components determine the shape of the characters being output. In addition, precision describes the fidelity with which the other text aspects match those requested by an application program.

text-font definition entity
The entity used to define the appearance of characters in a text font. A character is defined by pairing its character code with a sequence of display strokes and positional information.

text path
An attribute of text that indicates the direction, relative to the character-up vector, in which the text is drawn within the text plane.

text position
A point defined within the text primitive that together with the text reference points, defines the text plane. The text position also determines, together with text alignment, the location of the text string.

text precision
A workstation-dependent attribute of text that indicates the accuracy of interpretation of positioning and clipping a text primitive.

text preference points
Two points in the text function that together with the text position determine the text plane.

textual-numeric data base
This type of source data base contains records made up of a combination of textual and numeric data elements. See data base, source data base, numeric data base, and full-text data base.

texture
The degree of smoothness of the surface of an object. The texture of an object's surface can affect the way in which light is reflected from it and, therefore, the image brightness can be affected. A local variation in pixel values that repeats in a regular or random way across a portion of an image or object. A crosshatch, speckle, or other pattern that can be used to

fill an area in a drawing or diagram. On some graphics programs that allow on-screen drawing, brushes can be set to paint out a swatch in the selected texture. In others, the texture must be added afterward. In monochrome graphics (with no color), texture is the way of distinguishing surfaces.

t function
A code identifying a tool select command on a program tape.

thematic map
A map specifically designed to communicate geographic concepts such as the distribution of densities, relative magnitude, gradients, spatial relationships, movements, and all the required interrelationships and aspects among the distributional characteristics of the earth's phenomena. CAD systems allow the quick assignment and identification of both graphic and nongraphic properties, thus facilitating the use of thematic maps.

theorem
A proposition, or statement, to be proved based on a given set of premises.

theorem proving
A problem-solving approach in which a hypothesized conclusion (theorem) is validated using deductive logic.

thermal absorptivity
Fraction of the heat impinging on a body that is absorbed.

thermal conductivity
Rate of heat flow in a homogeneous material, under steady conditions, through unit area, per unit temperature gradient in direction perpendicular to the area. Thermal conductivity is usually expressed in English units as Btu/ft^2/hr/°F for a thickness of 1 inch.

thermal diffusivity
Rate at which temperature diffuses through a material. Ratio of thermal conductivity to product of density and specific heat, commonly expressed in ft^2/hr.

thermal emissivity
Ratio of heat emitted by a body to heat emitted by a blackbody at same temperature.

thermal expansion
Takes place when heated or cooled materials undergo a reversible change in dimensions that depends on the original size of the body and the temperature range studied. In addition to the change in dimensions, a change in shape may occur; that is, the expansion or contraction may be anisotropic.

thermal fatigue
Fracture resulting from the presence of temperature gradients that vary with time in a manner to produce cyclic stresses in a structure.

thermal-insulating efficiency
Ratio of heat saved by an insulation to the heat that would be lost without insulation.

thermal reflectivity
Fraction of heat impinging on a body that is reflected.

thermal shock
The development of a steep temperature gradient and accompanying high stresses within a structure.

thermal transmittance
Rate at which heat is transmitted through a material by combined conduction, convention, and radiation. Overall coefficient of heat transfer. Term used particularly for textile fabrics and batting, where heat transfer between opposite surfaces is not confined to conduction. For solid materials, thermal conductivity is a measure of thermal transmittance.

thermistors
Thermally sensitive resistors that change in electrical resistance with variations in temperature.

thermit welding
Welding with heat produced by the reaction of aluminum with a metal oxide. Filler metal, if used, is obtained from the reduction of the appropriate oxide.

thermochromic display
A display device consisting of materials that change to different colors when heated to different temperatures.

thermocouple
A device for measuring temperatures, consisting of two dissimilar metals that produce an electromotive force roughly proportional to the temperature difference between their hot and cold junction ends.

thin-film memory
A main memory device that uses a thin film of metal as its storage medium. Information is stored by magnetizing the thin-film material.

third generation
In the NC field, the period of technology associated with integrated circuits. In the com-

puter field, the period of technology utilizing integrated circuits, core memory, and advanced programming concepts.

thrashing
In virtual storage systems, a condition in which the system can do little useful work because of excessive paging.

threat
One or more events that may lead to intentional or unintentional modification, destruction, or disclosure of data.

three-dimensional view
A view of an object expressing its length, height, and depth. Also called 3-D.

three-plus-one address instruction
An instruction that contains three address parts, the plus-one address being that of the instruction that is to be executed next, unless otherwise specified.

threshold function
A two-valued switching function of one or more not-necessarily Boolean arguments that take the value of one if a specified mathematical function of the arguments exceeds a given threshold value, and zero otherwise.

thresholding
Converting a gray-scale image with three or more gray levels into a binary image (i.e., reassigning pixel gray levels to only two values) by separating regions of an image based on pixel values above or below a chosen intensity level. The comparison of an element value, such as pixel brightness, against a set point value or threshold. All elements whose values are above threshold are set to the binary value 1. All elements below threshold are set to the binary value 0. The process of defining a specific intensity level. If the pixel's brightness is above the threshold level, it will appear white in the image. If the brightness is below the threshold level, it will appear black.

threshold responsive
A control type that responds to a change in light intensity.

through-beam
A method of object detection in which a source is mounted on one side and a detector on the other. The object is detected when it "breaks" on the beam.

throughput
The rate at which information is processed through a computer. The number of units of work performed by a CAD/CAM system or a workstation during a given period of time. A quantitative measure of system productivity.

throw
The number of circuits that each individual pole of a switch can control. Also, the movement of a contact from one stationary point to another.

thumbwheel switch
A rotating numeric switch used to input numeric information to a controller.

tic mark
A mark made by some drawing programs on each page to allow the alignment of individual sheets to form multiple-page images.

tie line
A private-line communications channel of the type provided by communications common carriers for linking two or more points together.

tight standard
A time standard that provides a qualified worker with sufficient time to do a defined amount of work of specified quality when following the prescribed method, working at normal pace and experiencing normal fatigue and delays.

tilt
Orientation of a view, as with a video camera, in elevation. Motion in the elevation direction. The angle at which an object is turned. It is normally measured in degrees.

time bucket
In an MRP system, refers to the number of days summarized into one columnar display. A weekly time bucket would contain all of the relevant planning data for an entire week. Weekly time buckets are considered to be the largest possible (at least in the near and medium term) to permit effective MRP.

time constant
Any of a number of parameters of a dynamic function that have units of time. Parameters that particularly characterize the temporal properties of a dynamic function, such as the period of a period function or the inverse of the initial slope of a first-order exponential response to a step.

time-division multiplexing
Sharing a single facility among several data paths by dividing up the channel capacity into time slices. Transmission is in successive frames, each consisting of one bit from each path.

time fence
A policy or guideline established to note where various restrictions or changes in operating procedures take place. For example, changes to the master production schedule can be accomplished easily beyond the cumulative lead time but can become increasingly more difficult to a point where changes should be resisted. Time fences can be used to define these points.

time formula
A collection of standard time data arranged in the form of an algebraic expression for determining the time standard for an operation.

time out
A set time period for waiting before a terminal/system performs some action. Typical uses include poll release (assumes terminal cannot respond) and automatic discount (assumes terminal cannot transmit further or there is a line problem of some sort). It is the most convenient control technique to prevent difficulties with one terminal in a network from bringing the whole net to a stop.

time phased
A reference to the fact that requirements for resources have been dated as to the need.

time-phased contract
Refers to the practice of showing requirements, scheduled receipts, the projected available balance, and planned order releases in their proper time relationship to each other to the vendor.

time-phased order point (TPOP)
MRP for independent-demand items. Gross requirements come from a forecast, not by way of explosion. This technique can be used to plan warehouse inventories as well as planning for service (repair) parts, since MRP logic can readily handle items with dependent demand, independent demand, or a combination of both. See distribution requirements planning.

time-phased planning
Planning future operations based on a series of related sequential time segments or buckets.

timer
In relay-panel hardware, an electromechanical device that can be wired and preset to control the operating interval of other devices. In PC, a time is internal to the processor, which is to say it is controlled by a user-programmed instruction. A timer instruction has greater capability than any hardware timer. Therefore, PC applications do not require hardware timers.

time series
A sequence of quantitative data assigned to specific moments in time. These data are usually studied with regard to their distribution in time.

time-sharing
A computer environment in which multiple users can use the computer virtually simultaneously by means of a program that time-allocates the use of computer resources among the users in a near-optimum manner. The use of a common CPU memory and processing capabilities by two or more CAD/CAM terminals to execute different tasks simultaneously.

time slice
An interval of time on the central processing unit allocated for use in performing a task. Once the interval has expired, CPU time is allocated to another task; thus, a task cannot monopolize CPU time beyond a fixed limit.

time standard
A predetermined time allowance for performing a given operation, normally based on a desired level of efficiency.

toggle
Pressing a command key repeatedly turns the function on and off.

token
Special bit patterns or packet that travels on a token access network (i.e., token bus, token ring networks) from node to node, giving the node that currently has possession of the token exclusive access to the network for transmitting its message. The token is passed on if the node has nothing to transmit; the token is "held" if the node initiates a transmission and released when the node is finished transmitting. In this manner, conflicts between nodes wishing to transmit at the same time on the single line can be resolved.

token bus
A network architecture based on a bus topology and using a token-based media access method.

token ring
Network architecture using a ring topology and a token-based media access method.

tolerance
A specified allowance for error from a desired or measured quantity. A term used in con-

junction with the generation of a nonlinear tool path comprised of discrete points. The tolerance controls the number of these points. In general, the term denotes the allowed variance from a given standard, that is, the acceptable range of data.

tool
An object that contacts the workpiece and removes material from it. A term used loosely to define something mounted on the end of the robot arm; for example, a hand, a simple gripper, or an arc-welding torch.

tool center point
The robot arm is the means of moving the tool center point (TCP) from one programmed point to another in space. The control directs the movement of the TCP in terms of direction speed and acceleration along a defined path between consecutive points. This method of control, termed a controlled-path system, utilizes the mathematical ability of a computer to give the operator coordinated control of the predefined TCP in a familiar coordinate system during the teaching operation. It also controls the TCP in terms of position, velocity, and acceleration between programmed points in the reply or automatic modes of operation. The TCP is predefined by the operator, who enters a tool length into the control. This represents the distance between the end of the robot arm and the TCP. In the controlled-path system the coordinates of the TCP are stored as x, y, and z coordinates in space and not as robot axis coordinates. During the automatic mode of operation, the position operation the position of the TCP and the orientation of the end-effector relative to the TCP are known at all times by the control system.

tool-change time
Period of time required to replace one tool (or set of tools) with another; includes time to position tool for removal and time to reposition new tool.

tool, consumable, perishable
A tool that has significant consumption and is inventoried as a material or supply item.

tool design
That division of mechanical design that specializes in the design of tools.

tool function
A tape command identifying a tool and calling for its selection. The address is normally a "T" word and may be used in conjunction with a turret or automatic tool changer.

tool gaging
Gaging performed on tools to determine if they are within specified tolerance.

tool-handling system
System used to access and store tools and aggregates of tools.

tooling
A set of required standard or special tools for production of a particular part, including jigs, fixtures, gages, cutting tools, and others. The definition specifically excludes machine tools.

tooling design
A set of specifications used to construct tools for the system.

tooling engineer-computer language
A software system for drafting, design, and NC workpiece program generation. It has a number of characteristics that are useful in the design and fabrication of progressive stamping dies.

tool life
The anticipated life of a tool. It is usually expressed as either the number of pieces the tool is expected to make before it wears out or as the number of hours of use anticipated.

tool movement time
Period of time between the performance of elementary operations in which a tool moves from one position to another; does not include tool-change time.

tool offset
Employed in order to avoid the time-consuming requirement for setting the depth of the tool to an exact dimension in a turret. The tool may be set to an approximate dimension, generally within 0.1 inch of the required setting, and manual switches can be used to make up the difference. Found extremely helpful with turrets on a lathe or adjusting the "Z" or depth motion of a turret drill having three axes of control.

tool order
The authorization to design and produce tooling.

tool path
Centerline of the tip of an NC cutting tool as it moves over a part produced on a CAD/CAM system. Tool paths can be created and displayed interactively or automatically by a

Top-Down
Block Diagram of Hierarchical Top-Down Approach

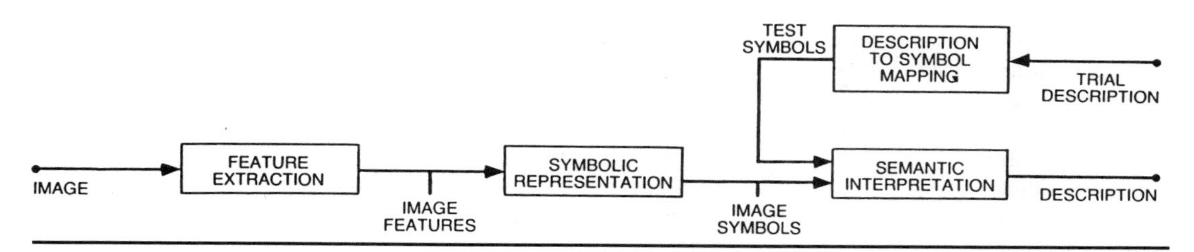

CAD/CAM system and reformatted into NC tapes by means of a postprocessor to guide or control machining equipment.

top-down approach
An approach in which the interpretation stage is guided in its analysis by trial or test descriptions of a scene. Sometimes referred to as *hypothesize and test.*

top-down logic
A problem-solving approach used in production systems, where production rules are employed to find a solution path by chaining backward from the goal.

top-down programming
The design and coding of computer programs using a hierarchical structure in which related functions are performed at each level of the structure.

topography
Variation in height (z axis, values) of elements or objects in a scene, usually inferred from changes in range values.

topology
An ideal TOP network is a hierarchical network consisting of various subnetworks interconnected to form a logical whole. A minimal TOP network is a collection of end systems attached to a physical medium to create a network segment. Network segments may be connected by repeaters, bridges, or intermediate systems to create a network. For example, the end systems on one floor create a network segment in a building. Each floor's network segment can be connected to create a building-wide network. Networks that provide direct data communications to end sys-

Topology
TOP Network Hierarchy Example. *Source*: Boeing Company.

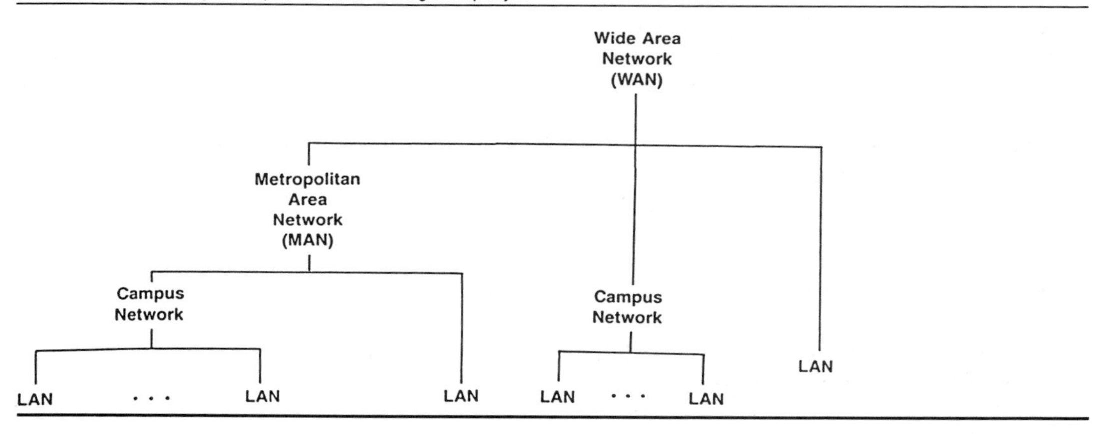

tems on a user's premises are commonly referred to as local area networks (LANs).

total-systems concept
The complete integration of all major operating systems within a business organization into the one functional organized system operating under the discipline of a data-processing facility.

total transmittance
Ratio of light transmitted by material to light incident on it. The ratio depends on spectral distribution of incident light energy.

total travel
The sum of pretravel and overtravel.

touch sensitive
Refers to the technology that enables a system to identify a point of contact on the screen by coordinates and transmit the information to a computer controller to be translated into some system action.

touch sensors
Used to obtain information associated with the contact between the finger(s) of a manipulator hand and objects in the workspace. They are normally much lighter than the hand and are sensitive to forces much smaller than those sensed by the wrist and contact sensors. Touch sensors may be mounted on the outer and inner surfaces of each finger. The outer sensor may be used to search for an object and possibly to determine its identity, position, and orientation. Outer sensors may also be used for sensing unexpected obstacles and stopping the manipulator before any damage can occur. The inner-mounted sensors may be used to obtain information about an object before it is acquired and about grasping forces and workpiece slippage during acquisition. Touch sensors may be classified into two types, binary and analog. A binary touch sensor is a contact device, such as a switch. Being binary, its output is easily incorporated into a computer controlling the manipulator. A simple binary touch sensor consists of two microswitches, one on the inner side of each finger. An analog touch sensor is a compliant device whose output is proportional to a local force. Analog touch sensors are usually mounted on the inner surface of the fingers to measure gripping force and to extract information about the object between the fingers.

toughness
Ability of a metal to absorb energy and deform plastically before fracturing. It is usually measured by the energy absorbed in a notch-impact test, but the area under the stress-strain curve in tensile testing is also a measure of toughness.

tpturn
A generalized two-axis turning postprocessor written in FORTRAN IV and compatible with the Manufacturing Software and Services' TOOLPATH NC processor or APT/ADAPT processors.

trace program
A computer program that performs a check on another computer program by exhibiting the sequence in which the instructions are executed and usually the result of executing the instruction.

tracer
Request upon a transportation line to trace a shipment for the purpose of expediting its movement or establishing delivery.

track
To move an object on screen using a light pen, mouse, or other cursor-positioning device. Also known as level or channel. A path parallel to the edge of the tape along which information may be stored by the presence or absence of holes or magnetized areas. The EIA standard 1-inch-wide tape has eight tracks. A channel on a direct-access device that contains data that can be read by a single reading head, without changing its position. It may refer to the track on a disk that rotates under a read head or the track on a magnetic stripe that is passed by a fixed read head. Network about which materials, tools, and/or gages travel.

track ball
A cursor-control device that looks like a ball held in an upside-down cup. Turning the ball moves the cursor, and some programs allow the ratio of turns to displacement to be set by the user. Normally, a track ball gives only information about desired movement but not absolute positioning.

tracking
Continuous position control response to a continuously changing input. Moving a predefined (tracking) symbol across the surface

Track Ball
Track Ball Graphic Input Device

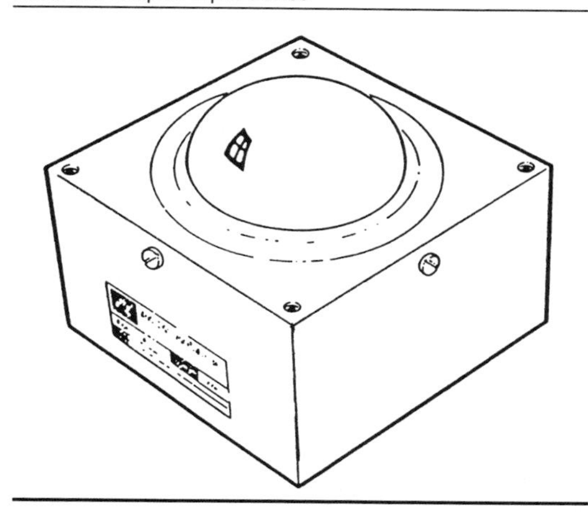

of the CRT with a light pen or an electronic pen. Processing sequences of images in real time to derive a description of the motion of one or more objects in a scene. A form of echoing that couples the measure of an input function to a displayed image. It is enabled or disabled as a mode.

tracking, abort branches
A special software routine called *abort branch* included in the controls for stationary baseline tracking robots. This function constantly monitors the position of the tracking window. If at any time during automatic operation the robot's TCP coordinate in the tracking direction coincides with the coordinate of either limit of the tracking window, the robot control immediately initiates an abort branch. The abort branch will direct the robot to exit from the part along a pretaught safe path relative to the part. The abort branch is taught by the operator during the teaching operation and, depending on the configuration of the part on which the robot is working, can contain a number of smaller branch programs. This will ensure that the TCP of the robot will always follow a safe path and a quick exit path from its operating position on the part, regardless of its position in the taught program.

tracking signal
The running sum of the forecast errors (RSFE) dividend by the mean absolute deviation (MAD) of a series of numbers.

traditional drafter
A drafter not trained in CAD, whose time is spent on traditional drafting techniques and tasks.

traffic
In communications, transmitted and received messages. The total information flow of a communications system.

trailer label
Information written after a file has been processed, indicating how many logical records make up the file, what is the batch total, and so on.

trailing zero
In positional notation, a zero in a less significant digit place than the digit place of the least-significant nonzero digit of a numeral.

transaction code
A code used to identify a specific type of transaction.

transaction file
A file containing update transactions to be processed against a master file.

transaction processing
A style of data processing in which files are updated and results generated immediately as a result of data entry.

transducer
A device for converting energy from one form to another. A device used to convert physical parameters, such as temperature, pressure, and weight, into electrical signals. Examples include an electric pressure gage that converts mechanical pressure into electrical signals and a photocell that converts light into electrical signals. A thermocouple that converts temperatures into millivolts is a type of transducer. In numerical control design, it usually applies to a device for converting rotary motion into a varying sine wave voltage. A resolver is a type of transducer.

transfer line
A manufacturing system in which individual stations, indirectly connected together, are used for dedicated purposes. The indirect connecting device performs no function other than the movement of workpieces from one station to another. Generally, a synchronous system. Group of workstations closely connected together by an automated material-handling system and designed for high-volume

production of a single part number or very similar part numbers. High-volume metal-working machines for environments, such as the automobile industry, that can tolerate inflexibility.

transfer machine
An appurtenance, machine, or device designed primarily to grasp a workpiece and move it from one state of a manufacturing operation to another.

transfer molding
A method of molding thermosetting materials, in which the plastic is first softened by heat and pressure in a transfer chamber, then forced through high pressure through suitable sprues, runners, and gates into a closed mold for final curing.

transfer rate
The speed at which data can be read/written into storage.

transfer time
Time to transfer pallet between queue and station or between queue and track.

transfer vector
A transfer table used to communicate between two or more programs. The table is fixed in relationship with the program for which it is the transfer vector. The transfer vector provides communication linkage between that program and any remaining subprograms.

transform
In computer graphics, to move an object or image by translation, rotation, and scaling.

transformation
A method of converting a point in one coordinate system to the same physical point in a different coordinate system. A mathematical conversion system used in providing industrial robots with line-tracking ability. A linear transformation that converts the coordinates of one representational system into another.

transformational grammar
A phrase-structure grammar that incorporates transformational rules to obtain the deep structure from the surface structure.

transformation matrix entity
An entity that allows translation and rotation to be applied to other entities. This is used to define alternate coordinate systems for the definition and viewing.

transformation pipeline
The series of mathematical operations that act on output primitives and geometric attributes to convert them from modeling coordinates to device coordinates.

transformation temperature
The temperature at which a change in phase occurs. The term is sometimes used to denote the limiting temperature of a transformation range.

transformation temperature range
Those ranges of temperature within which austenite forms during heating and transforms during cooling. The two ranges are distinct, sometimes overlapping, but never coinciding. The limiting temperatures of the ranges depend on the composition of the alloy and on the rate of change of temperature, particularly during cooling.

transformer coupling
One method of isolating I/O devices from the computer or robot controller.

transient
A value that changes, decays, or disappears in time. Momentary response of a dynamic system to a rapid input variation, such as a step or a pulse. General term referring to a value that changes in time. Response of a dynamic system to a transient input, such as a step or a pulse.

transient bill of material
Bill-of-material coding and structuring technique used primarily for transient (non-stocker) subassemblies. For the transient subassembly item, lead time is set to zero and lot sizing is on a lot-for-lot basis. This permits MRP logic to drive requirements straight through the transient item to its components but retains its ability to net against any occasional inventories of the subassembly. This technique also facilities the use of common bills of material for engineering and manufacturing.

transistor transistor logic
A signal processing system in which data in the form of low-level electrical signals is processed through circuits, either discretely or through integrated circuits comprised primarily of transistors.

transition
In a binary image, the point where the pixels change between light and dark.

transitional contact
A contact that closes for one program scan each time its internal coil is energized. This

transitional contact provides a one-shot function.

transitional fit
A fit that may have clearance of interference resulting from specified tolerances on hole and shaft, as given by six classes of transition locational fits of 13 nominal shaft sizes in ASA B4.1-1955.

translate
To move a graphic object without any rotation or scaling. In common use, it often implies moving the object in a straight line. To convert CAD/CAM output from one language to another. Also, by an editing command, to move a CAD display entity a specified distance in a specified direction. In data communication, the conversion of one code to another on a character-by-character basis. In programming, a type of language processor that converts a source program from one programming language to another.

translation
Movement of a body such that all axes remain parallel to what they were, that is, without rotation.

translational motion
Movement of a robot arm along one of three axes without rotation.

translation vector
A three-element vector that specifies the offsets (along the coordinate axes) required to move an entity linearly in space.

translucent
Light can pass through. With reference to drafting media, the use of translucent material for drawings allows for reproducible prints. Where light is dispersively transmitted.

transmission
The passage of light or other signal.

transmission speed
The rate at which data is passed through communication lines; usually measured in bits per second (bps).

transmit
To send data from one location and to receive it at another location.

transmittance
The ratio of transmitted energy to incident energy for a partially transmissive material, such as the transmissive mirror on a laser cavity. The percentage of incident light transmitted.

transparency
The property of independence from codes, media, or processing equipment. Program transparency means that any of a variety of computer equipment can be used to run the program. Transparency is the opposite of virtual in the sense that something virtual appears to exist but does not; while something transparent appears not to exist but in fact does.

transparent
Where light is transmitted without dispersion.

transportation inventory
Inventories that exist because materials must be moved. For example, if it takes 2 weeks to replenish a branch warehouse on the other side of the country, inventory equivalent to approximately 2 weeks of sales will normally be in transit and, therefore, this will be extra inventory.

transportation method
A linear programming model concerned with the minimization of costs involved in supplying requirements at several locations from several sources, with different costs related to the various combinations of source and requirement locations.

transport loaded
The basic element employed to move a part or object with the hand or other transporting device to a desired location.

transverse flow
The flow of a gas in a direction transverse to the laser-cavity axis that results in a higher output power.

trap
An unprogrammed conditional jump to a known location, automatically activated by hardware, with the location from which the jump occurred.

trapped
The ability to stop a controller from scanning; can be exercised only from a computer. The controller can still communicate to the computer but will have all outputs off.

traveling purchase requisition
A purchase requisition designated for repetitive use. After a purchase order has been prepared for the goods requisitioned, the form is returned to the originator who holds it until a repurchase of the goods is required. The name is derived from the repetitive travel between the originating and purchasing departments.

tree structure

A graph in which one node, the root, has no predecessor node, and all other nodes have exactly one predecessor. For a state space representation, the tree starts with a root node, representing the initial problem situation. Each of the new states that can be produced from this initial state by application of a single operator is represented by a successor node of the root node. Each successor node branches in a similar way until no further states can be generated or a solution is reached. Operators are represented by the directed arcs from the nodes to their successor nodes.

trend-forecasting models

Methods for forecasting sales data when a definite upward or downward pattern exists. Models include double exponential smoothing, regression, and additive trend.

triac

A solid-state component capable of switching alternating current.

triangulation

A method of determining distance by forming a right triangle consisting of a light source, camera, and sample. Distance can be calculated if the distance between the camera and light source is known as well as with the angle between the incident and reflected light beam.

trigger

A physical input device or set of devices that an operator can use to indicate significant moments in time. A means by which the operator indicates the current measure value is valid in a request or event input operation. A process that corresponds to such a mechanism is also capable of notifying measure processes in the VDI of significant events.

Triangulation

Techniques for Measuring Distance with Machine Vision. Reprinted with Permission of McGraw-Hill for Tech Tran Corp., from *Machine Vision Systems: A Summary and a Forecast.*

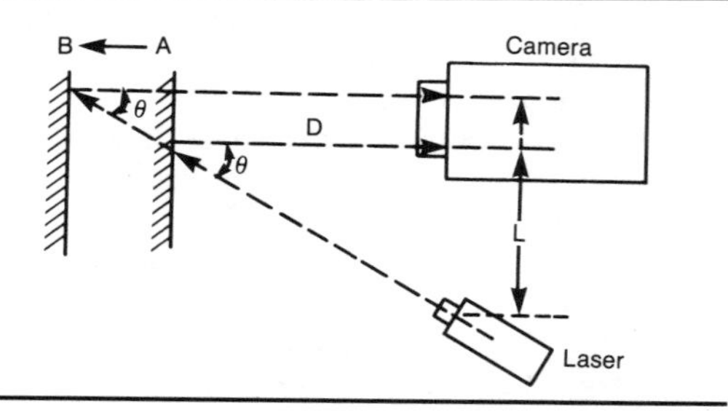

trigger process

A process that notifies a recipient logical input device when a trigger is fired. The trigger process only exists when a device is in event mode or is in request mode with a request pending.

trimming

The operation of removing excess material on the ends or edges of articles, resulting from some kind of forming operation.

trip force

Force required to cause the snap dome to actuate.

triple length register

Three registers that function as a single register.

triple precision

Pertaining to the use of three computer words to represent a number in accordance with the required precision.

triplet

A byte composed of three binary elements.

triplex fixture

A fixture designed to simultaneously hold three parts of the same type (number).

true density

Alternate term for density, as opposed to apparent density.

truncate

To terminate a computational process in accordance with some rule, for example, to end the evaluation of a power series at a specified term. The removal of one or more digits, characters, or bits from one end of an item of data when a string length or the required precision of a target variable has been exceeded. To cut off at a specified spot (as contrasted with round).

truth maintenance

(A misnomer.) The task of preserving consistent beliefs in a reasoning system whose beliefs change over time. A method of keeping track of beliefs (and their justifications) developed during problem-solving so that if contradictions occur, the incorrect beliefs or lines of reasoning and all conclusions resulting from them can be retracted.

truth table
An operation table for a logic operation. A table that describes a logic function by listing all possible combinations of input values and indicating for each combination the true output values.

truth value
One of the two possible values—true or false—associated with a proposition in logic.

t-stop
A system for rating lenses for sensitivity purposes that provides an equivalent aperature of a lens having 100 percent transmission efficiency. This system is based on actual light transmission and is considered as more realistic than the f-stop system.

tumbling
Finishing operation for small articles by which gates, flash, and fins are removed and/or surfaces are polished by rotating them in a barrel together with wooden pegs, sawdust, and polishing compounds.

tumbling barrel
The operation of removing burrs, fins, scale, and roughness by an abrasive material falling against finished piece parts in a revolving barrel.

tungsten-arc welding
Inert-gas shielded-arc welding using a tungsten electrode.

tungsten inert-gas welding
A technique using a tungsten electrode, generally without a filler material. Popular for light metal gages and high-precision work. Now known as gas tungsten-arc welding (GTAW).

tuning
The process of adjusting system control variables to make the system divide its resources most efficiently for the workload.

turnaround time
The elapsed time between submission of a job to a computing center and the return of results. In communications, the actual time required to reverse the direction of transmission from send to receive or vice versa when using a half-duplex circuit. For most communications facilities, there will be time required by line propagation and line effects, modem timing, and machine reaction. A typical time is 200 milliseconds on a half-duplex telephone connection. The elapsed time between the moment a task or project is input into the CAD/CAM system and the moment the required output is obtained.

turning center
A lathe, that is, a machine in which the piece being worked on rotates on an axis so the tools can work on it. When a lathe becomes sufficiently intelligent, with the addition of programmable numerical controls, it becomes a *turning center.*

turnkey
A CAD/CAM or robotic system for which the supplier/vendor assumes total responsibility for building, installing, and testing both hardware and software, and the training of user personnel. Also loosely, a system that comes equipped with all the hardware and software required to do a specific application or applications. Usually implies a commitment by the vendor to make the system work and to provide preventive and remedial maintenance of both hardware and software. Sometimes used interchangeably with stand-alone, although stand-alone applies more to system architecture than to terms of purchase.

turnkey system
A system in which the manufacturer takes full responsibility for complete system design and installation and supplies all necessary hardware, software, and documentation elements.

turnover
The number of times inventory is replaced during a time period; in other words, a measurement of investment inventory to support a given level of sales. It is found by dividing the cost of goods sold for the period by the average inventory for the period.

tutorial
A characteristic of CAD/CAM systems. If the user is not sure how to execute a task, the system will show how. A message is displayed to provide information and guidance.

twist
The angular displacement of a moving body around the principle axis of its motion.

two-circuit switch
In one position, contacts complete one circuit; in the other position, contacts complete a separate circuit after opening the original circuit.

two-out-of-five code

A binary coded decimal notation in which each decimal digit is represented by a binary numeral consisting of five bits of which two are of one kind, conventionally 1s, and three are of the other kind, conventionally 0s. The usual weights are 0-1-2-3-6, except for the representation of zero, which is then 01100.

two-plus-one address

An instruction that contains three address parts, the plus-one address being that part of the instruction that is to be executed next unless otherwise specified.

two-wire channel

Half-duplex transmission facility, characterized by a single wire pair.

U

UCC-APT
A version of APT developed by University Computing Company that contains special programming features: ALL FORTRAN functions, FORTRAN DO loops, logical IF, expanded solid and plane geometry capabilities, bounded FMILL, formatted printing, rough turning, and turning and threading subsets.

ultimate strength
The maximum conventional stress (tensile, compressive, or shear) that a material can withstand.

ultrasonic beam
A beam of acoustical radiation with a frequency higher than the frequency range for audible sound.

ultrasonic cleaning
Immersion cleaning aided by ultrasonic waves that cause microagitation.

ultrasonic frequency
A frequency, associated with elastic waves, that is greater than the highest audible frequency, generally regarded as being higher than 15 kilocycles per second.

ultraviolet
The region of the electromagnetic spectrum adjacent to the visible portion of the spectrum but with wavelengths between 0.01 and 0.04 microns.

unattended operation
The automatic features of a terminal station's operation that permit the transmission and reception of messages on an unattended basis.

unavoidable delay
An occurrence that is essentially outside the worker's control or responsibility that prevents him or her from doing productive work.

unavoidable delay allowance
Time included in the production standard to allow for time lost that is essentially outside the worker's control.

unbundled
The services, programs, training, and so forth sold independently of the computer hardware by the computer hardware manufacturer. Thus, a computer system that includes all products and services in a single price is said to be *bundled*.

unconditional transfer
An instruction that always causes a branch in program control away from the normal sequence of executing instructions.

undercut
A groove melted into the base metal adjacent to the toe of a weld and left unfilled. A machine cut where the cutter failed to arrive at a programmed point, following a command change in direction (in contrast to overshoot).

undershoot
The degree to which a system response to step changes in reference input falls short of the desired value.

undersizing, oversizing
CAD automatic editing tools for the systematic reduction or enlargement, respectively, of the areas in an IC design layout. This is typically performed to compensate for processing effects that may shift the edges of the layout areas either out or in.

UNIAPT
The minicomputer implementation and extension of the APT part programming language for small- and medium-size computers.

unibus
A bus that is a set of electrical conductors that carry specific signals to several other electrical circuits.

uniform-flow convention
In economy studies, the practice of assuming that receipts and disbursements are made at a constant rate throughout the time period of concern so as to facilitate calculation or analysis.

UNIGRAPHICS
Interactive graphics system that creates a dimensionally accurate data base accessible to all phases of design and production. It handles all design and drafting tasks and provides a direct link between computer-aided design and computer-aided manufacturing.

uniprocessing
Sequential execution of instructions by a processing unit or independent use of a processing unit in a multiprocessing system.

UNIS
Acronym for Sperry Univac Industrial System. UNIS is an interactive manufacturing control package that is data base oriented and

provides support for all phases of manufacturing and production control.

unit
A major building block for a set or system, consisting of a combination of basic parts, subassemblies, and assemblies packaged together as a physically independent entity.

unit of measure (purchasing)
The unit used to purchase an item. This may or may not be the same unit of measure used in the internal systems. Purchasing buys steel by the ton, but it may be issued and used in square inches.

unit record equipment
Conventional punched-card data-processing machines operated by control panel, as differentiated from stored-program computers.

unit screen
The logical display address space within which all drawing operations are executed and alphanumeric characters are deposited. The dimensions of the unit screen are 0 (inclusive) to 1 (noninclusive) in the horizontal, vertical, and depth dimensions.

universal product code (UPC)
A bar code (like those printed on a supermarket package) that uniquely identifies the product.

universe
The population or large set or data from which samples are drawn. Usually assumed to be infinitely large or at least very large relative to the sample. In speaking of computer graphic colors, the total group of possible colors that a display circuit can select from. More commonly, it is called the palette.

UNIX
A multiuser operating system developed by Bell Laboratories.

unpacked decimal
Representation of a decimal value by a single digit in one byte.

unpost
The act of removing a structure from display on a particular workstation.

unrecoverable error
An error that results in abnormal termination of a program.

unstable state
In a trigger circuit, a state in which the circuit remains for a finite period of time at the end of which it returns to a stable state without the application of a pulse.

unstratified
A language that can be used as its own metalanguage, for example, most natural language. See also metalanguage.

unwind
To state explicitly and in full, without the use of modifiers, all the instructions that are involved in the execution of a loop.

update
To modify a master file with current transaction information, according to a specified procedure.

upper arm
That portion of a jointed-arm robot that is connected to the shoulder.

upstream, downstream
Refers to belt travel in which the belt is moving relative to a fixed frame, and so forth.

uptime
The time during which equipment is either operating or available for operation, as opposed to downtime when no productive work can be accomplished.

urnary operator
An arithmetic operator having only one term. The urnary operators that can be used in absolute, relocatable, and arithmetic expressions are positive (+) and negative (−).

use
The basic element employed to perform the activity that is the purpose of the operation, other than the assembly.

use as is
Material that has been dispositioned as unacceptable per the specifications, however, the material can be used within acceptable tolerance levels.

user
A company, a business, or a person who uses advanced manufacturing technology and who contracts, hires, or is responsible for the personnel associated with its operation.

user address
A unique address for each communicating entity on the local network(s) where local control of addressing is maintained.

user documentation
A set of documentation relating to installation, operation, and maintenance of the system.

user identification
A one-to-eight character symbol identifying a system user. Abbreviated USERID.

user friendly
Computer program that anticipates user questions, problems, anomalies, and the like, and provides effective and useful responses.

user group
Any organization made up of users (as opposed to vendors) of various computing systems, software packages, and so on, that gives the users an opportunity to share knowledge they have gained in using a particular system, to exchange programs they have developed, and to jointly influence vendor software and hardware support and policy.

user interface (or *human interface*)
The component of an expert system that allows bidirectional communication between the expert system and its user. Most user interfaces utilize natural-language processing techniques.

user programs
Programs that have been written by the user, as contrasted to those supplied by the manufacturer.

utility-branch tracking
If a robot needs to take some form of corrective action in response to a signal that indicates the occurrence of some malfunction of the peripheral equipment, such action may be taken with the computer-controlled robot by using a function called *utility branch*. A utility branch is taught in a similar manner to abort branch and, if necessary, it can also contain a number of smaller branches. The utility branch is, however, initiated by an external signal from the peripheral equipment, rather than by an internal signal from the control. As an example, if, during a spot-welding sequence, the gun tip sticks to a flange, immediate correcting action is required. Upon receipt of the signal indicating the "stuck tip" condition, the control will immediately initiate the utility branch to take corrective action, regardless of where the robot is in its sequence of operations. In this case, the action taken may be, for example, a twisting motion to break the tip of the gun away from the flange.

utility function
One of a group of functions provided to create a matrix from the parameters that are natural to the application program. Functions are provided to assist with the creation of both modeling and viewing matrices.

utility-function-state list
A defined data structure that contains the parameters used by the utility functions that support the creation of viewing matrices.

utility program
A specialized program performing a frequently required everyday task. Examples include sorting, report generation, file updating, file dump and backup (maintaining backup files in case a master working file is fouled up), and so on. Those programs are usually supplied by the manufacturer of the equipment.

utility routine
Software used to perform some frequently required process in the operation of a computer system, for example, sorting, merging, and the like.

utility software
Programs that aid systems operations. They usually perform services frequently needed by users.

utilization of machines
Percentage of time a machine is actually being productive.

V

vacuum deposition
Condensation of thin metal coatings on the cool surface of work in a vacuum.

vacuum forming
Method of sheet forming in which the plastic sheet is clamped in a stationary frame, heated, and drawn down by a vacuum into a mold. In a loose sense, it is sometimes used to refer to all sheet-forming techniques, including drape forming involving the use of vacuum and stationary molds.

vacuum melting
Melting in a vacuum to prevent contamination from air, as well as to remove gases already dissolved in the metal; the solidification may also be carried out in a vacuum or at low pressure.

vacuum metallizing
Process in which surfaces are thinly coated with metal by exposing them to the vapor of metal that has been evaporated under vacuum (one-millionth of normal atmospheric pressure).

VAL
Manipulator-oriented programming language for robot programming. A product of Unimation/Westinghouse.

validation
An attempt to find errors by executing a program in a given real environment.

validation environment
A collection of tools used to validate a system element.

validation plan
A strategy for validation testing that identifies the items to be tested, the testing to be performed, test schedules, personnel requirements, reporting requirements, and so on.

validation procedure
A set of detailed instructions for the setup, operation, and evaluation of results for a given validation test.

validity check
Verification that each element of data is actually a valid character of the particular code in use.

valuator
An input class that returns a scalar value. The term used in the Core graphics standards for a device that is used to supply a continuous one-dimensional value, such as a dial or joystick.

valuator device
A logical input device providing a real number as its input measure.

value added
A company's sales minus purchased parts, raw materials, and outside services.

value-added carriers
A new class of communications common carrier authorized to lease raw communication trunks from the transmission carriers, augment these facilities with computerized switching, and provide enhanced or *value-added* communications service. Telenet Communications Corporation and others are now employing packet switching to provide value-added data communications services.

value-added service
A communications service utilizing communications common-carrier networks for transmission and providing added data services with separate additional equipment. Such added service features may be store-and-forward message switching, terminal interfacing, and host interfacing.

value analysis
An arrangement of techniques that make clear precisely the functions that the customer wants, establishes the appropriate cost for each function by comparison, and causes required knowledge, creativity, and initiative to be used to accomplish each function for that cost.

value servomotor
Controlled by feedback systems so that when the ultimate position is reached, the valve or motor stops operating.

vapor degreasing
Degreasing work in vapor over a boiling liquid solvent, the vapor being considerably heavier than air. At least one constituent of the soil must be soluble in the solvent.

vapor plating
Deposition of a metal or compound upon a heater surface by reduction or decomposition of a volatile compound at a temperature below

the melting points of the deposit and the basis material. The reduction is usually accomplished by a gaseous reducing agent, such as hydrogen. The decomposition process may involve thermal dissociation or reaction with the basis material. Occasionally used to designate deposition on cold surfaces by vacuum evaporation.

variable
A quantity that can assume any of a given set of values.

variable block format
A tape format that allows the number of words in successive blocks to vary.

variable costing
An inventory valuation method in which only variable production costs are applied to the product; fixed factory overhead is not assigned to the product. Variable production costs are direct labor, direct material, and variable overhead costs. Costing can be helpful for internal management analysis but is not widely accepted for external financial reporting. For inventory order quantity purposes, however, the unit costs must include both the variable and allocated fixed costs in order to be compatible to the other terms in the order quantity formula.

variable costs
An operating cost that varies directly with production volume, for example, materials consumed, power, direct labor, and sales commissions.

variable element
An element for which the leveled or normal time, under the same methods and working conditions, will change because of the varying characteristics of the parts being worked upon, as size, weight, shape, density, hardness, viscosity, tolerance requirements, finish, and so on.

variable expense
Expenditures that vary in proportion to the volume of production.

variable length
A record having a length independent of the length of other records with which it is logically or physically associated.

variable-mission manufacturing
Concept introduced by Cincinnati Milacron in the early 1970s; synonymous with FMS.

variable-point representation
A positional representation in which the position of the radix point is explicitly indicated by a special character at that position.

variance
The difference between any standard or expected value and an actual value.

variant process planning
Process planning based on the retrieval of existing plans of similar components.

VDC extent
A rectangular region of interest contained within the virtual device coordinate (VDC) range.

VDC range
A rectangular region within VDC space, consisting of the set of all coordinates representable in the declared coordinate type, precision, and encoding format of the metafile or VDI.

VDC space
A two-dimensional, Cartesian coordinate space of infinite precision and extent. Only a subset of VDC space, the VDC range, is realizable in a metafile or VDI.

VDM generator
The process or equipment that produces a metafile.

VDM interpreter
The process or equipment that reads a metafile and interprets the contents needed in order to drive a VDI or other device interface to obtain a picture that resembles the intended picture as closely as possible.

vector
A straight line segment that has both magnitude and direction. In some cases, only the direction of the vector, not its magnitude, is actually used by the graphics system (e.g., character-up direction).

vector graphics
A method of creating images by forming them out of line segments.

veitch diagram
A means of representing Boolean functions in which the number of variables determines the number of possible states, that is, 2 raised to a power determined by the number of variables.

velocity
A measure of speed or rate of motion.

velocity error
In a servomechanism that operates a manipulator joint, the difference between the rate of change of the actual position of that joint and the rate of change of the commanded position.

vendor
A company that supplies material.

vendor alternative
Other than the primary vendor. The alternate vendor may or may not supply a percentage of the items purchased but is usually approved to supply the items.

vendor lead time
The elapsed time between the placement of a purchase order to a vendor and when the goods are delivered and available for inspection and/or subsequent use.

vendor measurement
The act of measuring the vendor's performance to the contract. Measurements usually cover delivery, quality, and price.

vendor number
A numerical code used to identify one vendor from another vendor.

vendor scheduler
An individual whose main responsibility is ensuring vendor performance to the schedule. The position usually consists of taking the MRP output reports, communicating directly with the vendor in terms of ordering and rescheduling the required purchased material, and assuming the responsibility for delivery on the promised date. This includes communicating in advance with the master scheduler when parts will not arrive on time to support the schedule. By using the vendor scheduler approach, the buyer is then freed from day-to-day order placement and expediting and thus has the time to do cost reduction, negotiation, vendor selection, alternate sourcing, and the like.

Venn diagram
A schematic representation of the universal set and its subsets. A rectangle is usually used to represent the universal set, points to designate elements of the set, and circles to depict subsets of the universal set. The diagram is sometimes referred to as a Euler diagram.

verification
A system-generated message to a workstation, acknowledging that a valid instruction or input has been received. The process of checking the accuracy, viability, and/or manufacturability of an emerging design on the system.

verification plan
A strategy for verification testing that identifies the items to be tested, the testing to be performed, test schedules, personnel requirements, reporting requirements, and so forth.

verification process
A set of detailed instructions for the setup, operation, and evaluation of results for a given verification test.

verifier
A device similar to a card punch used to check the inscribing of data by rekeying.

version number
A means for uniquely designating one specification definition or translator implementation from a preceding or subsequent one.

vertex
The coordinate points that the program uses to keep track of objects. The function of a mouse or digitizer actually is to create vertex points. We see them as lines and shapes, but the program knows them only as vertices. Drawing a rectangle, for example, requires the setting of five vertex points. One corner will be both a start and end point to draw the lines. The point on a polyhedron common to three or more sides.

vertical display
A method of displaying or printing output from an MRP system where requirements, scheduled receipts, projected balance, and the like are displayed vertically, that is, down the page. Vertical displays are often used in conjunction with bucketless systems.

vertical integration
A method of manufacturing a product, such as a CAD/CAM system, whereby all major modules and components are fabricated in-house under uniform company quality control and fully supported by the system vendor.

vertical machining
Vertical machining center—similar to the horizontal machining center in that it consists of a robot and a milling machine working together. The vertical machining center differs in that the cutting tool rotates on a vertical axis and can only be moved vertically (Y axis). It can drill or cut only on the top surface of the workpiece.

vertical resolution
The number of horizontal lines that can be seen in the reproduced image.

vertical stroke
The amount of vertical motion of the robot arm from one elevation to the other.

vertical tabulation character
A format-effector that causes the print or display position to move to the corresponding position in the next of a series of predetermined lines.

vertice
The intersection of two surfaces of an object.

very-large-scale integration
The process of combining several hundred thousand electronic components into a single integrated circuit.

V format
A data set format in which logical records are of varying length (includes a length indicator), and in which V-format logical records may be blocked, with each block containing a block-length indicator.

via
A means of passing from one layer or side of a PC board to the other. A feed through. A CAD system can automatically optimize via placement by means of special application software.

via hole
A plated-through hole used as a through connection but in which there is no intention to insert a component lead or other reinforcing material.

vibration test
A test to determine the ability of a device to withstand physical oscillations of specified frequency, duration, and magnitude.

video
The analog time-varying output signal from an image sensor that conveys the image data. The data displayed on the screen of a CRT.

video disk
A rigid storage medium for analog or digital data written/read by a laser. A main feature is random access to a large selection of information.

videotex
An interactive information network that uses telephone lines, a decoder, and a television to connect a home user with a mainframe computer to display both text and graphic information.

Video Disk
Sony Videodisc System Consists of Video Disk Player, Monitor, and Computer Control

vidicon
A small television camera originally developed for use in closed-circuit television monitoring. It provides an analog voltage output corresponding to the intensity of the incoming light. An image pickup tube in which a change density pattern is formed by photoconduction and stored on that surface of the photoconductor that is scanned by an electron beam.

vidicon camera
An image-sensing device that uses an electron gun to scan a photosensitive target on which a scene is imaged.

view
The description of a viewing operating that produces a displayable picture.

view entity
A structure entity used to provide the definition of a human-readable representation of a two-dimensional projection of a selected subset of the model and/or nongeometry information.

viewer-centered viewing
A viewing situation in which the viewer's position is consider fixed and the most likely dynamic behavior is for the viewer to look around at the world.

viewing box
The clipping box used to define a view.

viewing coordinates
A three-dimensional coordinate system used for defining windows and projection param-

eters. The origin of the viewing coordinate system is the intersection with the view plane of a line through the view reference point parallel to the view plane normal vector.

viewing operation
An operation that maps positions in world coordinates to positions in normalized device coordinates. In addition, it specifies the portion of the world coordinate space that is to be visible.

viewing parameters
The parameters that may be used to define the viewing coordinate system and other aspects of a view. The parameters are maintained in the utility-function-state list.

viewing transformation
A normalization operation that maps positions in world coordinates to positions in viewing coordinates. In addition, a viewing operation specifies the portion of world coordinate space that is to be displayed.

view mapping
The transformation from viewing coordinates to logical device coordinates. This transformation includes a projection mapping and optional clipping. Also called the window/viewport transformation.

view matrix
A 4 × 4 matrix used to specify the viewing transformation. The view matrix may be composed using utility functions.

view plane
A two-dimensional plane through which three-dimensional objects are projected. The view plane is established by giving a view reference point, a view plane normal, and a view plane distance.

view plane distance
The distance of the view plane from the view reference point measured along the view plane normal vector in world coordinates.

view plane normal
A vector normal to the view plane used to orient the plane.

viewport
Rectangle on the view surface within which a view of the world coordinate object is shown and into which windows are mapped. The position or direction from which the scene is observed.

view reference point
A convenient world coordinate point on or near the object being viewed. All of the viewing parameters are specified relative to the view reference point. The view reference point becomes the origin of the viewing coordinate system.

view surface
A two-dimensional logical output surface. Images on a view surface are drawn on a corresponding physical output surface (e.g., plotter surface or display screen) in a device-dependent way by the device driver for that output device.

view table
The workstation-dependent table of views found on each workstation and referenced by the workstation-independent view index attribute.

view-up vector
A vector in world coordinates relative to the view reference point that, if it were within the window, would appear upright on the display surface. The view-up vector is projected onto the view plane in the direction of the view plane normal. The projection is the U axis in the UV viewing coordinate system.

view volume
A pyramid (possibly truncated) for perspective projection or a parallelepiped for parallel projection. The boundaries of the view volume are determined by the window defined in the view plane, the front and back clipping planes, and projectors from the projection reference point through the window corners for perspective projection, or projectors in the direction of projection through the window corners for parallel projection.

virtual
Conceptual or appearing to be rather than actually being. An adjective implying that data, structures, or hardware appear to the programmer or user to be different from their real condition, the conversion being performed by software.

virtual address
In virtual storage systems, an address that refers to virtual storage and must, therefore, be translated into a real storage address when it is used.

virtual address area
The area of virtual storage whose addresses are greater than the highest address of the real address area.

virtual address space
In virtual storage systems, the virtual storage assigned to a job, terminal user, or system task.

virtual cell
A software implementation of a manufacturing cell, where workstations or support systems are dynamically assigned or allocated to a cell controller that is responsible for the manufacture of a family of parts. The cell controller acts as a project manager that obtains control of required stations for limited time periods to accomplish certain steps in the production process. Manufacturing resources are allocated to the cell only when it is ready to use them and are relinquished when the particular processing step is completed so that other cells may utilize that same resource.

virtual circuit
A connection between a source and a sink in a network that may be realized by different circuit configurations during transmission of a message.

virtual device
An idealized graphics device that presents a set of graphics capabilities to graphics software or systems by means of a VDI.

virtual device coordinates
The absolute two-dimensional coordinates used to specify position in VDC space.

virtual device interface
An interface between the device-independent and the device-dependent levels of a graphics system. The interface may be implemented in either a device driver or a device.

virtual device metafile
A mechanism for retaining and transporting graphics data and control information at the level of the virtual device interface.

virtual image
In computer graphics, the complete visual representation of an encoded image that would be displayed if a display surface of sufficient size were available.

virtual screen
A section of memory that is used to save an image for screen display. Normally, the virtual screen is larger than the actual amount that can be shown at one time, so the display acts as a window showing a selected part of the virtual image. Using a virtual screen takes more memory and processing power, but it allows fast screen updates and smooth scrolling.

virtual storage
Addressable space that appears to the user as real storage, from which instructions and data are mapped into real storage locations. The size of virtual storage is limited by the addressing scheme of the computing system (or virtual machine) and by the amount of auxiliary storage available, rather than by the actual number of real storage locations. This procedure leaves the programmer free to address total storage without concern as to whether primary or secondary storage is actually being addressed, and effectively includes the large inexpensive capacity of secondary storage in the system. Optimally, the computer should be able to operate either with or without virtual storage without major software modification. Benefits of virtual storage operation are enhanced when it is implemented by hardware that carries out the data-swapping algorithms.

virtual storage access method (VSAM)
An access method for direct or sequential processing of fixed- and variable-length records on direct-access devices. The records in a VSAM data set or file can be organized in logical sequence by a key field (key sequence), in the physical sequence in which they are written on the data set or file (entry sequence), or by relative record number.

virtual storage management
Routines that allocate address spaces and virtual storage areas within address spaces and that keep a record of free and allocated storage within each address space.

virtual telecommunications access method (VTAM)
A set of programs that control communications between terminals and application programs under DOS/VS, OS/VSI, and OS/VS2.

virtual terminal
Terminal, possibly emulating a different type of terminal, connected through the network as a terminal to a system connected as a node.

virtual unit address
In a mass-storage system (MSS) an address for a virtual drive. The virtual unit address can be assigned to any staging drive group. Each staging drive can have more than one virtual unit address but only one real address.

viscosity
Resistance to flow exhibited within a material. Can be expressed in terms of relationship be-

tween applied shearing stress and resulting rage of strain in shear. Viscosity usually means *Newtonian viscosity*, in which case the ratio of shearing stress to rate of shearing strain is constant. In non-Newtonian behavior (which occurs with plastics), the ratio varies with the shearing stress. Such ratios are often called the apparent viscosities at the corresponding shearing stresses.

viscosity coefficient
Shearing stress necessary to induce a unit velocity flow gradient in a material. The viscosity coefficient of a material is obtained from ratio of shearing stress to shearing rate.

viscous friction
The resistive force on a body moving through a fluid. Ideally, a resistive force proportional to relative velocities of a body and a fluid.

visibility
An attribute that indicates whether a part of an image (primitive or segment) is actually visible on the display surface.

visible light
The portion of the electromagnetic spectrum at wavelengths between 0.38 and 0.75 microns.

vision
The process of understanding the environment based on sensing image data, the light level or reflectance of objects. The ability to scan and/or look at a given piece or part, such as welding, and determine where the part is in relation to where it should be. Then, that information is digested and the logic of the system tells the system what to do.

vision addressability
The number of positions (pixels) in the X axis and in the Y axis that can be displayed on the CRT. A measure of picture display quality or resolution.

vision optical system
A device, such as a camera, that is designed, constructed, and installed to detect intrusion by a person into the robot-restricted work envelope and that could also serve to restrict a robot work envelope.

vision sensor
Sensor that identifies the shape, location, orientation, or dimensions of an object through visual feedback, such as a television camera.

vision system
A device that collects data and forms images that can be interpreted by a robot computer to determine the position or to "see" an object.

visual inspection
A term generally used to indicate inspection performed without the aid of test instruments.

visual inspection systems
By storing a statistical image of parts in the control processor, these systems can be used to reject parts due to defects or missing components and to control certain manufacturing processes.

voice activated
Systems that can recognize and respond to spoken words.

voice-grade channel
Typically, a telephone circuit normally used for speech communication and accommodating frequencies from 300 to 3000 Hz. Up to 10,000 Hz can be transmitted.

voice numerical
Voice numerical control is a voice-activated three-axis point-to-point two-axis contouring system.

voice-operated device
A device used on a telephone circuit to permit the presence of telephone currents to effect a desired control. Such a device is used in most echo suppressors.

voiceprint
A technique for verifying an individual's identity by the pattern produced by his or her voice.

voice recognition
A system of sound sensors that translates human voice tones into computer commands.

voice recognition accuracy
The degree to which voice recognition equipment will correctly encode the input utterance without repeated inputs or corrections; higher recognition accuracy provides faster data input.

voice recognition equipment vocabulary size
The number of utterances (words or short phrases) that can be distinctly recognized and digitally encoded. The larger the vocabulary, the more complex the source data that can be handled without special software.

voice response unit
A device that can accept a coded request for data, compose a coded response, interpret the coded response into locations of stored vocabulary, and produce speech as output.

Vision System

Machine Vision Process Block Diagram. Reprinted with Permission of McGraw-Hill for Tech Tran Corp., from *Machine Vision Systems: A Summary and a Forecast.*

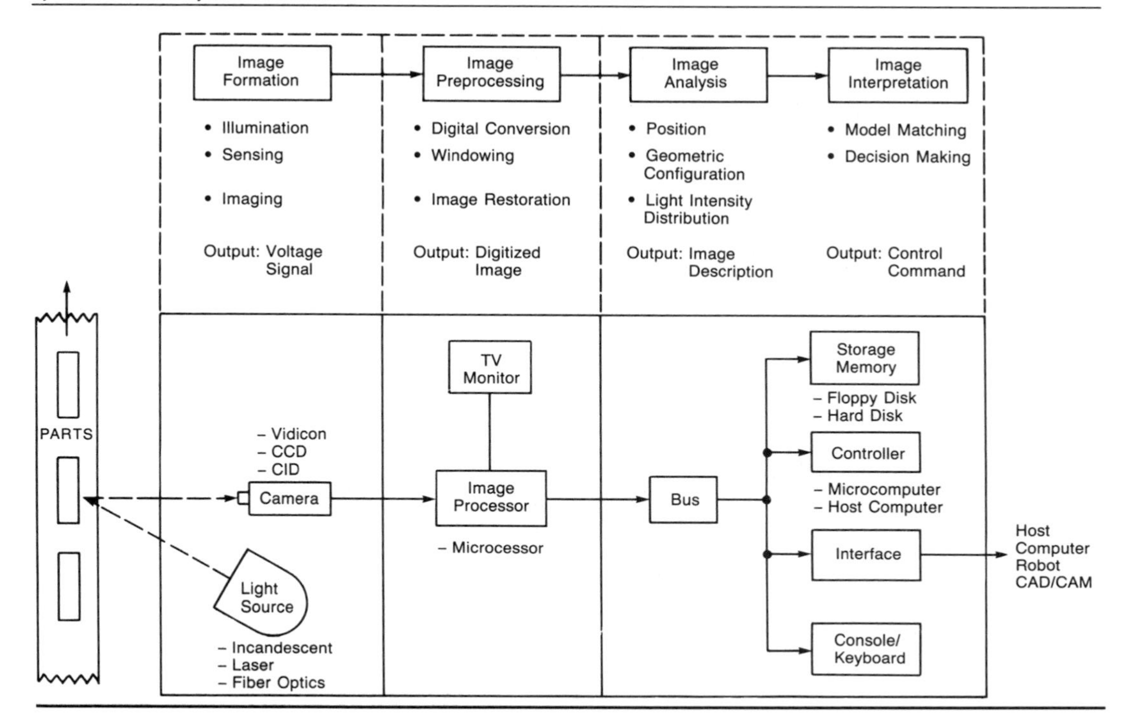

voice syntheses
Computer-generated sounds that simulate the human voice.

volatile memory
A memory system in a computer or control system that requires a continual source of electric current to maintain the data it is storing intact. Removal of power from a volatile memory system results in the loss of the data being stored.

volatile storage
A storage medium in which information cannot be retained without continuous power dissipation (semiconductor).

volatility
The percentage of records on a file that are added or deleted in a run.

voltage rating
The maximum voltage at which a given device may be safely maintained during continuous use in a normal manner. It is also called working voltage.

voltage ratio
Alternate term for surface-breakdown rate.

volume
A recording medium that is mounted and demounted as a unit, for example, a reel of magnetic tape, a disk pack, or a data cell.

volume element
A simple solid shape such as a sphere, cube, or polyhedron used as a component of a more complex computer graphics model.

volume model
The representation of three-dimensional objects as geometric solids, or it might be represented by the edges of those solids.

volume resistivity
A measure of the ability of material to resist flow of an electric current. The ratio of electrical potential gradient parallel to current in

a material to current density. Although volume resistivity is of interest, knowledge of its changes with temperature and humidity are even more important, and curves plotted against temperature are sometimes particularly desirable.

volume table of contents
A table or directory on a direct-access volume that describes each data set on the volume. Abbreviated VTOC.

von Neumann architecture
The current standard computer architecture that uses sequential processing.

wafer
A thin round slice of semiconductor material, usually silicon, on which hundreds of chips are made at once.

Wagner-Whitin algorithm
A mathematically complex dynamic lot-sizing technique that evaluates all possible ways of ordering to cover net requirements in each period of the planning horizon, to arrive at the optimum ordering strategy for the entire net requirements schedule.

waiting line
A line formed by units waiting for service; a queue.

wait state
The condition of a central processing unit when all operations are suspended.

wait station
A station at which a part or pallet waits, for example, a berth in a queue.

walkthrough programming
A method of programming a robot by physically moving the manipulator arm through a complete operating cycle. This is typically used for continuous path robots.

wall-to-wall inventory
Term for the technique where production material, parts, and assemblies enter the plant at one end and are processed through the plant to end product without ever having entered a formal stock area.

warm boot
Reloading the operating system or initializing the computer after it has been on and without turning the computer off.

warm start
The restart activity appropriate when a temporary failure has not disturbed backup storage. Current conversations can be continued by reading in the last filed version of the conversation control records, although the interrupted exchanges may be lost.

warning device
An audible or visual safety device such as a bell, horn, flasher, and the like used to alert persons to expect robot movement.

warning message
An indication during program compilation that a possible error has been detected.

waste-handling system
System used to collect, move, and store process waste.

watchdog
In control systems, a combination of hardware and software that acts as an interlock scheme, disconnecting the system's output from the process in event of system malfunction.

water-vapor permeability
Water-vapor transmission through a homogeneous body under unit vapor pressure difference between two surfaces, per unit thickness.

water-vapor transmission
Steady-state time rate of water-vapor flow through unit area of humidity at each surface.

waveguide laser
A sealed tube laser with a tube constructed in the form of a dielectric waveguide.

wavelength
The length of the light wave, measured from crest to crest. The wavelength determines the color of the light and also has an impact on the way in which electromagnetic radiation will interact with materials.

weight
In a positional representation, the factor by which the value represented by a character in the digit place is multiplied to obtain its additive contribution in the representation of a real number. A nonzero real number that appears in the numerator and denominator of the expression for a rational B-spline curve or surface. Increasing the weight associated with a particular control point will tend to draw the resulting curve or surface toward that control point.

weighted average
An averaging technique where the data to be averaged is multiplied by different factors. For example, a regular average is equivalent to a 50–50 weighted average. An average could be made up by taking 90 percent of one figure and 10 percent of another figure. This would then be a weighted average. Note that the weights must always be equal to 100 percent.

weighted value
The numerical value assigned to any single bit as a function of its position in the code word.

weight resistivity
A measure of the ability of a material to resist flow of electric current. Also called mass resistivity. Electrical resistance of a material per unit length and unit weight. A term generally reserved for metallic materials, that is, conductors. Weight resistivity in ohm-lb/mi^2 or ohm-gm/m^2 is equal to WR/L_1L_2 where W is the weight of the specimen in pounds or grams, R is the measured resistance in ohms, L_1 is the gage length in miles or meters used to determine R, and L_2 is the length of the specimen.

welding
Joining two or more pieces of material by applying heat, pressure, or both, with or without filler material, to produce a localized union through fusion or recrystallization across the interface. The thickness of the filler material is much greater than the capillary dimensions encountered in brazing. May also be extended to include brazing.

welding current
The current flowing through a welding circuit during the making of a weld. In resistance welding, the current used during preweld or postweld intervals is excluded.

welding cycle
The complete series of events involved in making a resistance weld. Also applies to semiautomatic mechanized fusion welds.

welding generator
A generator used for supplying current for welding.

welding lead
A work lead or an electrode lead.

welding machine
Equipment used to perform the welding operation; for example, spot-welding machine, arc-welding machine, seam-welding machine.

welding procedure
The detailed methods and practices, including joint-welding procedures, involved in the production of a weldment.

welding rod
Filler metal in rod or wire form used in welding.

welding schedule
A record of all welding machine settings plus identification of the machine for a given material, size, and finish.

welding sequence
The order of welding the various component parts of a weldment or structure.

welding stress
Residual stress caused by localized heating and cooling during welding.

welding technique
The details of a welding operation that, within the limitations of a welding procedure, are performed by the welder.

welding tip
A replaceable nozzle for a gas torch that is especially adapted for welding. A spot-welding or projection-welding electrode.

weld line
The junction of the weld metal and the base metal or the junction of the base-metal parts when filler metal is not used.

weldment
An assembly whose component parts are joined by welding.

weld metal
That portion of a weld that has been melted during welding.

wetting
A phenomenon involving a solid and a liquid in such intimate contact that the adhesive force between the two phases is greater than the cohesive force within the liquid. Thus, a solid that is wetted, on being removed from the liquid bath, will have a thin continuous layer of liquid adhering to it. Foreign substances, such as grease, may prevent wetting. Addition agents, such as detergents, may facilitate wetting by lowering the surface tension of the liquid.

wetting agent
A surface-active agent that produces wetting by decreasing the cohesion within the liquid.

what-if analysis
The process of evaluating alternate strategies. Answering the consequences of changes to forecasts, manufacturing plans, inventory levels, and so forth. Some companies have the capability of submitting various plans as a "trial fit" in order to find the best one.

where-used report
A report showing every part in the company's product line in which a certain part number is utilized and the level on the part's bill of materials at which it is utilized.

wideband
A communications channel having a bandwidth characterized by data transmission speeds of 10,000 to 500,000 bits per second.

Winchester disk
Rigid nonremovable magnetic oxide-coated random-access disk sealed in a filtered enclosure along with the read/write heads and head actuator. Heads fly only about 20 microinches from disk surface, allowing very dense data storage. Common disk sizes are $5\frac{1}{4}$ inches and 8 inches in diameter. Capacity ranges from 5 Mb (325 to 16,000 pages) for the $5\frac{1}{4}$-inch disk and from 6.5 to 635 Mb (1625 to 158,750 pages) for the 8-inch disk, calculated at 4000 characters per single-spaced page.

window
A selected portion, usually square or rectangular, of an image. A two-dimensional or three-dimensional region surrounding the object to be viewed. In three dimensions, a window is defined by a rectangle on the view plane, a front plane, and a back plane. The window is a frustum in a perspective projection. A portion of a drawing created by zooming in. Often used as a verb to indicate the displayed result of zooming either in or out. A glass or crystal used to transmit the emerging laser beam out of the cavity.

window clipping
Clipping against the sides of the view volume defined by a window.

windowing
A technique for reducing data-processing requirements by electronically defining only a small portion of the image to be analyzed. All other parts of the image are ignored.

window system
A system that divides a terminal's screen into pieces, as when various-sized pieces of paper are arranged on a desk.

windup
Colloquial term describing the twisting of a shaft under torsional load, so called because the twist usually unwinds, sometimes causing vibration or other undesirable effects.

wire chart
The primary end product of a CAD/CAM wiring diagram system, which generates it automatically. A wire chart lists all devices, connections, and properties in a wiring diagram indicating physical locations of devices and connections, as well as the optimum wiring order for each connection to be made.

wire frame
A method of geometric modeling in which a two- or three-dimensional object is represented by curve segments that are edges of the object. In the context of this specification, *wire-frame* and *edge-vertex* models are considered as the same technique and terms are used interchangeably. A type of display used in computer graphics that shows three-dimensional objects as only their boundaries, without removing hidden lines or showing shading and surfaces.

wire-frame model
A 3-D model, similar to a wire frame, in which the object is defined in terms of edges and vertices.

wire list
Wire-run list containing only two connections in each wire. It may also be called a from-to list.

wire net
A set of electrical connections in a logical net, having equivalent characteristics and a common identifier. No physical order of connections is implied. A wire net can be generated automatically by a CAD system.

wire-net list
Report that can be generated automatically by a CAD system; lists the electrical connections of wire net without reference to their physical order.

wire tags
Tags containing wire or net identification numbers and wrapped around both ends of each connection in an electrical system. Tags can be generated by a CAD/CAM system.

wire wrap
An interconnection technology that is an alternative to conventional etched printed circuit board technology. Wires are wrapped around the pins—a process normally done semiautomatically.

wiring diagram
Diagram containing components, wire runs, wires, miscellaneous graphic and nongraphic information, and text annotation.

wiring elementary
A wiring diagram of an electrical system in which all devices of that system are drawn be-

tween vertical lines that represent the power source. Can be generated by a CAD system. Also called a ladder diagram.

witness line
An annotation entity consisting of line segments and used in engineering drawings to indicate the beginning or the end of a measurement.

word
A group of characters occupying one storage location in a computer. It is treated by the computer circuits as an entity, by the control unit as an instruction, and by the arithmetic unit as a quantity. An ordered set of characters that may be used to cause a specific action of a machine tool. A grouping or a number of bits in a sequence that is treated as a unit. The basic storage unit of a computer's operation; a sequence of bits—commonly from 8 to 32—occupying a single storage location and processed as a unit by the computer.

word address format
Addressing each word in a block by one or more characters that identify the meaning of the word.

word length
The number of bits in a word. Longer word lengths increase efficiency and accuracy but add complexity and cost. Word length limits the number of memory locations that can be directly accessed using single-word addresses.

word time
In a storage device that provides serial access to storage locations, the time interval between the appearance of corresponding parts of successive words.

work breakdown structure (WBS)
A structured index to all elements of work and all end items produced by a product program.

work cell
A manufacturing unit consisting of one or more workstations.

work center
An administrative or accounting subdivision of a department. A specific production facility that may consist of one or more persons or machines.

work-center efficiency
The ratio of standard or allowed hours divided by actual hours worked on an operation.

work content
Time required by a module to perform a specific operation or set of operations on a given part.

work coordinates
The coordinate system referenced to the workpiece, jig, or fixture.

work cycle
A pattern of motions or processes that is repeated with negligible variation each time an operation is performed. A succession of operations or processes that is repeated with negligible variation each time a unit of production is completed.

work element
In process planning, it is a single task to be performed that cannot be subdivided.

work envelope
The set of points representing the maximum extent or reach of the robot hand or working tool in all directions. The work envelope can be reduced or restricted by limiting devices that establish limits that will not be exceeded in the event of any foreseeable failure of the robot or its controls. The maximum distance that the robot can travel after the limit device is actuated will be considered the basis for defining the restricted (or reduced) work envelope.

work factor
A system of predetermined motion time standards employing the work factor as an index of motion difficulty. This system is used for determining efficient methods and setting performance time standards.

work file
In sorting, an intermediate file used for temporary storage of data.

working coordinates
The coordinate system referenced to the earth or the shop floor.

working drawing
In plant design, a detailed layout of a component with complete dimensions and notes, approved for production. It can be generated by a CAD/CAM system.

working envelope
The set of points representing the maximum extent or reach of the robot hand or working tool in all directions.

working range
The volume of space that can be reached by maximum extensions of the robot's axis. The

Working Envelope
Working Envelope for Industrial Robot

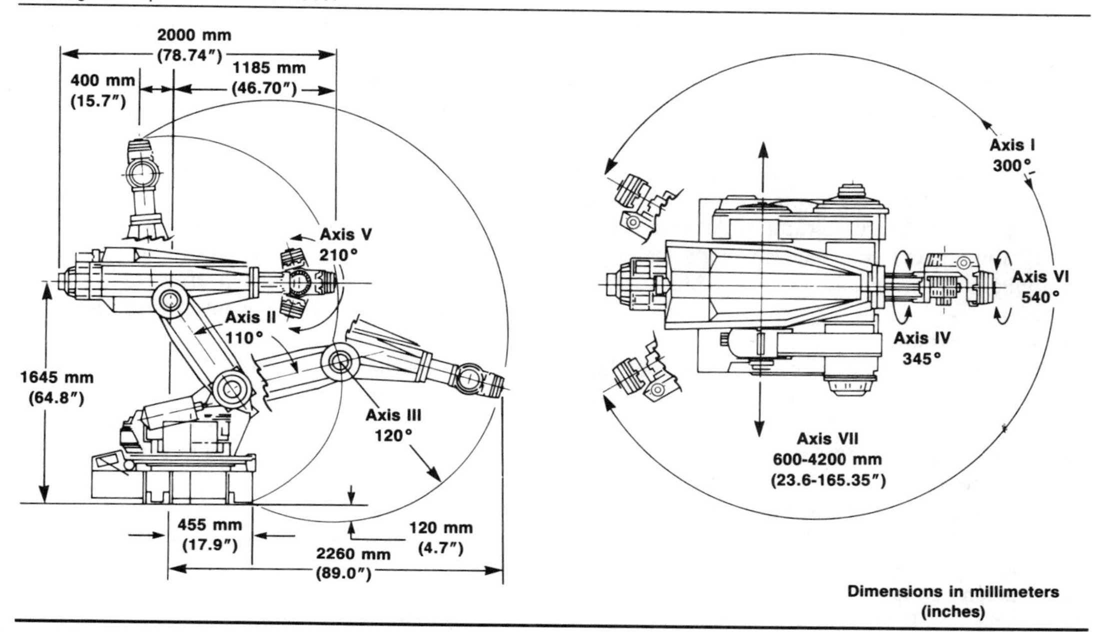

range of any variable within which the system normally operates.

working set
The set of a user's program pages in a virtual storage system that must be active in order to avoid excessive paging.

working space or volume
The physical space bounded by the working envelope in physical space.

working storage
That part of the system's internal storage reserved for intermediate results (i.e., while a computer program is still in progress). Also called temporary storage.

work in process
All materials, parts, and subassemblies that lie between release of the raw material to make a part and the finished product in finished-goods inventory.

work measurement
An analytical approach to establishing labor time standards, in which the work content or each work element is "measured" by timing and judgment.

work order
An authorization document to allow labor to be performed on a part.

workpiece
Any part in any stage of manufacture prior to its becoming a finished part. The piece of metal that is being worked.

workplace
An assigned location where a worker performs fabrication or assembly of materials or parts through manual or automatic means.

work queue
A queue of work that accumulates at a work center.

workstation
That section of a production center where the worker performs assigned tasks; includes the space required for his or her auxiliary equipment such as tools, a workbench, or a machine with any stands, containers, conveyors, and the like for the material being worked on. A manufacturing unit generally consisting of one numerical control machine tool or measuring machine, usually tended by a robot. The work

area and equipment used for CAD/CAM operations. It is where the user interacts (communicates) with the computer. Frequently consists of a CRT display and an input device as well as, possibly, a digitizer and a hardcopy device. In a distributed processing system, a workstation would have local processing and mass-storage capabilities. Also called a terminal or design terminal. An abstraction of physical graphics devices providing the logical interface through which the application program controls physical devices.

workstation attribute
Attributes that have values that are specific to a workstation. Workstation attribute values are workstation-dependent.

workstation attribute registers
Conceptual registers in which the current value of each workstation attribute is kept during structure traversal. These values determine the actual appearance of the image.

workstation attribute table
A table of all workstation attribute table entries for a given table-driven attribute, including bundled attributes. The attribute tables are referenced indirectly through an index. Not all workstation attributes are table-driven.

workstation category
Determines whether a particular workstation type can process graphics input only, graphics output only, or both output and input.

workstation-dependent
Entities that can vary from workstation to workstation.

workstation description table
A defined data structure that gives static (fixed) information for a particular workstation type.

workstation identifier
Uniquely identifies a particular workstation.

workstation-independent
Entities that are the same across all workstations.

workstation layout
The arrangement of the tools, fixtures, bins, chutes, and other equipment at a specific workstation.

workstation mandatory
Describes a property that must be realized identically on all workstations of a standard implementation.

workstation-state list
A defined data structure that gives dynamic (variable) information for a particular workstation.

workstation transformation
An isotropic transformation that maps the boundary and interior of a workstation window into the boundary and interior of a workstation viewport, performing translation and scaling. The transformation maps NDC coordinates (GKS) or LDC coordinates (PHIGS) to device coordinates. The effect of preserving aspect ratio is that the interior of the workstation window may not map to the whole of the workstation viewport.

workstation type
A type or class of actual workstation, sharing common characteristics and a single workstation-description table.

workstation viewport
A portion of display space currently selected for output of graphics.

workstation window
A rectangular region within the NDC coordinate system (GKS) or LDC coordinate system (PHIGS) that is represented on a display surface and is mapped to a workstation viewport.

world coordinate
A device-independent Cartesian coordinate system used by the application program to organize two-dimensional or three-dimensional modeled objects for display. The effect of applying the composite modeling transformation to modeling coordinates is to produce world coordinates.

world knowledge
In artificial intelligence, knowledge about the world (or domain of interest).

world model
A representation of the current situation.

worm gear
A short screw that mates to a gear whose axis of rotation is perpendicular to and offset from that of the worm screw. When the screw is turned, it drives the gear in rotation.

wraparound
The continuation from the maximum addressable location in storage to the first addressable location. On a CRT display device, the continuation from the last character position in the display buffer to the first position in the display buffer.

wrist

A set of rotary joints between the arm and robot end-effector that allow the end-effector to be oriented to the workpiece.

wrist force sensor

Consists of a structure with some compliant sections and transducers that measure the compliant sections along three orthogonal axes, as a result of the applied force and torque. There are different types of force transducers, such as strain gage, piezoelectric, magnetostrictive, magnetic, and others. A robot wrist force sensor measures the three components of force and three components of torque between the hand and the terminal link of the manipulator.

wrist movement

A robot ability that can make a minor contribution to the shape and size of the work envelope. However, the main significance of wrist movement is the ability to orient the gripper or any other end-of-arm tooling. Pitch refers to wrist movement in the vertical plane; yaw represents movement in the horizontal plane (swing); and the ability to rotate is denoted by roll.

write

The process of loading information into computer or robot controller memory. Process of inserting data into memory. This is a destructive process, in that any data already in a particular memory location are destroyed when new data are written into that location.

write-once, read-many (WORM)

A type of optical drive that uses a special pregrooved media in which the data bits are written using a diode laser in the drive. Optical drives are similar to OROMs in that once the data is written, it can be read many times but never erased.

write protect

A security feature in a CAD/CAM data storage device that prevents new data from being written over existing data.

written standard practice

A standard practice that has been recorded and approved by the proper authority or authorities.

Xenix
A multiuser operating system developed by Microsoft, a subset of Unix.

xeroradiography
A process utilizing a layer of photoconductive material on an aluminum sheet upon which an electrical charge is placed. After X-ray exposure, the electrical potential remaining on the plate in the form of a latent electrical pattern is developed by contact with a cloud of finely dispersed powder.

XL/NC
An all-purpose NC programming system for mills, lathes, grinders, and wire EDM machines. XL/NC requires no vocabulary; all programming is done in response to computer prompts and menus. Parts are described in "sections" so that each pattern, contour, or pocket may be mirrored, scaled, rotated, reversed, and roughed-out independently.

X-ray
Electromagnetic radiation, of wavelength less than about 500 angstrom units, emitted as the result of deceleration of fast-moving electrons (bremsstrahlung, continuous spectrum) or decay of atomic electrons from excited orbital states (characteristic radiation). Specifically, the radiation produced when an electron beam of sufficient energy impinges upon a target of suitable material.

***x*, *y* coordinates**
The horizontal (row) and vertical (column) designation of a position or dot in a matrix.

***xy* recorder**
A recorder that traces on a chart the relationship between two variables, neither of which is time. Sometimes the chart moves and one of the variables is controlled so that the relationship does increase in proportion to time.

Y

y

Used in either small- or capital-letter form for the vertical direction on or forward and back direction on graphs, displays, machine motion, and so on. Referring to video signals, the intensity component of the signal.

YAG

Yttrium aluminum garnet, a synthetic crystal used as a laser rod host material in an YAG:ND laser.

YAG:ND laser

A solid-state laser that uses a crystal of yttrium aluminum garnet (YAG), doped with neodymium as the lasing medium.

yaw

Rotation of the end-effector in a horizontal plane around the end of the manipulator arm. The angular displacement of a moving body about an axis that is perpendicular to the line of motion and to the top side of the body. Side-to-side motion at an axis.

yield

The ratio of usable output from a process to the materials of value input to the process. Yield is usually expressed as a percentage and may be in terms of total input or of a specific raw material. In IC manufacture, the number of good chips obtainable per wafer. It is inversely proportional to the area of the chip, that is, as the area of a chip increases, the number of good chips on that wafer will decrease proportionately.

yield factor

The precentage of material output from a process to the material that went into the process.

yield point

The first stress in a material, usually less than the maximum attainable stress, at which an increase in strain occurs without an increase in stress. Only certain metals exhibit a yield point. If there is a decrease in stress after yielding, a distinction may be made between upper and lower yield points.

yield strength

The stress at which a material exhibits a specified deviation from proportionality of stress and strain. An offset of 0.2 percent is used for many metals.

yon

Another name for the back-clipping plane, the imaginary surface that limits how far back objects are shown in a computer graphic showing three-dimensional objects. Another name for this plane is far.

Z

z axis
The vertical direction in three-dimensional space.

z-clipping
Ability to specify depth parameters for a three-dimensional drawing, such that all elements that are above or below the specified depth(s) become invisible. No change is made to the data base of the part or drawing. Useful in viewing cluttered or complex part geometry.

zero
The origin of all coordinate dimensions, defined in an absolute system as the intersection of the baselines of the x,y,z axes.

zero-address instruction
An instruction that contains no address part and is used when the address is implicit or when no address is required.

zero offset
A means of shifting the coordinate zero point from a fixed known zero point. The zero offset "shift" is normally accomplished by a series of switches or dials. On an NC unit, this feature allows the zero point on an axis to be relocated anywhere within a specified range, thus temporarily redefining the coordinate frame of reference.

zero point
The origin of a coordinate system.

zero reset
The ability to automatically realign each axis slide to the zero or target point by pressing a button when the slides are brought within close proximity of the target point.

zero suppression
The elimination of nonsignificant zeros, either before or after the significant figures in a tape command. The purpose is to reduce the number of characters that are required to be read.

zircon sand
A very refractory mineral, composed chiefly of zirconium silicate of extreme fineness, having low thermal expansion and high thermal conductivity.

zoom
To enlarge or reduce the size of an image. It may be done electronically or optically. A CAD capability that proportionately enlarges or reduces a figure displayed on a CRT screen.

zooming
The process of scaling all elements of a picture to give the appearance of having moved toward or away from a point or object of interest.

zoom ratio
On a graphics display that can magnify a selected portion of the image, the number of times the image is magnified, measured along an edge.

Points of Contact

INDUSTRIAL ROBOT ORGANIZATIONS

Robotics Associations

Association Française de Robotique Industrieele (AFRI)
89 Rue Falgueire
75015 Paris
France

British Robot Assoc.
35-39 High St.
Kempston
Bedford MK42 7BT
England

Japan Industrial Robot Assoc. (JIRA)
Kikai Shinko Kaikan Bldg.
3-5-8 Shiba-koen
Minato-ku
Tokyo 105
Japan

Robotics International of SME
One SME Dr.
P.O. Box 930
Dearborn, MI 48128
(313) 271-0778

Robot Institute of America
One SME Dr.
P.O. Box 930
Dearborn, MI 48128
(313) 271-0778

Societa Italiana Robotica Industriale (SIRI)
Instituto di Elettrotechnica ed Elettronics
Politechmico di Milano
Piazza Leonardo da Vinci 32
20133 Milano
Italy

Swedish Industrial Robot Assoc. (SWIRA)
Box 5506
Storgatan 19
S-14 85 Stockholm
Sweden

Robot Manufacturers

Accumatic Machinery Corp.
3537 Hill Ave.
Toledo, OH 43607
(419) 535-7997

Acrobe Positioning Systems, Inc.
3219 Doolittle Dr.
Northlake, IL 60062
(312) 273-4302

Advanced Robotics Corp.
Newark Ohio Industrial Park
Bldg. 8, Rt. 79
Hebron, OH 43025
(614) 929-1065

Ameco Corp.
P.O. Box 385
W158 N9335 Nor-X-Way Ave.
Menomonee Falls, WI 53051
(414) 262-2085

American Robot Corp.
354 Hookstown Rd.
Clinton, PA 15026
(412) 262-2085

Anorad Corp.
110 Oser Ave.
Hauppauge, NY 11788
(516) 231-1990

Armax Robotics, Inc.
38700 Grand River Ave.
Farmington Hills, MI 48018
(313) 478-9330

Automatix, Inc.
217 Middlesex Tpk.
Burlington, MA 01803
(617) 273-4340

Automaton Corp.
23996 Freeway Park Dr.
Farmington Hills, MI 48024
(313) 471-0554

Binks Manufacturing Co.
9201 W. Belmont Ave.
Franklin Park, IL 60131
(312) 671-3000

Cincinnati Milacron
215 S. West St.
South Lebanon, OH 45036
(513) 932-4400

Comet Welding Systems
900 Nicholas Blvd.
Elk Grove Village, IL 60007
(312) 956-0126

Control Automation, Inc.
P.O. Box 2304
Princeton, NJ 08540
(609) 799-6026

Cybotech Corp.
P.O. Box 88514
Indianapolis, IN 46208
(317) 292-7440

The DeVilbiss Co.
300 Phillips Ave.
P.O. Box 913
Toledo, OH 43692
(419) 470-2169

Elicon
273 Viking Ave.
Brea, CA
(714) 990-6647

Expert Automation
40675 Mound Rd.
Sterling Heights, MI 40878
(313) 977-0100

Fleximation Systems Corp.
53 Second Ave.
Burlington, MA 01803
(617) 229-6670

General Electric Co.
Automatic Systems
1285 Boston Ave.
Bridgeport, CT 06602
(203) 382-2876

General Numeric Corp.
390 Kent Ave.
Elk Grove Village, IL 60007
(312) 640-1595

GMFanuc Robotics Corp.
5600 New King St.
Troy, MI 48098
(313) 641-4100

Graco Robotics, Inc.
12898 Westmore Ave.
Livonia, MI 48150
(313) 261-3270

Hobart Brothers Co.
600 W. Main St.
Troy, OH 45473
(513) 339-6011

Hodges Robotics
International Corp.
3710 N. Grand River
Lansing, MI 48906
(517) 323-7427

IBM
P.O. Box 1328
Boca Raton, FL 33432
(305) 998-2000

Ikegai America Corp.
770 W. Algonquin Rd.
Arlington Heights, IL 60005
(312) 437-1488

Industrial Automates, Inc.
6123 W. Mitchell St.
Milwaukee, WI 53214
(414) 327-5656

Intelledex
33840 Eastgate Cir.
Corvallis, OR 97333
(503) 758-4700

International Intelligence/Robomation
6353 El Camino Real
Carlsbad, CA 92008
(714) 438-4424

I.S.I. Manufacturing, Inc.
31915 Groesbeck Hwy.
Fraser, MI 48026
(313) 294-9500

Lamson Corp.
P.O. Box 4857
Syracuse, NY 13221
(315) 432-5500

Mack Corp.
3695 E. Industrial Dr.

Flagstaff, AZ 86001
(602) 526-1120

Microbot, Inc.
453-H Ravendale Dr.
Mountain View, CA. 94043
(415) 968-8911

Mobot Corp.
980 Buenos Ave.
San Diego, CA 92110
(714) 275-4300

Nordson Corp.
555 Jackson St.
Amherst, OH
(216) 988-9411

Nova Robotics
262 Prestige Park Rd.
East Hartford, CT 06108
(203) 528-9861

Pickomatic Systems
37950 Commerce
Sterling Heights, MI 48077
(313) 939-9320

Positech Corp.
Rush Lake Rd.
Laurens, IA 50554
(712) 845-4548

Prab Robots, Inc.
5944 E. Kilgore Rd.
Kalamazoo, MI 49003
(616) 349-8761

Reis Machines
1426 Davis Rd.
Elgin, IL 60120
(312) 741-9500

Rob-Con, Ltd.
12001 Globe
Livonia, MI 48150
(313) 591-0300

Sandhu Machine Design, Inc.
308 S. State St.
Champaign, IL 61820
(217) 352-8485

Schrader Bellows/Scovill, Inc.
200 W. Exchange St.
Akron, OH 44309
(216) 375-5202

Seiko Instruments, Inc.
2990 W. Lomita Blvd.
Torrance, CA 90505
(213) 330-8777

Sigma
6505C Serrano Ave.
Annaheim, CA 92807
(714) 974-0166

Sormel/Black & Webster
281 Winter St.
Waltham, MA 02254
(617) 890-9100

Sterling Detroit Co.
261 E. Goldengate Ave.
Detroit, MI 48203
(313) 366-3500

Swanson-Erie Corp.
814 E. 8th St.
P.O. Box 1217
Erie, PA 16512
(814) 453-5841

TecQuipment, Inc.
P.O. Box 1074
Acton, MA 01720
(617) 263-1767

Thermwood Corp., Inc.
P.O. Box 436
Acton, MA 01720
(812) 937-4476

Unimation/Westinghouse, Inc.
Shelter Rock Ln.
Danbury, CT 06810
(203) 744-1800

United States Robots
650 Park Ave.
King of Prussia, PA 19406
(215) 768-9210

Westinghouse Electric Corp.
Industry Automation Div.
400 High Tower Office Bldg.
Pittsburgh, PA 15205
(412) 778-4349

Yaskawa Electric America, Inc.
305 Era Dr.
Northbrook, IL 60062
(312) 564-0770

Robot Manufacturers (Europe)

A.O.I.P. Kremlin Robotique
6 rue Maryse Bastie
9100 Evry
France

ASEA AB
S-72183 Vasteras
Sweden

British Federal Welder & Machine Co., Ltd.
Castle Mill Works
Dudley
West Midlands, DY1 4DA
United Kingdom

Camel Robot SRL
Palozzolo Milanese
Italy

Digital Electronics Automation SpA
Co Torino 70
Moncalieri, Piemonte 10024
Italy

Electrolux AB
Industrial Systems
S-105 45 Stockholm
Sweden

Fiat Auto SpA
CSO Agnelli 200
Torino, Piemonte
Italy

Hall Automation, Ltd.
Colonia Way
Watford
Herts, WD2 4FG
United Kingdom

Jungheinrich Unternelmensverwaltung
Friedrich-Ebert-Dabb 129
2000 Hamburg 70
West Germany

KUKA
Schweissanlagen & Roboter GmbH
P.O. Box 431280
Zugspitzstr. 140
D-8900
Augsburg 43
West Germany

Mouldmation Limited
2 Darwin Close
Burntwood, Walsall
Staffs, WS7 9HP
United Kingdom

Nimak
Werkstrabe
Postfach 86
5248 Wissen/Sieg
West Germany

Olivetti SpA
Controllo Numerico
Fr S Bernardo
V Torino 603
Ivrea, Piemonte
Italy

Pendar
Bridgwater
Somerset
United Kingdom

Regie Nationale des
Usines Renault SA
66Av Edouard Vaillaut
Boulogne-Billancourt
France

R. Kaufeldt AB
P.O. Box 42139
S-126 Stockholm
Sweden

Sormel
rue Becquerel
25009 Besanicon Cedex
France

Trallfa
Paint-Welding Robot Systems

P.O. Box 113
4341 Bryne
Norway

Unimation, Inc.
Units A3/A4
Stafford Park 4
Telford, Salop
United Kingdom

Volkswagenwerk AG
Abt. Industrieverkauf
3180 Wolfsburg
West Germany

Robot Manufacturers (Japan)

Dainichi Kiko Co., Ltd.
Kosai-cho
Nakakomagun Yeamanshi Pref.
400-04
Japan

Fanuc, Ltd.
3-5-1
Asahigoaka, Hino City
Tokyo
Japan

Hitachi, Ltd.
Shin-Maru Bldg.
1-5-1
Marunouchi, Chiyoda-ku
Tokyo
Japan

Kawasaki Heavy Industries, Ltd.
World Trade Center Bldg.
2-4-1
Hamamatsucho, Minato-ku
Tokyo
Japan

Matsushita Industrial Equipment Co., Ltd.
3-1-1 Inazumachi
Toyonaka City Osaka Pref.
Japan

Mitsubishi Heavy Industries, Ltd.
2-5-1
Marunouchi, Chiyoda-ku
Tokyo
Japan

Sankyo Seiki Manufacturing Co., Ltd.
1-17-2
Shunbashi, Minati-ku
Tokyo 105
Japan

Tokico, Ltd.
1-6-3
Funta, Kawasaki-ku
Kawasaki City
Kanagaw Pref.
Japan

Yaskawa Electric Mfg. Co., Ltd.
Ohtemachi Bldg.
1-6-1
Ohtemachi, Chiyoda-ku
Tokyo
Japan

Robot Rental/Lease Firms

Hi-Tech Assembly
8130 N. Knox
Skokie, IL 60076
(312) 676-0080

Rob-Con, Ltd.
12001 Globe Rd.
Livonia, MI 48150
(313) 591-0300

Thermwood Machinery Manufacturing Co., Inc.
P.O. Box 436
Dale, IN 47523
(812) 937-4476

Robot Consulting/Applications Firms

Automation Systems/American Technologies
1900 Pollitt Dr.
Fair Lawn, NJ 07410
(201) 797-8200

Blanarovich Engineering
Box 292
Don Mills, Ontario M3C 2S2
Canada
(416) 438-6313

Franklin Institute Research Laboratory, Inc.
The Benjamin Franklin Pkwy.
Philadelphia, PA 19103
(215) 448-1000

Productivity Systems, Inc.
21999 Farmington Rd.
Farmington Hills, MI 48024
(313) 474-5454

RMT Engineering, Ltd.
P.O. Box 2333, Sta. B
St. Catherines, Ontario L2M 7M7
Canada
(416) 937-1550

Robot Systems, Inc.
50 Technology Pkwy.
Norcross, GA 30092
(404) 448-4133

Technology Research Corp.
8328-A Traford Ln.
Springfield, VA 22152
(703) 451-8830

U.S. Robotics Research Organizations

Carnegie-Mellon University
The Robotics Institute
Schenley Park
Pittsburgh, PA 15213
(412) 578-2597

Charles Stark Draper Laboratory, Inc.
555 Technology Sq.
Cambridge, MA 02139
(617) 258-1000

Environmental Research Institute of Michigan
Robotics Program
P.O. Box 8618
Ann Arbor, MI 48107
(313) 994-1200

George Washington University
725 23rd St. N.W.
Washington, DC 20052
(202) 676-6083

Jet Propulsion Laboratories
Robotics Group
4800 Oak Grove Dr.
Pasadena, CA 91103
(213) 354-6101

MIT
Artificial Intelligence Laboratory
545 Technology Sq.
Cambridge, MA 02139
(617) 253-6218

National Bureau of Standards
Bldg. 220, Rm. A123
Washington, DC 20234
(301) 921-2381

Naval Research Laboratory
Code 7505
Washington, DC 20375
(202) 545-6700

North Carolina State University
Raleigh, NC 27650
(919) 737-2336

Purdue University
School of Electrical Engineering
West Lafayette, IN 47906
(317) 749-2607

Rensselaer Polytechnic Institute
Rm. 5304 JEC
Troy, NY 12181
(518) 270-6724

Robotics Institute
Carnegie-Mellon University
Schenley Park
Pittsburgh, PA 15213
(412) 578-3611

SRI International
Artificial Intelligence Center

Menlo Park, CA 94025
(415) 859-2311

Stanford University
Artificial Intelligence Laboratory
Dr. John McCarthy, Director
Stanford, CA 94305
(415) 497-2797

Texas A&M University
Dept. of Industrial Engineering
College Station, TX 77840
(713) 845-5531

United States Air Force
AFWAL/MLTC (USAF ICAM)
Wright Patterson AFB, OH 45433
(513) 255-2232

University of Central Florida
College of Engineering
P.O. Box 25000
Orlando, FL 32816
(305) 275-2236

University of Cincinnati
Institute of Applied Interdisciplinary Research
Location 72
Cincinnati, OH 45221

University of Florida
Institute for Intelligent Machines & Robotics
Rm. 300, Mechanical Engineering
Gainesville, FL 32601
(904) 392-0814

University of Michigan
Robotics Program
ECE Dept.
Ann Arbor, MI 48109
(313) 764-7139

University of Rhode Island
College of Engineering
102 Bliss Hall
Kingston, RI 02881
(401) 792-2187

University of Wisconsin
1513 University Ave.
Madison, WI 53706
(608) 262-3543

VISION SYSTEM ORGANIZATIONS

System Manufacturers

Adaptive Technologies, Inc.
600 W. North Market Blvd., No. 1
Sacramento, CA 95834
(916) 920-9119

Adept Technology, Inc.
1212 Bordeaux Dr.
Sunnyvale, CA 94089
(408) 747-0111

Advanced Robotics Corp.
777 Manor Park Dr.
Columbus, OH 43228
(614) 870-7778

American Robot
121 Industry Dr.
Pittsburgh, PA 15275
(412) 787-3000

Analog Devices
3 Technology Way
Norwood, MA 02062
(617) 329-4700

Applied Intelligent Systems
110 Parkland Plaza
Ann Arbor, MI 48103
(313) 995-2035

Applied Scanning Technology
1988 Leghorn St.
Mountain View, CA 94043
(415) 967-4211

ASEA Robitics, Inc.
16250 W. Glendale Dr.
New Berlin, WI 53151
(414) 875-3400

Autoflex, Inc.
25880 Commerce Dr.
Madison Heights, MI 48071
(313) 398-9911

Automatic Inspection Devices
One SeaGate
Toledo, OH 43666
(419) 247-5000

Automatic Vision
1300 Richard St.
Vancouver, BC V6B 3G6
Canada

Automation Intelligence
1200 W. Colonial Dr.
Orlando, FL 32804-7194
(305) 237-7030

Automation Systems, Inc.
1106 Federal Rd.
Brookfield, CT 06804
(203) 775-2581

Automatix
1000 Tech Park Dr.
Billerica, MA 01821
(617) 667-7900

Beeco, Inc.
4175 Millersville Rd.
Indianapolis, IN 46205
(317) 547-1717

Cambridge Instruments, Inc.
Subsidiary of Cambridge Instruments, Ltd.
40 Robert Pitt Dr.
Monsey, NY 10952
(914) 356-3331 or 356-3877

CBIT Corporation
Horse Shoe Trail, RD 2
Chester Springs, PA 19425
(215) 469-0358

Cochlea
2284 Ringwood Ave., Unit C
San Jose, CA 95131
(408) 942-8228

Cognex
72 River Park St.
Needham, MA 02194
(617) 449-6030

Computer Systems Co.
26401 Harper Ave.
St. Clair Shores, MI 48081
(313) 779-8700

Contrex
47 Manning Rd.
Billerica, MA 01821
(617) 273-3434

Control Automation
Princeton-Windsor Industrial Park
P.O. Box 2304
Princeton, NJ 08540
(609) 799-6026

Cosmos Imaging Systems, Inc.
30100 Crown Valley Pkwy.
Suite 32
Laguna Niguel, CA 92677
(714) 495-2662

CR Technology
1701 Reynolds Ave.
Irvine, CA 92714
(714) 751-6901

Cybotech
P.O. Box 88514
Indianapolis, IN 46208
(317) 298-5890

DCI Corp.
110 S. Gold Dr.
Robbinsville, NJ 08691
(609) 587-9132

Diffracto, Ltd.
6360 Hawthorne Dr.
Windsor, Ontario N8T 1J9
Canada
(519) 945-6373

Digital/Analog Design Assoc.
530 Broadway
New York, NY 10012
(212) 966-0410

Eaton Corp.
4201 N. 27th St.
Milwaukee, WI 53216
(414) 449-6345

Eigen/Optivision
P.O. Box 848
Nevada City, CA 95959
(916) 272-3461

Electro-Optical Information Systems
710 Wilshire Blvd.
Suite 501
Santa Monica, CA 90401
(213) 451-8566

Everett/Charles Test Equipment, Inc.
2887 N. Towne Ave.

Pomona, CA 91767
(714) 621-9511

Federal Products Corp.
Boice Div.
P.O. Box 12-185
Albany, NY 12212
(518) 785-2211

Gallaher Enterprises
P.O. Box 10244
Winston-Salem, NC 27108-0318
(919) 725-8494

General Electric
P.O. Box 17500
Orlando, FL 32860-7500
(305) 889-1200

General Numeric Corp.
390 Kent Ave.
Elk Grove Village, IL 60007
(312) 640-1595

GMF Robotics Corp.
Northfield Hills Corp. Ctr.
5600 New King St.
Troy, MI 48098
(313) 641-4242

Hamamatsu Systems
40 Bear Hill Rd.
Waltham, MA 02154
(617) 890-3440

Ham Industries
835 Highland Rd.
Macedonia, OH 44056
(216) 467-4256

Hitachi America, Ltd.
50 Prospect Ave.
Tarrytown, NY 10591-4698
(914) 332-5800

Honeywell Visitronics
P.O. Box 5077
Englewood, CO 80155
(303) 850-5050

Image Data Systems
315 W. Huron, Suite 140
Ann Arbor, MI 48103
(313) 761-7222

Industrial Technology and Machine Intelligence
1 Speen St., Suite 240
Framingham, MA 01701
(617) 620-0184

Industrial Vision Systems, Inc.
Victory Research Park
452 Chelmsford St.
Lowell, MA 01851
(617) 459-9000

Integrated Automation
2121 Allston Way
Berkeley, CA 94704
(415) 843-8227

Intelledex, Inc.
33840 Eastgate Cir.
Corvallis, OR 97333
(503) 758-4700

International Imaging Systems
1500 Buckeye Dr.
Milpitas, CA 95035
(408) 262-4444

International Robomation/Intelligence
2281 Las Palmas Dr.
Carlsbad, CA 92008
(610) 438-4424

IRT Corp.
3030 Callan Rd.
P.O. Box 85317
San Diego, CA 92138-5317
(619) 450-4343

Itek Optical Systems/Litton Industries
10 Maguire Rd.
Lexington, MA 02173
(617) 276-2000

ITRAN
670 N. Commercial St.
P.O. Box 607
Manchester, NH 03105
(603) 669-6332

Key Image Systems
20100 Plummer St.
Chatsworth, CA 91311
(213) 993-1911

KLA Instruments
2051 Mission College
Santa Clara, CA 95054
(408) 988-6100

L.N.K. Corp.
P.O. Box 1619
College Park, MD 20740
(301) 927-3223

Machine Intelligence Corp.
330 Potrero Ave.
Sunnyvale, CA 94086
(408) 737-7960

Machine Vision International
Burlington Center
325 Eisenhower
Ann Arbor, MI 48104
(313) 996-8033

Mack Corp.
3695 East Industrial Dr.
P.O. Box 1756
Flagstaff, AZ 86002
(602) 526-1120

Micro-Poise
P.O. Box 88512
Indianapolis, IN 46208
(317) 298-5000

Object Recognition Systems
1101-8 State Rd.
Princeton, NJ 08540
(609) 924-1667

Octek
7 Corporate Pl.
South Bedford St.
Burlington, MA 01803
(617) 273-0851

Opcon
720 80th St., S.W.
Everett, WA 98203
(206) 353-0900

Optical Gaging Products
850 Hudson Ave.
Rochester, NY 14621
(716) 544-0400

Optical Specialties
4281 Technology Dr.
Fremont, CA 94538
(415) 490-6400

Optrotech, Inc.
Suite 206
111 S. Bedford St.
Burlington, MA 01803
(617) 272-4050

Pattern Processing Technologies
511 Eleventh Ave. S.
Minneapolis, MN 55415
(612) 339-8488

Penn Video
929 Sweitzer Ave.
Akron, OH 44311
(216) 762-4840

Perceptron
23855 Research Dr.
Farmington Hills, MI 48024
(313) 478-7710

Photonic Automation, Inc.
3633 W. McArthur Blvd.
Santa Ana, CA 92704
(714) 546-6651

Photo Research Vision Systems
Div. of Kollmorgen
3099 N. Lima St.
Burbank, CA 91504
(818) 954-0104

Prothon
Div. of Video Tek
199 Pomeroy Rd.
Parsippany, NJ 07054
(201) 887-8211

Quantex Corp.
252 N. Wolfe Rd.
Sunnyvale, CA 94086
(408) 733-6730

Rank Videometrix
9421 Winnetka
Chatsworth, CA 91311
(818) 343-3120

Recognition Concepts, Inc.
924 Incline Way
P.O. Box 8510
Incline Village, NV 89450
(702) 831-0473

Robotic Vision Systems
425 Rabro Dr. E.
Hauppauge, NY 11788
(526) 273-9700

Selcom
P.O. Box 250
Valdese, NC 28690
(704) 874-4102

SILMA, Inc.
1800 Embarcadero
Palo Alto, CA 94301
(415) 493-0145

Spatial Data Systems
420 S. Fairview Ave.
P.O. Box 978
Goleta, CA 93117
(805) 967-2383

Syn-Optics
1225 Elko Dr.
Sunnyvale, CA 94089
(408) 734-8563

Synthetic Vision Systems
2929 Plymouth Rd.
Ann Arbor, MI 48105
(313) 665-1850

Technical Arts Corp.
180 Nickerson St., No. 303
Seattle, WA 98109
(206) 282-1703

Testerion, Inc.
9645 Arrow Hwy.
P.O. Box 694
Cucamonga, CA 91730
(714) 987-0025

3M, Vision Systems
Suite 300
8301 Greensboro Dr.
McLean, VA 22102
(703) 734-0300

Time Engineering
1630 Big Beaver Rd.
Troy, MI 48083
(313) 528-9000

Unimation
Shelter Rock Ln.
Danbury, CT 06810
(203) 744-1800

United Detector Technology
12525 Chadron Ave.
Hawthorne, CA 90250
(213) 978-0516

Vektronics
5750 El Camino Real
P.O. Box 459
Carlsbad, CA 92008
(619) 438-0992

Vicom Systems
2520 Junction Ave.
San Jose, CA 95134
(408) 946-5660

Videk
Div. of Eastman Technology
343 State St.
Rochester, NY 14650
(800) 445-6325, Ext. 15

View Engineering
1650 N. Voyager Ave.
Simi Valley, CA 93063
(805) 522-8439

Visionetics
P.O. Box 189
Brookfield Center, CT 06805
(203) 775-4770

Vision System Technologies
1532 S. Washington Ave.
Piscataway, NJ 08854
(201) 752-6700

Visual Matic
2171 El Camino Real
Oceanside, CA 92054
(619) 722-8299

Vuebotics Corp.
6086 Corte Del Cedro
Carlsbad, CA 92008
(619) 438-7994

MACHINE VISION CAMERA MANUFACTURERS

AFP Imaging
Video Products Group
50 Executive Blvd.
Elmsford, NY 10523
(914) 592-6100

Artek Systems
170 Finn Ct.

Farmingdale, NY 11735
(516) 293-4420

Circon
749 Wood Dr.
Santa Barbara, Ca 93111
(805) 967-0404

Cohu, Inc./Elec... Div.
5725 Kearny ... Rd.
Box 85623 ... 2138
San Dieg...
(619) 2...

...on, Inc.
...r Rd.
Cy... 44313
...68293

...en Systems Corp.
...Walsh Ave.
...a Clara, CA 95050
...8) 727-6766

Datacopy Corp.
1215 Terra Bella
Mountain View, CA 94043
(415) 965-7900

EG&G Reticon
345 Potrero Ave.
Sunnyvale, CA 94086
(408) 738-4266

Eikonix
23 Crosby Dr.
Bedford, MA 01730
(617) 275-5070

Fairchild CCD Imaging
3440 Hillview Ave.
Palo Alto, CA 94304
(415) 493-8001

General Electric Company
Electronic Camera Operation
890 7th North St.
Liverpool, NY 13088
(315) 456-2834

ITT
Electro-Optical Products Div.
3700 E. Pontiac St.
Fort Wayne, IN 46801
(219) 423-4341

Javelin Electronics
19831 Magellan Dr.
P.O. Box 2033
Torrance, CA 90502
(213) 327-7440

Micron Technology, Inc.
2805 E. Columbia Rd.
Boise, ID 83706
(208) 383-4000

MII
P.O. Box 395
Birdsboro, PA 19508
(215) 562-5361

Optron Corp.
30 Hazel Terr.
Woodbridge, CT 06525
(203) 389-5384

Panasonic Industrial Co.
Electronic Components Div.
One Panasonic Way
Secaucus, NJ 07094
(201) 348-5277

Periphicon
P.O. Box 324
Beaverton, OR 97075
(503) 222-4966

Pulnix America, Inc.
453 Ravendale Dr.
Mountain View, CA 94043
(415) 964-0955

RCA
Closed-Circuit Video Equipment
New Holland Ave.
Lancaster, PA 17604
(717) 397-7661

Sierra Scientific
2598 Bayshore Frontage Rd.
Mountain View, CA 94043
(415) 969-9315

Sony Corp. of America
15 Essex Rd.
Paramus, NJ 07652
(201) 368-5000

Thomson–CSF
Components Corp.
Electron Tube Div.

301 Route 17 N.
Rutherford, NJ 07070
(201) 438-2300

Video Logic Corp.
597 N. Mathilda Ave.
Sunnyvale, CA 94086
(408) 245-8622

Video Measurements
10 Havens St.
Elmsford, NY 10523
(914) 592-2025

VSP Laboratories
670 Airport Blvd.
Ann Arbor, MI 48104
(313) 769-5522

Xybion Electronic Systems
7750-A Convoy Ct.
San Diego, CA 92111
(619) 277-8220

MACHINE VISION COMPONENT MANUFACTURERS

American Volpi
26 Aurelius Ave.
Auburn, NY 13021
(315) 255-1105

Chorus Data Systems
P.O. Box 810
27 Proctor Hill Rd.
Hollis, NH 03049
(603) 465-7100

Colorado Video Inc.
P.O. Box 928
Boulder, CO 80306
(303) 444-3972

Datacube
4 Dearborn Rd.
Peabody, MA 01960
(617) 535-6644

Data-Sud Systems/U.S.
2219 S. 48th St., Suite J
Tempe, AZ 85282
(602) 438-1492

Digital Graphics Systems
2629 Terminal Bl.
Mountain View, CA 94043
(415) 962-0200

Dolan-Jenner Industries
Blueberry Hill Industrial Park
P.O. Box 102
Woburn, MA 01801-[illegible]
(617) 935-7444

General Scanning
500 Arsenal St.
P.O. Box 307
Watertown, MA 02172
(617) 924-1010

Imaging Technology
600 W. Cummings Park
Woburn, MA 01801
(617) 938-8444

I.T.M.I.
1 Speen St., Suite 240
Framingham, MA 01701
(617) 576-2585

Microtex
80 Trowbridge St.
Cambridge, MA 02138
(617) 491-2874

Poynting Products, Inc.
P.O. Box 1227
Oak Park, IL 60304
(312) 489-6638

Toko America, Inc.
5520 W. Touhy Ave.
Skokie, IL 60077
(312) 677-3640

Videtics, Ltd.
258 King St. N.
Waterloo, Ontario N2J 2Y9
Canada
(519) 885-6852

CONSULTANTS AND SYSTEM INTEGRATORS

Anorad Corp.
110 Oser Ave.
Hauppauge, NY 11788
(516) 231-1990

Farmingdale, NY 11735
(516) 293-4420

Circon
749 Wood Dr.
Santa Barbara, Ca 93111
(805) 967-0404

Cohu, Inc./Electronics Div.
5725 Kearny Villa Rd.
Box 85623
San Diego, CA 92138
(619) 277-6700

Cyberanimation, Inc.
4621 Granger Rd.
Akron, OH 44313
(216) 666-8293

Cybergen Systems Corp.
2070 Walsh Ave.
Santa Clara, CA 95050
(408) 727-6766

Datacopy Corp.
1215 Terra Bella
Mountain View, CA 94043
(415) 965-7900

EG&G Reticon
345 Potrero Ave.
Sunnyvale, CA 94086
(408) 738-4266

Eikonix
23 Crosby Dr.
Bedford, MA 01730
(617) 275-5070

Fairchild CCD Imaging
3440 Hillview Ave.
Palo Alto, CA 94304
(415) 493-8001

General Electric Company
Electronic Camera Operation
890 7th North St.
Liverpool, NY 13088
(315) 456-2834

ITT
Electro-Optical Products Div.
3700 E. Pontiac St.
Fort Wayne, IN 46801
(219) 423-4341

Javelin Electronics
19831 Magellan Dr.
P.O. Box 2033
Torrance, CA 90502
(213) 327-7440

Micron Technology, Inc.
2805 E. Columbia Rd.
Boise, ID 83706
(208) 383-4000

MII
P.O. Box 395
Birdsboro, PA 19508
(215) 562-5361

Optron Corp.
30 Hazel Terr.
Woodbridge, CT 06525
(203) 389-5384

Panasonic Industrial Co.
Electronic Components Div.
One Panasonic Way
Secaucus, NJ 07094
(201) 348-5277

Periphicon
P.O. Box 324
Beaverton, OR 97075
(503) 222-4966

Pulnix America, Inc.
453 Ravendale Dr.
Mountain View, CA 94043
(415) 964-0955

RCA
Closed-Circuit Video Equipment
New Holland Ave.
Lancaster, PA 17604
(717) 397-7661

Sierra Scientific
2598 Bayshore Frontage Rd.
Mountain View, CA 94043
(415) 969-9315

Sony Corp. of America
15 Essex Rd.
Paramus, NJ 07652
(201) 368-5000

Thomson–CSF
Components Corp.
Electron Tube Div.

301 Route 17 N.
Rutherford, NJ 07070
(201) 438-2300

Video Logic Corp.
597 N. Mathilda Ave.
Sunnyvale, CA 94086
(408) 245-8622

Video Measurements
10 Havens St.
Elmsford, NY 10523
(914) 592-2025

VSP Laboratories
670 Airport Blvd.
Ann Arbor, MI 48104
(313) 769-5522

Xybion Electronic Systems
7750-A Convoy Ct.
San Diego, CA 92111
(619) 277-8220

MACHINE VISION COMPONENT MANUFACTURERS

American Volpi
26 Aurelius Ave.
Auburn, NY 13021
(315) 255-1105

Chorus Data Systems
P.O. Box 810
27 Proctor Hill Rd.
Hollis, NH 03049
(603) 465-7100

Colorado Video Inc.
P.O. Box 928
Boulder, CO 80306
(303) 444-3972

Datacube
4 Dearborn Rd.
Peabody, MA 01960
(617) 535-6644

Data-Sud Systems/U.S.
2219 S. 48th St., Suite J
Tempe, AZ 85282
(602) 438-1492

Digital Graphics Systems
2629 Terminal Bl.
Mountain View, CA 94043
(415) 962-0200

Dolan-Jenner Industries
Blueberry Hill Industrial Park
P.O. Box 1020
Woburn, MA 01801-0820
(617) 935-7444

General Scanning
500 Arsenal St.
P.O. Box 307
Watertown, MA 02172
(617) 924-1010

Imaging Technology
600 W. Cummings Park
Woburn, MA 01801
(617) 938-8444

I.T.M.I.
1 Speen St., Suite 240
Framingham, MA 01701
(617) 576-2585

Microtex
80 Trowbridge St.
Cambridge, MA 02138
(617) 491-2874

Poynting Products, Inc.
P.O. Box 1227
Oak Park, IL 60304
(312) 489-6638

Toko America, Inc.
5520 W. Touhy Ave.
Skokie, IL 60077
(312) 677-3640

Videtics, Ltd.
258 King St. N.
Waterloo, Ontario N2J 2Y9
Canada
(519) 885-6852

CONSULTANTS AND SYSTEM INTEGRATORS

Anorad Corp.
110 Oser Ave.
Hauppauge, NY 11788
(516) 231-1990

Automated Vision Systems
1590 La Pradera Dr.
Campbell, CA 95008
(408) 370-0229

Automation Unlimited
10 Roessler Rd.
Woburn, MA 01801
(617) 933-7288

Bahr Technologies, Inc.
1842 Hoffman St.
Madison, WI 53704
(608) 244-0500

Hansford Manufacturing
3111 Winton Road S.
Rochester, NY 14623
(716) 427-8150

Image Technology Methods Corp.
103 Moody St.
Waltham, MA 02154
(617) 894-1720

Key Technology, Inc.
517 N. Elizabeth
P.O. Box 8
Milton-Freewater, OR 97862

Medar, Inc.
38700 Grand River Ave.
Farmington Hills, MI 48018
(313) 477-3900

Multicon
7035 Main St.
Cincinnati, OH 45244
(513) 271-0200

Raycon Corp.
77 Enterprise
Ann Arbor, MI 48103
(313) 769-2614

Robotic Objectives, Inc.
60 Church St., The Forum
Yalesville, CT 06492
(203) 269-5063

Spectron Engineering
800 W. 9th Ave.
Denver, CO 80204
(303) 623-8987

TAU Corp.
10 Jackson St., Suite 101
Los Gatos, CA 95030
(408) 395-9191

Technology Research Corp.
8328-A Traford Ln.
Springfield, VA 22152
(703) 451-8830

Tech Tran Corp.
134 N. Washington St.
Naperville, IL 60540
(312) 369-9232

Vision Systems International
3 Milton Dr.
Yardley, PA 19067
(215) 736-0994

Visual Intelligence Corp.
Amherst Fields Research Park
160 Old Farm Rd.
Amherst, MA 01002
(413) 253-3482

RESEARCH ORGANIZATIONS

Carnegie-Mellon University
The Robotics Institute
Pittsburgh, PA 15213
(412) 578-3826

Environmental Research Institute of Michigan
Robotics Program
P.O. Box 8618
Ann Arbor, MI 48107
(313) 994-1200

George Washington University
725 23rd Street, N.W.
Washington, DC 20052
(202) 676-6919

Jet Propulsion Laboratories
Robotics Group
4800 Oak Grove Dr.
Pasadena, CA 91103
(213) 354-6101

Massachusetts Institute of Technology
Artificial Intelligence Laboratory
545 Technology Sq.
Cambridge, MA 02139
(617) 253-6218

National Bureau of Standards
Industrial Systems Div.
Bldg. 220, Rm. A123
Washington, DC 20234
(301) 921-2381

Naval Research Laboratory
Code 2610
Washington, DC 20375
(202) 767-3984

North Carolina State University
Dept. of Electrical Engineering
Raleigh, NC 27650
(919) 737-2376

Purdue University
School of Electrical Engineering
West Lafayette, IN 47906
(317) 749-2607

Rensselaer Polytechnic Institute
Center for Manufacturing Productivity
Jonsson Engineering Center
Troy, NY 12181
(518) 270-6724

SRI International
Artificial Intelligence Center
Menlo Park, CA 94025
(425) 497-2797

Stanford University
Artificial Intelligence Laboratory
Stanford, CA 94022
(415) 497-2797

University of Central Florida
IEMS Dept.
Orlando, FL 32816
(305) 275-2236

University of Cincinnati
Institute of Applied Interdisciplinary Research
Location 42
Cincinnati, OH 45221
(513) 475-6131

University of Maryland
Computer Vision Laboratory
College Park, MD 20742
(301) 454-4526

University of Rhode Island
Dept. of Electrical Engineering
Kingston, RI 02881
(401) 792-2187

University of Southern California
School of Engineering
University Park
Los Angeles, CA 90033

University of Texas
Austin, TX 78712
(512) 471-1331

University of Washington
Department of Electrical Engineering
Seattle, WA 98195
(206) 543-2056

U.S. Air Force
AFWAL/MLTC
Wright Patterson AFB, OH 45437
(513) 255-6976

CAD/CAM ORGANIZATIONS

Adage, Inc.
1 Fortune Dr.
Billeria, MA 01821
(617) 667-7070

Applicon, Inc.
32 Second Ave.
Burlington, MA 01803
(617) 272-7070

Auto-trol Technology Corp.
12500 N. Washington St.
Denver, CO 80233
(303) 452-4919

Avera Corp.
200 Technology Cir.
Scotts Valley, CA 95066
(408) 438-1401

CADAM, Inc.
1935 N. Buena Vista
Burbank, CA 91504
(213) 841-9470

CAD/CAM, Inc.
2844 East River Rd.
Dayton, OH 45439
(513) 293-3381

Cadlinc, Inc.
700 Nicholas Blvd.
Elk Grove Village, IL 60007
(312) 228-7300

CGX Corp.
42 Nagog Park
Alton, MA 01720

Cincinnati Milacron
Machine Tool Div.
4701 Marburg Ave.
Cincinnati, OH 45209
(513) 841-8100

Computervision Corp.
201 Burlington Rd.
Bedford, MA 01730
(617) 275-1800

Control Data Corp.
Manufacturing Industry Marketing
P.O. Box O
Minneapolis, MN 55440
(612) 853-8100

Design Aids, Inc.
27822 El Lazo Blvd.
Laguna Niguel, CA 92677
(714) 831-5611

GE Calma Co.
5155 Old Ironsides Dr.
Santa Clara, CA 95050
(408) 727-0121

General Electric CAE International, Inc.
300 TechneCenter Dr.
Milford, OH 45150

Gerber Scientific Instrument (GSI)
83 Gerber Road W.
South Windsor, CT 06074
(203) 644-1551

Gerber Systems Technology, Inc.
40 Gerber Rd. E.
South Windsor, CT 06074
(203) 644-2581

Giddings and Lewis
305 W. Delavan Dr.
Jamesville, WI 53547
(608) 756-2363

Graphics Technology Corp.
1777 Conestoga St.
Boulder, CO 80301
(303) 449-1138

IBM Corp.
1133 Westchester Ave.
White Plains, NY 10604
(914) 696-1960

Intergraph Corp.
One Madison Industrial Park
Huntsville, AL 35807
(205) 772-3411

Kearney and Trecker Corp.
West Allis, WI 53214

Manufacturing and Consulting Services, Inc.
2960 South Daimler Ave.
Santa Ana, CA 92705
(714) 540-3921

The Manufacturing Productivity Center
IIT Research Institute
10 W. 35th St.
Chicago, IL 60616
(312) 567-4800

Manufacturing Software and Services
Div. of LeBlond Makino Machine Tool
7667 Wooster Pike
Cincinnati, OH 45227

Matra Datavision, Inc.
Corporate Place I
99 South Bedford St.
Burlington, MA 01803

McDonnell Douglas Automation Co.
P.O. Box 516
St. Louis, MO 63166
(314) 233-2299

Mentor Graphics Corp.
10200 S.W. Nimbus Ave., G7
Portland, OR 97223
(503) 620-9817

Prime Computer, Inc.
1 Speen St.
Framingham, MA 01701
(617) 872-4770

Spectragraphics Corp.
3333 Camino Del Rio S.
San Diego, CA 92108
(714) 584-1822

St. Onge, Ruff & Associates, Inc.
617 W. Market St.
P.O. Box 309M
York, PA 17405
(717) 854-3061

Summagraphics Corp.
35 Brentwood Ave.
P.O. Box 781
Fairfield, CT 06430
(203) 384-1344

Summit CAD Corp.
5222 FM 1960 W. 102
Houston, TX 77069
(713) 440-1468

Synercom Technology, Inc.
500 Corporate Dr.
Sugar Land, TX 77478
(713) 491-5000

Technology Research Corp.
8328-A Traford Ln.
Springfield, VA 22152
(703) 451-8830

Tektronix, Inc.
P.O. Box 500
Berverton, OR 97077
(503) 682-3411

Telesis Corp.
21 Alpha Rd.
Chelmsford, MA 01824
(617) 256-2300

T&W Systems, Inc.
7372 Prince Dr., No. 106
Huntington Beach, CA 92647
(714) 847-9960

Valid Logic Systems, Inc.
650 North Mary Ave.
Sunnyvale, CA 94086
(408) 773-1300

VG Systems, Inc.
21300 Oxnard St.
Woodland Hills, CA 91367
(213) 346-3410

ARTIFICIAL INTELLIGENCE ORGANIZATIONS

Advanced Decision Systems
201 San Antonio Cir., No. 286
Mountain View, CA 94040
(415) 941-3912

AI Decision Systems
8624 Via del Sereno
Scottsdale, AZ 85258
(602) 991-0599

Aion Corp.
101 University Ave., 4th Floor
Palo Alto, CA 94301
(415) 328-9595

Apollo Computer, Inc.
330 Billerica Rd., Dept. A1
Chelmsford, MA 01824
(617) 256-6600

Applied Expert Systems, Inc. (APEX)
5 Cambridge Center
Cambridge, MA 02142
(617) 492-7322

Applied Intelligent Systems
110 Parkland Plaza
Ann Arbor, MI 48103
(313) 995-2035

Arity
358 Baker Ave.
Concord, MA 01742
(617) 371-2422

Artelligence, Inc.
14902 Preston Rd., Suite 212-252
Dallas, TX 75240
(214) 437-0361

Arthur D. Little
25 Acorn Park
Cambridge, MA 02140
(617) 864-5770

Artificial Intelligence Corp.
100 Fifth Ave.
Waltham, MA 02254
(617) 890-8400

Artificial Intelligence, Inc.
P.O. Box 81045
Seattle, WA 98108
(206) 271-8633

Artificial Intelligence Research Group
921 N. La Jolla Ave.
Los Angeles, CA 90046
(213) 656-7368

Automated Reasoning Corp.
290 W. 12th St., Suite 1-D
New York, NY 10014
(212) 206-6331

Bolt Beranek and Newman, Inc.
10 Moulton St.
Cambridge, MA 02238
(617) 491-1850

Brattle Research Corp.
55 Wheeler St.
Cambridge, MA 02138
(617) 492-1982

Carnegie Group, Inc.
Commerce Court at Station Square
Pittsburgh, PA 15219
(412) 642-6900

Cognitive Systems, Inc.
234 Church St.
New Haven, CT 06510
(203) 773-0726

Cray Research
608 Second Ave.
S. Minneapolis, MN 55402
(612) 333-5889

Digital Equipment Corp.
77 Reed Rd.
Hudson, MA 01749-2895
(617) 568-7100

Expertelligence, Inc.
559 San Ysidro Rd.
Santa Barbara, CA 93108
(805) 969-7874

Expert-Knowledge Systems, Inc.
6313 Old Chesterbrook Rd.
McLean, VA 22101
(703) 734-6966

Expert Systems, Inc.
686 West End Ave., Suite 3A
New York, NY 10025
(212) 662-7206

Expert Systems International
1150 First Ave.
King of Prussia, PA 19406
(215) 337-2300

Expert Technologies, Inc.
2600 Liberty Ave.
Pittsburgh, PA 15230
(412) 355-0900

Exsys, Inc.
P.O. Box 75158, Con. Sta. 14
Alburquerque, NM 87194
(505) 836-6676

Franz, Inc.
1141 Harbor Bay Pkwy.
Alameda, CA 94501
(415) 769-5656

General Research Corp.
7655 Old Springhouse Rd.
McLean, VA 22102
(703) 893-5915

Gold Hill Computers, Inc.
163 Harvard St.
Cambridge, MA 02139
(617) 492-2071

Human Edge Software
2445 Faber Pl.
Palo Alto, CA 94303
(415) 493-1593

IBM Corp.
P.O. Box 218
Yorktown Heights, NY 10598
(914) 945-3036

Inference Corp.
5300 W. Century Blvd.
Los Angeles, CA 90045
(213) 417-7997

Intellicorp
1975 El Camino Real W.
Mountain View, CA 94040
(415) 323-8300

Interstate Voice Products
1849 W. Sequoia Ave.
Orange, CA 92668
(714) 937-9010

Kurzweil Applied Intelligence
411 Waverly Oaks Rd.
Waltham, MA 02154
(617) 893-5151

Level Five Research
4980 South A1A
Melbourne Beach, FL 32951
(305) 729-9046

The LISP Co.
P.O. Box 487
Redwood Estates, CA 95044
(408) 354-3668

LISP Machine, Inc.
6033 West Century Blvd.
Los Angeles, CA 90045
(213) 642-1116

Logical Business Machines
264 Santa Ana Ct.
Sunnyvale, CA 94086
(408) 737-1911

Lucid
707 Laurel St.
Menlo Park, CA 94025
(415) 329-8400

McDonnell Douglas–Knowledge Engineering
20705 Valley Green Dr., VG2-BO1
Cupertino, CA 95014
(408) 446-6553

Microelectronics & Computer Technology Corp. (MCC)
9430 Research Blvd.
Austin, TX 78759
(512) 343-0860

Mitre Corp.
Burlington Rd.
Bedford, MA 01730
(617) 271-2000

Palladian
4 Cambridge Center
Cambridge, MA 02142
(617) 661-7171

Perceptronics
21111 Erwin St.
Woodland Hills, CA 91367
(818) 884-7572

Reasoning Systems
1801 Page Mill Rd.
Palo Alto, CA 94303
(415) 494-6201

Schlumberger-Doll Research
Old Quarry Rd.
Ridgefield, CT 06877
(203) 431-5000

Software Architecture & Engineering, Inc.
1500 Wilson Blvd., Suite 800
Arlington, VA 22209
(703) 276-7910

Speech Systems, Inc.
18356 Oxnard St.
Tarzana, CA 91356
(818) 881-0885

SRI International
333 Ravenswood Ave.
Menlo Park, CA 94025
(415) 326-6200

Sun Microsystems
2550 Garcia Ave.
Mountain View, CA 94043
(800) 821-4643

Symbolics, Inc.
11 Cambridge Center
Cambridge, MA 02142
(617) 577-7500

Technology Research Corp.
Springfield Professional Park
8328-A Traford Ln.
Springfield, VA 22152
(703) 451-8830

Teknowledge, Inc.
525 University Ave.
Palo Alto, CA 94301-1982
(415) 327-6600

Tektronix, Inc.
700 Professional Dr., P.O. Box 6026
Gaithersburg, MD 20877
(301) 948-7151

Thinking Machines Corp.
245 First St.
Cambridge, MA 02142
(617) 876-1111

Thoughtware
2699 S. Bayshore Dr., Suite 1000A
Coconut Grove, FL 33133
(305) 854-2318

Voice Control Systems, Inc.
16610 Dallas Pkwy.
Dallas, TX 75248
(214) 248-8244

Xerox Corp.
250 N. Halstead St.
Pasadena, CA 91109
(818) 351-2351

Acronyms and Abbreviations

ABM	Asynchronous Balanced Mode
AC	Alternating Current
ACK	ACKnowledge
A/D	Analog-to-Digital
ADJ	ADJusted
AFI	Address Format Identifier
AGVS	Automated Guided-Vehicle System
AI	Artificial Intelligence
AID	Auto-Interactive Design
ALGOL	ALGOrithmic Language
ALU	Arithmetic-Logic Unit
AM	Amplitude Modulation
AMF	Advanced Manufacturing Facility
AML	Advanced Manipulator Language
AMT	Advanced Manufacturing Technology
ANSI	American National Standards Institute
AP	Access Primitives
APAR	Authorized Program-Analysis Report
APL	A Programming Language
APT	Automatically Programmed Tools
APU	Auxiliary Power Unit
AQL	Acceptable Quality Level
AROM	Alterable Read-Only Memory
ARPANET	Advanced Research Project Agency NETwork
ART	Average Response Time
ASA	American Standards Association
ASCII	American Standard Code for Information Interchange
ASN.1	ISO Abstract Syntax Notation One
ASR	Automatic Send Receive
AWG	American Wire Gauge
BAL	Basic Assembly Language
BAS	Basic Activity Subset
BCD	Binary Coded Decimal
BCS	Basic Combined Subset
BIU	Bus Interface Unit
BNA	Boeing Network Architecture
BNF	Backus-Naur Form
BOM	Bill Of Materials
BPI	Bits Per Inch
BPS	Bits Per Second
BSC	Binary Synchronous Communications
BSS	Basic Synchronized Subset
BTAM	Basic Telecommunications Access Method
CAD	Computer-Aided Design
CAD/CAM	Computer-Aided Design/ Computer-Aided Manufacturing
CADD	Computer-Aided Design and Drafting
CAI	Computer-Aided Instruction
CAM	Computer-Aided Manufacturing
CAM-I	Computer-Aided Manufacturing International
CAPOSS-E	CApacity Planning and Operation Sequence System Extended program
CAPP	Computer-Aided Process Planning
CASE	Common Application Service Elements

CASPA	Computer-Aided Sculptured Pre-Apt system
CAT	Computer-Aided Testing
CAV	Constant Angular Velocity
CBI	Computer-Based Instruction
CBX	Computerized Branch Exchange
CCD	Charge-Coupled Device
CCITT	Comite Consultatif Internationale de Telegraphie et Telephonie
CCR	Commitment, Concurrency, and Recovery
CID	Charge-Injection Device
CIL	Computer-Independent Language
CIU	Communications Interface Unit
CIM	Computer-Integrated Manufacturing
CL	Cutter Location
CLF	Cutter Location File
CLI	Cutter Location Information
CLS	CLear Screen
CMOS	Complementary MOS
CNC	Computer Numeric Control
CNR	Carrier-to-Noise Ratio
COBOL	Common Business-Oriented Language
COPICS	Communications-Oriented Production Information and Control System
CP	Continuous Path
CPM	Critical Path Method
CPS	Characters Per Second or Controlled Path System
CPU	Central Processing Unit
CRT	Cathode-Ray Tube
CSMA/CD	Carrier-Sense Multiple Access with Collision Detect
CV	ComputerVision Corporation
CW	Continuous Wave
DA	Design Automation or Destination Address
DAA	Data Access Arrangement
DART	Data Accumulation and Retrieval of Time
DAS	Data Acquisition System
DASD	Direct-Access Storage Device
DAT	Dynamic Address Translation
DCE	Data Circuit-terminating Equipment
DBMS	Data Base Management System
DBX	Digital Branch eXchange
DDAS	Digital Data Acquisition System
DDC	Direct Digital Control
DDCMP	Digital Data Communications Message Protocol
DDL	Data Definitions Language
DDS	Dataphone Digital Service
DIP	Dual-on-Line Package
DIS	Draft International Standard
DLC	Data Link Control
DMA	Direct Memory Access
DML	Data-Manipulation Language
DNC	Direct Numerical Control or Distributed Numerical Control

DO	Dark Operated
DOS	Disk Operating System
DP	Draft Proposal
DRAW	Direct Read After Write
DRC	Design-Rules Checking
DRDW	Direct Read During Write
DRO	Digital ReadOut
DRP	Distribution Requirements Planning
DTE	Data Terminal Equipment
DVST	Direct View Storage Tube
EAM	Electrical Accounting Machine
EAN	European Article Numbering System
EAROM	Electrically Alterable Read-Only Memory
EB	End-of-Block Character
EBCDIC	Extended Binary Coded Decimal Interchange Code
ECMA	European Computer Manufacturers Association
EEROM	Electrically Erasable Read-Only Memory
EM	End-Medium character
ENQ	ENQuiry character
EOA	End-of-Address code
EOF	End of File
EPROM	Erasable Programmable Read-Only Memory
ES	Expert System or Electrical Schematic
ESC	Escape Character
FADU	File Access Data Unit
FAS	Final Assembly Schedule
FCS	Frame Check Sequence
FDM	Frequency Division Multiplexing
FEA	Finite Element Analysis
FEM	Finite Element Modeling
FIFO	First-In, First-Out
FIPS	Federal Information Processing Standards
FMC	Flexible Manufacturing Cell or Flexible Manufacturing Center
FMD	Frequency Multiplexing Division
FORTRAN	FORmula TRANslating system
FPDU	FTAM Protocol Data Unit
FTAM	File Transfer, Access, and Management
FTP	File Transfer Protocol
GCLISP	Golden Common LISP
GDP	Generalized Drawing Primitive
GKS	Graphics Kernal System
GMAW	Gas Metal Arc Welding
GPS	General Problem Solver
GT	Group Technology
GTAW	Gas Tungsten-Arc Welding
HDLC	High-Level Data Link Control
HEM	Hostile Environment Machines
HLS	Hue, Lightness, Saturation
HSV	Hue, Saturation, Value
Hz	Hertz
IC	Integrated Circuit
ICAM	Integrated Computer-Aided Manufacturing (USAF)
ICOT	Institute for new generation COmputer Technology
$IDEF_0$	ICAM DEFinition method, version zero

IDSS	ICAM Decision Support System
IDU	Interface Data Unit
IEEE	Institute of Electrical and Electronic Engineers
IGES	Initial Graphics Exchange Specification
IGS	Interactive Graphics Systems
IHS	Intensity, Hue, Saturation
IMS	Information Management System
I/O	Input/Output
IP	Internet Protocol
IPDU	Internet Protocol Data Unit
IPL	Initial Program Load
IPS	Inches Per Second
IS	International Standard
ISDN	Integrated Services Digital Network
ISO	International Standards Organization
JCL	Job Control Language
JCS	Job Control Statement
JTM	Job Transfer and Manipulation
KE	Knowledge Engineer
KIPS	Knowledge Information Processing System
KR	Knowledge Representation
KWIC	KeyWord In Context
LAN	Logical Area Network or Local Area Network
LAN-WAN	Local Area Network to Wide Area Network communications
LAPB	Link Access Protocol Balanced
LCB	Line Control Block
LCD	Liquid Crystal Display
LDC	Logical Device Coordinate
LED	Light-Emitting Diode
LHS	Left-Hand Side
LID	Logical Input Device
LIFO	Last-In, First-Out
LIPS	Logical Inferences Per Second
LIS	Large Interactive Surface
LISP	LISt Processing language
LLC	Logical Link Control or Link Layer Control
LO	Light Operated
LPM	Lines Per Minute
LPDU	Link layer Protocol Data Unit
LRC	Longitudinal Redundancy Check
LSAP	Link layer Service Access Point
LSB	Least-Significant Bit
LSDU	Link layer Service Data Unit
LSI	Large-Scale Integration
LUT	LookUp Table
MAC	Medium Access Control
MAP	Manufacturing Automation Protocol
MAU	Media Access Unit
MCC	Microelectronics and Computer technology Corporation
MHS	Master Handling System
MICR	Magnetic Ink Character Recognition
MIG	Metal Inert-Gas welding
MIS	Management Information System
MMFS	Manufacturing Message Format Standard
MOS	Metal-Oxide Semiconductor

MPS	Master Production Schedule or MultiProcessing System
MRP	Material Requirements Planning
MRPS	Manufacturing Resource Planning System
MSB	Most-Significant Bit
MSS	Mass-Storage System
MSI	Medium-Scale Integration
MTBF	Mean Time Between Failures
MTP	Machine Tool Program
MTTR	Mean Time To Repair
MUF	Machine Utilization Factor
MUM	Methodology for Unmanned Manufacturing
NAK	Negative AcKnowledgment
NANS	National Automated Nesting System
NAPLPS	North American Presentation-Level Protocol Syntax
NBS	National Bureau of Standards
NC	Numerical Control
NDC	Normalized Device Coordinate
NEC	National Electrical Code
NGS	Numerical Geometry System
NISL	Network Interface SubLayer
NIU	Network Interface Unit
NL	Natural Language
NM	Network Management
NLC	New Line Character
NLI	Natural-Language Interface
NLP	Natural-Language Processing
NLU	Natural-Language Understanding
NOVRAM	NOnVolatile Random-Access Memory
NPAI	Network Protocol Address Information
NRM	Normal Response Mode
NSAP	Network Service Access Point
NSDU	Network Service Data Unit
NTIS	National Technical Information Service
OCR	Optical Character Recognition
ODR	Optical Data Recognition
OEM	Original Equipment Manufacturer
OMR	Optical Mark Recognition
OS	Operating System
OSHA	Occupational Safety and Health Act
OSI	Open System Interconnection
PBX	Private Branch Exchange
PC	Personal Computer, Program Counter, Programmable Controller, or Printed Circuit
PCM	Pulse Code Modulation
PD	Programmable Device
PDN	Public Data Network
PDU	Protocol Data Unit
PEP	Parametric Element Processor
PERT	Program Evaluation and Review Technique
PGI	Parameter Group Identifier
PI	Parameter Identifier

P&ID	Piping and Instrumentation Diagram
PID	Proportional, Integral, Derivative control
PIP	Peripheral Interchange Program
PL/1	Program Language 1
PLA	Programmed Logic Arrays
PLC	Programmable Logic Controller
PM	Phase Modulation
POL	Problem-Oriented Language
PPB	Part Period Balancing
PTP	Point-To-Point control system
PTT	national Post, Telephone, and Telegraph entities
RAA	Remote Axis Admittance
RAM	Random-Access Memory
RCC	Remote Center Compliance
RGB	Red, Green, Blue
RHS	Right-Hand Side
RJE	Remote Job Entry
RNR	Receiver Not Ready
ROM	Read-Only Memory
ROS	Read-Only Storage
RPC	Remote Procedure Calls
RPG	Report Program Generator or Random Program Selection
SA	Source Address
SAP	Service Access Point
SASE	Specific Application Service Element
SCARA	Selective Compliance Assembly Robot Arm
SDA	Source Data Automation
SDLC	Syncronous Data Link Communication
SDU	System Data Unit
SEAP	Service Element Access Point
SIGGRAPH	Special Interest Group on computer GRAPHICS
SKU	StockKeeping Unit
SM	System Manager
SMT	Surface Mount Technology
SNA	System Network Architecture
SNDCF	SubNetwork Dependent Convergence Function
SNICP	SubNetwork Independent Convergence Protocol
SNPA	SubNetwork Point of Attachment
SPDU	Session Protocol Data Unit
SRAM	Static RAM
SSAP	Session Service Access Point
SSI	Small-Scale Integration
SVC	Switched Virtual Circuit
SYSGEN	SYStem GENeration
SYSLOG	SYStem LOG
TCP	Tool Center Point
TEM	Transverse Electromagnetic Mode
TIG	Tungsten Inert-Gas welding
TOP	Technical and Office Protocol
TPDU	Transport Protocol Data Unit
TPOP	Time-Phased Order Point
TRC	Technical Review Committee
TSAP	Transport Service Access Point
TSDU	Transport Service Data Unit
VDC	Virtual Device Coordinate

VDI	Virtual Device Interface
VDM	Virtual Device Metafile
VDT	Video Display Terminal
VLSI	Very-Large-Scale Integration
VMM	Variable Mission Manufacturing
VRC	Vertical Redundancy Checksum
VSAM	Virtual Storage Access Method
VTAM	Virtual Telecommunications Access Method
VTOC	Volume Table of Contents
WAN	Wide Area Network
WBS	Work Breakdown Structure
WD	Working Draft
XID	Exchange IDentification

Reference Documents

ISO documents may be obtained from:
Ms Frances Schrotter
ANSI
ISO TC97/SC6 Secretariat
1430 Broadway
New York, NY 10018
(212) 354-3300

National Bureau of Standards documents may be obtained from:
National Bureau of Standards
Gaithersburg, MD 20899

IEEE documents may be obtained from:
IEEE Standards Office
345 E. 47th St.
New York, NY 10017

FIPS documents may be obtained from:
National Technical Information Service (NTIS)
U.S. Department of Commerce
5285 Port Royal Rd.
Springfield, VA 22161

MAP V2.1 document may be obtained from:
General Motors Corp.
Manufacturing Engineering and Development
Advanced Product and Manufacturing Engineering Staff
APMES A/MD-39
GM Technical Center
Warren, MI 48090-9040

CCITT documents may be obtained from:
International Telecommunication Union
Place des Nations
CH-1211
Geneve 20
Switzerland

TOP V1.0 document may be obtained from:
Boeing Computer Services
Network Services Group
P.O. Box 24346, M/S 7C/16
Seattle, WA 98124-0346

Bibliography

BOOKS, REPORTS

Aleksander, I. 1983. *Artificial vision for robots*. New York: Chapman and Hall.*

Ballard, D. H., and C. M. Brown. 1982. *Computer vision*. New York: Prentice-Hall.*

Baxes, G. A. 1984. *Digital image processing*. New York: Prentice-Hall.

Brady, J. M. 1981. *Computer vision*. New York: North-Holland.

Castleman, K. R. 1979. *Digital image processing*. New York: Prentice-Hall.

Department of Commerce, International Trade Administration. 1985. *A competitive assessment of the U.S. flexible manufacturing systems industry*.

Department of Commerce, Office of Industry Assessment, Industry Analysis Division. 1984. *A competitive assessment of the U.S. manufacturing automation equipment industries*.

Dodd, G. G., and L. Rossol. 1979. *Computer vision and sensor-based robots*. New York: Plenum.

Dun & Bradstreet, Inc. 1983. *An analysis of the robotics industry*. New York.

Engelberger, J. F. 1980. *Robotics in practice*. New York: Amacom Division of American Management Associations.

Faugeras, O. 1983. *Fundamentals in computer vision*. Cambridge: Cambridge Univ. Press.

Fisk, J. D. 1981. *Industrial robots in the United States: Issues and perspectives*. Congressional Research Service, The Library of Congress, Report No. 81-78 E, 30 March.

Fu, K. 1984. *VLSI for pattern recognition and image processing*. NY: Springer-Verlag.

Gomersall, A. 1984. *Machine intelligence: An international bibliography with abstracts on sensors in automated manufacturing*. NY: Springer-Verlag.*

Goos, G., and J. Hartmanis. 1984. *Digital image processing systems*. New York: Springer-Verlag.*

* Available from The Manufacturing Technology Bookstore, 134 N. Washington St., Naperville, Illinois 60540 (312) 369-9232.

Hollingum, J. 1984. *Machine vision: The eyes of automation*. New York: Springer-Verlag.*

Holland, S. W., L. Rossol, and M. R. Ward. 1979. *CONSIGHT-I: A vision-controlled robot system for transferring parts from belt conveyors*. General Motors Research Publication GMR-2912, February. Warren, MI.

Hunt, V. D. 1986. *Artificial intelligence and expert systems sourcebook*. New York: Chapman and Hall.

———. 1983. *Industrial robotics handbook*. New York: Industrial Press.

———. 1985. *Smart robots*. New York: Chapman and Hall.

———. 1987. *Mechatronics—Japan's Newest Threat*. New York: Chapman and Hall.

National Electrical Manufacturers Association (NEMA). 1985. *U.S. manufacturers five-year industrial automation plans for automation machinery and plant communication systems*. NEMA, 15 March.

Office of Technology Assessment, U.S. Congress. 1984. *Computerized manufacturing automation: Employment, education, and the workplace*. U.S. Government Printing Office, OTA-CIT-235, April.

Onoe, M., et al. 1981. *Real-time/parallel computing: Image analysis*. New York: Plenum.

Pao, Y. H., and G. W. Ernst. 1982. *Context-directed pattern recognition and machine intelligence techniques for information processing*. New York: IEEE.

Parsons, H. M., and G. P. Kearsley. 1981. *Human factors and robotics: current status and future prospects*. Human Resources Research Organization, October. New York.

Pratt, W. K. 1978. *Digital image processing*. New York: John Wiley.*

Pugh, A. 1983. *Robot vision*. New York: Springer-Verlag.*

Rifkin, S. B. 1982. *Industrial robots: A survey of foreign and domestic U.S. patents*. U.S. Department of Commerce, National Technical Information Service, August.

Robot Institute of America. 1982. *Robot institute of America worldwide robotics survey and directory*. Detroit, Michigan, RIA.

Stucki, P. 1979. *Advances in digital image processing*. New York: Plenum.*

Susnjara, K. 1982. *A manager's guide to industrial robots*. New York: Corinthian.

Toepperwein, L., M. T. Blacknow, et al. 1980. *ICAM robotics application guide*. Report AFWAL-TR-80-4042, vol. II. Air Force Wright Aeronautical Laboratories, Materials Laboratory, Wright-Patterson Air Force Base, OH.

———. 1984. *Machine Vision*. Detroit, Michigan: Society of Manufacturing Engineers.*

ARTICLES

Albert, M. 1984. Fixturing keeps FMS on track. *Modern Machine Shop* (March).

Arai, J. 1982. Robot growth in Japan. *Manufacturing Productivity Frontiers* (October).

Berger, C. 1982. Selecting robots for assembly operations. *Manufacturing Productivity Frontiers* (September).

Brooks, T. L. 1982. Visual and tactile feedback handle tough assembly task. *Robotics Today*, (October).

Brosilow, R. 1982. How to step up to robotic arc welding. *Welding Design and Fabrication* (August).

Catalano, F. 1983. Robotics, vision advances emerge from industry, university cooperation. *Mini-Micro Systems* (February).

Coates, V. T. 1983. The potential impact of robotics. *The Futurist* (February).

Deciding on an FMS. 1983. *American machinist* (May).

DiMaria, E. 1982. In Europe, France pushes robot research. *American Metal Market/Metalworking News* (November).

Eastwood, Dr. M. A. 1983. Introduction to robot control. *Tooling & Production* (February).

Flexible manufacturing's future. 1985. *Production*, newsbits (May).

Goebel, K. 1983. A guide to coating robot applications. *Robotics World* (February).

Gupta, P. 1982. Multiprocessing improves robotic accuracy and control. *Computer Design* (November).

Hughes fabricating precision parts with automated system. 1983. *Aviation Week & Space Technology*, 11 April, 66.

Kinnucan, P. 1981. How smart robots are becoming smarter. *High Technology*, 1, no. 1 (September/October).

Klein, A. 1982. Vision systems boost robot productivity. *Appliance Manufacturer* (August).

Lundquist, E. 1982. Robotic vision systems eye factory applications. *Mini-Micro Systems* (November).

Merritt, R. 1982. Industrial robots: Getting smarter all the time. *Instruments & Control Systems* (July).

Morley, R. E. 1982. An overview: Programmable controllers and robots in automated factories. *Design Engineering* (December).

Morris, H. M. 1983. Adding sensory inputs to robotic systems increases manufacturing flexibility. *Control Engineering* (March).

Moskal, B. S. 1985. Material handling breaches the factory of the future. *Industry Week*, 24 June, 46–52.

Rheinhold, A. G., and G. Venderbrug. 1980. Robot vision for industry: The autovision system. *Robotics Age* 2, no. 3 (Fall).

Sanderson, R. J. 1983. A survey of the robotic vision industry. *Robotics World* (February).

Senia, A. 1984. Inside apple's macintosh factory. *California Business* (September).

Shunk, D. L., et al. 1982. Applying the systems approach and group technology to a robotic cell. *Robotics Today* (October).

Stauffer, R. N. 1982. IBM advances robotic assembly in building a word processor. *Robotics Today* (October).

Thomas, R. 1983. Sensing devices extend applications of robotic cells. *IE Magazine* (March).

CONFERENCE PROCEEDINGS

Brady, M. and R. Paul. 1984. *Robotics research: The first international symposium*. Boston, MA: MIT Press.

* Available from The Manufacturing Technology Bookstore, 134 N. Washington St., Naperville, Illinois 60540 (312) 369-9232.

Casasent, D. P. 1982. *Proceedings of conference on industrial applications of machine vision.* IEEE.

———. 1981. *Proceedings of conference on pattern recognition and image processing.* IEEE.

———. 1981. *Proceedings of 1st international conference on robot vision and sensory control.* IFS (Publications).

———. 1982. *Proceedings of 2nd international conference on robot vision and sensory controls.* IFS (Publications).

———. 1982. *Proceedings of SPIE conference on robotics and industrial inspection.* SPIE.

———. 1983. *Proceedings of 3rd international conference on robot vision and sensory controls.* North-Holland.*

———. 1984. *Robotics research: The next five years and beyond.* Robotics International of SME.*

———. 1984. *Robots 8th conference proceedings.* SME.*

———. 1984. *The 3rd annual applied machine vision conference proceedings.* Robotics International of SME.*

Robotics International of SME. *Robots conference proceedings.* Robotics International of SME.

Rosenfeld, A. 1982. *Proceedings of SPIE conference on robot vision.* SPIE.

Society of Manufacturing Engineers, *Conference Proceedings for Robotics and Machine Vision, 1985–1987,* Dearborn, Michigan.

DIRECTORIES

AI Trends '86
DM Data, Inc.
Scottsdale, AZ 85251

CAD/CAM, CAE: Survey, Review and Buyers Guide
DARATECH, Inc.
Cambridge, MA 02138

CAD/CAM; ROBOTICS; and Industry Directory
Technical Database Corp.
Conroe, TX

Industrial Robots (1983)
Society of Manufacturing Engineers
Marketing Services Div.
One SME Dr.
P.O. Box 930
Dearborn, MI 48128

RELATED PERIODICALS

Assembly Engineering
Hitchcock Publishing Co.
New York, NY

CIM
CADLINC Inc.
Elk Grove Village, IL

CIM Technology
CASA/Society of Manufacturing Engineers
Dearborn, MI

Electronic Imaging
Morgan-Grampian Publishing Co.
Berkshire Common
Pittsfield, MA 01201
(413) 499-2550

High Technology Magazine
High Technology Publishing Corp.
Boston, MA

Image and Vision Computing
Butterworth Scientific, Ltd.
P.O. Box 63
Westbury House, Bury St.
Guildford, Surrey GU1 5BH
United Kingdom

The Industrial Robot
IFS (Publications), Ltd.
35–39 High St.
Kempston, Bedford
MK42 7BT England

Industrial Robots International
158 Linwood Pl.
P.O. Box 13304
Fort Lee, NJ 07024

The International Journal of Robotics Research
The MIT Press
28 Carleton St.
Cambridge, MA 02142

* Available from The Manufacturing Technology Bookstore, 134 N. Washington St., Naperville, Illinois 60540 (312) 369-9232.

Le nouvel Automatisme
41 rue de la Grange-aux Belles
75483 Paris Cedex 10
France

Managing Automation
Thomas Publishing Co.
New York, NY

Manufacturing Engineering
Society of Manufacturing Engineers
Dearborn, MI

Manufacturing Technology Horizons
Tech Tran Corp.
134 N. Washington St.
Naperville, IL 60540
(312) 369-9232

Material Handling Engineering
Penton/IPC
Cleveland, OH

Photonics Spectra
Optical Publishing Co., Inc.
P.O. Box 1146
Berkshire Common
Pittsfield, MA 01202

Production Engineering
Penton/IPC
Cleveland, OH

Robomatics Reporter
EIC/Intelligence
38 W. 38th St.
New York, NY 10018

Robotech Japan
Topika, Inc.
Nagatani Bldg.
7-17-4, Ginza
Chuko-ku, Tokyo 104
Japan

Robotics Age
P.O. Box 725
La Canada, CA 91011

The Robotics Report
Washington National News Reports, Inc.
Suite 400
7620 Little River Tpk.
Annandale, VA 22003

Robotics Technology Abstracts
Cranfield Press
Management Library
Cranfield Institute of Technology
Cranfield Bedford MK43 OAL
United Kingdom

Robotics Technology Abstracts
Tech Tran Corp.
134 N. Washington St.
Naperville, IL 60540
(312) 369-9232

Robotics Today
Society of Manufacturing Engineers
One SME Dr.
P.O. Box 930
Dearborn, MI 48128

Robotics World
Communication Channels, Inc.
6255 Barfield Rd.
Atlanta, GA 30328

Robot Insider
Fairchild Publications
7 E. 12th St.
New York, NY 10003

Robot News International
IFS (Publications), Ltd.
35–39 High St.
Kempston, Bedford
MK42 7BT England

Robot/X News
Robotics Publications, Inc.
P.O. Box 450
Mansfield, MA 02048

Sensors
North American Technology, Inc.
174 Concord St.
Peterborough, NH 03458
(603) 924-7261

Vision
Society of Manufacturing Engineers
One SME Dr.
P.O. Box 930
Dearborn, MI 48128

Appendix A

FMS Suppliers and Users in the United States

FMS Supplier FMS User Location/Date	Machine Station	Material Handling	Features No. Type	Control Mfg's	Description
Cincinnati Milacron Vought Aero Products Dallas, TX (1984)	8 MC	4 CW	S, P 2-I B, C	DEC/1124 DEC/1170 DEC/1144	540 parts for B-1 bomber, aft and intermediate fuselage
Cincinnati Milacron FMC Corp. Aiken, SC (1984)	4MC	CW	I, C	DEC/1144	19 components, Army's Bradley fighting vehicle
Cincinnati Milacron Cincinnati Milacron Plastics Machinery Cincinnati, OH (1985)	4 MC	CW 1 SP	P		Toggle line, plastic injection-mold equipment
Cincinnati Milacron General Dynamics Convair Lynburg Field, CA (1982)	4 MC	PS	P	Allen-Bradley	65,000 bulkheads, side frames, seat, tracks, supporting arms
Cincinnati Milacron General Dynamics Convair Lynburg Field, CA (1982)	2 MC	PS	P	Allen-Bradley	Same as above
Cincinnati Milacron General Dynamics Convair Lynburg, Field, CA (1982)	6 MC	PS	P	Allen-Bradley	Same as above
Cincinnati Milacron General Dynamics Fort Worth, TX (1985)	6 MC	A			
Cincinnati Milacron General Electric Evandale, OH (1985)	4 NT	R I			
Cincinnati Milacron Caterpillar Davenport, IA (1985)	3 MC 1 HI 1 NB	PT	P A C	Milacron	4 large track-type loader frames, 100 machine steps
Cincinnati Milacron Caterpillar Aurora, IL (1986)	2 MC	PS	P, I, A B, C	Milacron	Excavator sticks, booms
Cincinnati Milacron New York Air Brake (1984)					
Dearborn Caterpillar Decatur, IL (1984)	3 NB 5 MC 1 ND 1 NM	PT	C	Allen-Bradley	Truck-axle banjo housing
Giddings & Lewis Caterpillar Aurora, IL (1979)	2 NM 4 ND 1 HI	PT	I C B	G&L	Wheel-type loader frames
Gidding & Lewis General Electric Erie, PA (1984)	3 MC 3 NB 3 NM 1 SP	CT R	I B C	DEC/1144	6 motor frames and gear boxes for locomotives

FMS Supplier FMS User Location/Date	Machine Station	Material Handling	Features No. Type	Control Mfg's	Description
Giddings & Lewis McDonnell Douglas Astronautics St. Louis, MO (1985)	3 NB 2 NB 2 NT	CW R	I, P B, X A	G&L	Missile parts (high-speed machining)
Ingersoll Milling J.I. Case Components Div. Racine, WI (1978)	2 MC 2 HC 1 NM	PS Fixed Path 1 NB	I C Head changer		2 1300 lb. castings for tractor transmission cases
Ingersoll Milling Ingersoll Milling Rockford, IL (1984)	5 MC	CW	I S P	Allen-Bradley	2500 heads, gearboxes, brackets
Ingersoll Milling Ingersoll Milling Rockford, IL (1987)	3 MC	CNC but no automatic loading until 1987.			
Ingersoll Milling Ingersoll Milling Rockford, IL (1987)	4 MV	CNC but no automatic loading until 1987.			
Kearney & Trecker Allis-Chalmers WI (1971)	5 MC 4 HI 1 NM	CT	C	Inter Data Bendix	Tractor transmission components
Kearney & Trecker Deere & Company Waterloo, IA (1979)	11 MC 5 HI	CT	P C	DEC KT, CNC	8 transmission and clutch housings
Kearney & Trecker Avco-Lycoming Williamsport, PA (1975)	9 MC 4 HI	CT	C B	Inter Data	2 families of aircraft engine components
Kearney & Trecker Avco-Lycoming Stratford, CT (1979)	7 MC 2 NV	CT	B,C	Inter Data Bendix	Turbine components
Kearney & Trecker Rockwell International Newark, OH (1970)	8 MC 1 NV	CT	C, P I	Bendix	Carrier-drive axles
Kearney & Trecker Hughes Aircraft El Segundo, CA (1982)	9 MC	CT	I, A C, B	DEC/1144 KT, CNC	Castings for aircraft and missile parts
Kearney & Trecker Mack Truck Hagerstown, MD (1983)	4 MC 2 HC 1 NB	CT	C, I B, A	DEC/1144 Gemini	7 transmission casings
Kearney & Trecker Onan Minneapolis, MN (1984)	1 NV 1 MC	PT (CW in 3 yr)	C, B A	KT, CNC	12 generator frames
Kearney & Trecker Mercury Marine Fond du Lac, WI (1984)	9 MC 1 HI 1 NB	RG	B C A	DEC GEMINI	6 parts for outboard engine, engine block, crankcase assy
Kearney & Trecker Georgetown Mfg. Georgetown, KY (1984)	6 MC	X stack crane	A B C	DEC GEMINI	150 manifolds, spindle-housing castings

FMS Supplier FMS User Location/Date	Machine Station	Material Handling	Features No. Type	Control Mfg's	Description
Kearney & Trecker Cummins Engine Columbus, IL (1984)	6 MC	CW	A B C	DEC GEMINI	Cummings brake components
Kearney & Trecker Warner Ishi Shelbyville, IL (1984)	2 MC 2 NT	R	I, A B, C	DEC GEMINI	Turbocharger housings
Kearney & Trecker Sundstrand Aviation Rockford, IL (1984)	2 MC	CW AS/RS	I, B C, A	DEC GEMINI	Aircraft components
Comau (Italy) Buick Gear & Axle Hamtramck, MI (1985)	3 MC	PR	I P C	Allen-Bradley	6 differential-housing axle tubes
Mazak (Japan) Caterpillar Peoria, IL (1984)	3 MC 1 X	R	I C B, D	Mazatrol	Sprocket segments
Mazak (Japan) Mazak, Line A Florence, KY (1981)	4 MC	PS CW	C S B		Parts for machine tools
Mazak (Japan) Mazak, Line B Florence, KY (1981)	4 MC 1 NM 1 NV 1 NT	PS CW	S		Parts for machine tools
Oerlikon-Motch Rockwell Motch New Castle, PA (1985)	2 NV 1 ND		2-I		22 steering knuckles
Westinghouse/DeVlieg General Dynamics Fort Worth, TX (1986)	6 MC	CW R	I X		Aircraft components
Shin Nippon Koki Boeing Aircraft Seattle, WA (1985)	5 MC	CW AS/RS	C	DEC	
White Sundstrand Boeing Aerospace (1984)	3 MC	PR	P S	OMNI	Missile components
White Sundstrand Watervliet Arsenal NY (1985)	8 MC 2 NV	CW	I P	OMNI	6 gun tubes
White Sundstrand Buick, Detroit Diesel Allison Hamtramk, MI (1985)	8 MC	CT	C B	OMNI	40 heavy transmission parts
White Sundstrand Caterpillar E. Peoria, IL (1983)	8 MC 3 NT 1 SP	PT	P I B	Swinc	Transmission cases, covers, and assemblies
Acme-Cleveland Vickers Omaha, NE (1984)	5 MC 3 NT 2 SP	R PR	2-I	Westinghouse	Blocks
Harris Press Harris Press Fort Worth, TX (1981)	3 MC 3 NM 2 SP	RG 4 GR	I P		Cylindrical parts

FMS Supplier FMS User Location/Date	Machine Station	Material Handling	Features No. Type	Control Mfg's	Description
Comau (Italy) Borg Warner York, PA (1984)	4 MC 1 NV	PR AS/RS	P C	Allen-Bradley	80 reciprocating compressor parts

Machining stations
MC = machining centers
NM = NC milling machine
NV = NC vertical lathe
ND = NC drilling machine
NB = NC boring machines
NG = NC gear cutting
NT = NC turning machine
GR = NC grinding machine
HI = head indexer
HC = head changer
SP = special machines
X = other

Material handling
PT = pallet transfer
PS = pallet shuttle
CT = cart towline
CW = cart, wire guided
PR = roller conveyors
RG = rail-guided carts
R = robots
X = other

Features
I = inspection
P = parts cleaning
A = adaptive control
B = probing
C = central coolant
S = self-contained coolant
X = other